FIDIC Users' Guide

publishing

FIDIC Users' Guide

Third edition

A Practical Guide to the Red, Yellow, MDB Harmonised and Subcontract Books

Brian Barr and Leo Grutters

Originally created by Brian Totterdill

Foreword by Gwyn Owen

Published by ICE Publishing, One Great George Street, Westminster, London SW1P 3AA.

Full details of ICE Publishing sales representatives and distributors can be found at: www.icevirtuallibrary.com/info/printbooksales

First published 2006

Other titles by ICE Publishing:

The NEC Compared and Contrasted.
F. Forward. ISBN 978-0-7277-3115-9
ICE Manual of Construction Law.
V. Ramsey, J. Baster, A. Minogue and M. O'Reilly (eds).
ISBN 978-0-7277-4087-8
NEC3: A User's Guide.
J. Broome. ISBN 978-0-7277-4109-7

www.icevirtuallibrary.com

A catalogue record for this book is available from the British Library

ISBN 978-0-7277-5856-9

Commissioning Editor: Rachel Gerlis
Production Editor: Imran Mirza
Market Specialist: Catherine de Gatacre

Typeset by Academic + Technical, Bristol
Index created by Indexing Specialists (UK) Ltd, Hove, East Sussex
Printed and bound by CPI Group (UK) Ltd, Croydon CRO 4YY

Contents

Part 3

The FIDIC Conditions of Contract for Plant and Design Build for Electrical and Mechanical Plant, and for Building and Engineering Works, designed by the Contractor (Yellow Book, YLW Book) 353

Foreword

It gives me great pleasure to write a few words here for you to consider before you read this excellent third edition. My friend and colleague Brian Totterdill who passed away in 2011 after he had also completed and published its second edition initially wrote the FIDIC Users' Guide. It is a sign of the times that contract forms, precedent and practice develop so quickly that there is a need for a third edition of what has become a standard reference guide. Who better to undertake the work of such a publication than the present authors Brian Barr and Leo Grutters? Both authors are seasoned practitioners in the field of contract management and administration and also in the world of dispute resolution in its many forms. They are both industry recognised within their fields, are FIDIC President Listed and have taken on the significant role of the development of this new edition of the book, no doubt knowing that from somewhere Brian Totterdill is looking over their shoulder with a critical, but approving gaze.

The FIDIC form of contract is one, if not the, leading form of contract used on internationally funded development projects throughout the world. It is used by a cross section of industry, from developing nations through to sophisticated promoters, who all need a reliable, industry proven form of contract incorporating logical risk allocation leading to certainty in both time and cost. FIDIC has taken a global lead in the development of various forms of contract which provide such a platform and it is inevitable that these contract forms continue to be developed and modified as required by the industry and its users.

The first and second edition of the book guided the readers through the practical aspects of the use of the various standard FIDIC forms of contract and provided guidance to the understanding of the various clauses and provisions found therein. The books followed the seemingly simple presentation of language found in the 1999 suite of FIDIC contracts and enabled the reader, be they legally trained or not, to gain a better understanding of the meaning behind those simple words. However, a few years is a long time in the development of legal understanding and various practitioners have tested those standard forms of contract by litigation and arbitration and, together with their ever widening use within the development industry, a need has been created for the various standard forms to be amended and modified to reflect the progression of that legal understanding. Indeed, the need for further and more focused forms of contract has also arisen and FIDIC has revised, or is in the process of revising, the currently used forms and has published new forms of contract to fulfil such needs.

The present authors, as they are practitioners in this very field, have their fingers on the pulse of such changes. They have accordingly revised and edited the earlier editions of the FIDIC User's Guide and have produced an excellent third edition which incorporates the current industry's thinking and understanding of the FIDIC terms and conditions. They have also incorporated into the book the new FIDIC forms of contract, making the book a more all-encompassing volume for readers. This third edition contains an updated view of the meaning and understanding of the various clauses found within the FIDIC suite of contracts while maintaining the previous editions' layout format. New and additional flow charts have been incorporated which make the book a thorough source of materials for those who need a ready reference guide on their bookshelves.

As with Brian Totterdill's earlier editions, this third edition provides a thorough and up-to-date practical and reliable reference book for users of the various FIDIC forms of contract. I have no doubt that this new edition will be used by many practitioners throughout the world and will provide a uniform understanding of the administration of good practice.

Gwyn Owen

Chartered Arbitrator, Adjudicator & Mediator,
Past President DRBF

Acknowledgements

This book would not have been possible without the tireless support and assistance of the following people: Silke Mittelstädt, Rachel Gerlis, Axel Haider and his team at GibConsult GmbH, Matthew Barr, Lynne Barr, Nicole Mittelstädt and all the other unnamed people who have supported and assisted in many ways.

Appreciation must also be extended to Dr. Cyril Chern, Siobhan Fahey, Gwyn Owen, Nicholas Gould and Giovanni di Folco as professional colleagues for their valuable contributions and reviews of the content of this book.

The authors would also like to extent their appreciation to Enrico Vink and François Baillon of FIDIC for their assistance and support concerning the required copyrights, without which this book would not have been possible.

Preface to the third edition

The third edition comes at a time where the international construction industry is coming to terms with the loss of one of its great contributors. Brian W. Totterdill, the author and creator of the first and second edition of this book sadly passed away, leaving behind a legacy of great importance that will stretch over many years. His books on dispute resolution for the international construction industry are widely known and accepted and have become the standard of reference for many practitioners and users.

Not only is this third edition a welcome update reflecting current construction market requirements, it is also a tribute to Brian W. Totterdill in that his work is continued, so that the industry may experience the benefits of his life's passion and dedication for many more years.

We, Brian Barr and Leo Grutters, as authors of this third edition, are fully aware of our responsibility in preserving that legacy. Nevertheless, and in keeping with current construction industry needs and requirements, we have tried not to just copy from the previous editions, but to provide additional valuable information, facts and figures through in-depth research and to discuss in detail some of the more controversial topics the industry has raised since publication of the last edition. Our combined practical experience in excess of 70 years is used to produce a work of reference that we are convinced will serve the industry for years to come.

This third edition has been extended to include a more in-depth analysis of the Multilateral Development Banks Harmonised *Conditions of Contract for Construction*, commonly known as the MDB Edition or PNK Book as it has come to be known as well. This Contract has been prepared in conjunction with the World Bank and a group of Multilateral Development Banks, and it is understood that all of these banks adopt this edition of the FIDIC document in their standard bidding documents. A first version of the General Conditions was released to participating banks in May 2005, followed by an amended second version in March 2006 and a third amended version in June 2010. The MDB includes some significant changes to the *Conditions of Contract for Construction* and is reviewed in Part 2 of this book.

Over the years, the use of the FIDIC 1999 suite of Contracts has seen a steady increase in use. New contract forms have been published such as the 2008 *Conditions of Contract for Design, Build and Operate Projects*, more commonly referred to as the DBO or GOLD Book. Another new contract was published in 2011: *Conditions of Subcontract for Construction for building and engineering works designed by the Employer* (2011). This latter Contract form is specifically designed for use with the *Conditions of Contract for Construction for Building and Engineering Works Designed by the Employer* (1999), or commonly known as the 'RED Book'. The interaction between the RED Book and the Subcontract or SUB Book is reviewed in Part 4.

Abbreviations

DAB	Dispute Adjudication Board
YLW	Conditions of Contract for Plant and Design Build for Electrical and Mechanical Plant, and for Building and Engineering Works, Designed by the Contractor (1999)
PNK	Conditions of Contract for Construction MDB Harmonised Edition for Building and Engineering Works Designed by the Employer (2010)
SUB	Conditions of Sub Contract for Construction for Building and Engineering Works Designed by the Employer (2011)
SLV	Conditions of Contract for EPC/Turnkey Projects (1999)
GLD	Conditions of Contract for Design, Build and Operate Projects (2008)
MDB	Multilateral Development Banks
DB	Dispute Board
FIDIC	Fédération Internationale des Ingénieurs Conseils
RED	Conditions of Contract for Construction for Building and Engineering Works Designed by the Employer (1999)
ICE	Institution of Civil Engineers (UK)
ICC	International Chamber of Commerce
DBF	Dispute Board Federation
BoQ	Bill of Quantities

About the authors

Brian Barr is a Chartered Civil Engineer and Adjudicator with over forty years of experience in the public and private sectors specialising in the management of International Development Construction projects with particular emphasis on the avoidance and resolution of disputes. During his career he has been involved in the procurement, administration and supervision of major infrastructure works particularly development projects supported by International Financing Institutions.

He is a Fellow of three major professional bodies – The Institution of Civil Engineers and The Institution of Highways and Transportation in the United Kingdom, and The Dispute Board Federation in Switzerland. His name is also included on the FIDIC President's List of Dispute Adjudicators.

He has lived and worked in many developing countries in Africa, Central/South/South-East Asia, Eastern Europe, the Caribbean and Central America for more than twenty years. During this period he has held key positions including; Engineer, Engineer's Representative, Project Manager, Representative on Dispute Boards, Dispute Review Expert, Dispute Advisor, Claims Advisor, Technical Auditor.

Leo Grutters is a Civil Engineer and a Dispute Resolution Expert with nearly 30 years of professional experience. He holds an MBA and is a Fellow of the Chartered Institute of Arbitrators. As a Member of FIDIC's President's List of Approved Dispute Adjudicators he has been involved as an independent neutral expert in many large prestigious international engineering projects involving such wide-ranging sectors as roads, power generation and distribution, off-shore, marine, renewable energies, etc. He is a forensic analyst focused on the logical and fair resolution to contractual disputes. Due to his extensive participation in international contractual dispute resolution procedures such as Adjudication, Dispute Boards and Arbitration, he is a project-focused expert that understands the interaction between complex engineering methodologies and the unavoidable legal and commercial implications when seen in a contractual environment. He strongly believes in a common sense approach and will always strive to achieve the optimum result acceptable by all involved. He also regularly lectures on Dispute Resolution Methodologies, Proactive Contract Administration Systems and Procedures and he is a regular speaker at related international events. He also routinely conducts training and coaching events on the entire FIDIC Rainbow Suite of Contracts, as well as other construction management related topics.

Part 1

Introduction to Part 1

The book uses the masculine gender throughout which follows the convention used within the FIDIC forms of contract. We trust this is perceived as a deficiency in the English language rather than bias on the part of the authors.

FIDIC Users' Guide
ISBN 978-0-7277-5856-9

ICE Publishing: All rights reserved
http://dx.doi.org/10.1680/fug.58569.003

Chapter 1
Introduction to the FIDIC Conditions of Contract

1.1. Construction contracts

Construction projects are governed, in some form, by a contract. Any such construction Contract is a legally binding agreement between two Parties – the Owner, who is generally referred to as the Employer, and the Contractor. In its most basic form the Employer initiates the project, decides what he wants, makes the site available for construction, gives instructions, supervises the construction, pays for the project and occupies and/or uses the completed project. The Contractor builds the project and is paid for his work.

The Contract controls the relations between the Employer and the Contractor and includes a substantial bundle of documents. These documents include details of the work which the Employer wishes to have constructed, and which the Contractor has offered to construct. It also includes pertinent details on the payments that will be made from the Employer to the Contractor. The Contract includes Conditions of Contract, which lay down the responsibilities and obligations of both Parties. When unexpected problems arise, which often result in delays and additional costs, the Conditions of Contract determine which of the Parties must bear the consequences.

A typical construction contract will include the following.

- *The Contractor's Tender,* which is the Contractor's offer to carry out the work for a certain price. The Tender will be based on information and documents provided by the Employer, and includes the Contractor's proposal for the rates of payment for the different items of work.
- *The Contract Agreement*, which is the document that confirms the Contractor's offer and the Employer's acceptance, and establishes the formal Contract. Under some jurisdictions the Contract Agreement is required in order to establish a legally binding agreement between the Parties.
- *The Employer's Letter of Acceptance,* which is the letter from the Employer to the Contractor in which the Employer accepts the Contractor's offer to carry out the work. This offer and acceptance constitute a legally binding agreement, within the terms that are stated in the Letter of Acceptance. The Letter of Acceptance will refer to the Contractor's Tender and any subsequent agreements.
- *The Conditions of Contract*, which is generally based on a standard document, modified to suit the requirements of the Employer for the particular project.

- *The Technical Documents,* which include drawings, specifications, bills of quantities and other schedules and give the Employer's technical requirements for the project, together with the detailed calculations that make up the Accepted Contract Amount. When the design is to be prepared by the Contractor the drawings and specifications are replaced by the Employer's Requirements and Contractor's Proposals.
- *The Time Schedule,* which outlines the basic milestones the Contractor must achieve over the duration of the project. This is replaced soon after the project has started by a more detailed time schedule. It is prepared by the Contractor, outlining in much greater detail, when the various activities making up the project will be executed.

1.2. The FIDIC Conditions of Contract

The most commonly used Conditions of Contract for international construction projects are published by the Fédération Internationale des Ingénieurs-Conseils (FIDIC), the International Federation of Consulting Engineers. The traditional FIDIC Contract for civil engineering construction is the FIDIC *Conditions of Contract for Works of Civil Engineering Construction,* commonly known as 'The RED Book'. In 1999, this was superseded by the FIDIC *Conditions of Contract for Construction,* which are the subject of this book, together with the *Conditions of Contract for Plant and Design-Build.* In addition, this book will analyse in detail the interrelationship between the RED Book and the newly published FIDIC *Conditions of Subcontract for Construction for building and engineering works designed by the Employer* (first edition, 2011).

The FIDIC Conditions of Contract include

- the General Conditions
- the Particular Conditions.

The General Conditions were drafted with care by some of the most experienced professionals in the industry and is thus intended to be used unchanged for every project. The Particular Conditions are prepared for the individual project requirements and include any changes or additional clauses that the Parties have decided to include suiting the local and specific conditions under which the project is to be executed. Some employers have printed their own versions of the General Conditions, with changes to suit their own, often one-sided, requirements. This procedure cannot be recommended. One of the prime advantages of using standard Conditions of Contract is that contractors tendering for the project and contract administrators are familiar with the standard Conditions and are aware of their responsibilities and of the consequences of any failure to meet their obligations. Any changes or additions are in a separate document and so everyone's attention is drawn to the changes. When the General Conditions have been reprinted with small but significant changes, then the changes may be overlooked and the project will suffer as a consequence.

The General Conditions also include the Appendix to Tender, which is a schedule of essential information, most of which must be completed by the Employer before he issues the Tender documents and a few items which are completed by the Tenderer.

In any project, there will be problems and to overcome these problems it may be necessary to carry out additional work. This inevitably will take time and money. In an ideal world, the Parties will endeavour to agree on the extra amounts of money to be spent before they are incurred. However, the most common situation is that the Contractor first spends the money and then claims it back from the Employer. When this situation arises, it is necessary to decide whether the Employer must reimburse the Contractor, or whether the Contractor must bear the additional cost. If this cannot be agreed by the Parties' representatives on the Site then an initial decision will be made by the Engineer or other Employer's Representative. This is an interim decision and is subject to appeal to an arbitrator or the Courts. However, reference to arbitration or the Courts is a slow and expensive procedure. Under the 1999 FIDIC Conditions of Contract, if either Party is not satisfied with the Engineer's initial acceptance or rejection of a claim, then the dispute can be referred immediately, on site, for a decision by a Dispute Adjudication Board (DAB). This decision must be implemented, but the dispute may then be referred to arbitration for a final decision.

The basis on which all such decisions must be made is laid down in the Conditions of Contract. The Conditions of Contract give the procedural rules and lay down the rights and obligations of the Parties to the Contract. Successive revisions to the standard Conditions of Contract have increased the complexity of these rules and procedures so that the current Conditions of Contract are virtually manuals of good project management practice, rather that purely legal statements of party rights and obligations.

1.3.   The Traditional FIDIC Conditions of Contract

FIDIC publishes a family of different Conditions of Contract to suit the requirements of different types of construction projects. Since 1999 there have been two separate and distinct sets of FIDIC Conditions of Contract available for use – the Traditional FIDIC Conditions of Contract and the new 1999 FIDIC Conditions of Contract.

The Traditional FIDIC Conditions of Contract are as follows.

- *Conditions of Contract for Works of Civil Engineering Construction* (fourth edition, 1987, reprinted 1992 with further amendments), known as 'The Old Red Book'. A supplement to the fourth edition was published in 1996.
- *Conditions of Contract for Electrical and Mechanical Works* (third edition, 1987), known as 'The Old Yellow Book'.
- *Conditions of Contract for Design-Build and Turnkey* (first edition, 1995), known as 'The Orange Book'.
- *Client/Consultant Model Services Agreement* (third edition, 1998), known as 'The White Book'.

Other FIDIC publications include

- *Conditions of Subcontract for Works of Civil Engineering Construction* (first edition, 1994)
- *Tendering Procedure* (second edition, 1994)
- Guides to the Use of the Different FIDIC Conditions of Contract

- *Amicable Settlement of Construction Disputes* (first edition, 1992)
- *Mediation; Explanation and Guidelines* (first edition, 1993)
- *Insurance of Large Civil Engineering Projects* (1981, 1997)
- other publications on different aspects of the construction of engineering projects and the work of FIDIC.

All the FIDIC publications use English as the official and authentic text. Translations into some languages have been published by FIDIC. A list of all their publications is available from FIDIC (FIDIC Bookshop, Box 311, CH-1215 Geneva 15, Switzerland) or can be found at their website (www.fidic.org).

1.4.   The 1999 FIDIC Conditions of Contract

A publication such as a standard Conditions of Contract needs to be revised every few years. Construction procedures develop with changes to the size and complexity of projects and with experience of the use of the Contract procedures to overcome problems. These changes are reflected in the different editions of the FIDIC Conditions of Contract. However, around 1997, FIDIC decided that the time had come for a major review of all its Conditions of Contract. FIDIC decided that they would not just revise the existing editions, or even publish new editions, but would change the basic purpose of each of the Contracts and completely rearrange the layout of the Conditions of Contract.

The original FIDIC Conditions of Contract was for Works of Civil Engineering, followed by the Contract for Electrical and Mechanical Works. The principle of different Conditions of Contract to suit different types of engineering was modified in 1995 by the publication of the *Conditions of Contract for Design-Build and Turnkey*. The Design-Build and Turnkey Contract included detailed provisions for design by the Contractor, but was also written in a different style and the layout of Clauses and Sub-Clauses were different to the earlier Contracts.

The style and layout of the Design-Build Contract was developed further when, following test editions in 1998, FIDIC published the first edition of a new set of Conditions of Contract in 1999. These are listed below.

- *The Conditions of Contract for Construction.* This Contract is for Building and Engineering Works that have been designed by the Employer and replaces the traditional FIDIC Old Red Book (fourth edition, 1987) (as amended) (The RED Book).
- *The Conditions of Contract for Plant and Design-Build.* This Contract is for Electrical and Mechanical Plant and for Building and Engineering Works, designed by the Contractor and replaces the traditional Old Yellow and Orange Books. The publication is yellow in colour (The YLW Book).
- *The Conditions of Contract for EPC/Turnkey Projects.* This Contract is for Engineering, Procurement and Construction or Turnkey Projects where the Contractor takes total responsibility for the design and execution of the project, providing a completed project ready for occupation. This was at the time a new FIDIC Contract and the publication is silver in colour (The Silver Book).
- *The Short Form of Contract.* This Contract is for Building or Engineering Works of relatively small capital value or time period or for relatively simple works where a

much shorter form of contract is suitable. It is also used for projects incorporating relatively simple activities or procedures that are of a highly repetitive nature. This was, at the time, a new FIDIC Contract and is green in colour (The Green Book).

After the publication of this new set of contract forms, FIDIC set upon a task to develop more bespoke contracts for varying typical construction type of projects. They include the following contracts.

- *The Form of Contract for Dredging and Reclamation Works.* This Contract is specifically drafted for Dredging and Reclamation Works projects. This is a relatively new FIDIC Contract (first edition, 2006) and it is turquoise in colour (The Blue Book).
- *The Form of Contract for Design, Build and Operate Projects* (first edition) published in 2008. This Contract is for projects where the Contractor is to design, build and operate a project for a considerable period after construction (typically ca. 20 years). This is a new FIDIC Contract and the publication is gold in colour (The GOLD Book).
- *The Conditions of Subcontract for Construction for building and engineering works designed by the Employer* (first edition, 2011). This Contract is specifically drafted for use in conjunction with the 1999 RED Book. This is a new FIDIC Contract and the cover of the book has a red outline with a white background. (The SUB Book).

At the time of publication of this book, FIDIC have formed a number of additional task groups that are considering the need for further specialised contracts. In particular the follow-on for the Gold Book is being considered as important. Experience from users of the Gold Book has shown that it is a very good publication for so-called 'green-field' projects (i.e. new plants or buildings). However, modern times demand that existing plants also need significant rehabilitation to make them more eco-friendly as well as more efficient. The Gold Book does not cater very well for such 'brown-field' projects. It is probable that the new book for such existing plant/building projects will be called the FIDIC Brown Book.

FIDIC was also keen to be seen to guide the users towards good practices in developing projects that incorporate any of the Standard Forms of Contract. Thus, in 1982, FIDIC first published its *Tendering Procedure*. This dealt primarily with procedure for awarding civil engineering contracts. It was updated in 1994 to take account of the then current 1987 fourth edition of the RED Book and the third edition of the 1987 Old Yellow Book. Recently in 2011, FIDIC published its *Procurement Procedures Guide*, a book that is much more than a further edition of the *Tendering Procedures*. It takes into account not only the new rainbow suite of contracts published in 1999, but also the ever increasing acceptance and use of novel approaches to the procurement of engineering projects. This publication is light blue in colour.

In 2000, FIDIC published *The FIDIC Contracts Guide*, which gives detailed guidance on the use of the three principal 1999 Contracts (RED, YLW and Silver). Peter L. Booen wrote *The Contracts Guide* and was also the principal drafter of the 1999 Contracts. Hence, it gives an authoritative guide to the intentions of the panels who drafted these Contracts. The Contracts Guide includes the complete texts of these three contracts, printed side-by-side for easy comparison, together with comments on the background to each Sub-Clause.

1.5. The MDB Harmonised Construction Contract

During May 2005, FIDIC published the *Multilateral Development Bank Harmonised Edition of the Conditions of Contract for Construction* (the MDB Edition). This Contract was prepared and agreed with a group of development banks that have been licensed to include it in their procurement procedures. It was updated in March 2006, with the current version published in June 2010. We refer to it as the Pink Book (PNK).

1.6. Social clauses

The Conditions of Contract give the rights and obligations of the Parties to the Contract. The Parties' rights are expressed in detail in order to facilitate the allocation of liability in the event of any delays or additional costs being incurred by either Party. The detailed obligations, particularly the obligations of the Contractor, are aimed at delivering the completed Works within the requirements for quality, time and cost. In recent years standard Conditions of Contract, particularly FIDIC, have incorporated additional clauses that are intended to avoid, or restrict, corruption and the impact of the construction work on the environment and the local community, as well as improving disease avoidance and the care and facilities for the workers on the Site. The PNK Book includes a number of additional clauses for this purpose.

This trend is increasing and some organisations are now talking about including additional clauses in the Particular Conditions that would require the Contractor to execute the Works in a manner that may not be the most economic method, but would bring additional benefits to the country of the project. Some examples of these clauses are listed below.

- Dividing the project into smaller Contracts, thus enabling them to be carried out by local Contractors, which would assist in the development of local industry.
- The construction of labour camps and other temporary works in such a way that they can be handed over to the local community on completion of the project.
- Clauses requiring the use of local Subcontractors and local labour, even if this means providing training and other requirements.
- Clauses discouraging the import of machinery, in order to oblige the Contractor to use local labour rather than machines.

Any such clauses will add to the cost of the project, and possibly require a longer time period for construction, so they will only be included on the initiative of the Employer, with the agreement of any funding institution. However, construction projects are built for the benefit of the community, therefore, if additional benefits can be obtained for a reasonable additional cost, this may be a worthwhile investment.

1.7. Contents of this book

A further review of each of these Contracts is given in Chapter 3. Parts 2 and 3 include a Clause-by-Clause review of the *Conditions of Contract for Construction* [RED and Pink (PNK) Books] and for the *Conditions of Contract for Plant and Design-Build* [Yellow (YLW) Book] respectively. Part 4 gives a review and Clause-by-Clause comparison between the Subcontract (SUB) Book and the Construction Contract (RED Book).

FIDIC Users' Guide
ISBN 978-0-7277-5856-9

http://dx.doi.org/10.1680/fug.58569.009

Chapter 2
Comparisons between the different FIDIC Conditions of Contract

2.1.   The Traditional and the 1999 Conditions of Contract

Prior to 1999, the most popular form of FIDIC contract was the 1987 Conditions of Contract for Civil Engineering Construction (referred to here as the Old Red Book). In 1999, this was replaced by the *Conditions of Contract for Construction* or 'The RED Book'. The other traditional FIDIC Conditions of Contract have also been replaced by the new 1999 Conditions of Contract. With the increasing use of FIDIC Contracts, many users of the 1999 Rainbow Suite of Contracts will be using FIDIC for the first time. In the past these people may have had an advantage over those of us who were used to the traditional Contracts and had to convert to the 1999 Contracts. This perceived disadvantage, if any, is rapidly diminishing as the current 1999 Contracts gain momentum throughout the world. However, there are many similarities between the new and the old FIDIC Conditions and some Sub-Clauses are identical but have been rearranged and renumbered. There are also many changes, some of which are significant new requirements and others that constitute just a minor change of wording, possibly to clarify the previously intended meaning. It is very easy, but very dangerous, to assume that a Sub-Clause that looks familiar is unchanged from the traditional Conditions of Contract.

2.2.   The Traditional Old Red Book and the Conditions of Contract for Construction

Part 2 of this book refers to the 1999 *Conditions of Contract for Construction* and does not attempt to make frequent comparisons between the new and the old contracts. Comparisons of individual Sub-Clauses are particularly difficult, because some Sub-Clauses have changes in wording that may not be immediately apparent. Also, each of the Sub-Clauses must be read in conjunction with other Sub-Clauses and the rearrangement of the layout adds to the difficulty in identifying all the Sub-Clauses that relate to a particular subject. While the previous Contracts have remained in use, and will continue to do so for some time, when the 1999 Contracts are used it is essential to concentrate on the Contract for the particular project rather than to make constant comparisons with the Contract that was used for a previous project.

The reduction from the 'old' 72 to the 'new' 20 main Clauses, with the consequent rearrangement and increase in the number of Sub-Clauses, makes a direct comparison impossible. While it is not possible to list all the significant changes from the Traditional to the 1999 Contract, the most important changes include the following.

9

(*a*) *The General Conditions* and the *Guidance for the Preparation of Particular Conditions* are now included in the same publication and are no longer referred to as Part I and Part II. While this was certainly the cause of some significant confusion during the early life of the 1999 Contracts, it has become apparent over the subsequent years that to have the entire contractual and commercial conditions in one book, rather than having to page through, search and review between two (or more) separate volumes is certainly easier when executing the project to which the contract refers. Nevertheless, the 1999 Contracts still require an Appendix to Tender, Appendices and other add-ons to make the project contract complete for effective and efficient use by the Parties.

(*b*) New terms have been introduced and defined at Clause 1, including the following.

(*i*) 'Schedules', which is defined at Sub-Clause 1.1.1.7, is used to describe the Bills of Quantities and other Schedules submitted by the Contractor with the Letter of Tender and included in the Contract.

(*ii*) 'Base Date', at 1.1.3.1, is the date 28 days prior to the latest date for submission of the Tender. This is an important addition, as it clearly defines the date before which the Contractor is deemed to have included all aspects that may have an effect on his Tender.

(*iii*) 'Defects Notification Period', at 1.1.3.7, replaces the Defects Liability Period providing a more accurate description.

(*iv*) 'Accepted Contract Amount', at 1.1.4.1, is the amount accepted in the Letter of Acceptance, whereas 'Contract Price' includes adjustments.

(*v*) 'Goods', at 1.1.5.2, includes the Contractor's Equipment, Materials, Plant and Temporary Works.

(*vi*) A procedure has been introduced at Sub-Clause 2.4 [*Employer's Financial Arrangements*] enabling the Contractor to request from the Employer adequate proof of his ability to pay for the project.

(*vii*) A procedure has been introduced at Sub-Clause 2.5 [*Employer's Claims*] for use when the Employer wishes to claim against the Contractor.

(*viii*) The role of the Engineer has been changed, at Clause 3 and in various other Sub-Clauses. Most notably the traditional Contract requirement for the Engineer to be impartial has been removed!

(*ix*) The term 'Engineer's Representative' has been abolished and is apparently replaced by a reference, at Sub-Clause 3.2 [*Delegation by the Engineer*], to a possible resident engineer.

(*x*) Provisions for Quality Assurance have been introduced at Sub-Clause 4.9 [*Quality Assurance*].

(*xi*) There is provision, at Sub-Clause 4.20 [*Employer's Equipment and Free-Issue Material*], for the Employer to provide equipment and free-issue materials for the use of the Contractor.

(*xii*) The Contractor is required to submit detailed monthly progress reports, at Sub-Clause 4.21 [*Progress Reports*], the content of which is clearly defined.

(*xiii*) The delay causes listed at Sub-Clause 8.4 [*Extension of Time for Completion*], which entitle the Contractor to an extension of time, have changed. Delay caused by the Employer's other Contractors on the site and unforeseeable shortages in the availability of personnel or goods caused by epidemic or government actions have been added. Other special circumstances that may occur have been omitted. However, the provision for extensions as a result of delays caused by authorities at Sub-Clause 8.5 [*Delays Caused by Authorities*] and the Force Majeure provision at Clause 19 may cover some of the situations that would previously have been regarded as special circumstances.

(*xiv*) Provisions for Value Engineering have been introduced at Sub-Clause 13.2 [*Value Engineering*].

(*xv*) Sub-Clause 13.8 [*Adjustments for Changes in Cost*] includes a complex formula for calculating adjustments for changes in Cost.

(*xvi*) Provision for Force Majeure is included at Clause 19.

(*xvii*) Provision for a Dispute Adjudication Board (DAB) has been included at Sub-Clauses 20.2–20.4, together with changes to the role of the Engineer and to the procedures for dealing with Contractor's claims.

The fact that the role of the Engineer has changed is immediately apparent with the removal of the traditional Clause 67 reference of any dispute for an Engineer's decision, but the changes also affect a large number of other Sub-Clauses. The impact on the actual performance of the Engineer's duties may be less than would be expected from an initial study of the new conditions. When carrying out his duties the Engineer is deemed to act for the Employer, as Sub-Clause 3.1(a), and is, by definition at Sub-Clause 1.1.2.6, considered part of the 'Employer's Personnel'. However, the Engineer is still required to be 'fair' as discussed later in this chapter. The Contractor is no longer required to send copies of some notices direct to the Employer but other notices and information must still be sent direct to the Employer, with a copy to the Engineer.

While the Clause 67 reference of disputes to the Engineer has been replaced by the Dispute Adjudication Board, a new requirement has been introduced for claims to be determined by the Engineer. Sub-Clause 3.5 [*Determinations*] gives the procedure for the Engineer to decide matters such as Contractor's claims and requires him to consult with each Party '*in an endeavour to reach agreement*' and then '*make a fair determination in accordance with the Contract, taking due regard of all relevant circumstances*'. This suggests that the Engineer will not necessarily take the action that would be preferred by the Employer. This task is to be carried out by the Engineer and as per Sub-Clause 3.2 [*Delegation by the Engineer*] cannot be delegated to an assistant. Similarly, Sub-Clause 14.6 [*Issue of Interim Payment Certificates*] requires Interim Payment Certificates to show the amount that the Engineer '*fairly determines to be due*'.

Many improvements and other changes have been introduced but some small changes in wording can only be identified by a detailed comparison of Sub-Clauses. For example, the definition of 'Cost', at Sub-Clause 1.1.4.3 is based on '*all expenditure reasonably incurred*' compared with the previous reference to '*all expenditure properly incurred*'. The full implication of this change has become especially apparent when viewed in the light of Contractor's Claims. Whereas under the traditional wording the term '*properly incurred*' seemed to suggest that a simple proof of incurring the costs would suffice, it has now become more onerous on the Contractor to prove the quantum of his claim. The term '*reasonably incurred*' is now accepted as meaning that the Contractor not only has to prove that he actually incurred the claimed costs, but he must also show that these costs were unavoidable for the claimed event and would not have been incurred under normal circumstances.

2.3. Comparison between the different 1999 Conditions of Contract

The traditional FIDIC Conditions of Contract appear to have been written by different committees, at different times, with very little attempt at co-ordination. However, the 1999 Conditions of Contract have been properly co-ordinated, the style and layouts are identical and many of the provisions (where possible and/or logical to do so) are the same throughout the family of Contracts. This co-ordination will bring considerable benefits to the users of the Contracts. Anyone who is familiar with one Contract will be able to convert to a different Contract with the minimum of effort.

2.3.1 *The Conditions of Contract for Construction* (The RED Book)

This is the basic 1999 FIDIC Conditions of Contract and is reviewed Clause-by-Clause in Part 2, together with the MDB Harmonised Edition (the PNK Book).

As a change from the previous traditional Old Red Book, the 1999 Conditions of Contract are also intended for use for Electrical and Mechanical Works, which are designed by the Employer. Provision has been added, at Clause 9, for Tests on Completion, which are likely to be required for mechanical plant, and follows the previous Old Yellow Book for Electrical and Mechanical Works. A definition of 'Tests after Completion' has also been added at Sub-Clause 1.1.3.6 although details are to be provided in the Particular Conditions and Specification.

Although this Contract is intended for use when the Works are designed by the Employer, it recognises that some design may be required from the Contractor. The requirements are given at Sub-Clause 4.1 [*Contractor's General Obligations*], but rely on further details in the Specification. Clause 5 of the YLW Book provides useful guidance for additional requirements that could be included in the Particular Conditions.

2.3.2 *The Conditions of Contract for Plant and Design-Build* (The YLW Book)

This Contract is intended for use for Electrical and Mechanical Plant and for Building and Engineering Works, designed by the Contractor. It has been developed from the previous Orange Book for Design-Build and Turnkey. A Clause-by-Clause review and comparison with the *Conditions of Contract for Construction* is included in Part 3.

Due to the coordinated re-structuring of the traditional books, the majority of Clauses in this new YLW Book Contract are identical to Clauses in the *Conditions of Contract for Construction* (new RED Book). This will be a considerable help to the users of FIDIC Contracts. Anyone who is familiar with the *Conditions of Contract for Construction* should have no difficulty in adapting to the Plant and Design-Build Conditions of Contract. The main differences derive from the Contractor being responsible for the design of the Works and the Plant and Design-Build Conditions being for a lump sum contract with payments from Schedules rather than by re-measurement of the quantities of work that have been executed.

The main differences can be summarised as follows.

- The Definitions at Sub-Clause 1.1 [*Definitions*] change. '*Employer's Requirements*' and '*Contractor's Proposal*' replace '*Specification*' and '*Drawings*'. '*Schedule of Guarantees*' and '*Schedule of Payments*' replace '*Bill of Quantities*' and '*Daywork Schedule*'. However, Sub-Clause 13.6 [*Daywork*] still refers to a '*Daywork Schedule*'.
- At Sub-Clause 1.5 [*Priority of Documents*] the priority of documents changes to suit the different documents as defined at Sub-Clause 1.1 [*Definitions*].
- Sub-Clause 1.9 [*Delayed Drawings or Instructions*] is replaced by [*Errors in the Employer's requirements*], which is the equivalent terminology for the different situations.
- A procedure has been introduced at Sub-Clause 2.4 [*Employer's Financial Arrangements*] enabling the Contractor to request from the Employer adequate proof of his ability to pay for the project.
- A procedure has been introduced at Sub-Clause 2.5 [*Employer's Claims*] for use when the Employer wishes to claim against the Contractor.
- At Sub-Clause 4.1 [*Contractor's General Obligations*] the Contractor's obligations change to suit different situations.
- Sub-Clause 4.5 [*Assignment of Benefit of Subcontract*] is omitted and replaced by a very short Sub-Clause on [*Nominated Subcontractors*].
- Clause 5 [*Nominated Subcontractors*] is cancelled. The Clause number has been used for a detailed Clause [*Design*].
- Sub-Clause 8.3 [*Programme*], subparagraph (a) requires the programme to include trial operation.
- Sub-Clause 9.1 [*Contractor's Obligations*] for [*Tests on Completion*] includes additional requirements for pre-commissioning tests, commissioning tests and trial operation.
- Clause 12 [*Measurement and Evaluation*] is not required. The Clause number has been used for a new Clause for [*Tests after Completion*].
- Sub-Clauses 13.1 [*Right to Vary*] and 13.2 [*Value Engineering*] are reduced in scope to suit the different circumstances.
- Clause 14 is changed to suit the different payment procedures for the lump sum contract.
- Sub-Clause 20.2 [*Appointment of the Dispute Adjudication Board*] is changed so that the DAB is only appointed after a party gives notice to the other party of its intention to refer a dispute to the DAB. A so-called Ad-Hoc DAB! The *Guidance for the Preparation of Particular Conditions* suggests that for certain types of projects, particularly those involving extensive work on Site, it would be appropriate for the DAB to visit the Site on a regular basis. A permanent or Standing DAB should then be appointed as in the *Conditions of Contract for Construction*. Obviously a DAB that has to be appointed after the dispute has arisen will not be available immediately and will take longer to consider the problem and reach a decision.

The reason for this difference is not fully clear but it seems that on a design-build scenario the task-group entrusted with the drafting of the Contract considered that different people with different backgrounds and skill-sets might be appropriate for different design and construction disputes. If this is the case, then the problem could also be overcome in the composition of the DAB. For example, on a very large project it may be appropriate to appoint additional people to the DAB and for the Chairman to select the people who will deal with a particular dispute.

These changes are generally a consequence of the change to a lump sum contract with all the design being provided by the Contractor. Work, which has been designed by the Contractor, is not normally measured for payment because payment for measured quantities could lead to uneconomic design. Also, Contractor design is generally specified on a performance basis and is subject to testing after completion to check the operation and performance. At first, there seems no good reason why FIDIC published two separate Contracts. The alternative provisions could surely have been incorporated into a single document, either as alternative Clauses, or in the *Guidance for the Preparation of Particular Conditions*. However, on further reflection the distinction between separate books for design by the Employer and Design by the Contractor makes sense; especially for projects where the subject of construction involves significant portions of Contractor's in-house or proprietary technology necessitating the contract to be performance driven rather than quantity based. Such performance criteria are defined in detail before the parties enter into an agreement. The Particular Conditions are meant to accommodate project related requirements in addition to the standard conditions; they are to be kept simple so as to avoid chaotic changes that are impossible to control. To change a re-measurable contract into a High-Tec performance driven contract by 'simply' changing a couple of core standard clauses is tantamount to create chaos if done by people not familiar with the intimate way the standard Contracts have been drafted. It is for this reason that the differentiation at the standard Contract level makes a lot of sense, and this again is confirmed by the industry after some 10–12 years of intensive use.

The *Conditions of Contract for Construction* may be used when there is some design to be carried out by the Contractor. It also may sometimes be desirable to import the more detailed Contractor design Clauses and procedures into those contracts that combine Contractor design with Employer design.

The problems of preparing Employer's Requirements and of evaluating Tenders for design work must not be underestimated. Some Employers use a two-stage tender procedure for evaluating different design proposals but this may not be permitted by the funding agency.

2.4.   *The Conditions of Contract for EPC/Turnkey Projects (The Silver Book)*

The Conditions of Contract for EPC (Engineering, Procurement and Construction) *and Turnkey Projects* are intended for use when the Employer gives his overall requirements and the Contractor takes a much greater share of the risks, carries out all the design and construction and hands over the completed project, ready for operation 'at the turn of a key'. FIDIC state, in the Introduction to these Conditions, that they may be suitable when the Employer wants a greater degree of certainty in price and time and is prepared to pay a higher price in order to achieve that certainty. The Contractor takes virtually all the risks and must be paid a premium for taking the risks. This procedure could be appropriate where the EPC Contract is a critical part of a larger commercial venture. The Introduction also refers to situations where this Contract would not be suitable. An EPC/Turnkey Contract will involve more negotiation than an Employer Design or Design-Build Contract, before the Contract terms are agreed. Contractors must consider very carefully the implications of this allocation of risk before agreeing to the Contract Price.

In the Introductory Note to the Silver Book, FIDIC emphasises the additional requirements at tender stage if the Contractor is being required to take additional risks, and continues:

> *'Thereafter the Contractor should be given freedom to carry out the work in his chosen manner, provided the end result meets the performance criteria specified by the Employer. Consequently the Employer should only exercise limited control over and should in general not interfere with the Contractor's work.'*

The layout and content of the Clauses and Sub-Clauses is similar to *The Plant and Design-Build Contract* (YLW Book). The question to be considered is whether these Clauses are appropriate for a Contract under which the Contractor is being paid to take more of the risk. If the Contractor is to achieve the certainty of price and time stipulated then the involvement of the Employer during construction must be reduced to an absolute minimum. While the opportunities for Employer involvement, or interference, have been reduced from the *Conditions of Contract for Construction*, the Contract still allows more Employer involvement than is consistent with the above requirement, or is desirable in order to achieve the overall requirement for the project. This becomes a bigger problem in projects where the Employer is advised or forced to select the Silver Book Contract but is not sufficiently aware of and skilled in the intimate workings of such a Contract. His natural disposition will always be to 'get involved' because it is 'his' project!

Comparing this Contract with the Plant and Design-Build Contract (YLW Book), most of the changes reflect the greater allocation of risk to the Contractor. The wording of some clauses, such as 4.12, must be queried as to whether the Contractor should be expected to allow for risks that could not be foreseen when he prepared his Tender. In general, the Employer transfers risks to the Contractor but maintains his rights to be involved in administration and supervision. The changes include the following.

- The Appendix to Tender and Letter of Acceptance are not used. The information given in the Appendix to Tender under other Contracts must be included in the Employer's Requirements. The Letter of Acceptance is incorporated into the Contract Agreement that, by definition at Sub-Clause 1.1.1.2, can include annexed memoranda.
- Sub-Clause 1.1 [*Definitions*] includes changes to reflect the different requirements.
- Sub-Clauses 1.5 [*Priority of Documents*] and 1.6 [*Contract Agreement*] are changed because there is no Letter of Acceptance.
- Sub-Clause 1.9 [*Errors in the Employer's Requirements*] is deleted.

- A new Sub-Clause 1.9 [*Confidentiality*] has been introduced.
- Sub-Clause 1.12 [*Confidential Details*] reduces the information that the Contractor is obliged to disclose.
- Clause 3 [*The Engineer*] is replaced by [*The Employer's Administration*]. Throughout the Contract it is the Employer who carries out the duties allocated to the Engineer in other Contracts.
- Sub-Clauses 4.7 [*Setting Out*], 4.10 [*Site Data*], 4.11 [*Sufficiency of the Contract Price*], 4.12 [*Unforeseeable Difficulties*], 7.2 [*Samples*], 7.3 [*Inspection*] and 8.1 [*Commencement of Works*] are all reduced in length or modified to reflect the different risk allocation, often removing the Contractor's right to claim.
- Sub-Clause 8.4 [*Extension of Time for Completion*] is changed to remove the right to an extension of time in the event of exceptionally adverse climatic conditions or unforeseeable shortages. The other extension situations remain, which implies that it is still necessary to allow for the situation that the Contractor does not complete the Works on time, but is not responsible for the delay.
- The provisions of Sub-Clause 10.2 [*Taking Over of Parts of the Works*] are greatly reduced and only permitted by agreement.
- Sub-Clause 10.4 [*Surfaces Requiring Reinstatement*] is deleted.
- Sub-Clause 12.2 [*Delayed Tests after Completion*] is changed.
- The provisions in Sub-Clause 13.8 [*Adjustments for Changes in Costs*] are removed and it just refers to the Particular Conditions.
- Clause 14 [*Contract Price and Payment*] has major changes to reflect the greater risk taken and additional guarantees that have been given by the Contractor.
- Sub-Clause 17.3 [*Employer's Risks*] has been reduced.

The FIDIC Contracts Guide includes a note that a two-stage tender procedure may be adopted for some tenders that include Contractor design. The first stage would comprise technical proposals, on which the Employer may comment before he invites one or more of them to submit final priced offers. This is particular appropriate for EPC/Turnkey projects and will need to allow time for investigations and discussion on a more detailed design proposal. Some Employers make a payment to tenderers for the additional work in preparing a second stage tender. It may be necessary to discuss and modify the Particular Conditions in order to satisfy the tenderer.

The Silver Book is becoming popular with Employers, who see the advantages of a firm price and time. However, the Tender price will be higher than for a YLW Book contract, because the Contractor is taking more of the risks and has to allow for this in his Tender. In order to maintain these benefits, it is essential that the Employer does not interfere with the Contractor's work. FIDIC have published a warning that the EPC/Turnkey Contract is not advised when there is high unforeseen risk.

2.5.   *The Short Form of Contract* (The Green Book)

According to the FIDIC Foreword to the *Short Form of Contract*, it is suitable not only for work of relatively small capital value, but also for '*fairly simple or repetitive work or work of short duration without the need for specialist sub-contracts*'. In other words, it is suitable for projects that will not have too many problems. However, it should only be used after proper consideration of the consequences of using a shorter Contract. If, or when, problems do arise the Parties must be ready to negotiate an agreed solution, rather than rely on their interpretation of the Conditions of Contract. The role of an Adjudicator will be important.

The Short Form will serve a useful purpose but should be used with some caution. In particular, Employers who are not accustomed to using FIDIC Conditions of Contract must resist the temptation to regard the Short Form as being a simple Contract that will be easy to use. The other FIDIC Contracts, being longer and more complex, include more detailed procedures and give more guidance to Employers who encounter problems. The Short Form is more suitable for use by experienced Employers. While it does not include provision for an Engineer, it will be necessary to designate experienced people as 'Authorised Person' under Sub-Clause 3.1 and 'Employer's Representative' under Sub-Clause 3.2 [*Delegation by the Engineer*].

The Guidance Notes for the Short Form confirm that the Employer's Representative acts for and in the interests of the Employer, without any dual-role requirement to be impartial. However, the Employer's Representative will be making immediate decisions on any problems or claims and these decisions will be subject to review by the Adjudicator. The Employer's Representative should therefore act fairly and in accordance with the provisions of the applicable law and the Contract.

The Short Form of Contract is, as might be expected, a shorter form of *the Conditions of Contract for Construction*. The Short Form has only 15 Clauses and 52 Sub-Clauses instead of 20 Clauses and 163 Sub-Clauses. This is not just a matter of some Sub-Clauses being omitted and others retained, but also involves a rearrangement of the essential requirements with some requirements being omitted.

A detailed analysis and comparison with the *Conditions of Contract for Construction* would be longer than the Short Form of Contract itself and would serve no useful purpose here. The Short Form must be studied and judged on its own and the FIDIC publication includes a very useful section 'Notes for Guidance'. However, the following points of comparison are important.

- There are no Particular Conditions. If changes or additional Clauses are required then Particular Conditions must be added. Particular Conditions are already included in the priority list in the Appendix to the Agreement, with a high priority.
- The Appendix to Tender and Letter of Acceptance are not used and the Contract Agreement is replaced by a more general 'Agreement'. The Agreement refers to the Employer's request for Tender, the Contractor's Tender offer and the Employer's acceptance in a single document. The essential information that is normally in the Appendix to Tender is included in the Appendix, which is issued with the request for Tender and becomes part of the Agreement.
- Sub-Clause 6.1 [*Engagement of Staff and Labour*] gives a list of Employer's Liabilities, which would entitle the Contractor to an extension of time under Sub-Clause 7.3 [*Inspection*] and/or reimbursement of Cost under Sub-Clause 10.4

[*Surfaces Requiring Reinstatement*]. While this list appears to repeat the similar claims causes that are scattered throughout the RED Book, the use of the word 'liability' and the absence of detail means that it will be more difficult for claims to be rejected and the Employer could find that he has an increased liability under the Short Form.

■ The Dispute Adjudication Board is replaced by a single 'Adjudicator'. The Adjudicator is only appointed after the dispute has arisen. The shorter contract with fewer provisions could make the task of the Adjudicator more difficult. The Parties should consider selecting and appointing a suitable person at the start of the project so that that person is available to visit the Site if required before a claim develops into a dispute.

Employers considering the use of the Short Form should also look at the *FIDIC Form of Contract for Dredging and Reclamation Works* (*Dredgers Contract*; first edition, 2006), the Blue Book. This is similar to the Short Form but includes a number of improvements. Some of these improvements could usefully be introduced into the Short Form.

2.6. *The Contract for Dredging and Reclamation Works* (The Blue Book)

A test edition of the FIDIC *Form of Contract for Dredging and Reclamation Works* (The Blue Book) was published in 2001. This was used extensively within the industry until the first edition was published in April 2006. Only minor technical changes were made to the original text. The layout and content is similar to the FIDIC's Green Book, modified to suit the particular requirements for dredging and reclamation works.

It also incorporates several changes and improvements, presumably with the benefit of experience of use of the Green Book. However, unlike the Green Book, the Blue Book incorporates an Engineer (not party to the Contract) who must perform several duties including instructing variations, approving Contractor's designs, issuing payment certificates, issuing Taking-Over certificates and so on. The Engineer must act fairly but is not impartial hence disputes may be settled through a 'standing' DAB procedure followed, if necessary, by arbitration.

There is no defect notification period within the Blue Book due to the high cost of remobilising dredging equipment to correct any defects. This places higher than normal risks on the Employer who may consider additional indemnity requirements when preparing such contracts.

2.7. *The MDB Harmonised Conditions of Contract for Construction* (The PNK Book)

FIDIC licensed Multilateral Development Banks (MDB) to use a revised version of the *Conditions of Contract for Construction*. This is known as the *Multilateral Development Banks Harmonised Conditions of Contract for Construction*, and is referred to here as the PNK Book. The first version was released in May 2005 followed by the second in March 2006 and the third (current) in June 2010. The PNK Book may only be used for Contracts that are financed by one of the participating Banks. The PNK Book includes a number of changes to the 1999 RED Book. Some of these changes incorporate Clauses from previous

MDB Contracts and the intention is to standardise the contract and reduce the Particular Conditions.

The participating Banks are listed and the changes to the RED Book are reviewed in Part 2. Some of these changes relate specifically to actions that involve the Bank. Others include improvements that could usefully be incorporated into the Particular Conditions for other projects. The complete wording must be consulted, but some of the more important changes can be summarised as follows.

(*a*) General changes of layout and wording.
  (*i*)   The Appendix to Tender is renamed Contract Data and becomes Part A of the Particular Conditions.
  (*ii*) The DAB has become the Dispute Board (DB).
  (*iii*) The Bank and Borrower are defined and named in the Contract Data.
(*b*) Changes to Sub-Clauses related to the Bank.
  (*i*)   The Contractor must permit the Bank or its appointee to inspect or audit the Contractor's accounts (Sub-Clause 1.15 [*Inspections and Audit by the Bank*]).
  (*ii*) If the Bank suspends disbursements under its loan, the Employer must notify the Contractor (Sub-Clause 2.4 [*Employer's Financial Arrangements*]).
  (*iii*) All equipment, material and services to be incorporated in or required for the Works shall have their origin in any eligible source country as defined by the Bank (Sub-Clause 4.1 [*Contractor's General Obligations*]).
  (*iv*) The conditions to be fulfilled prior to issuing an instruction to Commence Works have changed (Sub-Clause 8.1 [*Commencement of Works*]).
  (*v*)   Payment must be made if the loan is suspended (Sub-Clause 14.7 [*Payment*]).
  (*vi*) Suspension and termination are possible if the loan is suspended or the Engineer fails to issue an instruction recording the agreement that all precedent conditions under Sub-Clause 8.1 [*Commencement of Works*] have been fulfilled (Sub-Clause 16.1 [*Contractor's Entitlement to Suspend Work*] and Sub-Clause 16.2 [*Termination by Contractor*]).
(*c*) Other changes.
  (*i*)   Profit (described as '*reasonable profit*' in the RED Book) is set at 5% (Sub-Clause 1.2 [*Interpretation*]).
  (*ii*)   Constraints on the Engineer's authority are listed (Sub-Clause 3.1 [*Engineer's Duties and Authority*]).
  (*iii*) The Contractor shall give fair and reasonable opportunity for contractors from the Country to be appointed as Subcontractors (Sub-Clause 4.4 [*Subcontractors*]).
  (*iv*)   The Contractor is encouraged to employ staff and labour from within the Country (Sub-Clause 6.1 [*Engagement of Staff and Labour*]).
  (*v*)   Additional Health and Safety, welfare and similar clauses (Sub-Clause 6.7 [*Health and Safety*], 6.12–6.22).
  (*vi*)   Includes a Retention Money guarantee (Sub-Clause 14.9 [*Payment of Retention Money*]).
  (*vii*) A Clause is added against Corrupt or Fraudulent Practices (Sub-Clause 15.6 [*Corrupt or Fraudulent Practices*]).

(*viii*) Procedures for the appointment of the Dispute Board have been improved (Sub-Clause 20.2 [*Appointment of Dispute Board*]).

2.8.   *Conditions of Contract for Design, Build and Operate Projects* (The Gold Book)

FIDIC published a new form of contract in 2008 for design build and operate contracts known as the Gold Book. This is to be used where the employer not only wishes the Contractor to design and build a facility but also requires him to operate and maintain such facility for a number of years after completion. The Contract is based on the YLW Book with new clauses covering the operation service. Typical projects would include power generation plants, factories, water treatment plants and so on.

FIDIC acknowledged that the ideal contract form varies for different project requirements and therefore various scenarios were considered. Would one or more contracts cover the design, build and operational phases? Would the contract form be suitable for green or brown field sites? What length of operational period would the contract form cover? FIDIC opted for a scenario that included a single contract covering the design build and operational phases under one contractor, a green field site and a 20 year operational period. FIDIC states in the Foreword that '*The document, as written, is not suitable for contracts which are not based on the traditional Design-Build-Operate sequence, or where the Operation Period differs significantly from the 20 years adopted.*'

Under such a contract, the Contractor is likely to be a large consortium or joint venture covering the multitude of skills and experience required for such a project. The Contractor is not responsible for funding the work or for the commercial success of the project as this responsibility remains with the Employer. Similar to the Silver Book, there is no Engineer with supervision being undertaken by the Employer's Representative.

2.9.   *Conditions of Subcontract for Construction* (The Subcontract Book)

FIDIC published *the Conditions of Subcontract for Construction* for use with the FIDIC *Conditions of Contract for Construction* (the 1999 RED Book) in 2011 following the release of a test edition in 2009. The Task Group appointed to draft the Subcontract Book had two main objectives. These were to create a contract that could be used back to back with the RED Book and would also ensure that any rights and obligations of the Employer under the Main Contract would not be diluted.

The use of the term 'back to back' is frequently used when discussing the interaction between the Subcontract and RED Books. It means the SUB Book adopts a similar structure and format of the RED Book, they have common terms and definitions with cross references, they can only be read together and the Subcontractor is generally under an obligation to perform his duties in a similar manner as the Contractor under the Main Contract.

Due to applicability of some of the Sub-Clauses within certain jurisdictions, the drafting task group have included alternative clauses within the guidance and further useful guidance is contained within the flow charts at the beginning of the book. Refer to Chapter 3 for details of the content and Part 4 for a detailed review.

2.10. FIDIC *Procurement Procedures Guide*

The Procurement Procedures Guide is new (first edition, 2011) and provides best practice procedures for procuring construction works. It provides guidance on how to achieve competitive tenders and thereby deliver satisfactory project objectives. It has its ancestry in FIDIC's *Tendering Procedure* Book which provided procedures for awarding contracts based on the traditional Old Red Book and the 1987 Old Yellow Book. This new guide not only takes into account the 1999 suite of contracts but also new management techniques currently used by the construction industry and financial institutions.

The Procurement Procedures Guide describes the basic considerations required for procurement management and explains how to develop a project strategy. It covers prequalification techniques, consultancy appointments and how to obtain tenders for all the contract forms including: Green, Blue, RED, YLW and Silver Books. It also provides advice on consultancy appointments and the award of works contracts.

FIDIC Users' Guide
ISBN 978-0-7277-5856-9

ICE Publishing: All rights reserved
http://dx.doi.org/10.1680/fug.58569.023

Institution of Civil Engineers

publishing

Chapter 3
The 1999 FIDIC Red, Pink, Yellow and Subcontract Books

3.1. Contents of the FIDIC publication

The 1999 FIDIC *Conditions of Contract for Construction* (the Red Book or RED), *The Multilateral Development Banks Harmonised Edition* (the Pink Book or PNK), the 1999 *Conditions of Contract for Plant and Design-Build* (the Yellow Book or YLW) and the *Conditions of Subcontract for Construction* (the Subcontract Book or SUB) have similar layouts and many of the Clauses are identical, or have minor changes. Clauses 5 and 12 of the YLW Book are completely different to the other books to allow for dedicated main clauses on Design and Testing after Completion.

The FIDIC publications (RED, YLW and SUB) are arranged in three sections – General Conditions, Guidance and Forms. The PNK Book also contains three sections; however the Guidance is replaced by Particular Conditions. The books contain far more than just the Conditions of Contract. The information contained within these sections is important for the Employer or Consultant when preparing tender documents. While one could say that only the General Conditions and other Contract documents are important for the Contractor and the staff on the Site who will administer the Contract, it is nevertheless logical for any prudent contractor to fully understand the way in which the FIDIC Standard Forms of Contracts have been compiled and drafted, if at least to efficiently put together his Tender/ Offer with a minimum potential for future disputes and/or conflict.

The contents of the FIDIC publications are shown in Appendix 4, 5 6 and 7 for RED, PNK, YLW and SUB respectively. A summary is provided in Table 3.1.

These documents are discussed in detail in Part 2 for RED and PNK, Part III for YLW and Part 4 for SUB but, in general, they cover the following subjects.

(*a*) The Foreword (RED and YLW) includes a brief review of the different 1999 Contracts with introductory comments on the use of the FIDIC Conditions of Contract. This is followed by three charts giving the typical sequence of events during the Contract, for payment and for disputes. The charts are useful, but have been over-simplified. Expanded and additional charts are provided in Chapter 5.
The Foreword (SUB) includes introductory comments on the predecessor to the book and stating that it has been prepared for use with the RED and PNK. Useful charts follow which cover typical sequences including principal events, payment and disputes.

Table 3.1 Contents of FIDIC Publications (RED, PNK, YLW and SUB)

RED and YLW	PNK	SUB
Foreword	Introduction	Foreword
General Conditions of Contract	General Conditions of Contract	General Conditions of Subcontract
– Contents	– Contents	– Contents
– The General Conditions, Clauses 1 to 20	– The General Conditions, Clauses 1 to 20	– The General Conditions of Subcontract, Clauses 1 to 20
– Appendix: General Conditions of Dispute Adjudication Agreement	– Appendix: General Conditions of Dispute Board Agreement	– Index and Index of Sub-Clauses
– Annex: Procedural Rules	– Annex: Procedural Rules	
– Index of Sub-Clauses	– Index of Sub-Clauses	
Guidance	Particular Conditions	Guidance
– Introduction	– Part A: Contract Data	– Introduction
– Notes on the preparation of Tender documents	– Part B: Specific Provisions	– Notes on the preparation of Tender documents
– Annex A: Example Form of Parent Company Guarantee		– Annex A: Particulars of Main Contract
– Annex B: Example Form of Tender Security		– Annex B: Scope of Subcontract Work and Schedule of Subcontract Documents
– Annex C: Example Form of Performance Security – Demand Guarantee		– Annex C: Incentive(s) for Early Completion, Taking-Over by the Contractor and Subcontract Bill of Quantities
– Annex D: Example Form of Performance Security – Surety Bond		– Annex D: Equipment, Temporary Works, Facilities, and Free-Issue Materials to be provided by the Contractor
– Annex E: Example Form of Advance Payment Guarantee		– Annex E: Insurances
– Annex F: Example Form of Retention Money Guarantee		– Annex F: Subcontract Programme
– Annex G: Example Form of Payment Guarantee by Employer		– Annex G: Other Items
Forms	Forms	Forms
– Letter of Tender	– Annex A: Letter of Bid	– Letter of Subcontractor's Offer

Table 3.1 Continued

RED and YLW	PNK	SUB
– Appendix to Tender	– Annex B: Letter of Acceptance	– Appendix to the Subcontractor's Offer
– Contract Agreement	– Annex C: Contract Agreement	– Contractor's Letter of Acceptance
– Dispute Adjudication Agreement (for one-person DAB)	– Annex D: Dispute Board Agreement (Sole Member DB)	– Subcontract Agreement
– Dispute Adjudication Agreement (for each member of a three-person DAB)	– Annex E: Dispute Board Agreement (each Member of a three-member DB)	
	– Annex F: Performance Security: Demand Guarantee	
	– Annex G: Performance Security: Performance Bond	
	– Annex H: Advance Payment Security: Demand Guarantee	
	– Annex I: Retention Money Security: Demand Guarantee	
	– Annex J: Parent Company Guarantee	
	– Annex K: Bid Security Bank Guarantee	

The Introduction (PNK) (there is no Foreword) provides reasons for publishing a separate book and the problems they hope to solve by its use (i.e. Incorporating MDB changes into the General Conditions, standardising bidding documents and reduce the number of additions and amendments within the Particular Conditions). This is followed by the Terms of Conditions of Use that describe the license agreement regulating the terms and conditions of use. It also provides a list of participating Banks.

(b) The General Conditions (RED, YLW, PNK and SUB) comprise Clauses 1 to 20 together with the Appendix and Annex for the Dispute Adjudication Board agreements (with the exception of SUB). These are the General Conditions that should be included unchanged in the Contract.

(c) The Guidance (RED, YLW and SUB) includes brief comments on the preparation of Tender documents, together with detailed guidance and proposals for changes and additional Sub-Clauses which may be required for a particular project and will form the

Particular Conditions. This is followed by examples of forms for the securities and guarantees which are referred to in the General and Particular Conditions. Copies of any forms that are required for the particular Contract should be included with the Particular Conditions. The Guidance is of interest to the Employer or Consultant who is preparing the Contract documents, but is also useful for the people who ultimately have to administer the Contract during project execution, in the sense that they need to know how, and why a particular clause is drafted or constructed the way it is. This will help them to better understand and in consequence to administer the contract more efficiently.

The Particular Conditions (PNK) (there is no Guidance) comprises two parts; Part A Contract Data and Part B Specific Provisions. The Contract Data replaces the Appendix to Tender but with modified terms and additional items (e.g. Bank and Borrower). All information must be provided by the Employer unlike the Appendix to Tender where some items are input by the Contractor. Part B contains the Specific Provisions that are required to suit the particular requirements of the project.

(*d*) The Forms (RED, YLW, PNK and SUB) include appropriate examples of the letters and agreements referred to in the General Conditions. Copies should be included with the Particular Conditions as appropriate. The Appendix to Tender (Contract Data in the case of PNK) is particularly important because this document gives facts, figures and information that are essential to complete the Clauses in the Conditions of Contract, as well as to properly administer the contract during project execution.

For the efficient administration of a FIDIC Contract, the pages that make up the General Conditions, Particular Conditions and Appendix to Tender/Contract Data may be combined and used as a single document, dependent on the individual preferences of the user.

3.2. Preparation of the Contract Documents

The Conditions of Contract do not just give the rights and obligations of the Parties as a legal framework to be used for the resolution of disputes. They also include project management procedures that are essential for the administration of the project, whether or not there are any problems. For example, the procedures for the Engineer to issue additional drawings and for the Contractor to receive payments are laid down in the Conditions of Contract. The proper use of the procedures in the Conditions of Contract can avoid problems becoming disputes and assist the fast and economic resolution of any disputes that do arise.

It is also necessary to draw the user's attention to the fact that the General Conditions combined with the Particular Conditions and other Contract Documents consist of numerous Sub-Clauses that can only be interpreted by taking into account many interconnections and cross-references. Whereas such interconnections and cross-references are sometimes stated in the relevant Sub-Clause, this is however the exception rather than the rule. Proper admin-istration requires an intimate knowledge of these mechanisms and should never be underestimated.

The Conditions of Contract on it's own is not complete. Certain information must be provided in other documents in order to complete the Contract documents. Lists of the Sub-Clauses that refer to information being provided in other parts of the documents that make up the

Contract are given in Tables 3.2–3.14. If, for any reason, this information is being provided in a different document then the Sub-Clause should be amended in the Particular Conditions. This information must be coordinated with other documents in order to ensure that the Contract, as a whole, will serve its intended purpose.

In order to prepare the Contract documents for a project using the FIDIC procedures it is necessary to do the following.

■ Incorporate the FIDIC General Conditions unchanged, together with the General Conditions of Dispute Adjudication Agreement/Dispute Board Agreement and the Annex: Procedural Rules. The General Conditions include numerous references to information given in other documents, referring to
For the RED/YLW/PNK
  □ the Appendix to Tender/Contract Data (Table 3.2)
  □ the Particular Conditions (Table 3.4)
  □ the Specification/Employer's Requirements (Table 3.6 and 3.8)
  □ the Schedules (Table 3.10)
  □ the Contract (Table 3.12).
For the SUB Book
  □ the Appendix to Subcontractors Offer (Table 3.3)
  □ the Particular Conditions (Table 3.5)
  □ the Specification (Table 3.7 and 3.9)
  □ the Schedules (Table 3.11)
  □ the Subcontract (Table 3.13)
  □ the Main Contract (Table 3.14).
If this information is not provided then, due to wording such as '*If there is no such table ... this Sub-Clause shall not apply*' the particular Sub-Clause may have been effectively deleted, possibly unintentionally.
■ Insert essential information in the Appendix to Tender/Contract Data/Appendix to Subcontractor's Offer, which is printed near the end of the FIDIC publication. The Employer must insert most of the information required (all data in the case of PNK) in order to complete the documents for calling Tenders. The Tenderer will insert other information. Each item in the Appendix to Tender/Contract Data/Appendix to Subcontractor's Offer, with the exception of Contract Name and Identification Number, relates to a Sub-Clause in the General Conditions and the requirements in relation to particular Sub-Clauses are considered in Part 2 (RED and PNK), Part 3 (YLW) and Part 4 (SUB). A list of the Sub-Clauses that refer to information in the Appendix to Tender/Contract Data is given in Table 3.2 and Appendix to Subcontractor's Offer in Table 3.3.
■ Prepare Particular Conditions to suit the Employer's requirements for the particular project. The preparation of the Particular Conditions is considered later in this chapter, with discussion of the relevant Sub-Clauses in Parts 2, 3 and 4.
■ Many of the Sub-Clauses in the General Conditions rely on other Sub-Clauses, both for their application and to maintain a fair balance between the interests of the Parties. When considering changes and additions to the FIDIC Sub-Clauses it is important to consider the relationship between different Sub-Clauses and to maintain this balance.

Table 3.2 Sub-Clauses that refer to the Appendix to Tender (RED and YLW)/Contract Data (PNK)

Sub-Clause	
1.1.2.2	Definition of 'Employer'
1.1.2.3	Definition of 'Contractor'
1.1.2.4	Definition of 'Engineer'
1.1.2.11	Banks Name (PNK only)
1.1.2.12	Borrower's Name (PNK only)
1.1.3.3	Definition of 'Time for Completion'
	Time for Completion of Sections
1.1.3.7	Definition of 'Defects Notification Period'
1.1.5.6	Definition of 'Section'
1.2	Profit (PNK only)
1.3(a)	Systems of electronic transmission
1.3(b)	Addresses for communications
1.4	Governing Law
	Ruling Language
	Language for Communications
1.6	Time for Parties to enter into the Contract Agreement (PNK only)
2.1	Times for access and possession of Site
3.1(B)(ii)	Engineer's Duties and Authority (% Variations) (PNK only)
4.2	Amount of Performance Security
5.1	Period for notice of errors (YLW only)
6.5	Normal working hours
8.1(c)	Effective Access to the Site (PNK only)
8.7	Delay damages, Works and Sections
13.5(b)	Percentage for Provisional Sums
13.8	Table of Cost adjustment data
14.2	Advance payment, currencies, proportions, repayment provisions
14.3(c)	Retention percentage and limit
14.5	Lists for Plant and Materials payments
14.6	Minimum Interim Payment Certificate
14.8	Publishing sources of commercial interest rates (PNK only)
14.9	Percentage value of Sections (YB only)
14.15	Currencies of payment
17.6	Maximum total liability (PNK only)
18.1	Submission of insurance information
18.2(d)	Maximum insurance deductible
18.3	Minimum limit per occurrence for insurance
20.2	Date for appointment of DAB/DB (RED and PNK only)
	List of DB sole members (PNK only)
	Number of DAB members
20.3	DAB/DB appointing authority

Table 3.3 Sub-Clauses that refer to the Appendix to Subcontractor's Offer for SUB

Sub-Clause	
1.1.3	Appendix to the Subcontractor's Offer
1.1.4	Contractor's name and address
1.1.7	Contractor's Subcontract Representative name and address
1.1.20	Subcontract DAB name and address
1.1.28	Subcontract Section
1.1.31	Subcontract Time for Completion
1.1.34	Subcontractor's name and address
1.1.37	Subcontractor's Offer
1.1.39	Subcontractor's Representative
1.4	Subcontract Communications addresses of recipients
1.11	Definition of Subcontract Sections
8.7	Subcontract Damages for Delay
13.5	Subcontract Adjustments for Changes in Cost
14.2	Total Subcontract Advance Payment, Number and timing of instalments and currency and proportions
14.6	Minimum amount of subcontract payment
14.7	Payment of Retention Money under the Subcontract
14.11	Subcontract Currencies of Payment
20.5	Appointment of a three person DAB

Any attempt to change this balance by amending sub-clauses that appear to favour any Party over the other generally causes problems and additional costs in either the Tender or in the resolution of the resulting claims and disputes. A list of the Sub-Clauses that refer to the Particular Conditions is given in Table 3.4 and 3.5.

■ Check the Sub-Clauses that refer to information in the Specification (RED, PNK and SUB) or Employer's Requirements (YLW) and ensure that they include the necessary information. A list of these Sub-Clauses is given in Table 3.6 and 3.7. In addition, there are Sub-Clauses which may be amplified or require additional information in the Specification (RED, PNK and SUB) or Employer's Requirements (YLW) and a list of these Sub-Clauses is given in Table 3.8 and 3.9. If a RED or PNK Book contract includes significant Contractor's design then some of the YLW Book clauses may be included in the Particular Conditions.

■ Check the Sub-Clauses that refer to the Schedules and ensure that these Schedules are provided. The Schedules are prepared by the Employer, completed by the Tenderer, and included in the Contract. A YLW Book Contract may include additional Schedules for information that is required to be included in the Tenderer's proposal. A list of these Sub-Clauses is given in Table 3.10 and 3.11.

■ Check the Sub-Clauses that refer to information which may be included elsewhere in the Contract/Subcontract and ensure that any necessary information is included in the appropriate document. The General Conditions do not state in which document this information should be included. Some of this information may have been included in supplementary documents that were agreed after the Tender was submitted and are

Table 3.4 Sub-Clauses that refer to the Particular Conditions (RED, PNK and YLW)

Sub-Clause	
1.1.1.10	Definition of Contract Data (PNK only)
1.1.3.6	Definition of 'Tests after Completion' (RED only)
1.5	Priority of Documents
1.6	Contract Agreement
1.13	Compliance with Laws
3.1	Engineer to obtain Employer's approval
4.1	Procedures for design by Contractor (RED and PNK only)
4.2	Form for Performance Security
4.4	Procedures for consent to Subcontractors
4.16	Procedures for delivery of Goods to Site
4.21	Procedures for progress reports
4.22	Security of the Site
8.1	Period for Commencement Date
12.1	Alternative details of tests after Completion (YLW only)
14.1	Evaluation of Contract Price
14.2	Form for advance payment guarantee
14.8	Percentage for financing charges
14.9	Retention Money Guarantee (PNK only)
17.6	Amount for limitation of liability
18.1	Insurance details (RED and YLW only)
18.2	Insurance details
18.3	Insurance procedures
20.3	DAB appointing authority

attached to the Letter of Acceptance, Contract Agreement, Contractor's Letter of Acceptance or Subcontract Agreement. A list of these Sub-Clauses is given in Table 3.12, 3.13 and 3.14.

- Consider the need for a Contract Agreement. The FIDIC standard agreement form may need to be modified to suit the Employer's standard agreement, but the information contained in the FIDIC form must be included as a minimum.
- Decide whether a one-person or three-person Dispute Adjudication Board (Dispute Board for PNK) is required and decide whether to nominate potential members in the Tender documents. Any amendments to the General Conditions of Dispute Adjudication Agreement or the Annex: Procedural Rules must be included in the Particular Conditions.
- Consider the use of FIDIC Annexes A to G for the Forms of Securities and Guarantees which are referred to in the General Conditions and the Appendix to Tender/Contract Data. The FIDIC standard forms may need to be modified to suit the Employer's standard forms and any applicable law.
- Consider the inclusion of a standard Letter of Tender for use by Tenderers when submitting their Tender. If the FIDIC standard form is not used then the Instructions to Tenderers should include the requirement that the Tenderer must confirm the appropriate information.

Table 3.5 Sub-Clauses that refer to or require information to be included in the Particular Condition of Subcontract (SUB)

Sub-Clause	
1.1.2	Annex
1.1.8	Employer – Employer's name and address
1.1.9	Engineer – Engineer's name and address
1.1.11	Main Contract – Brief particulars of Main Contract
1.1.18	Subcontract Bill of Quantities – Subcontractor's Bill of Quantities
1.1.27	Subcontract Programme
1.5	Priority of Subcontract Documents
1.9	Subcontract Agreement – Form of Subcontract Agreement annexed to the Particular Conditions
2.1	Subcontractor's Knowledge of Main Contract – list of confidential parts
2.2	Compliance with Main Contract – Exclusions from compliance with Main Contract as set out under subparagraph (xi) of this Sub-Clause
4.1	Subcontractor's General Obligations – exclusions from providing personnel, superintendence, labour, Subcontract Plant, Subcontractor's Equipment, Subcontractor's Documents etc.
7.1	Subcontractor's Use of Equipment, Temporary Works, and/or Other Facilities – specified in Annex D for use by the Subcontractor
7.2	Free-Issue Materials – specified in Annex D for use by the Contractor
8.3	Subcontract Programme – requirements set out in Part A of Annex F
9.1	Subcontract Tests on Completion – as specified in the Subcontract
9.2	Main Contract Tests on Completion – as specified in the Subcontract and the Main Contract
10.2	Taking-Over Subcontract Works – provision for taking over the Subcontract Works before taking over by the Employer in Annex C
10.3	Taking-Over by the Contractor – provision for taking over the Subcontract Works before taking over by the Employer in Annex C
14.6	Interim Subcontract Payments – percentage and limit of retention/minimum amount
17.3	Subcontract Limitation of Liability
18.1	Subcontractor's Obligation to Insure – risks, sums and names of insurance
18.2	Insurance arranged by the Contractor and/or the Employer – Insurance details under the Main Contract

- Check that all relevant data in the Employer's possession on sub-surface and hydrological conditions has been made available to the Tenderers for their information, as per Sub-Clause 4.10. Note that any additional information that came into the Employer's possession after the Base Date shall also be made available to the Tenderer or Contractor (if after Letter of Acceptance) and even during execution of the project.
- Review the FIDIC Tender procedures as given in The FIDIC Contracts Guide and the FIDIC Procurement Procedures Guide. Adopt or modify as appropriate.

Table 3.6 Sub-Clauses that refer to the Specification (RED and PNK) or Employer's Requirements (YLW)

Sub-Clause	
1.1.3.6	Definition of 'Tests after Completion' (PNK only)
1.1.6.3	Definition of 'Employer's Equipment'
1.13(a)	Permissions obtained by Employer
2.1	Possession of foundation, structure, plant or means of access
4.1(b)	Contractor's Drawings (RED and PNK only)
4.1(d)	Operation and maintenance manuals (RED and PNK only)
4.6	Opportunities for work by others
4.18	Values for emissions and discharges
4.19	Details and prices of services available on the Site
4.20	Details of Employer's Equipment and free-issue materials
5.1	Criteria for designers (YLW only)
5.2	Technical documents to be included in Contractor's Documents (YLW only) Language for Contractor's Documents (YLW only)
5.3	Contractor's Documents to be submitted for review and/or approval (YLW only) Review periods (YLW only)
5.4	Other standards for compliance (YLW only)
5.5	Training to be provided for Employer's Personnel (YLW only)
5.6	Numbers and types of copies of as-built drawings (YLW only)
5.7	Manuals for operation and maintenance (YLW only)
6.1	Arrangements for staff and labour
6.6	Facilities for staff and labour
6.13	Provision of suitable food for Contractor's Personnel (PNK only)
7.8	Payment of royalties (a) natural materials from outside the Site (b) disposal of demolition, excavation and surplus materials
17.1	Details of care for Employer's accommodation and facilities (PNK only)

3.3.   The layout of the General Conditions

The General Conditions include 20 main Clauses with 163 Sub-Clauses RED, 179 Sub-Clauses PNK, 167 Sub-Clauses YLW and 94 Sub-Clauses SUB. The arrangement of information in Sub-Clauses is far from perfect. When the user requires information to help resolve a particular problem it may be necessary to consult several different Sub-Clauses.

Table 3.7 Sub-Clauses that refer to the Specification (SUB)

Sub-Clause	
1.1.29	Definition of Subcontract Specification
1.5	Priority of Subcontract documents
10.1	Requirements for 'as-built' documents and operation and maintenance manuals

Table 3.8 Sub-Clauses that may usefully be amplified in the Specification (RED and PNK) or Employer's Requirements (YLW)

Sub-Clause	
1.8	Requirements for Contractor's Documents (RED and PNK only)
1.8	Number of copies of Contractor's Documents (YLW only)
1.13	Permissions being obtained by the Employer
2.1	Phased possession foundations, etc.
4.1	Contractor's designs (RED and PNK only)
4.1	Intended purposes for which the Works are required (YLW only)
4.6	Other contractors, etc. on the Site
4.7	Setting out points, etc.
4.14	Third parties
4.18	Environmental constraints
4.19	Services available on the Site
4.20	Employer's equipment and free issue material
5.1	Nominated Subcontractors (RED and PNK only)
5.1	Criteria for design personnel (YLW only)
5.2	Contractor's documents required and whether for approval (YLW only)
5.4	Technical standards and building regulations (YLW only)
5.5	Operational training for the Employer's Personnel (YLW only)
5.6	As-built drawings and other records (YLW only)
5.7	Operation and maintenance manuals (YLW only)
6.6	Facilities for personnel
6.18	Festivals and religious customs (PNK only)
6.22	Employment records of workers (PNK only)
6.23	Workers' organisations (PNK only)
7.2	Samples
7.4	Testing during manufacture and/or construction
9.1	Tests on Completion
9.4	Damages for failure to pass Tests on Completion (YLW only)
12.1	Tests after Completion (YLW only)
12.4	Damages for failure to pass Tests after Completion (YLW only)
13.5	Provisional Sums

The Index, which is located after the General Conditions, is a considerable help in locating references to a particular subject but detailed study and experience in the use of the General Conditions will often reveal important information in unexpected places. This is a typical characteristic of the FIDIC rainbow suite of Standard Forms of Contract and is often referred to as the 'interlinking dynamics' when having to work with or administering these type of contracts.

The main Clauses can be considered in groups of Clauses dealing with related subjects.

Table 3.9 Sub-Clauses that may usefully be amplified in the Specification (SUB)

Sub-Clause	
4.1	Subcontractor's General Obligations
6.1	Co-operation under the Subcontract
7.1	Subcontractor's Use of Equipment, Temporary Works, and/or Other Facilities
7.2	Free-Issue Materials
7.5	Subcontractor's Equipment and Subcontract Plant
8.5	Subcontract Progress Reports
9.1	Subcontract Tests on Completion
9.2	Main Contract Tests on Completion
10.1	Completion of Subcontract Works
10.3	Taking-Over by the Contractor
12.2	Quantity Estimated and Quantity Executed
13.2	Valuation of Subcontract Variations
13.6	Subcontract daywork
18.1	Subcontractor's Obligation to Insure
18.2	Insurance arranged by the Contractor and/or the Employer

Table 3.10 Sub-Clauses that refer to Schedules (RED, PNK and YLW)

Sub-Clause	
1.1.1.1	Contract includes schedules
1.1.1.6	Defines Schedules (YLW only)
1.1.1.7	Defines Schedules, which may include Bill of Quantities (RED and PNK only)
1.1.1.9	Daywork schedule and Schedule of Payment Currencies (PNK only)
1.1.1.10	Daywork schedule (RED only)
1.1.1.10	Schedule of Guarantees and Schedule of Payments (YLW only)
1.5	Schedules are the lowest priority (RED and PNK only)
1.5	Schedules have priority above Contractor's Proposals
12.2	Method of Measurement (RED and PNK only)
12.3	Quantity of an item (RED and PNK only)
13.5	Overhead charges and Profit
13.6	Daywork Schedules
13.8	Table of Cost adjustment data (RED and PNK only)
14.1	Refers to quantities and price data in Schedules (YLW only)
14.4	Schedule of Payments
14.5	Plant and Materials intended for the Works (PNK only)
14.15	Payment Currencies (PNK only)

Table 3.11 Sub-Clauses that refer to Schedules (SUB)

Sub-Clause	
1.5	Priority of Subcontract documents
2.1	Subcontractor's knowledge of main contract
12.2	Schedules of rates and prices
13.6	Subcontract daywork

- Clause 1: General Provisions covers subjects that apply to the Contract in general, such as definitions, the applicable language and law, the priority of the different documents that make up the Contract and the use of the different documents.
- Clauses 2 to 4: The Employer; The Engineer; The Contractor deal with the duties and obligations of the different organisations or entities that play a part in the execution of the Works. It is significant that the Contractor's Clause contains more Sub-Clauses than all the others added together. It is the Contractor who is responsible for executing the Works and so is the most active of the people who are involved with the project. Many other Sub-Clauses throughout the General Conditions refer to the Contractor's obligations.
- Clause 5 (RED, PNK and SUB): Nominated Subcontractors deals with the requirements for Subcontractors who have been nominated either in the Contract or as a variation under Clause 13. In the YLW Book nominated Subcontractors can only be instructed as a Variation and this is covered at Sub-Clause 4.5 [*Nominated Subcontractors*].
- Clause 5 (YLW): Design gives the requirements concerning the Contractor's design obligations.
- Clauses 6 and 7: Staff and Labour; Plant, Materials and Workmanship deal with the requirements for the items of workforce, materials and equipment/tools which the Contractor brings to the Site and uses to execute the project. The PINK Book expands Clause 6 by the addition of 13 new Sub-Clauses relating to staff and labour.
- Clauses 8, 9, 10 and 11: Commencement, Delays and Suspension; Tests on Completion; Employer's Taking Over; Defects Liability follow the sequence of events during the construction of the project, as well as defining certain milestones such as Commencement Date.
- Clause 12 (RED, PNK and SUB): Measurement and Evaluation gives the procedures when the work is to be paid by measurement of actual quantities.
- Clause 12 (YLW): Tests after Completion gives the procedures for any performance tests that are required after the Works have been handed over and occupied by the Employer
- Clauses 13 and 14: Variations and Adjustments; Contract Price and Payment give the procedures for the Employer to pay the Contractor for his work.
- Clauses 15 and 16: Termination by Employer; Suspension and Termination by Contractor refer to events that may occur at any time during the construction sequence and may bring the Contract to a close.
- Clause 17: Risk and Responsibility relates to the project as a whole and includes Sub-Clauses that are only used rarely, together with matters that are critical to the

Table 3.12 Sub-Clauses that refer to the Contract (RED, PNK and YLW)

Sub-Clause	
1.1.1.1	List of Contract documents
1.1.2.3	Contractor as named in Letter of Tender
1.1.2.5	Contractor's Representative as named in Contract
1.1.2.8	Subcontractors named in Contract
1.1.2.9	DAB members named in Contract
1.1.3.4	Tests on Completion specified in Contract
1.1.3.6	Tests after Completion specified in Contract
1.1.4.1	Accepted Contract Amount in Letter of Acceptance
1.1.4.10	Provisional Sum specified in Contract
1.1.6.7	Site as specified in the Contract
1.8	Number of copies of Contractor's Documents
4.1	Contractor's obligations to the extent in the Contract (RED and PNK only)
	Plant and Contractor's Documents specified in Contract
	Contractor's design as specified in Contract (RED and PNK only)
4.3	Contractor's Representative may be named in Contract
4.7	Setting out points specified in Contract
4.9	Quality assurance system stated in Contract
4.11	Obligations covered by Accepted Contract Amount
4.15	Access route arrangements unless otherwise stated in Contract
5.1(a)	Nominated Subcontractors stated in Contract (RED and PNK only)
6.5	Working hours stated otherwise in Contract
7.1(a)	Manner of execution specified in Contract
7.1(c)	Unless otherwise specified in Contract
7.2	Samples specified in Contract
7.4	Tests specified in Contract
7.6	Plant, Materials or other work not in accordance with Contract
8.2	Work for completion stated in Contract
8.3(c)	Inspections and Tests specified in Contract
9.4	Other obligations under the Contract
12.1	Measurement records by Engineer unless otherwise in Contract (RED and PNK only)
12.1	Requirements for Tests after Completion (YLW only)
12.2	Method of measurement except as otherwise stated in Contract (RED and PNK only)
12.3	Engineer's valuation except as otherwise stated in Contract (RED and PNK only)
13.4	Currencies provided for in Contract
13.6	Daywork Schedule included in Contract
14.4	Schedule of payments in Contract
14.7	Payment country specified in Contract
17.3(f)	Part of Permanent Works occupied by Employer specified in Contract
18.1	Insurance requirements
20.2	List of potential DAB/DB members in Contract

Table 3.13 Sub-Clauses that refer to the Subcontract (SUB)

Sub-Clause	
1.1.2	Annex attached to the Particular Conditions
1.1.16	Subcontract documents listed in Sub-Clause 1.5
1.1.17	Subcontract Agreement
1.1.22	Subcontract Drawings
1.1.23	Subcontract Goods
1.1.25	Subcontract Plant
1.1.26	Subcontract Price
1.1.29	Subcontract Specification
1.1.30	Subcontract Tests on Completion
1.1.33	Subcontract Works
1.1.35	Subcontractor's Documents
1.3	Subcontract Interpretation
1.4	Subcontract Communications
1.5	Priority of Subcontract Documents
1.6	Notices, Consents, Approvals, Certificates, Confirmations, Decisions and Determinations
1.7	Joint and Several Liability under the Subcontract
1.8	Subcontract Law and Language
1.9	Subcontract Agreement
1.10	No Privacy of Contract with Employer
2.1	Subcontractor's Knowledge of Main Contract
2.2	Compliance with Main Contract
2.5	Main Contract Documents
3.1	Contractor's Instructions
3.3	Contractor's Claims in connection with the Subcontract
3.4	Employer's Claims in connection with the Main Contract
3.5	Co-ordination of Main Works
4.1	Subcontractor's General Obligations
5.1	Assignment of Subcontract
5.2	Subcontracting
6.1	Co-operation under the Subcontract
6.3	Contractor's Subcontract Representative
6.4	Subcontractor's Representative
8.3	Subcontract Programme
8.4	Extension of Subcontract Time for Completion
8.7	Subcontract Damages for Delay
9.1	Subcontract Tests on Completion
9.2	Main Contract Tests on Completion
10.1	Completion of the Subcontract Works
11.1	Subcontractor's Obligations after Taking-Over
11.2	Subcontractor Defects Notification Period
12.2	Quantity Estimated and Quantity Executed

Table 3.13 Continued

Sub-Clause	
12.3	Evaluation under the Subcontract
13.2	Valuation of Subcontract Variations
13.4	Subcontract Adjustments for Changes in Legislation
13.6	Subcontract Adjustments for Changes in Cost
14.1	The Subcontract Price
14.2	Subcontract Advance Payment
14.4	Subcontractor's Statement at Completion
14.6	Interim Subcontract Payments
14.7	Payment of Retention Money under the Subcontract
14.8	Final Subcontract Payment
14.10	Cessation of the Contractor's Liability
15.1	Termination of Main Contract
15.2	Valuation at Date of Subcontract Termination
15.3	Payment after Termination of Main Contract
15.4	Termination of Main Contract in Consequence of Subcontractor Breach
15.5	Notice to Correct under the Subcontract
15.6	Termination of Subcontract by the Contractor
16.2	Termination by Subcontractor
16.3	Payment on Termination by Subcontractor
17.1	Subcontractor's Risks and Indemnities
17.2	Contractor's Indemnities
17.3	Subcontract Limitation of Liability
18.3	Evidence of Insurance and Failure to Insure
19.1	Subcontract Force Majeure
20.2	Subcontractor's Claims
20.4	Subcontract Disputes

Parties' responsibilities and overlap with the requirements of other important Sub-Clauses.

- Clause 18: Insurance includes important procedures which must be implemented at or before the commencement of the Works in addition to the procedures to be used when a problem occurs which will give rise to an insurance claim.
- Clause 19: Force Majeure is a general Clause that will only be used when the particular problem occurs. The final Sub-Clause refers to release from performance in a wider context than just due to Force Majeure.
- Clause 20: Claims, Disputes and Arbitration will probably be the most frequently used Clause in the whole Conditions of Contract. It includes procedures such as the submission and response to Contractor's claims, which must be used when a problem has arisen, as well as the procedures for the resolution of claims and disputes. Clause 20 also includes the procedures for the appointment of the Dispute Adjudication Board.

Table 3.14 Sub-Clauses that refer to the Main Contract (SUB)

Sub-Clause	
1.1.9	Engineer
1.1.11	Main Contract
1.1.12	Main Contract DAB
1.1.13	Main Contract Tests on Completion
1.1.14	Main Works
1.1.30	Subcontract Tests on Completion
1.3	Subcontract Interpretation
1.8	Subcontract Law and Language
2.1	Subcontractor's Knowledge of Main Contract
2.2	Compliance with Main Contract
2.3	Instructions and Determinations under Main Contract
2.4	Rights, Entitlements and Remedies under Main Contract
3.1	Contractor's Instructions
3.4	Employer's Claims in connection with the Main Contract
4.2	Subcontract Performance Security
4.4	Subcontractor's Documents
5.1	Assignment of Subcontract
7.4	Ownership of Subcontract Plant and Material
7.5	Subcontractor's Equipment and Subcontract Plant
8.3	Subcontract Programme
8.4	Extension of Subcontract Time for Completion
8.5	Subcontract Progress Reports
8.6	Suspension of Subcontract Works by the Contractor
8.7	Subcontract Damages for Delay
9.2	Main Contract Tests on Completion
10.2	Taking-Over Subcontract Works
11.1	Subcontractor's Obligations after Taking-Over
11.2	Subcontract Defects Notification Period
11.3	Performance Certificate
12.1	Measurement of Subcontract Works
12.3	Evaluation under the Subcontract
13.1	Variation of Subcontract Works
13.5	Subcontract Adjustments for Changes in Costs
13.6	Subcontract Daywork
14.1	The Subcontract Price
14.2	Subcontract Advance Payment
14.3	Subcontractor's Monthly Statements
14.4	Subcontractor's Statement at Completion
14.5	Contractor's Application for Interim Payment Certificate
14.6	Interim Subcontract Payments
14.7	Payment of Retention Money under the Subcontract
14.9	Delayed Payment under the Subcontract

Table 3.14 Continued

Sub-Clause	
14.10	Cessation of the Contractor's Liability
15.1	Termination of Main Contract
15.2	Valuation at Date of Subcontract Termination
15.3	Payment after Termination of Main Contract
15.4	Termination of Main Contract in Consequence of Subcontractor Breach
15.6	Termination of Subcontract by the Contractor
16.3	Payment on Termination by Subcontractor
17.1	Subcontractor's Risks and Indemnities
17.2	Contractor's Indemnities
17.3	Subcontract Limitation of Liability
18.1	Subcontractor's Obligation to Insure
18.2	Insurance arranged by the Contractor and/or the Employer
19.1	Subcontract Force Majeure
20.1	Notices
20.2	Subcontractor's Claims
20.3	Failure to Comply
20.4	Subcontract Disputes
20.5	Appointment of the Subcontract DAB
20.6	Obtaining Subcontract DAB's Decision
20.7	Subcontract Arbitration

3.4. The sequence of construction operations

The Sub-Clauses that refer to events during the construction period are generally arranged in the order in which the events occur

- Commencement
- Progress
- Completion
- Defects Period.

Other Clauses cover particular subjects, such as the duties of the Engineer, the submission of claims, the resolution of disputes and the provisions for payment. These matters can arise at any time during the construction period and are arranged in a reasonably logical order.

Due to the complexity of the General Conditions it is impossible to follow a logical sequence for all the Sub-Clauses. Many relevant Sub-Clauses will be found in unexpected places and some essential information is thus difficult to locate. Additional cross-references have been included in the review of Sub-Clauses in Part 2 of this book, in order to assist the user to locate the different sub-clauses that are relevant to a particular situation.

A series of flow charts (Figures 5.1 to 5.16) is included in Chapter 5 in order to show the sequence of events and the relevant Sub-Clauses for certain important operations. These flow charts must be read together with the relevant Sub-Clauses.

3.5.   Preparation of the Particular Conditions

The Conditions of Contract give the rights and obligations of the Parties, including project management obligations. The technical requirements are given in the other documents that make up the Contract. The FIDIC General Conditions are the standard document that must be included, unchanged, in every Contract. Any changes or additional obligations required by the Employer can be included in the Particular Conditions. Any changes must be kept to a minimum, must not disturb the balance between the Employer and the Contractor, and must be carefully checked to ensure that there is no conflict with any other Clause. Changes that are just a matter of personal preference should be avoided.

FIDIC, in the Guidance for the Preparation of Particular Conditions, give suggested wording for clauses to meet some of the most common requirements. This Guidance is extremely valuable and should be studied carefully when preparing Particular Conditions. If the FIDIC Clauses are used then there should not be a problem of conflicting Clauses, but there may be a need for additional cross-references and additional links in flow charts. Other FIDIC Contracts published since the 1999 Contracts should also be reviewed because they may include some improvements that could be included in the Particular Conditions.

These are the major matters that make further Clauses in the Particular Conditions essential.

■   Conflict with the applicable law and Laws that must be coordinated with the Contract.
■   The requirements of the financing authority.
■   Regulations and the Employer's procedures that must be coordinated with the Contract.
■   The need to import Clauses from another FIDIC Contract.

Examples of these requirements are given in the review of Sub-Clauses in Part 2 of this book. Examples of financing authority requirements are given in the review of the PNK Book in Part 2.

The need to co-ordinate the Contract with applicable law is particularly important in countries with a Civil Law jurisdiction. For example, the countries in Central Europe, which are making extensive use of the FIDIC Contracts, generally have detailed laws that govern all construction in that country. The full list of requirements varies from country to country, but some typical examples are shown below.

■   The need for the design to be approved before construction can start on Site, which causes additional problems with the use of Design-Build Contracts.
■   The power to give the Contractor possession of the Site may lie with a regional authority, rather than with the Employer.
■   The need for the Contractor to keep a 'Site Diary' – this is a legal document in which any authorised person or authority can make an entry.

- Certain authorities may have the power to issue instructions to the Contractor; this, of course, is in conflict with the FIDIC procedure under which only the Engineer can issue instructions to the Contractor.
- Legal restrictions on the power of any non-Government employee to authorise instructions and payments to the Contractor for a Government-financed contract, which may also affect Variations and the measurement of quantities.
- In certain countries applicable legislation may have the effect of making the time bar contained in Sub-Clause 20.1 [*Notification of Claims*] invalid.

If this study and co-ordination of the standard Contract with the Laws of the Country is carried out by a national authority or organisation, then the same provisions could be used for all Contracts in that country.

FIDIC Users' Guide
ISBN 978-0-7277-5856-9

ICE Publishing: All rights reserved
http://dx.doi.org/10.1680/fug.58569.043

Chapter 4
Claims and dispute procedures

4.1. Introduction

The procedures for the resolution of claims are arguably the most important part of any Conditions of Contract. Certainly the claims provisions are the most frequently used Clauses in any Contract. The FIDIC Conditions include provisions for the submission, consideration and resolution of claims and disputes in a number of different Clauses. Successive revisions to the traditional RED and the publication of the Orange Book, for Design-Build and Turnkey Contracts, introduced additional provisions and more complicated procedures. The 1999 Conditions of Contract have continued this process, introduced further provisions and even more complicated procedures. While Clause 20 is headed '*Claims, Disputes and Arbitration*', there are also a large number of other Clauses that include procedures that must be followed by the Employer, the Engineer and the Contractor, for the submission and response to claims.

The claims procedures in individual Clauses are reviewed in Parts 2, 3 and 4 of this book. This chapter gives a general review of the procedures for the submission, response and resolution of claims, draws attention to the need for co-ordination between different Clauses, and lists some of the Clauses which include similar procedures. These procedures may have been modified in the Particular Conditions and the actual Contract should be checked to ascertain the detailed procedures for the particular project.

Unexpected situations are an inevitable feature of every construction project. Delays and additional costs may result, leading to claims from either Party to the Contract. Claims may arise under any of the Contracts for a particular project. The procedures under all the 1999 FIDIC Contracts are similar, but may differ in detail from the procedures described in this chapter.

If the Particular Conditions include provision for management or steering committee meetings, then these may be utilised to discuss the situation and how to avoid further problems. Such meetings would be extremely valuable to resolve technical and contractual problems and prevent minor differences from developing into major disputes.

The 1999 FIDIC Conditions of Contract include respective procedures that must be followed for the Employer to claim against the Contractor. This helps with the efficient administration of the project and helps eliminate the cases where the Employer takes unilateral adverse actions against the Contractor. All the parties involved in the Contract (i.e. Employer, Engineer and the Contractor) should agree at an early stage in the Contract on the actual procedures to be followed.

4.2.   Claims and the Conditions of Contract

The Conditions of Contract give the rights and obligations of the Parties to the Contract, that is, the Employer and the Contractor. Other people, such as the Engineer, a Consultant or a Subcontractor may be involved in the preparation, analysis or administration of the claim but cannot be the principal who makes or receives the claim. While it may be legally possible for an outside person to claim that either the Employer or the Contractor has caused them damage by negligence or failing to comply with some legal obligation, any such claim would be outside the scope of this book. However, attention is drawn to Sub-Clause 17.1 [*Indemnities*] in which the Parties indemnify each other with respect to claims from third parties.

All claims that are made because of problems that arose under or in connection with a particular Contract must follow the procedures laid down in that Contract. The claims may be made because of the following reasons.

■   Made in accordance with a Clause, which states that the Contractor may be entitled to additional time and/or money in certain circumstances, or the Employer may be entitled to claim money and/or an extension to the Defects Notification Period from the Contractor.
■   One Party alleges that the other Party has failed to fulfil an obligation that is required by a Clause in the Contract.
■   One Party alleges that it is entitled to payment for some other reason, possibly because of some legal entitlement, which applies regardless of whether it is mentioned in the Contract.

Under the relevant detailed sections in Parts 2, 3 and 4 for each FIDIC Standard Form of Contract, that is, the RED, PNK, YLW and SUB contracts the particular claims procedures are discussed in detail. What follows is a generic summary of the core principles of making a claim under the terms and conditions of the FIDIC Standard Forms of Contract.

4.3.   Claims by the Contractor

Most claims are made by the Contractor and may be claims for an extension of time for completion of the Works, or for reimbursement of money that has been spent or will be spent, or both. If the claim is for money that is expected to be spent then, even if liability is agreed, the Engineer will not certify money until the Contractor has actually spent it. Claims for additional time frequently result in a claim for additional payment, which, under the FIDIC Conditions, must be submitted as a separate claim.

In general, the sequences of procedures for the submission of Contractor's claims, in accordance with Clause 20 [*Claim, Disputes and Arbitration*] and other clauses can be summarised as follows.

(*a*) In certain circumstances, the Contractor must give notice that he is aware of a situation, which may arise that could potentially have an adverse effect on the project in that it may increase costs and/or the time for completion. The Engineer may then take action to avoid, mitigate or resolve the issue.

(*b*) The Contractor gives notice that he considers himself to be entitled to additional time for completion and/or additional payment.

(*c*) The Contractor gives notice when he has actually suffered delay or incurred additional cost.

(*d*) The Contractor keeps contemporary records to support his claim and the Engineer may inspect the details. Factual records are checked at the time, to establish and agree the facts, regardless of any query or objections on liability.

(*e*) The Contractor submits his fully detailed claim with supporting particulars.

(*f*) The Engineer responds and approves or disapproves the claim. The Engineer must give his response on the principle, regardless of whether he has asked for further information.

(*g*) If and to the extent that the claim has been approved in principle, the Engineer will certify for interim payment any amount that has been substantiated.

(*h*) The Engineer proceeds in accordance with Sub-Clause 3.5 [*Determinations*], to determine any extension of time and/or additional payment.

(*i*) If the Contractor does not agree with the Engineer's determination and negotiation fails to achieve agreement between the Parties, then the claim becomes a dispute and the procedures of Sub-Clauses 20.4 [*Obtaining Dispute Boards Decision*], 20.5 [*Amicable Settlement*] and 20.6 [*Arbitration*] may be followed, basically: (i) a decision by the Dispute Adjudication Board, followed by (ii) an attempt at amicable settlement, followed by (iii) arbitration.

In the above list the Engineer's determination under Sub-Clause 3.5 [*Determinations*] has been included as point (*h*). This follows the order of events in Sub-Clause 20.1 [*Contractor's Claims*]. However, there are a number of situations when the Engineer may wish to make an earlier determination, or the Contractor may request an earlier determination. It should be noted that as per the wording of Sub-Clause 3.5 the Engineer has an obligation to make a determination whenever he is required to do so under the provisions of the Contract. The wording seems to suggest that these determinations are thus limited to situations that are explicitly governed by a relevant Sub-Clause according to which it is obligatory for the Engineer to make a determination. The FIDIC Conditions do not specifically allow the Contractor to request a determination, but action under Sub-Clause 3.5 may avoid the need to refer a dispute to the DAB under Sub-Clause 20.4 [*Obtaining Dispute Adjudication Board's Decision*].

Situations that may be suitable for an immediate determination under Sub-Clause 3.5 include the following.

- Any claim which the Resident Engineer, appointed under Sub-Clause 3.2 [*Delegation by the Engineer*], has stated will be rejected.
- Any claim the Engineer rejected in principle as point (*f*).
- The Engineer's review of Contractor's Document under Clause 5 (YLW) [*Design*].
- Adverse comments by the Engineer on the Contractor's programme as Sub-Clause 8.3 [*Programme*].
- A requirement to increase production under Sub-Clause 8.6 [*Rate of Progress*].
- A rejection under the second paragraph of Sub-Clause 20.1 [*Contractor's Claims*] (time-bar).
- Any other action by the Engineer that is disputed by the Contractor.

In practice, these procedures may be simplified and meetings may be held to try to resolve any problems.

All claims for additional time and/or money must follow the procedures of Sub-Clause 20.1 [*Contractor's Claims*], which requires a notice to the Engineer '*as soon as practicable, and not later than 28 days after the Contractor became aware, or should have become aware, of the event or circumstance*'. Failure to comply with this requirement may result in the Contractor losing his entitlement to the claim, under the second paragraph of Sub-Clause 20.1. Failure to comply with the requirements of other Sub-Clauses will result in the evaluation of the claim being reduced if the investigation of the claim was prevented or prejudiced by this failure, under the final paragraph of Sub-Clause 20.1.

It is thus important that all the appropriate notices are issued referring to all the relevant clauses. The people who are working on the Site will be subject to pressure not to delay the Works and to co-operate in resolving any claims in a non-adversarial manner. However, if a claim does escalate to DAB/DB level or even beyond to Arbitration then any failure to submit a notice or comply with a procedure will be used against the Party who failed to follow the correct procedure.

Notices are issued, as Sub-Clause 1.3 [*Communications*] in writing, to the Engineer or to a designated assistant, who is preferably resident on the Site. Notices may be required under several different Sub-Clauses for the same claim. The Sub-Clause 20.1 notice is a requirement for all claims, both for extension of time and for additional payment. Claims for an extension of time must also comply with the requirements of Sub-Clause 8.4 [*Extension of Time for Completion*].

Sub-Clause 8.3 [*Programme*] includes a general requirement for the Contractor to give notice of any '*specific probable events or circumstances which may adversely affect the work, increase the Contract Price or delay the execution of the Works*'.

Sub-Clause 20.1 [*Contractor's Claims*], which is reviewed in detail in Chapter 26 (RED), Chapter 49 (PNK) and Chapter 72 (YLW), also requires the Contractor to submit his claim in detail and to submit contemporary records. This information is essential for the Engineer, and any subsequent DAB/DB or arbitrator, to make a fair assessment of the claim. Many justifiable claims have been lost, or costs have escalated to a totally unreasonable extent, due to a failure by the Contractor to keep proper records at the time the claim occurred. Similarly, unjustified claims have been successful because the Engineer failed to keep proper records on behalf of the Employer.

Other Sub-Clauses that include specific requirements for claims are reviewed in Parts 2, 3 and 4 of this book and include the following.

■ Sub-Clauses that require the Contractor to give notice of an event that may cause delay or additional cost.
  1.9   (RED and PNK) [*Delayed Drawings or Instructions*]
  1.9   (YLW) [*Errors in the Employer's Requirements*]

4.12   [*Unforeseeable Physical Conditions*]

4.24   [*Fossils*]

5.4   (YLW) [*Change in technical standard*]

8.3   [*Programme*]

8.4   [*Extension of Time for Completion*]

16.1   [*Contractor's Entitlement to Suspend Work*]

17.4   [*Consequences of Employer's Risks*]

19.2   [*Notice of Force Majeure*]

19.4   [*Consequences of Force Majeure*]

■ Sub-Clauses that state an entitlement for the Contractor to claim an extension of time and/or additional payment.

1.9   (RED and PNK) [*Delayed Drawings or Instructions*]

1.9   (YLW) [*Errors in the Employer's Requirements*]

2.1   [*Right of Access to the Site*]

4.7   [*Setting Out*]

4.12   [*Unforeseeable Physical Conditions*]

4.24   [*Fossils*]

7.4   [*Testing*]

8.3   [*Programme*]

8.4   [*Extension of Time for Completion*]

8.5   [*Delays caused by Authorities*]

8.9   [*Consequences of Suspension*]

10.2   [*Taking-Over of Parts of the Works*]

10.3   [*Interference with Tests on Completion*]

11.8   [*Contractor to Search*]

12.2   (YLW) [*Delayed Tests after Completion*]

12.4   (YLW) [*Delay in giving access to investigate failure of Tests after Completion*]

13.7   [*Adjustments for Changes in Legislation*]

14.8   [*Delayed Payments*]

16.1   [*Contractor's Entitlement to Suspend Work*]

16.4   [*Payment on Termination*]

17.1   [*Indemnities*]

17.4   [*Consequences of Employer's Risks*]

18.1   [*General Requirements for Insurances*]

19.4   [*Consequences of Force Majeure*]

19.6   [*Optional Termination, Payment and Release*]

■ Sub-Clauses that involve valuation or similar requirements.

12.3   (RED and PNK) [*Evaluation*]

12.4   (RED and PNK) [*Omissions*]

15.3   [*Valuation at Date of Termination*]

16.4   [*Payment on Termination*]

18.1   [*General Requirements for Insurances*]

■ Sub-Clauses that provide for the Contractor to claim profit as well as cost.

1.9   (RED and PNK) [*Delayed Drawings or Instructions*]

1.9   (YLW) [*Errors in the Employer's Requirements*]

2.1   [*Right of Access to the Site*]

4.7   [*Setting Out*]

7.4   [*Testing*]

10.2   [*Taking-Over of Parts of the Works*]

10.3   [*Interference with Tests on Completion*]

11.8   [*Contractor to Search*]

12.2   (YLW) [*Delayed Tests*]

12.4   (YLW) [*Failure to Pass Tests after Completion*]

13.2   [*Value Engineering*]

16.1   [*Contractor's Entitlement to Suspend Work*]

16.4   [*Payment on Termination*]

17.4   [*Consequences of Employer's Risks*]

When submitting a claim the Contractor should include reference to all the Sub-Clauses that may be relevant. Some claims situations are covered by more than one Sub-Clause and the Contractor's entitlement may vary dependent on which Sub-Clause is used as the justification for the claim. Some Engineers may ask that a claim notice include reference to specific Sub-Clauses. However, this is frequently not possible and is not essential. Claims are submitted '*under any Clause of these Conditions or otherwise in connection with the Contract*', as stated at Sub-Clause 20.1 [*Contractor's Claims*]. Claims may also be made under the governing law, which is referred to Sub-Clause 1.4 [*Law and Language*] and stated in the Appendix to Tender/Contract Data.

4.4.   Claims by the Employer

Previous FIDIC Contracts did not include procedures for claims by the Employer. In practice, the Employer probably deducted any money he thought he was entitled to claim and the Contractor had to submit a claim to recover the deduction. This situation inevitably caused problems and disputes and procedures for claims by the Employer are now given at Sub-Clause 2.5 [*Employer's Claims*].

Under the RED and YLW Books, either the Employer or the Engineer must give notice to the Contractor of any claim as soon as practicable after he became aware of the event or circumstance. These procedures are from a timing point of view much less onerous than the procedures for Contractor's claims. The Employer must give such notice as soon as practicable after he became aware of the event giving rise to the claim. This requirement is not explicitly time-barred, and there is no equivalent to the '*or should have been aware*' condition as included under Sub-Clause 20.1 [*Contractor's Claims*].

Some interpretation of Sub-Clause 2.5 considers that the Employer is required to submit full particulars together with the notice. However, this is not specifically stated. The first paragraph states that notice and particulars shall be given when the Employer considers himself entitled to payment. The second paragraph states that the notice shall be given as soon as practicable after becoming aware (i.e. it specifies the time when the notice shall be given). The third paragraph states the claim shall include substantiation but does not mention when it shall be submitted.

The lack of a time-bar on the Employer frequently results in controversy with the Contractor who alleges that the Employer has waited far too long before initiating his claim. Employers sometimes perceive a benefit in delaying claims until such time as they have a clearer view on the extent of Contractor's claims. This may then be used as leverage for the subsequent negotiations.

The PNK Book goes someway to resolving this issue by introducing, in Sub-Clause 2.5 [*Employer's Claims*], a 28-day time limit imposed on the Employer after becoming aware or should have become aware. There are no sanctions if the Employer fails to meet this deadline and the lack of a time-bar on the Employer remains. Even though there is no sanction, the Employer would still be in breach when failing to meet the time-bar, which may entitle the Contractor to recovery of damages under Law.

After the Employer's claim has been submitted to the Contractor, the Engineer then makes a determination under Sub-Clause 3.5 [*Determinations*]. Either Party can then refer the matter to the DAB/DB if they are not satisfied with the Engineer's determination.

Sub-Clause 2.5 [*Employer's Claims*] is clear that if the Employer '*considers himself to be entitled to any payment under any Clause of these Conditions or otherwise in connection with the Contract, and/or to any extension of the Defects Notification Period*' then notice must be given and the Sub-Clause 2.5 procedure followed. Sub-Clause 2.5 lists specific deductions for which notice is not required. Since the wording of Sub-Clause 2.5 is specific in that the 'notice' is not required in this instance, it would imply that the Employer nevertheless must go through the procedure of Sub-Clause 2.5, that is, submit particulars under the Sub-Clause to which the Engineer shall make a determination. In other words there seems to be no direct and automatic entitlement to recovery of the amounts in question for which no notice is required. This has proven to be an issue in practice where employers frequently deduct such charges without the proper procedures having been followed.

There are a number of Sub-Clauses that refer to Sub-Clause 2.5, but other Sub-Clauses refer to claims or deduction due to the Employer and do not refer to Sub-Clause 2.5. Clearly, the Employer should follow these procedures before making a deduction under any Sub-Clause.

Sub-Clauses that refer to claims by the Employer are as follows.

- Sub-Clauses that require notice under Sub-Clause 2.5 [*Employer's Claims*].
  - 7.5   [*Rejection*]
  - 7.6   [*Remedial Work*]
  - 8.6   [*Rate of Progress*]
  - 8.7   [*Delay Damages*]
  - 9.4   [*Failure to Pass Tests on Completion*]
  - 11.3   [*Extension of Defects Notification Period*]
  - 11.4   [*Failure to Remedy Defects*]
  - 15.4   [*Payment after Termination*]
  - 18.1   [*General Requirements for Insurances*]
  - 18.2   [*Insurance for Works and Contractor's Equipment*]
- Sub-Clauses under which Employer can claim and notice is not required.
  - 4.2   [*Performance Security*]

4.19   [*Electricity, Water and Gas*]
4.20   [*Employer's Equipment and Free-Issue Material*]
5.4   [*Evidence of Payments to nominated Subcontractors*]
9.2   [*Delayed Tests*]
10.2   [*Taking Over of Parts of the Works*]
11.6   [*Further Tests*]
11.11   [*Clearance of Site*]

4.5.   The Dispute Adjudication Board (DAB)/The Dispute Board (DB)

If a dispute arises between the Employer and the Contractor then it may be referred to the DAB/DB under Clause 20.4 [*Obtaining Dispute Board's decision*]. The wording of Sub-Clause 20.4 is not restricted to disputes as a result of a claim being rejected, but includes disputes of any kind whatsoever, in connection with or arising out of the Contract or the execution of the Works.

The World Bank has, for several years, required major contracts to include procedures for a Dispute Review Board (DRB), which is similar to the DAB/DB. The difference is that the DAB/DB gives a decision, which must be implemented, whereas the DRB just gives a recommendation.

Detailed provisions and procedures for the DAB/DB are included at Sub-Clauses 20.2 to 20.4, the Appendix '*General Conditions of Dispute Adjudication Agreement/General Conditions of Dispute Board Agreement*' and the Annex: Procedural Rules. These provisions are reviewed in Part 5 of this book.

The DAB/DB comprises either one or three people. The Board is chosen by agreement at the start of the project. For the three-person Board each side proposes one person for the other side's agreement and the Chairman is chosen by agreement. This is also some-times referred to as the 'bottom-up' procedure, as opposed to the Parties appointing a chairman, who in turn selects his wingmen ('top-down'). Failing agreement by the Parties, the selection is made by an independent organisation, such as, for example, FIDIC, the Dispute Board Federation (DBF) or The Institution of Civil Engineers, named in the Appendix to Tender/Contract Data. The procedures under PNK differ slightly from RED with improvement to the wording for appointment and reappointment of members of the DB.

The DAB/DB procedure described in this section is the procedure from the RED and PNK Book. This is the preferred procedure, although the *Conditions of Contract for Plant and Design-Build* (YLW) include a procedure under which the DAB is only appointed after a dispute has arisen and is only appointed for that dispute. This procedure for a so-called Ad-Hoc DAB is described in Chapter 72.

The detailed procedures for the selection of the DAB/DB members are mentioned in a number of different documents. The FIDIC *Guidance for the Preparation of Particular Conditions* states the important principle that should govern the process as '*it is essential*

that candidates for this position are not imposed by either Party on the other Party'.
Sub-Clause 20.2 [*Appointment of the Dispute Adjudication Board*] and the Appendix to
Tender/Contract Data state that the Parties shall jointly appoint the DAB by the date 28
days after the Commencement Date (21 days before the date stated in the Contract Data for
PNK).

Under the RED Book, the fourth paragraph of Sub-Clause 20.2 and the FIDIC standard form
for the Letter of Tender to be submitted by the Contractor include reference to a list of
members for the DAB being included in the Contract. The list appears to have been prepared
by the Employer and the Contractor can accept, reject or add to the suggested list in his
Tender. For the Employer to suggest names in the Tender documents, to be accepted or
rejected by the Contractor before his Tender has even been considered will inevitably
imply, rightly or wrongly, that these people are being imposed on the Contractor by the
Employer. To ensure that the procedure is seen to be fair, and hence to establish confidence
in the DAB, it is preferable that names are not accepted or rejected until the Tender has
been accepted. The Parties can then exchange names and details of candidates and negotiate
on an equal basis.

Under the PNK Book, the procedure is different. The Contract encourages the Parties to first
consider who will serve as the DB and it is only if they have not jointly appointed the DB 21
days before the date stated in the Contract Data that each party proposes a member for a
three-person DB.

From a practical point of view, to insert the names of candidates in the documents at too early
a stage can cause problems. The work of a DAB/DB member requires a commitment to be
available to spend time for the project as and when required. Most potential DAB/DB
members also have other commitments and will only be able to accept a limited number of
DAB/DB appointments at any one time. To be asked to allow one's name to be proposed
involves some commitment and so the time between the name being put forward and the
appointment being confirmed should be kept to a minimum.

The members of the DAB/DB sign an Agreement and must be independent of both Parties.
They are appointed at the start of the project and visit the Site every three or four months.
The DAB/DB will have been sent copies of correspondence, notices and minutes of meetings
and so will be aware of progress and be familiar with problems which may have arisen.
During visits the DAB/DB will inspect the work in progress and meet the Parties. Any potential problems may be raised jointly by the Parties and the DAB/DB asked to express an
opinion. These visits have been found to assist with the resolution of claims and other
problems before they escalate into serious disputes.

Any dispute can be referred to the DAB/DB for its decision and the DAB/DB proceeds as
Sub-Clause 20.4. The DAB/DB decision must be implemented but if either Party is not
satisfied it can give a notice of dissatisfaction within 28 days. This notice prevents the
decision becoming final and enables the dispute to be referred to arbitration. The decision
remains binding on the Parties until changed by an amicable settlement or an arbitral
award.

FIDIC, the DBF and The Institution of Civil Engineers provide lists of potential members for Dispute Boards.

4.6.   Amicable Settlement

Sub-Clause 20.5 [*Amicable Settlement*] provides for a 56-day period after the issue of a notice of dissatisfaction before arbitration can commence. This is to allow time for the Parties to attempt to settle the dispute amicably. From a practical point of view, it can be questioned why this additional period was inserted into the procedure. The dispute is normally the result of an issue discussed at length between the parties at project level. Subsequent failure to reach agreement sent the issue to the DAB/DB, who in turn made a decision. The suggestion that after all this tension filled procedural manoeuvring the parties all of a sudden see the light of day is doubtful. On the other hand, there is something to be said for the notion that the party aggrieved by the DAB/DB decision (and thus the most likely one to issue the notice of dissatisfaction) will have been able to better understand the reasons for his aggrieved situation, also since it was now confirmed by an independent board of experts.

The Contract does not stipulate any particular procedure for the attempt at amicable settlement. It may consist of a further attempt by the Engineer, or a direct negotiation between senior managers of the Contractor and the Employer. Taking the dispute away from the people who are directly involved and who have established fixed positions may help to achieve a settlement that is acceptable to both sides.

Alternatively, an independent Mediator or Conciliator may be appointed to try to help the Parties reach an agreed settlement. FIDIC has published a report on Mediation, which gives a procedure for an independent person to help with the negotiation using a form of shuttle diplomacy. The Institution of Civil Engineers, several Arbitration Centres and the United Nations Commission for International Trade Law (UNCITRAL) have published procedures for Conciliation. These procedures are similar to Mediation, but the Conciliator or Tribunal may also give a recommendation, which is often used as a basis for further negotiation.

While the FIDIC Conditions do not require any particular procedure during the amicable settlement period, the Particular Conditions could require that conciliation, or some other procedure, should be tried.

4.7.   Arbitration

Sub-Clause 20.6 [*Arbitration*] gives the provisions for a dispute to be settled by arbitration. The FIDIC Contract refers to the Arbitration Rules of the International Chamber of Commerce in Paris, but the rules of a different arbitral organisation may have been included in the Particular Conditions. Any arbitration will be subject to the applicable arbitration law as well as the Arbitration Rules and the governing law as stated in the Contract.

Arbitration is a very thorough procedure that is outside the scope of this book. It has a legal basis and should achieve the correct legal decision on a dispute. However, legal procedures are never straightforward and several years of legal arguments may only produce the solution that was apparent to any reasonable person at the time the dispute started. During negotiation, both sides should remember that the legal and managerial costs of arbitration are considerable, are never fully recovered even by the winner, and so could logically be added to or subtracted from any offer to settle.

4.8.   The Courts

A Contract is, by definition, a legally binding agreement and the governing law is stated in the Contract. Hence, any disagreement or dispute may eventually be referred to the Courts of the country of the governing law.

If the provision for arbitration has been deleted in the Particular Conditions then any dispute which is not settled by the Engineer or the DAB/DB must be referred to the Courts. However, most international contractors prefer international arbitration because it is seen to be completely independent of the country of the project. Furthermore, arbitration is conducted by technical people or legal professionals with experience of construction law, and is generally conducted in the language of the project. The Courts may require that all the project paperwork be translated into the language of the country. Any dispute that may arise concerning the conduct or result of an arbitration may also be referred to the Courts.

FIDIC Users' Guide
ISBN 978-0-7277-5856-9

ICE Publishing: All rights reserved
http://dx.doi.org/10.1680/fug.58569.055

Chapter 5
Flow charts

During the administration of a project, any major activity or procedure will involve a sequence of events that are covered by different Clauses in the Contract. FIDIC have attempted to arrange the Sub-Clauses in a logical order so as to assist the user to identify the relevant sequence of Sub-Clauses. However, owing to the complexity of the situations that arise and the need to refer to certain Sub-Clauses in a variety of different situations, it has not been possible to achieve a sequence to suit all circumstances. The following flow charts (Figures 5.1 to 5.16) show the sequence of events and the relevant Sub-Clauses for certain important operations. These flow charts must be read together with the relevant Sub-Clauses.

The flow charts given in Figures 52.1 and 52.2 (in Part 3), and the reference to Clause 12 in Figure 5.7 only apply to the YLW Book. However, if the relevant Sub-Clauses have been included in the Particular Conditions for a RED Book contract then the flow charts will apply.

Figure 5.1a Time periods in the Conditions of Contract (RED)

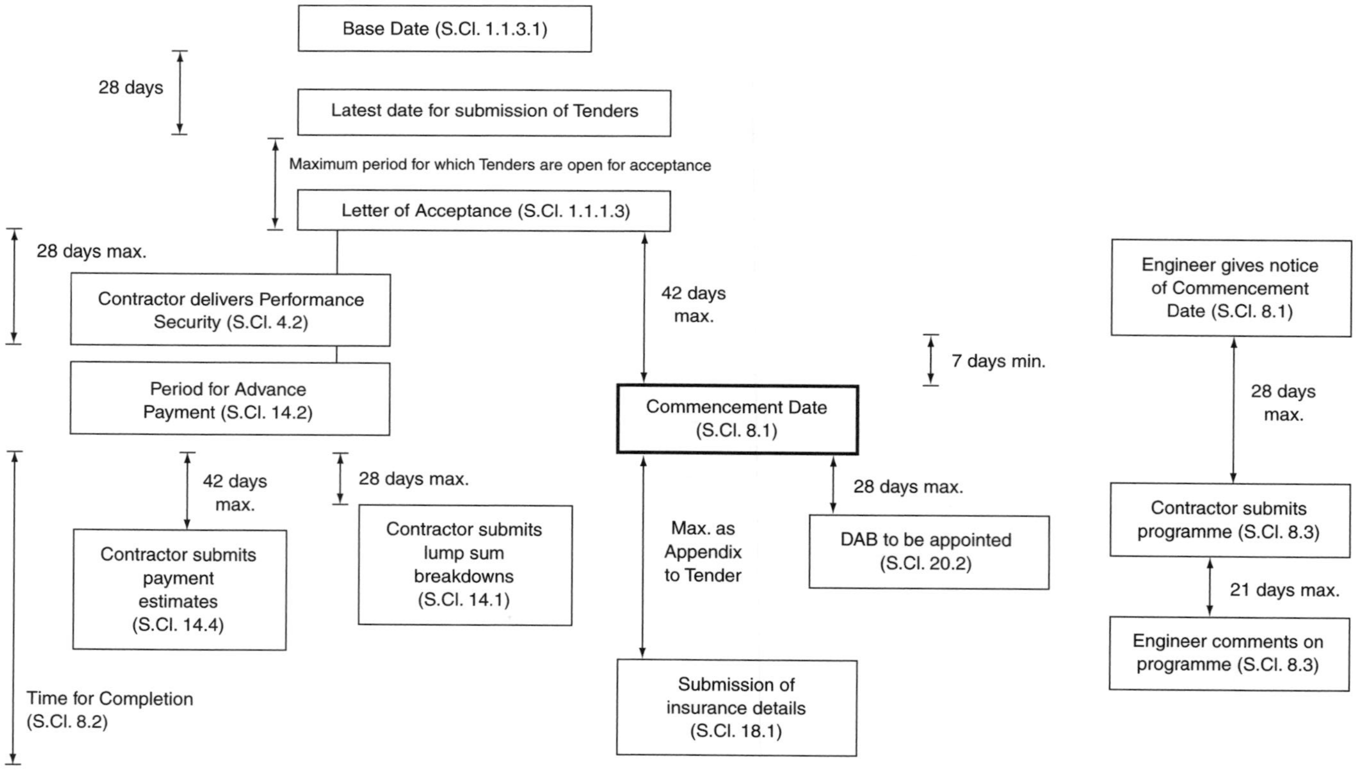

(*continued on Figure 5.1c*)

Figure 5.1b Time periods in the Conditions of Contract (PNK)

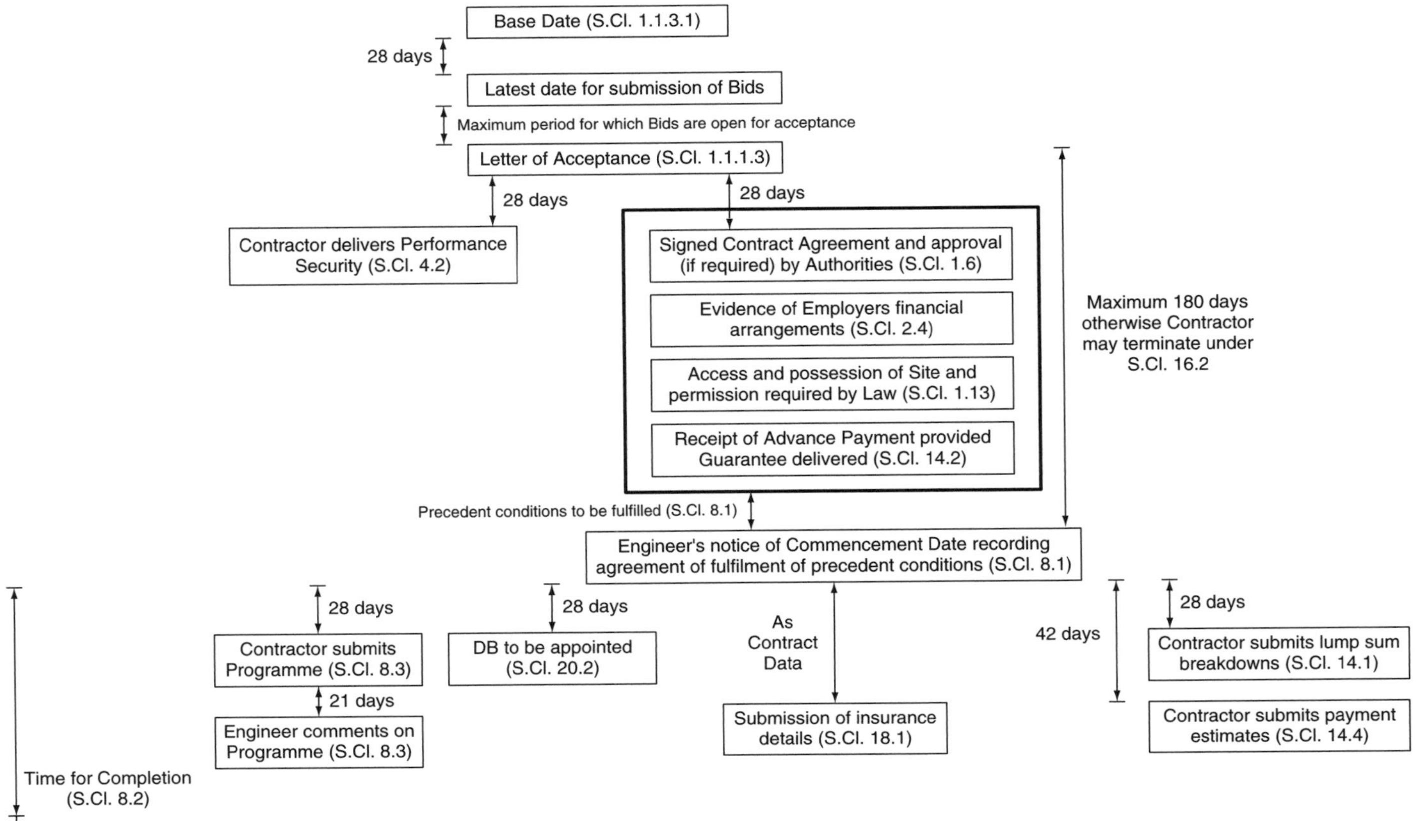

Base Date (S.Cl. 1.1.3.1)

28 days

Latest date for submission of Bids

Maximum period for which Bids are open for acceptance

Letter of Acceptance (S.Cl. 1.1.1.3)

28 days

Contractor delivers Performance Security (S.Cl. 4.2)

28 days

Signed Contract Agreement and approval (if required) by Authorities (S.Cl. 1.6)

Evidence of Employers financial arrangements (S.Cl. 2.4)

Access and possession of Site and permission required by Law (S.Cl. 1.13)

Receipt of Advance Payment provided Guarantee delivered (S.Cl. 14.2)

Maximum 180 days otherwise Contractor may terminate under S.Cl. 16.2

Precedent conditions to be fulfilled (S.Cl. 8.1)

Engineer's notice of Commencement Date recording agreement of fulfilment of precedent conditions (S.Cl. 8.1)

As Contract Data

Submission of insurance details (S.Cl. 18.1)

28 days

DB to be appointed (S.Cl. 20.2)

42 days

Contractor submits lump sum breakdowns (S.Cl. 14.1)

28 days

Contractor submits payment estimates (S.Cl. 14.4)

28 days

Contractor submits Programme (S.Cl. 8.3)

21 days

Engineer comments on Programme (S.Cl. 8.3)

Time for Completion (S.Cl. 8.2)

(continued on Figure 5.1c)

57

Figure 5.1c

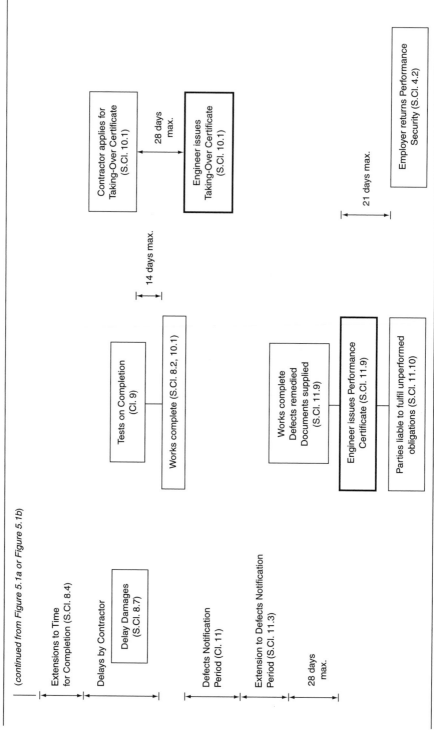

(continued from Figure 5.1a or Figure 5.1b)

Contractor applies for Taking-Over Certificate (S.Cl. 10.1)

28 days max.

Engineer issues Taking-Over Certificate (S.Cl. 10.1)

Employer returns Performance Security (S.Cl. 4.2)

21 days max.

14 days max.

Tests on Completion (Cl. 9)

Works complete (S.Cl. 8.2, 10.1)

Works complete Defects remedied Documents supplied (S.Cl. 11.9)

Engineer issues Performance Certificate (S.Cl. 11.9)

Parties liable to fulfil unperformed obligations (S.Cl. 11.10)

Extensions to Time for Completion (S.Cl. 8.4)

Delays by Contractor

Delay Damages (S.Cl. 8.7)

Defects Notification Period (Cl. 11)

Extension to Defects Notification Period (S.Cl. 11.3)

28 days max.

Figure 5.2 Progress requirements

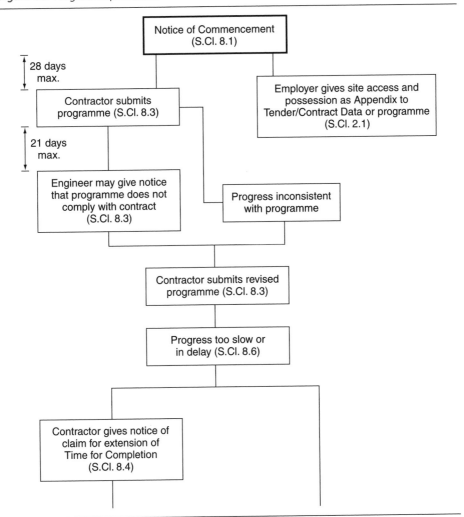

Notice of Commencement (S.Cl. 8.1)

28 days max.

Employer gives site access and possession as Appendix to Tender/Contract Data or programme (S.Cl. 2.1)

Contractor submits programme (S.Cl. 8.3)

21 days max.

Engineer may give notice that programme does not comply with contract (S.Cl. 8.3)

Progress inconsistent with programme

Contractor submits revised programme (S.Cl. 8.3)

Progress too slow or in delay (S.Cl. 8.6)

Contractor gives notice of claim for extension of Time for Completion (S.Cl. 8.4)

Figure 5.2 Continued

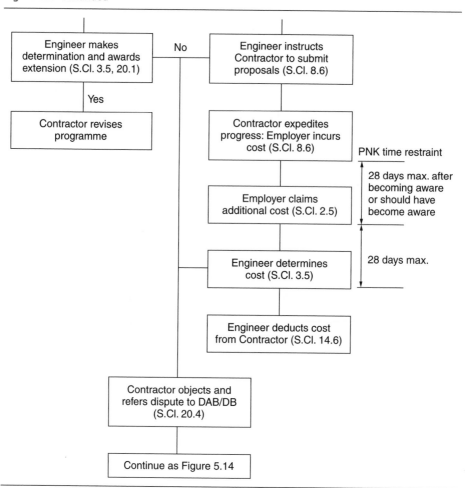

Figure 5.3 Workmanship procedures

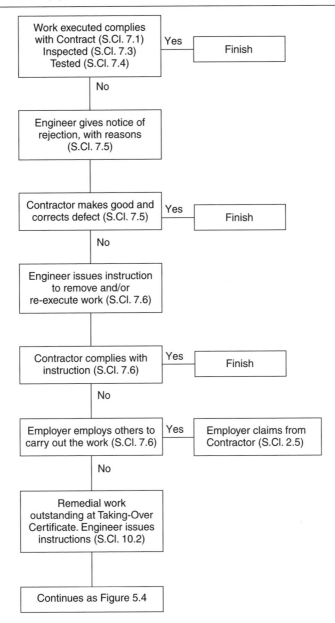

Notes:
1. If the Contractor questions an instruction by an assistant to the Engineer he may refer the matter to the Engineer (S.Cl. 3.2(b)).
2. The Contractor must comply with an instruction from the Engineer (S.Cl. 3.3), but may request payment as a Variation (Clause 13).
3. If a dispute arises it may be referred to the DAB/DB (S.Cl. 20.4).

Figure 5.4 Procedures at Completion of Works

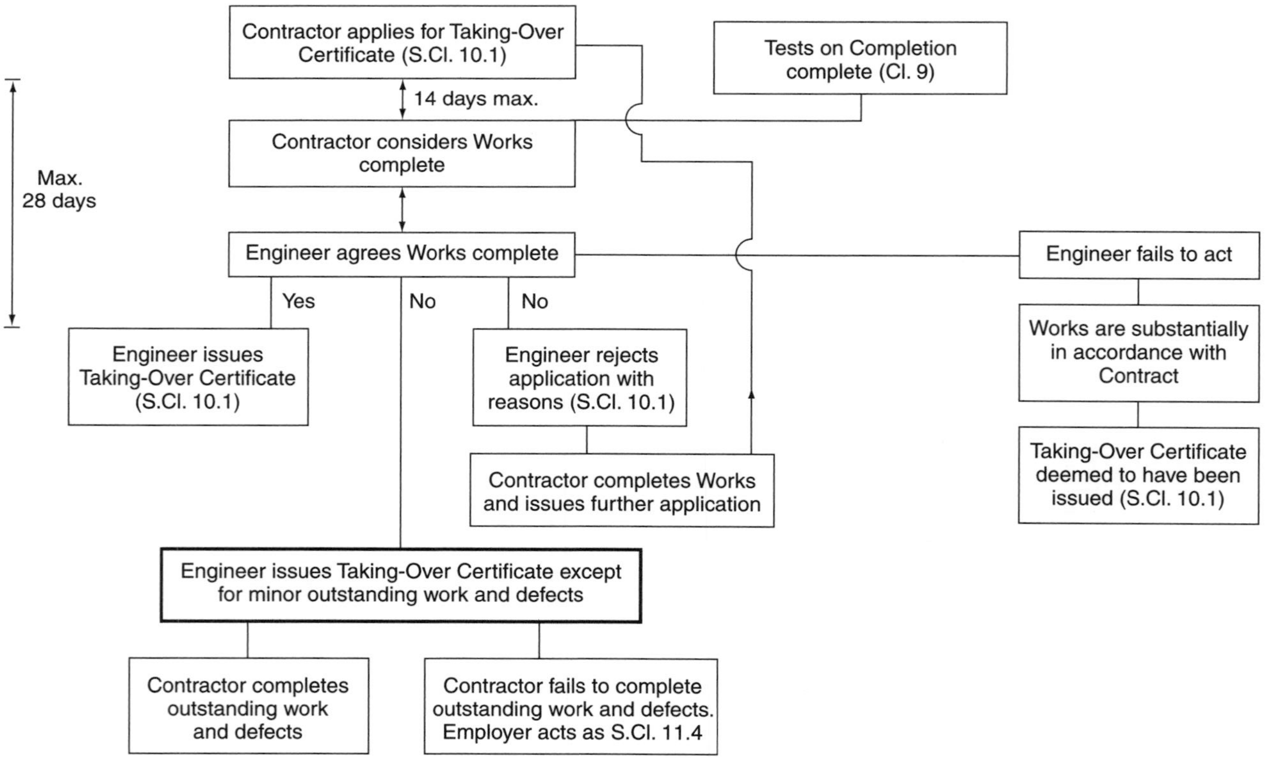

Figure 5.5 Procedures during Defects Notification Period

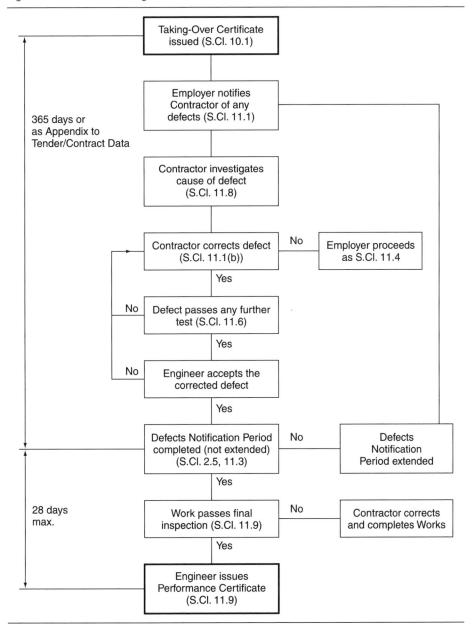

Figure 5.6 Procedures for advance payment

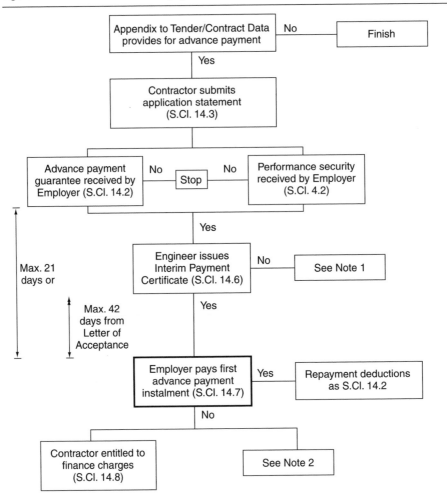

Note 1:
(i) If the Engineer fails to issue the Interim Payment Certificate within 28 days, the Contractor may give 21 days notice and then suspend or reduce the rate of work (S.Cl. 16.1).
(ii) If the Engineer fails to issue the Interim Payment Certificate within 56 days from the Contractor's application statement, the Contractor may terminate the Contract (S.Cl. 16.2(b)).

Note 2:
(i) If the Employer fails to make the payment, the Contractor may give 21 days notice and then suspend or reduce the rate of work (S.Cl. 16.1).
(ii) If the Contractor does not receive payment within 42 days of the due date, the Contractor may terminate the Contract (S.Cl. 16.2(c)).

Figure 5.7 Payment procedures

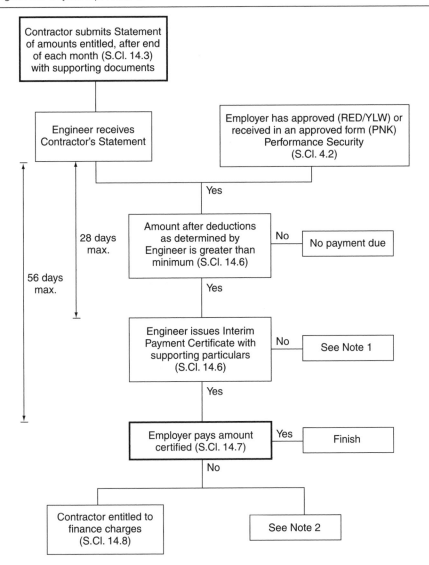

Note 1:
(i) If the Engineer fails to issue the Interim Payment Certificate, the Contractor may give 21 days notice and then suspend or reduce the rate of work (S.Cl. 16.1).
(ii) If the Engineer fails to issue the Interim Payment Certificate within 56 days from the Contractor's application statement, the Contractor may terminate the Contract (S.Cl. 16.2(b)).
Note 2:
(i) If the Employer fails to make the payment, the Contractor may give 21 days notice and then suspend or reduce the rate of work (S.Cl. 16.1).
(ii) If the Contractor does not receive payment within 42 days of the due date, the Contractor may terminate the Contract (S.Cl. 16.2(c)).

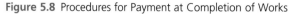

Figure 5.8 Procedures for Payment at Completion of Works

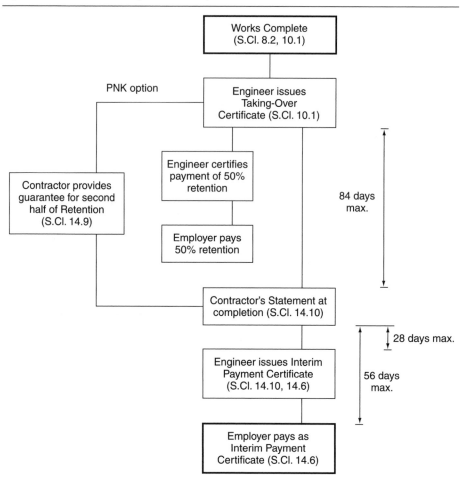

Note:
Sanctions for failure to certify or pay the Interim Payment Certificates are as Figure 5.4.

Figure 5.9 Procedures for final payment

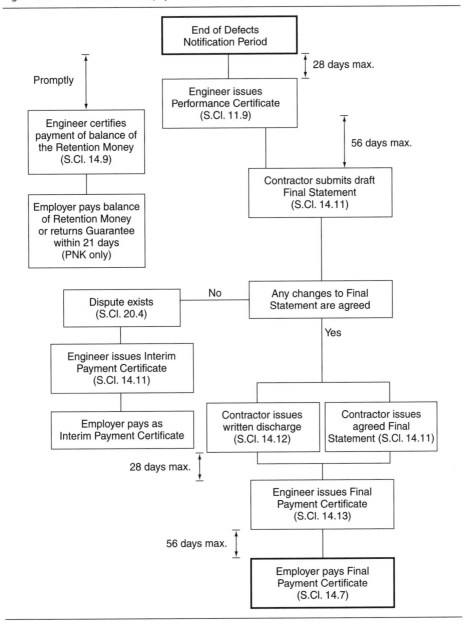

Figure 5.10 Claims by the Contractor

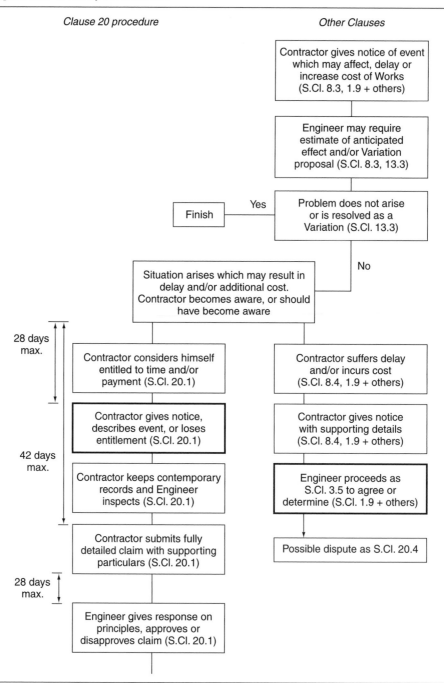

Figure 5.10 Continued

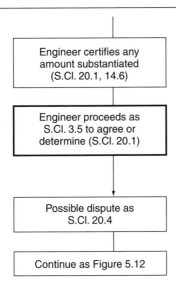

Notes:
1. S.Cl. 20.1 notice is obligatory for all claims.
2. Engineer's determination as S.Cl. 3.5 can be made at any time in the sequence as appropriate.
3. Notices etc. should be given as soon as possible within the stated period.
4. Consents and determinations must not be unreasonably withheld or delayed (S.Cl. 1.3).

Figure 5.11 Claims by the Employer

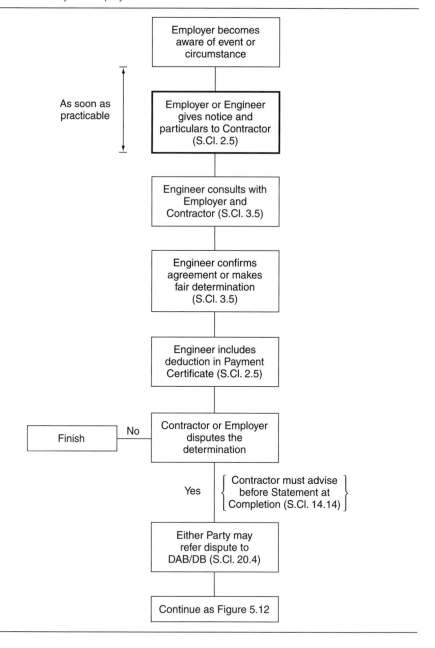

Figure 5.12 Procedures for disputes

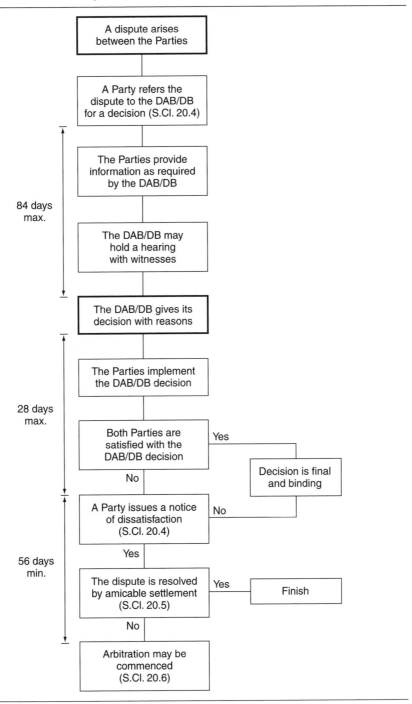

Figure 5.13 Variation Procedure

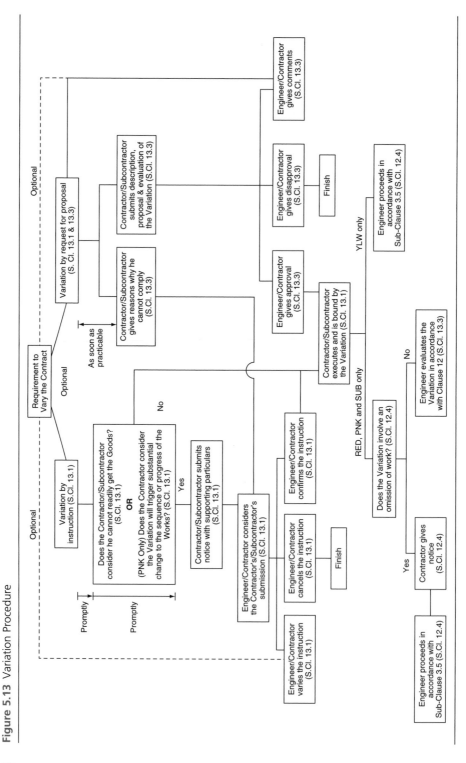

Figure 5.14 Suspension by the Engineer

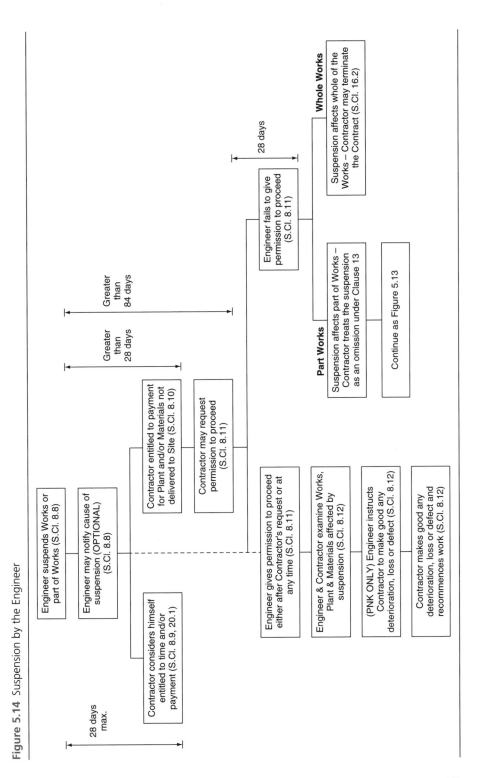

Figure 5.15 Force Majeure

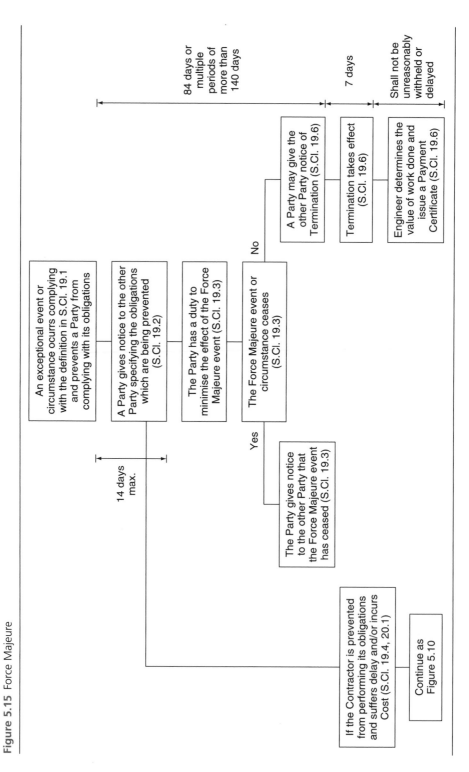

Figure 5.16 Tests on Completion

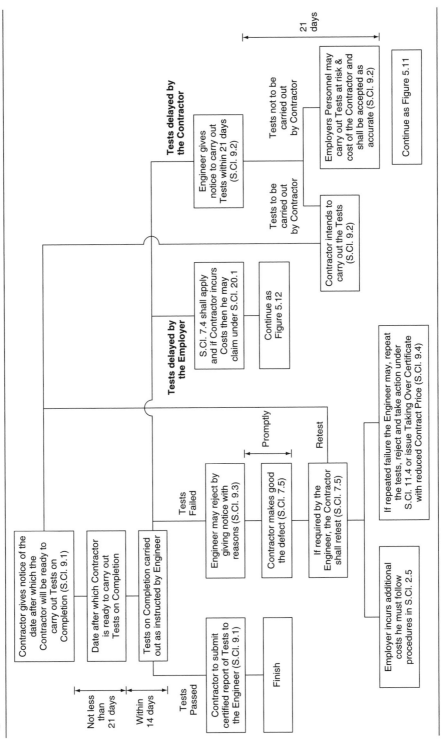

publishing

Part 2

The FIDIC Conditions of Contract for Construction For Building and Engineering Works designed by the Employer (RED and PNK Books)

FIDIC Users' Guide
ISBN 978-0-7277-5856-9

ICE Publishing: All rights reserved
http://dx.doi.org/10.1680/fug.58569.079

Chapter 6
Introduction to Part 2

The purpose for which this Contract is intended is stated on the title page:

The *Conditions of Contract for Construction* is for building and engineering works designed by the Employer, also referred to as the Red Book (RED) or as the Pink Book (PNK) for the MDB Harmonised Edition.

Part 2 of this book includes a detailed review and commentary for each of the Sub-Clauses in the FIDIC *Conditions of Contract for Construction,* RED and PNK Books. The commentary includes suggestions for possible changes that could be made in the Particular Conditions, together with cross-references to other Sub-Clauses that may be relevant.

The RED and PNK Books are very similar with many paragraphs unchanged. Only those changes have been considered in Chapters 29 to 51 relating to the PNK Book. Therefore, when considering the PNK Book, it is necessary to consult the chapters for both RED and PNK to obtain the full opinion.

This review does not attempt to give a detailed legal analysis of every Sub-Clause, but to draw readers' attention to problems that may arise in their interpretation and use. Potential strategies and/or solutions are discussed for such problems where appropriate. These comments, strategies and solutions and any references to the content of the Sub-Clauses must be read in relation to the complete Sub-Clause, which is reproduced in the text.

Part 1 of this book includes sixteen flow charts (Figures 5.1 to 5.16 in Chapter 5). These flow charts show the procedures required by the Conditions in certain sequences of events. For the RED and PNK Books the following flow charts should be consulted in connection with the Sub-Clauses as listed below.

- Figure 5.1a   Time periods in the Conditions of Contract (RED): Sub-Clauses 1.1.1.3, 1.1.3.1, 4.2, 8.1, 8.2, 8.3, 8.4, 8.7, 9, 10.1, 11, 11.3, 11.9, 11.10, 14.1, 14.2, 14.4, 18.1, 20.2
- Figure 5.1b   Time periods in the Conditions of Contract (PNK): Sub-Clauses 1.1.1.3, 1.1.3.1, 1.6, 1.13, 2.4, 4.2, 8.1, 8.2, 8.3, 8.4, 8.7, 9, 10.1, 11, 11.3, 11.9, 11.10, 14.1, 14.2, 14.4, 18.1, 20.2
- Figure 5.1c
- Figure 5.2   Progress Requirements: Sub-Clauses 2.1, 2.5, 3.5, 8.1, 8.3, 8.4, 8.6, 14.6, 20.1, 20.4

- Figure 5.3 Workmanship procedures: Sub-Clauses 2.5, 3.3, 7.1, 7.3, 7.4, 7.5, 7.6, 10.2, 13, 20.4
- Figure 5.4 Procedures at Completion of Works: Clause 9, Sub-Clauses, 10.1, 11.4
- Figure 5.5 Procedures during Defects Notification Period: Sub-Clauses 2.5, 10.1, 11.1, 11.3, 11.6, 11.8, 11.9
- Figure 5.6 Procedures for advance payment: Sub-Clauses 14.2, 14.3, 14.7, 14.8, 16.1, 16.2
- Figure 5.7 Payment procedures: Sub-Clauses 4.2, 14.3, 14.6, 14.7, 14.8, 16.1, 16.2
- Figure 5.8 Procedures for Payment at Completion of Works: Sub-Clauses 8.2, 10.1, 14.6, 14.9, 14.10
- Figure 5.9 Procedures for final payment: Sub-Clauses 14.7, 14.9, 14.11, 14.12, 14.13
- Figure 5.10 Claims by the Contractor: Sub-Clauses 1.3, 1.9, 3.5, 8.3, 8.4, 13.3, 14.6, 20.1, 20.4
- Figure 5.11 Claims by the Employer: Sub-Clauses 2.5, 3.5, 14.14, 20.4
- Figure 5.12 Procedures for disputes: Sub-Clauses 20.4, 20.5, 20.6
- Figure 5.13 Variation Procedure: Clause 12, Sub-Clauses 12.4, 13.1, 13.3
- Figure 5.14 Suspension by the Engineer: Clause 13, Sub-Clauses 8.8, 8.9, 8.10, 8.11, 8.12, 13, 16.2, 20.1
- Figure 5.15 Force Majeure: Sub-Clauses 19.1, 19.2, 19.3, 19.6
- Figure 5.16 Tests on Completion: Clause 9, Sub-Clauses 2.5, 7.5, 9.1, 9.2, 9.3, 9.4, 11.4, 20.1

The Red Book

FIDIC Users' Guide
ISBN 978-0-7277-5856-9

Chapter 7
Clause 1: General Provisions

Clause 1 covers general matters such as definitions, the law and language of the Contract and various matters concerning the documents. The FIDIC Forms of Contract follow a useful convention in that capitalised words in the individual Sub-Clauses are generally defined under Sub-Clause 1.1 [*Definitions*].

1.1 Definitions

In the Conditions of Contract ('these Conditions'), which include Particular Conditions and these General Conditions, the following words and expressions shall have the meanings stated. Words indicating persons or parties include corporations and other legal entities, except where the context requires otherwise.

1.1.1 The Contract

*1.1.1.1 '**Contract**' means the Contract Agreement, the Letter of Acceptance, the Letter of Tender, these Conditions, the Specification, the Drawings, the Schedules, and the further documents (if any) which are listed in the Contract Agreement or in the Letter of Acceptance.*

*1.1.1.2 '**Contract Agreement**' means the contract agreement (if any) referred to in Sub-Clause 1.6 [Contract Agreement].*

*1.1.1.3 '**Letter of Acceptance**' means the letter of formal acceptance, signed by the Employer, of the Letter of Tender, including any annexed memoranda comprising agreements between and signed by both Parties. If there is no such letter of acceptance, the expression 'Letter of Acceptance' means the Contract Agreement and the date of issuing or receiving the Letter of Acceptance means the date of signing the Contract Agreement.*

*1.1.1.4 '**Letter of Tender**' means the document entitled letter of tender, which was completed by the Contractor and includes the signed offer to the Employer for the Works.*

*1.1.1.5 '**Specification**' means the document entitled specification, as included in the Contract, and any additions and modifications to the specification in accordance with the Contract. Such document specifies the Works.*

*1.1.1.6 '**Drawings**' means the drawings of the Works, as included in the Contract, and any additional and modified drawings issued by (or on behalf of) the Employer in accordance with the Contract.*

1.1.1.7 'Schedules' means the document(s) entitled schedules, completed by the Contractor and submitted with the Letter of Tender, as included in the Contract. Such document may include the Bill of Quantities, data, lists, and schedules of rates and/or prices.

1.1.1.8 'Tender' means the Letter of Tender and all other documents which the Contractor submitted with the Letter of Tender, as included in the Contract.

1.1.1.9 'Appendix to Tender' means the completed pages entitled appendix to tender which are appended to and form part of the Letter of Tender.

1.1.1.10 'Bill of Quantities' and 'Daywork Schedule' mean the documents so named (if any) which are comprised in the Schedules.

1.1.2 Parties and Persons

1.1.2.1 'Party' means the Employer or the Contractor, as the context requires.

1.1.2.2 'Employer' means the person named as Employer in the Appendix to Tender and the legal successors in title to this person.

1.1.2.3 'Contractor' means the person(s) named as Contractor in the Letter of Tender accepted by the Employer and the legal successors in title to this person(s).

1.1.2.4 'Engineer' means the person appointed by the Employer to act as the Engineer for the purposes of the Contract and named in the Appendix to Tender, or other person appointed from time to time by the Employer and notified to the Contractor under Sub-Clause 3.4 [Replacement of the Engineer].

1.1.2.5 'Contractor's Representative' means the person named by the Contractor in the Contract or appointed from time to time by the Contractor under Sub-Clause 4.3 [Contractor's Representative], who acts on behalf of the Contractor.

1.1.2.6 'Employer's Personnel' means the Engineer, the assistants referred to in Sub-Clause 3.2 [Delegation by the Engineer] and all other staff, labour and other employees of the Engineer and of the Employer; and any other personnel notified to the Contractor, by the Employer or the Engineer, as Employer's Personnel.

1.1.2.7 'Contractor's Personnel' means the Contractor's Representative and all personnel whom the Contractor utilises on Site, who may include the staff, labour and other employees of the Contractor and of each Subcontractor; and any other personnel assisting the Contractor in the execution of the Works.

1.1.2.8 'Subcontractor' means any person named in the Contract as a subcontractor, or any person appointed as a subcontractor, for a part of the Works; and the legal successors in title to each of these persons.

1.1.2.9 ***'DAB'*** *means the person or three persons so named in the Contract, or other person(s) appointed under Sub-Clause 20.2 [Appointment of the Dispute Adjudication Board] or under Sub-Clause 20.3 [Failure to Agree Dispute Adjudication Board].*

1.1.2.10 ***'FIDIC'*** *means the Federation Internationale des Ingenieurs-Conseils, the international federation of consulting engineers.*

1.1.3 Dates, Tests, Periods and Completion

1.1.3.1 ***'Base Date'*** *means the date 28 days prior to the latest date for submission of the Tender.*

1.1.3.2 ***'Commencement Date'*** *means the date notified under Sub-Clause 8.1 [Commencement of Works].*

1.1.3.3 ***'Time for Completion'*** *means the time for completing the Works or a Section (as the case may be) under Sub-Clause 8.2 [Time for Completion], as stated in the Appendix to Tender (with any extension under Sub-Clause 8.4 [Extension of Time for Completion]), calculated from the Commencement Date.*

1.1.3.4 ***'Tests on Completion'*** *means the tests which are specified in the Contract or agreed by both Parties or instructed as a Variation, and which are carried out under Clause 9 [Tests on Completion] before the Works or a Section (as the case may be) are taken over by the Employer.*

1.1.3.5 ***'Taking-Over Certificate'*** *means a certificate issued under Clause 10 [Employer's Taking Over].*

1.1.3.6 ***'Tests after Completion'*** *means the tests (if any) which are specified in the Contract and which are carried out in accordance with the provisions of the Particular Conditions after the Works or a Section (as the case may be) are taken over by the Employer.*

1.1.3.7 ***'Defects Notification Period'*** *means the period for notifying defects in the Works or a Section (as the case may be) under Sub-Clause 11.1 [Completion of Outstanding Work and Remedying Defects], as stated in the Appendix to Tender (with any extension under Sub-Clause 11.3 [Extension of Defects Notification Period]), calculated from the date on which the Works or Section is completed as certified under Sub-Clause 10.1 [Taking Over of the Works and Sections].*

1.1.3.8 ***'Performance Certificate'*** *means the certificate issued under Sub-Clause 11.9 [Performance Certificate].*

1.1.3.9 ***'day'*** *means a calendar day and 'year' means 365 days.*

1.1.4 Money and Payments

*1.1.4.1 '**Accepted Contract Amount**' means the amount accepted in the Letter of Acceptance for the execution and completion of the Works and the remedying of any defects.*

*1.1.4.2 '**Contract Price**' means the price defined in Sub-Clause 14.1 [The Contract Price], and includes adjustments in accordance with the Contract.*

*1.1.4.3 '**Cost**' means all expenditure reasonably incurred (or to be incurred) by the Contractor, whether on or off the Site, including overhead and similar charges, but does not include profit.*

*1.1.4.4 '**Final Payment Certificate**' means the payment certificate issued under Sub-Clause 14.13 [Issue of Final Payment Certificate].*

*1.1.4.5 '**Final statement**' means the statement defined in Sub-Clause 14.11 [Application for Final Payment Certificate].*

*1.1.4.6 '**Foreign Currency**' means a currency in which part (or all) of the Contract Price is payable, but not the Local Currency.*

*1.1.4.7 '**Interim Payment Certificate**' means a payment certificate issued under Clause 14 [Contract Price and Payment], other than the Final Payment Certificate.*

*1.1.4.8 '**Local Currency**' means the currency of the Country.*

*1.1.4.9 '**Payment Certificate**' means a payment certificate issued under Clause 14 [Contract Price and Payment].*

*1.1.4.10 '**Provisional Sum**' means a sum (if any) which is specified in the Contract as a provisional sum, for the execution of any part of the Works or for the supply of Plant, Materials or services under Sub-Clause 13.5 [Provisional Sums].*

*1.1.4.11 '**Retention Money**' means the accumulated retention moneys which the Employer retains under Sub-Clause 14.3 [Application for Interim Payment Certificates] and pays under Sub-Clause 14.9 [Payment of Retention Money].*

*1.1.4.12 '**Statement**' means a statement submitted by the Contractor as part of an application, under Clause 14 [Contract Price and Payment], for a payment certificate.*

1.1.5 Works and Goods

*1.1.5.1 '**Contractor's Equipment**' means all apparatus, machinery, vehicles and other things required for the execution and completion of the Works and the remedying of any defects. However, Contractor's Equipment excludes Temporary Works, Employer's Equipment (if any), Plant, Materials and any other things intended to form or forming part of the Permanent Works.*

1.1.5.2 'Goods' means Contractor's Equipment, Materials, Plant and Temporary Works, or any of them as appropriate.

1.1.5.3 'Materials' means things of all kinds (other than Plant) intended to form or forming part of the Permanent Works, including the supply-only materials (if any) to be supplied by the Contractor under the Contract.

1.1.5.4 'Permanent Works' means the permanent works to be executed by the Contractor under the Contract.

1.1.5.5 'Plant' means the apparatus, machinery and vehicles intended to form or forming part of the Permanent Works.

1.1.5.6 'Section' means a part of the Works specified in the Appendix to Tender as a Section (if any).

1.1.5.7 'Temporary Works' means all temporary works of every kind (other than Contractor's Equipment) required on Site for the execution and completion of the Permanent Works and the remedying of any defects.

1.1.5.8 'Works' mean the Permanent Works and the Temporary Works, or either of them as appropriate.

1.1.6 Other Definitions

1.1.6.1 'Contractor's Documents' means the calculations, computer programs and other software, drawings, manuals, models and other documents of a technical nature (if any) supplied by the Contractor under the Contract.

1.1.6.2 'Country' means the country in which the Site (or most of it) is located, where the Permanent Works are to be executed.

1.1.6.3 'Employer's Equipment' means the apparatus, machinery and vehicles (if any) made available by the Employer for the use of the Contractor in the execution of the Works, as stated in the Specification; but does not include Plant which has not been taken over by the Employer.

1.1.6.4 'Force Majeure' is defined in Clause 19 [Force Majeure].

1.1.6.5 'Laws' means all national (or state) legislation, statutes, ordinances and other laws, and regulations and by-laws of any legally constituted public authority.

1.1.6.6 'Performance Security' means the security (or securities, if any) under Sub-Clause 4.2 [Performance Security].

1.1.6.7 'Site' means the places where the Permanent Works are to be executed and to which Plant and Materials are to be delivered, and any other places as may be specified in the Contract as forming part of the Site.

*1.1.6.8 '**Unforeseeable**' means not reasonably foreseeable by an experienced contractor by the date for submission of the Tender.*

*1.1.6.9 '**Variation**' means any change to the Works, which is instructed or approved as a variation under Clause 13 [Variations and Adjustments].*

Sub-Clause 1.1 [*Definitions*] defines the meaning of 60 words and expressions used in the Conditions of Contract. These are listed alphabetically on page vi in the RED Book and are defined in separate Sub-Clauses under six subheadings:

1.1.1 The Contract
1.1.2 Parties and Persons
1.1.3 Dates, Tests, Periods and Completion
1.1.4 Money and Payments
1.1.5 Works and Goods
1.1.6 Other Definitions

The definitions include technical and contractual terms, such as '*Final Payment Certificate*' and words that are in everyday use, such as '*Cost*' and '*Unforeseeable*'. Many of the words that are defined are well known. However, some words have definitions that are not obvious. For example, '*Employer's Personnel*' covers not only staff directly employed by the Employer, but also includes the Engineer and any other person who has been notified to the Contractor as being Employer's Personnel. Also, the equipment that is used by the Contractor for the construction of the Works is referred to as '*Contractor's Equipment*', and the word '*Plant*' is used for apparatus, machinery and vehicles that will be incorporated into the Permanent Works.

As stated before, words that are defined are printed with capital letters in the text. Exceptions are '*day*' and '*year*'. This is a useful convention and users are advised to check at Sub-Clause 1.1 [*Definitions*] to establish the meaning of a word in the context of FIDIC, if and when, such a word is printed with a capital letter, but is not at the beginning of a sentence.

Several of the terms defined at Sub-Clause 1.1 [*Definitions*] require further information, which must be given in the *Appendix to Tender*:

1.1.2.2 The name and address of the Employer
1.1.2.3 The name and address of the Contractor
1.1.2.4 The name and address of the Engineer
1.1.3.3 The number of days for completing the Works
        The number of days for the completion of Sections
1.1.3.7 The number of days in the period for notifying defects
1.1.5.6 Any part of the Works which is specified as a Section.

Other definitions refer to information that must be given elsewhere in the Contract documents.

7.1. Definitions that depend on the date for the submission of the Tender

1.1.3.1 The '*Base Date*' is fixed as 28 days before the latest date for the submission of the Tender.

1.1.6.8 Things which are 'Unforeseeable' means unforeseeable by the date for submission of the Tender.

It is noteworthy that, unlike PNK, the definition for 'Unforeseeable' is not linked to the Base Date, but rather to the submission date for the Tender. Any information a tenderer receives up to the tender date is therefore 'foreseeable' and will have a potential effect on the interpretation of Sub-Clause 4.12 [*Unforeseeable Physical Conditions*]. If such information is received a short time before the tender submission date, it may not be possible to incorporate necessary changes in the tender submission. A potential contractor must decide how best to reflect this new information into his risk management for the project.

Another potential problem is that the Invitation to Tender is not intended to be a Contract document under FIDIC procedures. It is therefore necessary to ensure that the date for submission of the Tender is recorded somewhere in the Contract. The FIDIC Guidance for the Preparation of Particular Conditions suggests at Sub-Clause 1.1 [*Definitions*] that the Base Date could be stated as a calendar date. The FIDIC Guidance also suggests at Sub-Clause 1.6 [*Contract Agreement*] that the Base Date could be stated in the Contract Agreement. This has the advantage that, if the date for submission of the Tender was delayed, then the Base Date should change and the actual date is known by the time the Contract Agreement is signed. Alternatively, the Appendix to Tender could be amended so as to allow the Contractor to insert the actual Tender date or Base Date.

7.2. Definition in the Letter of Tender that was accepted by the Employer

1.1.2.3 The name of the Contractor or his legal successors in title is given in the Letter of Tender which was accepted by the Employer.

The Letter of Tender is one of the documents that form the Contract and will be referred to in the Letter of Acceptance and Contract Agreement. The Contractor's name and address must also be added to the Appendix to Tender and will be stated in the Contract Agreement. It is obvious, but not always achieved in practice, that the same name and address should appear in all these documents.

7.3. Definitions that depend on the Letter of Acceptance or the Contract Agreement

1.1.1.1 The list of documents which make up the Contract must be given in the Contract Agreement or the letter of Acceptance.

1.1.2.5 The name of the Contractor's Representative is stated to be given in the Contract. Since the Contract is defined as a compilation of documents it is theoretically possible to include the name of such representative in any of the defined documents. However, common sense dictates that this is done either in the Contract Agreement, the Letter of Tender, or as a memorandum annexed to the Letter of Acceptance. Sub-

Clause 4.3 [*Contractor's Representative*] states that if the name is not stated in the Contract it must be submitted for the Engineer's approval before the Commencement Date.

1.1.2.9 Members of the DAB may be named in the Contract or appointed later, as discussed at Sub-Clause 20.2 [*Appointment of the Dispute Adjudication Board*].

1.1.3.2 The Commencement Date must be notified to the Contractor by the Engineer within 42 days after the Contractor receives the Letter of Acceptance, unless a different number of days are stated in the Particular Conditions. However the Engineer shall give the Contractor not less than 7 days notice.

1.1.4.1 The Accepted Contract Amount must be given in the Letter of Acceptance and may be repeated in the Contract Agreement.

The definition of the Letter of Acceptance, at Sub-Clause 1.1.1.3, allows it to include annexed memoranda, which should include the agreed confirmation of negotiations and letters from the Contractor covering matters such as *inter alia* the name of the Contractor's Representative. Both parties shall sign such annexed memoranda. Insurance arrangements should also have been discussed at this stage and any agreements annexed to the Letter of Acceptance, as reviewed at Clause 18 [*Insurance*].

7.4. Definitions that depend on the Contract documents, the Specification or Schedules

1.1.2.8 Subcontractors may be either named in the Contract, or appointed as a Subcontractor.

1.1.3.4 Tests on Completion are those tests that must be carried out before the Works or a Section (as the case may be) are taken over by the Employer, and may be specified in the Contract or agreed or instructed later.

1.1.3.6 Tests after Completion must be specified in the Contract with the procedures given in the Particular Conditions.

1.1.4.10 Any Provisional Sums must be specified in the Contract.

1.1.6.2 The Country in which (most of) the Site is located, should be apparent from the Contract. This is either defined in the Particular Conditions, or is readily determined from the Specifications and/or the Drawings. It appears from the definition that there can only be 1 (one) Country in which most of the Site is located.

1.1.6.3 Any Employer's Equipment must be stated in the Specification.

1.1.6.7 The Site must be defined in the Contract.

The names of Subcontractors and the details of Tests on Completion may be agreed or instructed during the construction. However, Tests after Completion are not considered to be a routine requirement for a project that is designed by the Employer and the procedures are not given in the FIDIC Conditions. If required these procedures must be given in the Particular Conditions with the technical details in the Specification. Guidance on these procedures can be obtained from Clause 12 [*Measurement and Valuation*] of the YLW Book.

The details of Provisional Sums and any Equipment that the Employer intends to provide for the Contractor's use must be given in the Specification or Schedules in order that the Contractor can plan his work.

The precise location and extent of the Site must obviously be given in the Contract. This may include areas that are not to be occupied by the Permanent Works but which the Employer intends to make available for the Contractor's use. If the Site crosses the border and is located in more than one country, for example a bridge spanning a river which forms the national boundary, then the identity of the country in which most of it is located will be critical for the definition of 'Country' under Sub-Clause 1.1.6.2. To avoid any confusion the name of the Country could be stated in the Particular Conditions.

The FIDIC *Guidance for the Preparation of Particular Conditions* suggests that definitions of Foreign Currency at 1.1.4.6 and Local Currency at 1.1.4.8 may need to be amended if the Employer intends to specify a particular currency for payment.

An additional definition, which would be useful and could be included in the Particular Conditions would be the 'Resident Engineer'. Previous FIDIC Contracts (prior to the 1999 series) included a specific provision for the Engineer to appoint an Engineer's Representative, who was normally resident on the Site. This role is not specifically envisaged under the 1999 Red Book (RED), although a resident engineer is mentioned at Sub-Clause 3.2 [*Delegation of the Engineer*] as a possible assistant to the Engineer. In practice, only small and uncomplicated projects may be managed sufficiently well without a dedicated person on the Site to handle the day-to-day issues that arise between the Employer and the Contractor. The majority of international construction projects require constant supervision so as to ensure that progress is not hampered. The Engineer may appoint a suitable assistant under Sub-Clause 3.2, to be resident on the Site thereby re-introducing the Resident Engineer. Suitable wording for such a definition might be:

1.1.2.11 '***Resident Engineer***' means an assistant to the Engineer who will be resident on the Site and is appointed from time to time by the Engineer under Sub-Clause 3.2 [*Delegation by the Engineer*] and notified to the Parties.

In the review of Clause 3, in Chapter 9, it is suggested that an additional Sub-Clause 3.6 [*Management Meetings*] could be added in the Particular Conditions. This would also require an additional definition:

1.1.6.10 '***Management Meeting***' means a meeting called by either the Engineer or the Contractor's Representative in accordance with Sub-Clause 3.6 [*Management Meetings*].

1.2 Interpretation
In the Contract, except where the context requires otherwise:

(a) *words indicating one gender include all genders;*
(b) *words indicating the singular also include the plural and words indicating the plural also include the singular;*
(c) *provisions including the word 'agree', 'agreed' or 'agreement' require the agreement to be recorded in writing, and*
(d) '***written***' *or '**in writing**' means hand-written, type-written, printed or electronically made, and resulting in a permanent record.*

91

The marginal words and other headings shall not be taken into consideration in the interpretation of these Conditions.

Sub-Clause 1.2 [*Interpretation*] states that headings and marginal words must not be taken into consideration in the interpretation of the Conditions. This statement also applies in reverse. The reader cannot rely on the heading when searching for the sub-clause that covers a particular situation. Many sentences within a Sub-Clause refer to a completely different subject to the heading. One of the problems with the administration of a FIDIC Contract is the need to collect information on any subject from a variety of different, and sometimes unexpected, Sub-Clauses. The Index of Sub-Clauses, which is located after the Appendix and Annex that follow Clause 20, is helpful but cannot cover all the possible subjects and Sub-Clauses.

Sub-Clause 1.2 [*Interpretation*] also confirms the usual conventions that one gender includes all genders, singular includes plural and *vice versa*. Other requirements that are important for efficient project management are that anything that is stated to be agreed must have been recorded in writing and *'written'* means a record *'resulting in a permanent record'*. The word *'permanent'* must be interpreted in the context of the contract, in order to cover the examples that are quoted, such as *'hand-written'* and *'electronically made'*. E-mail is now possibly the most widespread method of contractual communication. For an e-mail to comply with the contractual requirement to be a 'permanent record in writing' the Parties should agree on the format of such communications. Here, the applicable law may also have some relevant provisions. In this context there is also a requirement to define the acceptable Electronic Transmission Systems under Sub-Clause 1.3 [*Communications*] in the Appendix to Tender. To be safe, it is still recommended that for any communication of contractual significance the Parties may use an electronic media for speed, but thereafter forward an original document by traditional mailing systems.

The FIDIC *Guidance for the Preparation of Particular Conditions* suggests that Sub-Clause 1.2 [*Interpretation*] could be extended if a more precise explanation is required of the phrase 'Cost plus reasonable profit'. This phrase is included in Sub-Clauses such as Sub-Clause 1.9 [*Delayed Drawings or Instructions*], which refers to circumstances in which the Contractor may be entitled to additional payment. The FIDIC example states that reasonable profit would be one-twentieth (5%) of the Cost, that is the same as that used in the PNK Book. The phrase *'reasonable profit'* is certainly open to interpretation and argument, but it does not seem logical to stipulate a figure and then hide the figure at Sub-Clause 1.2 [*Interpretation*]. If a particular figure were imposed on the Contractor, it would be better if the figure were included in the Appendix to Tender. If the intention is to have a figure determined in the Contract, to avoid problems when claims are being valued, then it could be left open for completion by the Contractor.

1.3 Communications
Wherever these Conditions provide for the giving or issuing of approvals, certificates, consents, determinations, notices and requests, these communications shall be:

(a) in writing and delivered by hand (against receipt), sent by mail or courier, or transmitted using any of the agreed systems of electronic transmission as stated in the Appendix to Tender; and

(b) *delivered, sent or transmitted to the address for the recipient's communications as stated in the Appendix to Tender. However:*

    (i) *if the recipient gives notice of another address, communications shall thereafter be delivered accordingly; and*

    (ii) *if the recipient has not stated otherwise when requesting an approval or consent, it may be sent to the address from which the request was issued.*

Approvals, certificates, consents and determinations shall not be unreasonably withheld or delayed. When a certificate is issued to a Party, the certifier shall send a copy to the other Party. When a notice is issued to a Party, by the other Party or the Engineer, a copy shall be sent to the Engineer or the other Party, as the case may be.

Sub-Clause 1.3 [*Communications*] requires that the formal communications provided for in the Conditions, which are listed as approvals, certificates, consents, determinations, notices and requests, must be in writing and delivered by hand, mail, courier or electronic means. The acceptable electronic means are stated in the Appendix to Tender under '*Electronic transmission systems*'. The use of facsimile transmission is not listed but could be covered in the Appendix to Tender. The Sub-Clause requires hand delivery to be against a receipt, but does not require a receipt for the other means of delivery. Clearly it would often be prudent to use a system that requires the recipient to sign a receipt. Some actions, such as the payment provisions at Sub-Clause 14.7 [*Payment*] are required to be taken within a number of days following the receipt of another document so the date of receipt will be significant.

Communications must be sent to the address stated in the Appendix to Tender unless notice has been given of a different address. In practice, when the Parties and the Engineer have offices on Site they may wish to use these offices for most communications during the construction period and so must issue formal notices to this effect. Also, when a certificate or notice is issued to one Party it must be copied to the other Party and the Engineer. A particularly important statement is hidden within this Sub-Clause:

'*Approvals, certificates, consents and determinations shall not be unreasonably withheld or delayed.*'

In general, throughout the FIDIC Conditions of Contract, when the Contractor is required to take some action it must be carried out within a fixed time period. However, the same principle does not apply to many of the actions by the Engineer or the Employer. The consequences or compensation for a document being unreasonably withheld or delayed are not stated but clearly it would be a breach of Contract. However, the fact that documents shall not be unreasonably withheld or delayed does not apply to notifications! Under Sub-Clause 2.5 [*Employer's Claims*] where there is no time constraint for notification of a claim, other than it shall be given as soon as practicable, the issue is often the cause for heated debate or even dispute. If the Contractor feels that he has suffered delay or additional cost then he could submit a claim using the procedures of Sub-Clause 20.1 [*Contractor's Claims*]. The matter would then be considered by the Engineer, who would make a determination under Sub-Clause 3.5 [*Determinations*]. The DAB or Arbitration Tribunal could later review this.

In previous FIDIC Contracts the sub-clauses that required the issue of a notice to the Engineer frequently stated that a copy must be sent to the Employer. In the 1999 Contracts this requirement has been omitted from most of the Sub-Clauses but a general requirement has been included at Sub-Clause 1.3 [*Communications*]. Certificates must be copied to the other Party and notices to a Party must be copied to the other Party or the Engineer. However, notices that are sent to the Engineer do not have to be copied to the other Party. This means that notices of claims do not have to be copied to the Employer, although these are the notices that the Employer will probably wish to receive.

This requirement only refers to certificates and notices. However, the use of the word 'notice', which is not a defined term, could cause confusion. In other Sub-Clauses the use of words such as 'notify', 'submit' or 'instruct' appears to be a matter of word selection rather than a deliberate intention to either require, or not to require, a copy to the other Party. The requirement for copies of particular notices should be reviewed during the preparation of the Particular Conditions.

Certificates and notices include the following.

(*a*) Certificates issued by the Engineer to the Employer, with a copy to the Contractor.
   (*i*)   Payment Certificates as Sub-Clauses 14.2, 14.6, 14.13.
(*b*) Certificates issued by the Engineer to the Contractor, with a copy to the Employer.
   (*i*)   Taking-Over Certificate as Sub-Clause 10.1 or 10.2.
   (*ii*)   Performance Certificate as Sub-Clause 11.9.
(*c*) Notices issued by either Party to the other Party, with a copy to the Engineer.
   (*i*)   An error or defect in a document as Sub-Clause 1.8.
   (*ii*)   Infringement of intellectual or industrial property rights as Sub-Clause 17.5.
   (*iii*) Event or circumstances as Force Majeure as Sub-Clause 19.2.
   (*iv*)   Force Majeure ceases to affect as Sub-Clause 19.3.
   (*v*)   Termination due to Force Majeure as Sub-Clause 19.6.
   (*vi*)   Dissatisfaction with a DAB decision as Sub-Clause 20.4.
(*d*) Notices issued by the Employer to the Contractor, with a copy to the Engineer.
   (*i*)   A change to his financial arrangements as Sub-Clause 2.4.
   (*ii*)   Claim particulars as Sub-Clause 2.5.
   (*iii*) Details of a replacement Engineer as Sub-Clause 3.4.
   (*iv*)   Defects during the Defects Notification Period as Sub-Clause 11.1.
   (*v*)   Defects not at Contractor's Cost as Sub-Clause 11.2.
   (*vi*)   Date by which a defect is to be remedied as Sub-Clause 11.4.
   (*vii*) Termination by Employer as Sub-Clause 15.2.
   (*viii*) Release equipment after termination as Sub-Clause 15.2.
   (*ix*)   Termination for convenience as Sub-Clause 15.5.
(*e*) Notices issued by the Contractor to the Employer, with a copy to the Engineer.
   (*i*)   Suspend or reduce work as Sub-Clause 16.1.
   (*ii*)   Termination as Sub-Clause 16.2.
   (*iii*) Insurance cover ceases to be available as Sub-Clause 18.2.
(*f*) Notices issued by the Engineer to both Parties.
   (*i*)   Agreement or Engineer's determination as Sub-Clause 3.5.

(g) Notices issued by the Engineer to the Contractor, with a copy to the Employer.
- (i)   Does not require to inspect as Sub-Clause 7.3.
- (ii)   Intention to attend tests as Sub-Clause 7.4.
- (iii)   Rejection as Sub-Clause 7.5.
- (iv)   Commencement Date as Sub-Clause 8.1.
- (v)   Programme does not comply with the Contract as Sub-Clause 8.3.
- (vi)   Programme not consistent with actual progress as Sub-Clause 8.3.
- (vii)   Instruction to carry out delayed tests as Sub-Clause 9.2.
- (viii)   Tests on Completion after delay by the Employer as Sub-Clause 10.3.
- (ix)   Further tests as Sub-Clause 11.5.
- (x)   That part of the Works is to be measured as Sub-Clause 12.1.
- (xi)   Payment is to be withheld as Sub-Clause 14.6.
- (xii)   Correct a failure to carry out an obligation as Sub-Clause 15.1.

(h) Notices issued by the Contractor to the Engineer.
- (i)   Claims as Sub-Clauses 1.9, 4.12, 4.24, 16.1, 17.4, 19.4.
- (ii)   Requiring a drawing or instruction as Sub-Clause 1.9.
- (iii)   Date of Commencement of Subcontractor's Work as Sub-Clause 4.4.
- (iv)   Delivery to Site of Plant or other major items as Sub-Clause 4.16.
- (v)   Shortage or defect in free-issue materials as Sub-Clause 4.20.
- (vi)   Objection to a nominated Subcontractor as Sub-Clause 5.2.
- (vii)   Inspection before cover up as Sub-Clause 7.3.
- (viii)   Circumstances that may have an adverse effect as Sub-Clause 8.3.
- (ix)   Prolonged suspension as Sub-Clause 8.11.
- (x)   Date for delayed Tests on Completion as Sub-Clause 9.1.
- (xi)   Delayed Tests on Completion as Sub-Clause 9.2.
- (xii)   Application for a Taking-Over Certificate as Sub-Clause 10.1.
- (xiii)   Records are inaccurate as Sub-Clause 12.1.
- (xiv)   Damage due to Employer's risks as Sub-Clause 17.4.

(i) Notices issued by both Parties to each Member of a DAB.
- (i)   Dispute Adjudication Agreement takes effect as Appendix paragraph 2.
- (ii)   Termination of Dispute Adjudication Agreement as Appendix paragraph 7.

(j) Notices issued by a DAB Member to both Parties.
- (i)   Resignation as Appendix paragraph 2.
- (ii)   Termination of Dispute Adjudication Agreement as Appendix paragraph 7.
- (iii)   Notice of its decision, as Annex paragraph 5.
- (iv)   Notice of a hearing, as Annex paragraph 7.

There is no requirement for other correspondence or reports to be copied to a Party or to the Engineer although this may be done, by agreement or at the option of the Party concerned.

The fact that the word 'notice' is not a defined term has been the cause of many disputes over the years. This is particularly true in the context of what constitutes a 'proper notice' under Sub-Clause 20.1. In what may be seen as an acknowledgment of omission, FIDIC in its 2008 *Conditions of Contract for Design, Build and Operate Projects* (first edition) (The Gold Book) introduces for the first time a definition for a 'notice':

'Notice' means a written communication identified as a Notice and issued in accordance with the provisions of Sub-Clause 1.3. [Notices and Other Communications].

In comparison to the wording of Sub-Clause 1.3 [*Communications*] the only difference seems to be that a notice shall be identified as a 'Notice'. As such the definition falls short of addressing the real issue behind the omission, such as, what format (e.g. letter, meeting minutes, e-mail, etc.) such a notice shall have. To avoid these problems under the RED, YLW, PNK, SUB and Silver Books, the Parties are advised to agree on a format through either Particular Conditions or a sample Annex.

1.4 Law and Language
The Contract shall be governed by the law of the country (or other jurisdiction) stated in the Appendix to Tender.

If there are versions of any part of the Contract which are written in more than one language, the version which is in the ruling language stated in the Appendix to Tender shall prevail.

The language for communications shall be that stated in the Appendix to Tender. If no language is stated there, the language for communications shall be the language in which the Contract (or most of it) is written.

The Contractor, quite apart from any provisions in the Contract, is obliged to comply with the Laws of the country in which he is working and Sub-Clause 1.13 [*Compliance with Laws*] states that '*the Contractor shall, in performing the Contract, comply with applicable laws*'. If the Site is located in more than one country, as discussed under Sub-Clause 1.1 [*Definitions*], then the Contractor may have to comply with the Laws of more than one country. Work such as the manufacture of items of Plant in a different country will be subject to the Laws and regulations of that country.

Several other clauses, such as Sub-Clause 13.7 [*Adjustments for Changes in Legislation*], refer to the '*Laws of the Country*', with '*Laws*' and '*Country*' as defined at Sub-Clause 1.1. Sub-Clause 13.7 refers to changes in the Laws of the Country. The '*governing law*' as stated in the Appendix to Tender is the law which governs the Contract and is not necessarily the Law of the country in which the Site is located. However, if the governing law is not the Laws of the Country then there is likely to be confusion and problems due to any difference between the different legal requirements.

The '*ruling language*' is designated in the Appendix to Tender and is the language which has priority if any part of the Contract is written in more than one language. For example, the Contract Agreement is sometimes written in more than one language and technical documents such as manufacturers' brochures may have been written in a different language.

The '*language for communications*' should also be stated in the Appendix to Tender, but if it is not stated then it is the language in which most of the Contract is written. The separation of language for communications from the ruling language recognises that many FIDIC

Contracts have English as the language for daily use on the project, even when it is not the language of the Country of the project. English will then be the official language for documents and meetings, although internal documents and meetings may use a different language. Any assistants to whom the Engineer has delegated authority under Sub-Clause 3.2 [*Delegation by the Engineer*], together with the Contractor's Representative and any assistants to whom powers have been delegated under Sub-Clause 4.3 [*Contractor's Representative*], are required to be fluent in the language for communications.

If the original of any document was written in a different language and a translation takes priority then care is necessary to ensure that the translation truly gives the intended meaning of the document. It is not unknown for different translations of a document to exist that have different, or even opposite, meanings.

A problem sometimes arises in relation to the FIDIC General Conditions of Contract, which are available in several different languages, although it is stated in the Foreword that the English language version is the official and authentic text. If the FIDIC General Conditions are included in the Contract by reference it is important to state in the Particular Conditions exactly which language version is applicable. When a version of the General Conditions in the language for communications has been published by FIDIC then this version should be used for communications during the project. However, when FIDIC has not published an official translation it is possible that several different translations are available, which will inevitably differ in the translations of some words.

1.5 Priority of Documents
The documents forming the Contract are to be taken as mutually explanatory of one another. For the purposes of interpretation, the priority of the documents shall be in accordance with the following sequence:

(a) the Contract Agreement (if any),
(b) the Letter of Acceptance,
(c) the Letter of Tender,
(d) the Particular Conditions,
(e) these General Conditions,
(f) the Specification,
(g) the Drawings, and
(h) the Schedules and any other documents forming part of the Contract.

If an ambiguity or discrepancy is found in the documents, the Engineer shall issue any necessary clarification or instruction.

Sub-Clause 1.5 [*Priority of Documents*] gives the order of priority of the Contract documents and must be read in conjunction with any requirements of the applicable law. The FIDIC *Guidance for the Preparation of Particular Conditions* includes an alternative for use when the Contract does not give an order of precedence, but it is to be governed by the applicable law.

Under Sub-Clause 1.5 the first priority document is the Contract Agreement, followed by the Letter of Acceptance and Letter of Tender. These are the documents, in reverse order, which constitute the Contractor's offer, the Employer's acceptance of that offer and the Agreement signed by both Parties. This is logical because the Letter of Acceptance may include documents which have been agreed and which change the original Tender. The definitions at Sub-Clauses 1.1.1.2 and 1.1.1.3 suggest that either the Contract Agreement or the Letter of Acceptance might be omitted for a particular project, with the other document serving both purposes and including all the information that is required in both documents.

The Engineer can clarify any ambiguity or discrepancy between documents. However, he must bear in mind that he does not have authority to change the Contract (refer to Sub-Clause 3.1 [*Engineer's Duties and Authorities*]). His instruction will be a Variation if it meets the requirements of Sub-Clause 13.1 [*Right to Vary*] in which case the Contractor may be entitled to additional payment.

Particular requirements in a high-priority document may be overruled by a lower priority document when the changed priority is stated in the higher priority document. For example, Sub-Clause 7.1(c) [*Manner of Execution*] of these Conditions includes the statement '*except as otherwise specified in the Contract*'. This particular requirement in the specification would then take precedence over the requirements of Sub-Clause 7.1(c).

Where there is an ambiguity within a document legal clarification must be sought. For example, the principle of *contra proferentem* may apply. This is known as the 'interpretation against the draftsman'. It is a doctrine of contractual interpretation providing that, where a promise, agreement or term is ambiguous, the preferred meaning should be the one that works against the interests of the party who provided the wording.

1.6 Contract Agreement

The Parties shall enter into a Contract Agreement within 28 days after the Contractor receives the Letter of Acceptance, unless they agree otherwise. The Contract Agreement shall be based upon the form annexed to the Particular Conditions. The costs of stamp duties and similar charges (if any) imposed by law in connection with entry into the Contract Agreement shall be borne by the Employer.

Sub-Clause 1.6 [*Contract Agreement*] requires the Contract Agreement to be entered into within 28 days from the date the Contractor receives the Letter of Acceptance, unless the Parties have agreed otherwise. The FIDIC Conditions of Contract do not include a Sub-Clause or standard form for the Letter of Acceptance although it is an essential part of the sequence of documents at the start of the Contract. The standard form for the Letter of Tender states that it will be open for acceptance until a stated date and it is clear that the Letter of Tender and the Letter of Acceptance of that Tender, together with attached documents, will form an agreement between the Parties.

The Letter of Acceptance is defined at Sub-Clause 1.1.1.3 and is also referred to at the following Sub-Clauses.

- 1.5 as having priority only after the Contract Agreement (if any).
- 1.6 as starting the 28-day period for the Contract Agreement.
- 4.2 as starting a 28-day period for the Contractor to deliver the Performance Security to the Employer.
- 8.1 as starting a 42-day period for the Commencement Date.
- 18.1 as the end of a period for the Parties to agree terms for insurances.

The definition at Sub-Clause 1.1.1.2 and the reference at Sub-Clause 1.5(a) [*Priority of Documents*] confirm that there may not be a Contract Agreement document for a particular project. The Letter of Acceptance then becomes the formal, as well as the initial, acceptance by the Employer. However, there may be a legal requirement for a Contract Agreement document, particularly for government Contracts.

The FIDIC documents include a standard form for a Contract Agreement. If the Employer has his own standard form it should be checked to make sure that it includes all the details in the FIDIC Contract Agreement.

Sub-Clause 1.6 [*Contract Agreement*] requires the Employer to bear the cost of any charges imposed by law in connection with entry into the Contract Agreement. The Agreement must be 'based' on the FIDIC form that is annexed to the Particular Conditions. The Employer may wish to use his own standard form, which must then be checked to ensure that it is consistent with the FIDIC Conditions, includes the information that is given in the FIDIC form and complies with any requirements of the applicable law. The form that the Employer intends to use must be included with the Tender documents, although agreed modifications may be necessary following any Tender negotiations.

The FIDIC *Guidance for the Preparation of Particular Conditions* suggests that the Contract Agreement may record the Accepted Contract Amount, Base Date and/or Commencement Date. If this is intended then the various definitions, together with Sub-Clause 8.1 [*Commencement of Works*], should be amended in the Particular Conditions.

1.7 Assignment
Neither Party shall assign the whole or any part of the Contract or any benefit or interest in or under the Contract. However, either Party:

(a) *may assign the whole or any part with the prior agreement of the other Party, at the sole discretion of such other Party, and*
(b) *may, as security in favour of a bank or financial institution, assign its right to any moneys due, or to become due, under the Contract.*

Neither Party is permitted to assign the whole or any part of the Contract except with the agreement of the other Party, or for its limited use for financial security.

1.8 Care and Supply of Documents
The Specification and Drawings shall be in the custody and care of the Employer. Unless otherwise stated in the Contract, two copies of the Contract and of each subsequent

Drawing shall be supplied to the Contractor, who may make or request further copies at the cost of the Contractor.

Each of the Contractor's Documents shall be in the custody and care of the Contractor, unless and until taken over by the Employer. Unless otherwise stated in the Contract, the Contractor shall supply to the Engineer six copies of each of the Contractor's Documents.

The Contractor shall keep, on the Site, a copy of the Contract, publications named in the Specification, the Contractor's Documents (if any), the Drawings and Variations and other communications given under the Contract. The Employer's Personnel shall have the right of access to all these documents at all reasonable times.

If a Party becomes aware of an error or defect of a technical nature in a document which was prepared for use in executing the Works, the Party shall promptly give notice to the other Party of such error or defect.

Under Sub-Clause 1.8 [*Care and Supply of Documents*] either Party must give prompt notice to the other if they become aware of any error or defect of a technical nature in a document. While the Party who prepared a document must be responsible for its content, the other Party has an obligation to co-operate in minimising the consequences of any mistake. If there is a mistake on a Drawing then an experienced Contractor may notice the mistake when using the Drawing for planning or ordering materials. He can then ask for a corrected Drawing to be issued before the item of work is constructed. It may be difficult to prove that the Contractor failed to report a mistake immediately he became aware of the mistake, but any breach of the requirement could result in a claim by the Employer for the consequences of the failure to give notice, or a reduction in the value of the Contractor's claim for the consequences of the error or defect.

1.9 Delayed Drawings or Instructions

The Contractor shall give notice to the Engineer whenever the Works are likely to be delayed or disrupted if any necessary drawing or instruction is not issued to the Contractor within a particular time, which shall be reasonable. The notice shall include details of the necessary drawing or instruction, details of why and by when it should be issued, and details of the nature and amount of the delay or disruption likely to be suffered if it is late.

If the Contractor suffers delay and/or incurs Cost as a result of a failure of the Engineer to issue the notified drawing or instruction within a time which is reasonable and is specified in the notice with supporting details, the Contractor shall give a further notice to the Engineer and shall be entitled subject to Sub-Clause 20.1 [Contractor's Claims] to:

(a) an extension of time for any such delay, if completion is or will be delayed, under Sub-Clause 8.4 [Extension of Time for Completion], and

(b) payment of any such Cost plus reasonable profit, which shall be included in the Contract Price.

After receiving this further notice, the Engineer shall proceed in accordance with Sub-Clause 3.5 [Determinations] to agree or determine these matters.

However, if and to the extent that the Engineer's failure was caused by any error or delay by the Contractor, including an error in, or delay in the submission of, any of the Contractor's Documents, the Contractor shall not be entitled to such extension of time, Cost or profit.

Under Sub-Clause 1.9 [*Delayed Drawings or Instructions*] the Contractor is required to give notice if he requires an additional Drawing or instruction, the lack of which could delay the Works. The timing of this notice should give the Engineer reasonable time to prepare and issue the Drawing or instruction. However, the Contractor needs to know that the Engineer does not intend to issue the Drawing or instruction without being asked, so some discussion may precede the formal notice. The Engineer should have informed the Contractor of his intentions for the issue of further Drawings and instructions which he intends to issue under Sub-Clause 3.3 [*Instructions of the Engineer*].

The same circumstances should also be considered under Sub-Clause 8.3 [*Programme*]. The mere fact that the lack of an Instruction or Drawing if not issued by the stated time can delay the Works, is in itself a specific probable future event requiring the Contractor to give a notice. It might be wise to make the notice under Sub-Clause 1.9 [*Delayed Drawings or Instructions*] to include wording that will make it also compliant with Sub-Clause 8.3. This notice must be issued '*promptly*'. It would also enable the Engineer to ask the Contractor for a Variation proposal under Sub-Clause 13.3 [*Variation Procedure*] in order to overcome the problem. The existence of a potential problem and the measures being taken to overcome the problem would be included in the Contractor's monthly progress report under Sub-Clause 4.21(h) [*Progress Reports*].

Failure to issue the Drawing or instruction in a reasonable time might result in delay or additional Cost to the Contractor and require a further notice under Sub-Clause 1.9 and also under Sub-Clause 20.1 [*Contractor's Claims*]. This seems to be a double notification requirement especially when seen in the context of Sub-Clause 20.1 paragraph 3 where it is stated that the Contractor shall also submit any other notices that are required by the Contract. To avoid any debate or dispute on the validity and sufficiency of the notice requirements it might be prudent to make the second notice under Sub-Clause 1.9 to include a wording that will make it compliant with Sub-Clause 20.1. This notice would be included in the monthly progress report under Sub-Clause 4.21(f). What might be confusing is the requirement for the Engineer to proceed in accordance with Sub-Clause 3.5 [*Determination*]. What exactly is the Engineer to agree or determine after receiving a notice of intent to claim, which by implication is without the claim details and particulars itself? As discussed elsewhere in this book, the Engineer has a double obligation under Sub-Clause 3.5. He first has to attempt to get the parties to agree on a matter failing which he is to make a fair determination taking into account all relevant events and circumstances. The latter would certainly not be possible on the strength of a notice without particulars. However, his obligation to initially attempt mitigation between the parties would be plausible. The Engineer must therefore, upon receipt of a notice by the Contractor, immediately start the process of trying to resolve the potential dispute before it escalates. He can do this by, for example, starting discussions with the Contractor on what all parties involved would require in terms of supporting documents, or to evaluate mitigating possibilities to avert or reduce impact of the impending claim. Taken to the extreme it could be argued that the Engineer would be in breach of his obligations, if he did not do these things prior to actually receiving a notified claim. This principle is also found in the following Sub-Clauses.

- 1.9 [*Delayed Drawings or Instructions*]
- 2.1 [*Right of Access to Site*]
- 4.7 [*Setting Out*]
- 4.12 [*Unforeseeable Physical Conditions*]
- 4.24 [*Fossils*]
- 7.4 [*Testing*]
- 8.9 [*Consequences of Suspension*]
- 10.2 [*Taking Over of Parts of the Works*]
- 10.3 [*Interference with Tests on Completion*]
- 12.4 [*Omissions*]
- 13.7 [*Adjustments for Changes in Legislation*]
- 16.1 [*Contractor's Entitlement to Suspend Work*]
- 17.4 [*Consequences of Employer's Risks*]
- 19.4 [*Consequences of Force Majeure*]

The Contractor would claim additional time and Cost plus reasonable profit. The claim would be considered by the Engineer under Sub-Clause 3.5 [*Determinations*]. The Engineer would consider whether the delay in issuing the information delayed completion, as required by Sub-Clause 8.4 [*Extension of Time for Completion*], as distinct from any immediate delay to a particular part of the Works. The Engineer would also consider whether the Contractor was responsible for the delay, for example, by failing to provide some information that was required in order to complete the Drawings.

1.10 Employer's Use of Contractor's Documents
As between the Parties, the Contractor shall retain the copyright and other intellectual property rights in the Contractor's Documents and other design documents made by (or on behalf of) the Contractor.

The Contractor shall be deemed (by signing the Contract) to give to the Employer a non-terminable transferable non-exclusive royalty-free licence to copy, use and communicate the Contractor's Documents, including making and using modifications of them. This licence shall:

(a) *apply throughout the actual or intended working life (whichever is longer) of the relevant parts of the Works;*
(b) *entitle any person in proper possession of the relevant part of the Works to copy, use and communicate the Contractor's Documents for the purposes of completing, operating, maintaining, altering, adjusting, repairing and demolishing the Works; and*
(c) *in the case of Contractor's Documents which are in the form of computer programs and other software, permit their use on any computer on the Site and other places as envisaged by the Contract, including replacements of any computers supplied by the Contractor.*

The Contractor's Documents and other design documents made by (or on behalf of) the Contractor shall not, without the Contractor's consent, be used, copied or communicated to a third party by (or on behalf of) the Employer for purposes other than those permitted under this Sub-Clause.

This Sub-Clause enables the Employer to make use of documents that are the Contractor's copyright, for the purposes of the Contract.

1.11 Contractor's Use of Employer's Documents

As between the Parties, the Employer shall retain the copyright and other intellectual property rights in the Specification, the Drawings and other documents made by (or on behalf of) the Employer. The Contractor may, at his cost, copy, use, and obtain communication of these documents for the purposes of the Contract. They shall not, without the Employer's consent, be copied, used or communicated to a third party by the Contractor, except as necessary for the purposes of the Contract.

This Sub-Clause states that The Contractor may copy and use the Employer's documents, but only for the purposes of the Contract.

1.12 Confidential Details

The Contractor shall disclose all such confidential and other information as the Engineer may reasonably require in order to verify the Contractor's compliance with the Contract.

The Engineer has a duty to check that the Contractor is complying with the requirements of the Contract. For this the Contractor must provide any confidential information requested by the Engineer provided that this is reasonable. The extent of this information and any requirement to provide copies of documents will be limited by the Engineer's need to satisfy himself. Furthermore, he is not permitted to use this information for any other purpose. However, there may be information that the Contractor, or a Subcontractor, may wish to keep confidential. This would then be the subject of negotiation as to how the Engineer can comply with his obligations but permit sensitive information to remain confidential. The Employer may consider including in the Particular Conditions the Silver Book provision that the Contractor may state in his Tender that certain information is confidential.

The FIDIC *Guidance for the Preparation of Particular Conditions* includes an example Sub-Clause for use when the Employer requires the Contractor to keep the details of the Contract confidential and obtain the Employer's agreement before publishing any technical paper or other information concerning the Works.

1.13 Compliance with Laws

The Contractor shall, in performing the Contract, comply with applicable Laws. Unless otherwise stated in the Particular Conditions:

(a) the Employer shall have obtained (or shall obtain) the planning, zoning or similar permission for the Permanent Works, and any other permissions described in the Specification as having been (or being) obtained by the Employer; and the Employer

shall indemnify and hold the Contractor harmless against and from the consequences of any failure to do so; and

(b) *the Contractor shall give all notices, pay all taxes, duties and fees, and obtain all permits, licences and approvals, as required by the Laws in relation to the execution and completion of the Works and the remedying of any defects; and the Contractor shall indemnify and hold the Employer harmless against and from the consequences of any failure to do so.*

An overall requirement is given at Sub-Clause 1.13 [*Compliance with Laws*] that the Contractor shall comply with the applicable Laws, which includes giving notices and obtaining permits as required by the regulations. However, some permits should be obtained before the Employer calls for Tenders and the Contractor is not required to obtain planning permission, or any other permits which the Specification states will be obtained by the Employer. The phrase '*similar permission*' in subparagraph (a) could be ambiguous and should be clarified in the Particular Conditions. Each Party is required to indemnify and hold harmless the other Party against any failure to comply with this requirement.

The requirement that each Party shall indemnify and hold the other Party harmless against any failure to obtain permissions, and so on, is a severe sanction. The sums of money would be much larger than for a claim for Costs and reasonable profit. It is therefore essential that the responsibilities of the Parties are clearly stated. FIDIC provides that changes in principle must be given in the Particular Conditions. The Employer must obtain planning and similar permissions for the Permanent Works and, if the Employer intends to obtain other permissions then they will be described in the Specification or Employer's Requirements. All other permissions, approvals, and so on, will be obtained by the Contractor. However, if there is a change in the Laws which, as defined at Sub-Clause 1.1.6.5, includes regulation and by-laws, or '*in the judicial or official governmental interpretation of such laws*' then the Contractor will be entitled to time and costs under Sub-Clause 13.7 [*Adjustments for Changes in Legislation*]. If any authority causes disruption or a delay then the Contractor may be entitled to an extension of time under Sub-Clause 8.5 [*Delays Caused by Authorities*].

If the Contractor is required to carry out any design under the RED Book Sub-Clause 4.1 [*Contractor's General Obligations*] then the matter of obtaining approvals must be clarified in the Particular Conditions.

The FIDIC *Guidance for the Particular Conditions* also raises the point that the Employer may need to obtain import licences and various consents for Goods for a plant Contract, but will first require information from the Contractor.

1.14 Joint and Several Liability

If the Contractor constitutes (under applicable Laws) a joint venture, consortium or other unincorporated grouping of two or more persons:

(a) *these persons shall be deemed to be jointly and severally liable to the Employer for the performance of the Contract;*

(b) *these persons shall notify the Employer of their leader who shall have authority to bind the Contractor and each of these persons; and*

(c) *the Contractor shall not alter its composition or legal status without the prior consent of the Employer.*

If the Contractor is a joint venture, consortium or other grouping of two or more persons, then Sub-Clause 1.14 [*Joint and Several Liability*] requires that the separate persons are jointly and severally liable, but one of them must have the authority to bind them all and have been designated as the leader. The approval of the Employer is necessary if the persons in the joint venture wish to alter the composition or legal status of the Contractor. In this context the word '*person*' or '*persons*' is to be interpreted in its legal sense, such as, a legal entity such as a person or a company.

The applicable law may also include requirements for joint ventures and the Employer may have other requirements, which should be included in the Instructions to Tenderers. For example, a parent company guarantee may be required from each member. FIDIC includes an example form as Annex A to the *Guidance for the Preparation of Particular Conditions*.

Tenderers must have considered these requirements before they decide to tender for the Contract. The Employer will also have checked the details of the joint venture or grouping at pre-tender or tender stage, preferably during a pre-qualification procedure. The arrangements must be in place before the Employer accepts the Tender.

FIDIC Users' Guide
ISBN 978-0-7277-5856-9

Chapter 8
Clause 2: The Employer

The FIDIC standard form for the Contract Agreement includes the statement that the Employer covenants to pay the Contractor the Contract Price in consideration of the execution and completion of the Works and the remedying of defects therein. However, this does not mean that the Employer is only required to appoint an Engineer to administer the project and then sign the payment cheques. The development of Conditions of Contract over the years has imposed additional tasks on the Employer, involving both rights and obligations. In some countries the law also imposes duties on the Employer in a construction Contract.

The Conditions of Contract require the Employer, as distinct from the Engineer, to take certain actions during the construction period. Even though the Engineer is now classed as '*Employer's Personnel*' (refer to Sub-Clause 1.1.2.6) there are some tasks that are allocated to the Employer. While the Employer could delegate the paperwork to the Engineer, particularly if the designated Engineer is an employee of the Employer, the actual tasks require the Employer to be involved. It is important that the Employer designates a staff member, separate from the Engineer, to represent him whenever the Contract requires notice to, or action by, the Employer.

Clause 2 must be read in conjunction with other Clauses and covers

- possession of the Site
- assistance with permits, and so on
- Employer's Personnel, financial arrangements and claims.

2.1 Right of Access to the Site
The Employer shall give the Contractor right of access to, and possession of, all parts of the Site within the time (or times) stated in the Appendix to Tender. The right and possession may not be exclusive to the Contractor. If, under the Contract, the Employer is required to give (to the Contractor) possession of any foundation, structure, plant or means of access, the Employer shall do so in the time and manner stated in the Specification. However, the Employer may withhold any such right or possession until the Performance Security has been received.

If no such time is stated in the Appendix to Tender, the Employer shall give the Contractor right of access to, and possession of, the Site within such times as may be required to enable the Contractor to proceed in accordance with the programme submitted under Sub-Clause 8.3 [Programme].

If the Contractor suffers delay and/or incurs Cost as a result of a failure by the Employer to give any such right or possession within such time, the Contractor shall give notice to the Engineer and shall be entitled subject to Sub-Clause 20.1 [Contractor's Claims] to:

(a) an extension of time for any such delay, if completion is or will be delayed, under Sub-Clause 8.4 [Extension of Time for Completion]; and

(b) payment of any such Cost plus reasonable profit, which shall be included in the Contract Price.

After receiving this notice, the Engineer shall proceed in accordance with Sub-Clause 3.5 [Determinations] to agree or determine these matters.

However, if and to the extent that the Employer's failure was caused by any error or delay by the Contractor, including an error in, or delay in the submission of, any of the Contractor's Documents, the Contractor shall not be entitled to such extension of time, Cost or profit.

Sub-Clause 2.1 [Right of Access to the Site] refers to 'right of access to' and 'possession of the Site', but these terms are qualified at other Sub-Clauses. 'Access' refers to the right to enter the Site and must not be confused with 'access routes' that are the Contractor's responsibility as stated at Sub-Clause 4.15 [Access Route]. 'Possession' refers to the Contractor taking over control and responsibility for the Site. The Appendix to Tender states the number of days from the Commencement Date within which the Employer must give access to the Site. When the Employer issues the Letter of Acceptance he is fixing the latest calendar date for providing access to and possession of the Site. If this access and possession is to be restricted to different parts of the Site at different times then the restrictions must be stated in the Particular Conditions, dates being given in the Appendix to Tender, with full details in the Specification.

Sub-Clause 2.1 [Right of Access to the Site] states that possession does not necessarily mean exclusive possession, but shared possession requires clarification in the Particular Conditions. When the Contractor takes possession of the Site he assumes responsibility for matters such as safety, security and insurance. If the Contractor does not have full control of the Site and the activities on the Site, or if these powers are to be shared, then the extent of the Contractor's responsibilities must be clearly stated.

The 'Site' is defined at Sub-Clause 1.1.6.7 as including not just the area of land on which the Works are to be executed, but also any other places that are specified in the Contract. These may include areas that have been set aside for the Contractor to use for storage or for obtaining excavated materials or for any other purpose. It is important that the Site area is delineated clearly in the Contract Drawings or Specification.

If the Employer fails to give right of access to and possession of the Site within the stated period then the Contractor will be entitled to an extension of time, plus his Costs and a reasonable profit, subject to his following the correct procedures as detailed at Sub-Clause 2.1 [Right of Access to the Site] and Sub-Clause 20.1 [Contractor's Claims] and the delay not being attributable to a failure on the part of the Contractor.

2.2 Permits, Licences or Approvals

The Employer shall (where he is in a position to do so) provide reasonable assistance to the Contractor at the request of the Contractor:

(a) by obtaining copies of the Laws of the Country which are relevant to the Contract but are not readily available; and

(b) for the Contractor's applications for any permits, licences or approvals required by the Laws of the Country:

   (i)   which the Contractor is required to obtain under Sub-Clause 1.13 [Compliance with Laws],

   (ii) for the delivery of Goods, including clearance through customs, and

   (iii) for the export of Contractor's Equipment when it is removed from the Site.

Sub-Clause 2.2 [*Permits, Licences or Approvals*] obliges the Employer to assist the Contractor to obtain copies of the Laws of the Country and with applications for any permits, licences or approvals required by these Laws, in the circumstances that are listed in the Sub-Clause. This refers to the Laws of the Country, which are not necessarily the governing law as stated in the Appendix to Tender, but are documents to which the Employer can be assumed to have access.

The obligation is qualified as '*reasonable*' and the Employer being in the position to give assistance. The Contractor should also rely on his local partner, agent or representative before calling on the assistance of the Employer. Co-operation and assistance should always be given when possible, but it must be doubtful whether, if the assistance fails to achieve the desired result, this would reduce the Contractor's obligations or give grounds for a claim. However, a delay in any of these operations could result in a claim under Sub-Clause 8.5 [*Delays Caused by Authorities*] for Delays by Authorities.

The Sub-Clause does not refer to the Employer being able to charge the Contractor for providing this service. The provision of 'reasonable' assistance must be assumed to be free of any charge. However, if the Contractor is asking for excessive '*unreasonable*' assistance then the Employer might give notice under Sub-Clause 2.5 [*Employer's Claims*] that he intends to make a charge. The Contractor might then decide to withdraw his request and obtain the information elsewhere.

2.3 Employer's Personnel

The Employer shall be responsible for ensuring that the Employer's Personnel and the Employer's other contractors on the Site:

(a) co-operate with the Contractor's efforts under Sub-Clause 4.6 [Co-operation]; and

(b) take actions similar to those which the Contractor is required to take under subparagraphs (a), (b) and (c) of Sub-Clause 4.8 [Safety Procedures] and under Sub-Clause 4.18 [Protection of the Environment].

Employer's Personnel are defined at Sub-Clause 1.1.2.6 as being people so notified to the Contractor, including

- The Engineer and his assistants
- all staff, labour and employees of the Employer and the Engineer
- any other person who the Employer or the Engineer has decided to designate as Employer's Personnel.

The designation as Employer's Personnel must be used with discretion. Anyone so designated has rights under other Sub-Clauses including the Sub-Clause 17.1 [*Indemnities*]. The Employer also indemnifies the Contractor for certain actions by Employer's Personnel, including Sub-Clause 17.1.

Sub-Clause 4.6 [*Co-operation*] has a separate requirement for the Contractor to co-operate with other Contractors and public authority personnel. They are not normally Employer's Personnel, but presumably some individuals could be designated and notified to the Contractor as Employer's Personnel.

Sub-Clause 2.3 [*Employer's Personnel*] obliges the Employer to take responsibility for Employer's Personnel and for other Contractors' reciprocal co-operation under Sub-Clause 4.6 [*Co-operation*], Safety Procedures similar to Sub-Clause 4.8(a), (b), (c) [*Safety Procedures*] and Protection of the Environment as Sub-Clause 4.18 [*Protection of the Environment*]. It is important that the Employer includes similar Clauses in his Contracts with all the Contractors who will be working on the Site.

When two or more Contractors are working on the same Site the possibilities of delays and Costs from failures of co-operation, or other problems, can have serious consequences. If the Contractor incurs Costs as a consequence of an action by another Contractor then he may wish to claim against the Employer, but liability may be unclear depending on the applicable law. The FIDIC Clauses may be adequate when this Contractor is carrying out a high percentage of the total work on the Site. However, if the work is more evenly divided between two or more Contractors it will be necessary to review the provisions of the Contract.

2.4 Employer's Financial Arrangements

The Employer shall submit, within 28 days after receiving any request from the Contractor, reasonable evidence that financial arrangements have been made and are being maintained which will enable the Employer to pay the Contract Price (as estimated at that time) in accordance with Clause 14 [Contract Price and Payment]. If the Employer intends to make any material change to his financial arrangements, the Employer shall give notice to the Contractor with detailed particulars.

Sub-Clause 2.4 [*Employer's Financial Arrangements*] is a provision that provides some reassurance to Contractors with regard to the Employer's financial status. The Sub-Clause requires the Employer to provide evidence that he has the finance available to pay the Contractor in accordance with the Contract. The evidence must be provided within 42 days of a request by the Contractor. If the project is financed by an international development agency or similar organisation then it may be advisable to state this fact in the Particular Conditions to reassure tenderers and avoid the subsequent request. Any representative of the finance institution could then be declared as Employer's Personnel and visit the Site if necessary.

If the Employer fails to comply with this Sub-Clause then the Contractor can give 21 days' notice to suspend work, or reduce the rate of work, under Sub-Clause 16.1 [*Contractor's Entitlement to Suspend Work*]. If he does not receive reasonable evidence within 42 days of giving notice under Sub-Clause 16.1 then he is entitled to terminate the Contract under Sub-Clause 16.2 [*Termination by Contractor*]. The procedures of Sub-Clause 16.3 [*Cessation of Work and Removal of Contractor's Equipment*] and Sub-Clause 16.4 [*Payment on Termination*] will then apply.

The protracted time periods are necessary to give the Employer a reasonable time to make arrangements and satisfy the Contractor. However, they make a total period of 105 days or three and a half months from the initial request to the entitlement to terminate the Contract. During this period the Contractor may have spent substantial sums of money on the Contract, perhaps with little chance of recovery. Hence, it is useful to provide the name of any financing body in the Particular Conditions.

The Sub-Clause is not clear as to who provides the estimate of the Contract Price and what would constitute reasonable evidence that the Employer ultimately has the means to pay for the Contract. If the Contractor has already submitted claims, which have been rejected by the Engineer, then they may have very different estimates of the figure that should be taken as the estimated Contract Price. The matter of what constitutes reasonable evidence will depend on the circumstances and could require a statement from the bank, financing authority, Ministry of Finance, or whoever is providing the finance for the project. A payment guarantee by the Employer, on the standard form at Annex G at the back of the FIDIC document, would presumably meet the requirement. The consequences that can arise if the Employer fails to comply with the request are so serious that this situation could result in a dispute, which could be referred to the DAB.

2.5 Employer's Claims

If the Employer considers himself to be entitled to any payment under any Clause of these Conditions or otherwise in connection with the Contract, and/or to any extension of the Defects Notification Period, the Employer or the Engineer shall give notice and particulars to the Contractor. However, notice is not required for payments due under Sub-Clause 4.19 [Electricity, Water and Gas], under Sub-Clause 4.20 [Employer's Equipment and Free-Issue Material], or for other services requested by the Contractor.

The notice shall be given as soon as practicable after the Employer became aware of the event or circumstances giving rise to the claim. A notice relating to any extension of the Defects Notification Period shall be given before the expiry of such period.

The particulars shall specify the Clause or other basis of the claim, and shall include substantiation of the amount and/or extension to which the Employer considers himself to be entitled in connection with the Contract. The Engineer shall then proceed in accordance with Sub-Clause 3.5 [Determinations] to agree or determine (i) the amount (if any) which the Employer is entitled to be paid by the Contractor, and/or (ii) the extension (if any) of the Defects Notification Period in accordance with Sub-Clause 11.3 [Extension of Defects Notification Period].

This amount may be included as a deduction in the Contract Price and Payment Certificates. The Employer shall only be entitled to set off against or make any deduction from an amount certified in a Payment Certificate, or to otherwise claim against the Contractor, in accordance with this Sub-Clause.

Sub-Clause 2.5 [*Employer's Claims*] enables the Employer to give notice of any claim for an extension to the Defects Notification Period or for payments from the Contractor. This procedure must be followed for all deductions or claims for payment except payments for

- electricity, water and gas under Sub-Clause 4.19
- Employer's Equipment and free-issue material under Sub-Clause 4.20
- other services requested by the Contractor, for which the payment has presumably been agreed in advance.

The procedure also applies to any other claims for payment, including claims for amounts not insured or not recovered from the insurers as the final paragraph of Sub-Clause 18.1 [*General Requirements for Insurances*] or for the application of Sub-Clause 8.7 [*Delay damages*]. This procedure should prevent Employers from making unexpected deductions from payment certificates or raising unexpected claims or counterclaims.

The notice must be given '*as soon as practicable after the Employer became aware of the event or circumstances giving rise to the claim*', which should normally be a shorter period than the 28 days from the event which the Contractor is allowed under Sub-Clause 20.1 [*Contractor's Claims*]. However, the requirement omits the phrase '*or should have become aware*', which would be more usual in these circumstances. Any notice for extension must be given before the expiry of the Defects Notification Period, but other notices may presumably be given before or after the issue of the Performance Certificate. A notice of claim given after the issue of the Performance Certificate would need to overcome the statement at Sub-Clause 11.9 [*Performance Certificate*] that '*Only the Performance Certificate shall be deemed to constitute acceptance of the Works*'.

As required in the first paragraph, the Employer must give notice and detailed particulars. The second paragraph states that the notice must be given as soon as practicable after becoming aware of the event. The Contract is silent on when the detailed particulars must be provided although some interpretations have suggested these are to be given at the same time as the notice. This is not explicitly stated within the Sub-Clause. In practise, detailed particulars should be provided as soon as possible after the notice and must specify the Clause or basis of the claim and substantiate the amount and/or extension to which the Employer considers himself entitled.

Also, as seen under Sub-Clause 1.3 [*Communications*] the required notice is not included in the list of communications that shall not be unreasonably withheld or delayed. This may further delay the notice to the Contractor.

Any such claim by the Employer is then dealt with by the Engineer under Sub-Clause 3.5 [*Determinations*] for determinations. If the Engineer determines that the Employer is entitled to payment then the amount may be deducted from the Contract Price and from Payment Certificates. If either Party is not satisfied with the Engineer's determination then the resulting dispute could be referred to the Dispute Adjudication Board under Sub-Clause 20.4 [*Obtaining Dispute Adjudication Board's Decision*].

The actual procedure for recording the deduction will depend on the reason for the deduction. It is important that it is kept separate from the items of payments to the Contractor and the reasons for the deduction, or deductions, are clearly stated. A problem will arise if the Employer's claim occurs during the Defects Notification Period or when, for some reason, the Contractor is not asking for an Interim Payment Certificate. Even if the Contractor included the Engineer's determination in his monthly statement the time periods for payment, at Clause 14, could result in the Employer being entitled to claim against the Performance Security under Sub-Clause 4.2(b) [*Performance Security*], which would obviously be unreasonable.

FIDIC Users' Guide
ISBN 978-0-7277-5856-9

Chapter 9
Clause 3: The Engineer

The '*Engineer*' is defined at Sub-Clause 1.1.2.4 as the person appointed by the Employer and named in the Appendix to Tender. Under the introductory paragraph to Sub-Clause 1.1 [*Definitions*] the word '*person*' can mean a company, so the Engineer may, for example, be named as a firm of Consulting Engineers rather than an individual. If the Engineer is a company then the company should designate an individual to carry out the role of the Engineer. There are some significant advantages with this scenario, for example, the fact that any replacement of the designated person to carry out the role of the Engineer does not automatically invoke Sub-Clause 3.4 [*Replacement of the Engineer*]. Such a designated person shall have as a minimum the following responsibilities.

- Signing of all Communications under Sub-Clause 3.2 [*Delegation by the Engineer*]
- Signing of all determinations under Sub-Clause 3.5 [*Determinations*]

Alternatively, the Engineer may be an individual or a member of the Employer's own staff. If the Employer wishes to change the Engineer from the person named in the Appendix to Tender then he must follow the procedures of Sub-Clause 3.4 [*Replacement of the Engineer*].

One of the most important changes from the 1987 FIDIC fourth edition to the 1999 FIDIC Conditions of Contract is in the role of the Engineer. The 1987 FIDIC edition required the Engineer to exercise his discretion '*impartially within the terms of the Contract and having regard to all the circumstances*'. The 1999 Conditions state that the Engineer '*shall be deemed to act for the Employer*' but, when he is making a decision under Sub-Clause 3.5 [*Determinations*], '*shall make a fair determination in accordance with the Contract, taking due regard of all relevant circumstances*'.

Over the years it has become apparent that for projects where the Engineer is a reputable consultancy hired specifically by the Employer, there is no discernible difference between the decisions made by these Engineers, whether under the 'old' 1987 fourth edition or the 'new' 1999 Series. In contrast, it has been suggested that in projects where the Engineer is a staff member of the Employer, the decisions made by such Engineers do not always have due regard for all relevant circumstances.

Another consequence of this change is that Employers may now utilise a low-cost staff member rather than employing an expensive external professional. This is especially the case for projects not funded by any recognised institutions such as the World Bank or equivalent.

However the Engineer must remember that any decision may be overruled by the DAB within a very short period of time. A series of adverse decisions by the DAB may cause an Employer to question the competence of the Engineer and the true economic cost of such an appointment.

Furthermore, the previous FIDIC Contracts had a clear distinction between the tasks and procedures that are the duty of the Engineer and those that are matters for the Employer. The 1999 Conditions have omitted the requirement that some notices that are sent to the Engineer must be copied to the Employer. However, the distinction has been maintained in other provisions, although some of these are likely in practice to be carried out by the Engineer now that he is *'deemed to act for the Employer'*.

3.1 Engineer's Duties and Authority
The Employer shall appoint the Engineer who shall carry out the duties assigned to him in the Contract. The Engineer's staff shall include suitably qualified engineers and other professionals who are competent to carry out these duties.

The Engineer shall have no authority to amend the Contract.

The Engineer may exercise the authority attributable to the Engineer as specified in or necessarily to be implied from the Contract. If the Engineer is required to obtain the approval of the Employer before exercising a specified authority, the requirements shall be as stated in the Particular Conditions. The Employer undertakes not to impose further constraints on the Engineer's authority, except as agreed with the Contractor.

However, whenever the Engineer exercises a specified authority for which the Employer's approval is required, then (for the purposes of the Contract) the Employer shall be deemed to have given approval.

Except as otherwise stated in these Conditions:

(a) *whenever carrying out duties or exercising authority, specified in or implied by the Contract, the Engineer shall be deemed to act for the Employer;*

(b) *the Engineer has no authority to relieve either Party of any duties, obligations or responsibilities under the Contract; and*

(c) *any approval, check, certificate, consent, examination, inspection, instruction, notice, proposal, request, test, or similar act by the Engineer (including absence of disapproval) shall not relieve the Contractor from any responsibility he has under the Contract, including responsibility for errors, omissions, discrepancies and non-compliances.*

The Engineer has an extremely important role in the administration of the Contract and the way in which he carries out his duties will have a major impact on the work of the Contractor and the success of the project. The Contractor will have made an assumption concerning the likely performance of the Engineer (having been named in the Tender documents) and taken this into account when pricing his Tender.

This Sub-Clause recognises that having to obtain prior Employer's approval for some of his specified responsibilities is a constraint on the Engineer's authority and his freedom to make '*fair*' decisions. The Particular Conditions may include details of any requirements imposed on the Engineer for obtaining Employer's approval before exercising any authority given to him under the Contract. Any additional constraint imposed on the Engineer after the Contract has been agreed would be a breach of Contract by the Employer, for which the Contractor would be entitled to claim damages. It should be noted that there is no contractual obligation on the Contractor to check or verify if and to what extent the Engineer has complied with his obligation to obtain prior Employer's approval, when exercising his authority. If the Contractor suffers damage or is delayed under such a scenario, he would be entitled to claim.

The approvals that are required will vary dependent on whether the Engineer is an independent consultant or a member of the Employer's own staff. An Engineer who is a member of the Employer's staff may have limits on his authority such that additional expenditure, or certifying payment, requires approval from a senior person or a different department. If a named individual is the Engineer any limits on his authority may be stated. A consultant working for a private Employer may have been given the necessary authority, but a consultant working for a Government Employer will probably have limits placed on his authority. The Contractor does not see the contract between the Engineer and the Employer so a clear statement of his authority will be beneficial. These limitations may differ if the Engineer is not the designer for a RED Book Contract.

Approval could reasonably be required before issuing variations under Clause 13 [*Variations and Adjustments*] that would involve changes to the design or result in additional cost greater than a designated figure. Some Employers require the Engineer to obtain approval before certifying *any* additional cost or extension of time. However, under Sub-Clause 3.5 [*Determinations*], the Engineer is required to consult with the Employer before making '*a fair determination in accordance with the Contract*' on a claim for time or money. To require approval of the action after this consultation would imply that the Employer might wish to prevent the Engineer from giving a determination based on his technical and contractual assessment of the claim.

In accordance with Sub-Clause 3.1(a) the Engineer is deemed to act for the Employer '*except as otherwise stated in these Conditions*'. However, it is not clear which Sub-Clauses comply with this 'otherwise stated' exception. Several Sub-Clauses, such as Sub-Clause 3.5 and the payment provisions require the Engineer to be 'fair' and it must be assumed that he will always act in accordance with the requirements of the Contract. Furthermore, whether acting for the Employer or under Sub-Clause 3.5 he will presumably discuss any approval, certificate or action with the Employer if he wishes to obtain the Employer's point of view before deciding what action to take.

A problem will arise if the Employer persuades the Engineer to take some action that is clearly against the provisions of the Contract. This situation would be exposed if a dispute is referred to the DAB and their decision so clearly contradicts the action of the Engineer that no reasonable Engineer could have been expected to act in that way.

3.2 Delegation by the Engineer

The Engineer may from time to time assign duties and delegate authority to assistants, and may also revoke such assignment or delegation. These assistants may include a resident engineer, and/or independent inspectors appointed to inspect and/or test items of Plant and/or Materials. The assignment, delegation or revocation shall be in writing and shall not take effect until copies have been received by both Parties. However, unless otherwise agreed by both Parties, the Engineer shall not delegate the authority to determine any matter in accordance with Sub-Clause 3.5 [Determinations].

Assistants shall be suitably qualified persons, who are competent to carry out these duties and exercise this authority, and who are fluent in the language for communications defined in Sub-Clause 1.4 [Law and Language].

Each assistant, to whom duties have been assigned or authority has been delegated, shall only be authorised to issue instructions to the Contractor to the extent defined by the delegation. Any approval, check, certificate, consent, examination, inspection, instruction, notice, proposal, request, test, or similar act by an assistant, in accordance with the delegation, shall have the same effect as though the act had been an act of the Engineer. However:

(a) any failure to disapprove any work, Plant or Materials shall not constitute approval, and shall therefore not prejudice the right of the Engineer to reject the work, Plant or Materials;

(b) if the Contractor questions any determination or instruction of an assistant, the Contractor may refer the matter to the Engineer, who shall promptly confirm, reverse or vary the determination or instruction.

The Engineer, either as an individual or a designated member of a company, will require assistance to carry out all the duties assigned to him under the Contract. Details of the delegation must be sent in writing to both Parties. If a company is designated as Engineer then all the individuals who will be exercising authority by checking or instructing the Contractor, as resident engineer or as assistants to the Engineer, should be named and their authority confirmed.

Previous FIDIC Contracts included specific reference to the '*Engineer's Representative*', who was normally resident on the Site, acted as the representative of the Engineer and to whom most of the daily administration was delegated. Reference to this position has now been deleted and seems to be replaced by the phrase '*These assistants may include a resident engineer*'. By implication this may be interpreted as the Engineer no longer needs a deputy who is permanently on the Site. However, and from experience it is clear that this is not the case. In practice the Engineer will use Sub-Clause 3.2 [*Delegation by the Engineer*] extensively to delegate many of his authorities to specialist professionals that he can mobilise if and when required, for example, during earthworks a specialist soil and foundation professional.

In truth, for a major project, the Engineer will need a large team of engineers, inspectors and other specialists. A detailed organisation chart should be issued to the Contractor at the start of the project and updated whenever there are changes in personnel. All such people are designated by definition as Employer's Personnel.

Sub-Clause 4.3 [*Contractor's Representative*] requires the Contractor to designate a '*Contractor's Representative*' who will be employed full time on the Contract, so it would be logical for the Engineer to designate someone to act as resident engineer on the Site.

The Engineer is not permitted to delegate his authority under Sub-Clause 3.5 [*Determinations*], except with the agreement of both Parties. However, the duties under numerous other Sub-Clauses require immediate action on the Site. It is therefore convenient for the Engineer to delegate all authority to a deputy, except for specified Sub-Clauses.

3.3 Instructions of the Engineer

The Engineer may issue to the Contractor (at any time) instructions and additional or modified Drawings which may be necessary for the execution of the Works and the remedying of any defects, all in accordance with the Contract. The Contractor shall only take instructions from the Engineer, or from an assistant to whom the appropriate authority has been delegated under this Clause. If an instruction constitutes a Variation, Clause 13 [Variations and Adjustments] shall apply.

The Contractor shall comply with the instructions given by the Engineer or delegated assistant, on any matter related to the Contract. Whenever practicable, their instructions shall be given in writing. If the Engineer or a delegated assistant:

(a) gives an oral instruction,
(b) receives a written confirmation of the instruction, from (or on behalf of) the Contractor, within two working days after giving the instruction, and
(c) does not reply by issuing a written rejection and/or instruction within two working days after receiving the confirmation;

then the confirmation shall constitute the written instruction of the Engineer or delegated assistant (as the case may be).

Sub-Clause 3.3 [*Instructions of the Engineer*] requires the Contractor to comply with any instruction from the Engineer, or an assistant to whom authority has been delegated under Sub-Clause 3.2 [*Delegation by the Engineer*], with a procedure for the confirmation of oral instructions. If the Contractor considers that an instruction will result in additional Costs or delay to completion then he should confirm receipt as a Variation, in accordance with Sub-Clause 13.3 [*Variation Procedure*].

The Sub-Clause gives the Engineer the power to issue additional or modified Drawings. This is an important power because many Contracts under the FIDIC Conditions of Contract rely on a very small number of Drawings in the Tender documents. The majority of the detailed Drawings are issued as and when required during construction. It is for the Engineer to ensure that the Drawings are issued to suit the progress requirements and he will rely on the Contractor's programme as Sub-Clause 8.3 [*Programme*], the monthly progress reports as Sub-Clause 4.21 [*Progress Reports*], information on any design which is the responsibility of the Contractor as Sub-Clause 4.1(a) and (b) [*Contractor's General Obligations*] and any requests from the Contractor as Sub-Clause 1.9 [*Delayed Drawings or Instructions*].

3.4 Replacement of the Engineer

If the Employer intends to replace the Engineer, the Employer shall, not less than 42 days before the intended date of replacement, give notice to the Contractor of the name, address and relevant experience of the intended replacement Engineer. The Employer shall not replace the Engineer with a person against whom the Contractor raises reasonable objection by notice to the Employer, with supporting particulars.

The Employer is entitled to change the Engineer provided he gives 42 days' notice with details of the proposed replacement, and as long as the Contractor does not raise reasonable objections by notice to the Employer. There is no stated time limit for this objection by the Contractor, but common sense would dictate this to be as soon as possible after receiving the notification of the proposed change. A change to the named individual, when the Engineer is a company, does not explicitly require a notification by the Employer, although reasonable notice and discussion would assist in efficient administration by the Parties. The Employer does not have to give any reason for the change, but should take into account that to change the Engineer will have serious consequences for the administration of the Contract, whether the change is made at the start or during the progress of the Works.

The 42-day notice period must be considered in relation to the other time periods stated in the Conditions of Contract. For example, at the start of the project, if the Employer gave notice in the Letter of Acceptance, the Engineer named in the Appendix to Tender would still be obliged to give the notice of the Commencement Date, unless the 42 days stated at Sub-Clause 8.1 [*Commencement of Works*] had been changed in the Particular Conditions. Similarly, whenever the notice is given, the original Engineer will issue at least one monthly Interim Payment Certificate under Sub-Clause 14.6 [*Issue of Interim Payment Certificates*] during the notice period. A change of Engineer could also mean changes to the resident engineer and other assistants appointed under Sub-Clause 3.2 [*Delegation by the Engineer*]. The Contractor is thus advised to indicate as quickly as possible whether he intends to object to the replacement Engineer so that the 42-day period can be used as a changeover period as well as a notice period. If the Engineer should die, or otherwise cease to be available, then the notice period should be waived, by agreement between the Parties.

3.5 Determinations

Whenever these Conditions provide that the Engineer shall proceed in accordance with this Sub-Clause 3.5 to agree or determine any matter, the Engineer shall consult with each Party in an endeavour to reach agreement. If agreement is not achieved, the Engineer shall make a fair determination in accordance with the Contract, taking due regard of all relevant circumstances.

The Engineer shall give notice to both Parties of each agreement or determination, with supporting particulars. Each Party shall give effect to each agreement or determination unless and until revised under Clause 20 [Claims, Disputes and Arbitration].

Table 9.1 Sub-Clauses requiring the Engineer to proceed under Sub-Clause 3.5

Sub-Clause	Heading
1.9	[Delayed Drawings or Instructions]
2.1	[Right of Access to the Site]
2.5	[Employer's Claims]
4.7	[Setting Out]
4.12	[Unforeseeable Physical Conditions]
4.20	[Employer's Equipment and Free Issue Materials]
4.24	[Fossils]
7.4	[Testing]
8.9	[Consequences of Suspension]
9.4	[Failure to Pass Test on Completion]
10.2	[Taking Over of Parts of the Works]
10.3	[Interference with Tests on Completion]
11.4	[Failure to Remedy Defects]
11.8	[Contractor to Search]
12.3	[Evaluation]
12.4	[Omission]
13.2	[Value Engineering]
13.7	[Adjustments for Changes in Legislation]
14.4	[Schedule of Payments]
15.3	[Valuation at Date of Termination]
16.1	[Contractor's Entitlement to Suspend Work]
17.4	[Consequences of Employer's Risks]
19.4	[Consequences of Force Majeure]
20.1	[Contractor's Claims]

Throughout the Conditions of Contract, and certainly whenever the Contractor submits a claim for an extension of time or reimbursement of costs, the Engineer is required to proceed in accordance with Sub-Clause 3.5 [*Determinations*]. Table 9.1 gives a listing of all the Sub-Clauses under which the Engineer is required to proceed in accordance with Sub-Clause 3.5.

Sub-Clause 3.5 [*Determinations*] requires the Engineer to first consult with each Party in an endeavour to reach agreement. To comply with this requirement the Engineer must act as a mediator and try to help the Parties towards agreement.

If agreement is not achieved the Engineer must make '*a fair determination in accordance with the Contract, taking due regard of all relevant circumstances*'. The key phrase here is '*in accordance with the Contract*'. The determination must express the rights and obligations of the Parties, in accordance with the Contract and the applicable law, regardless of the preferences of either Party. Under Sub-Clause 3.2 [*Delegation by the Engineer*] the Engineer cannot delegate this task without the agreement of both Parties.

This Sub-Clause does not impose a time limit on the Engineer for making his determination, but the situation would be covered by the requirement of Sub-Clause 1.3 [*Communications*] that a determination '*shall not be unreasonably withheld*'. The timing of this determination is discussed further in Chapter 26, under Sub-Clause 20.1 [*Contractor's Claims*].

Sub-Clause 1.9 [*Delayed Drawings or Instructions*] is used here as an example to illustrate the complexity of determinations. The Contractor must give notice if he thinks it likely he will be delayed unless drawings or instructions are issued by a particular time. He must give a second notice if he suffers delay or additional costs due to a failure of the Engineer to issue such drawings or instructions. In such a case the Contractor may, subject to Sub-Clause 20.1, recover these costs or claim additional time through a procedure that includes the submission of detailed particulars followed by a Sub-Clause 3.5 [*Determination*]. However, according to Sub-Clause 1.9, the Engineer must proceed in accordance with Sub-Clause 3.5 as soon as he receives the second notice (i.e. without receiving any detailed particulars). In practice, these two separate determinations must be combined into one process. The Engineer's attempts at reaching agreement must commence upon receipt of the second Sub-Clause 1.9 notice but can only be concluded after receipt of detailed particulars under Sub-Clause 20.1.

9.1.　Additional Sub-Clause 3.6 [*Management Meetings*]

The FIDIC *Guidance for the Preparation of Particular Conditions* includes an example Sub-Clause for either the Engineer or the Contractor's representative to call a management meeting. This is a very useful provision and should be included in any FIDIC Contract. A definition of '*Management Meeting*' should be added to Sub-Clause 1.1.6. FIDIC procedures rely on the exchange of notices and written information whereas from a practical project management point of view a meeting to discuss a problem is far more effective than an exchange of paperwork. While most Contractors and Engineers arrange meetings when necessary without any provision in the Contract there may be occasions when a contractual provision is required. The FIDIC suggested wording is as follows:

> *The Engineer or the Contractor's Representative may require the other to attend a management meeting in order to review the arrangements for future work. The Engineer shall record the business of management meetings and supply copies of the record to those attending the meeting and to the Employer. In the record, responsibilities for any actions to be taken shall be in accordance with the Contract.*

The purpose of the Sub-Clause is presumably to encourage good management procedures and enable the Contractor to demand a meeting to discuss an important problem or proposal. Management meetings would be particularly useful in a claim situation or to prevent a claim developing into a serious conflict. When the Contractor gives an initial notice of a potential claim, a management meeting could serve as an early warning meeting at which the Engineer and the Contractor could discuss alternative ways to avoid or overcome a problem.

Traditionally if the Engineer wants the Contractor to attend a meeting, for any purpose, then he has always been able simply to tell the Contractor to do so. The significance including a provision for Management Meetings is that it enables the Contractor to require the Engineer to attend a meeting. The right for the Contractor to demand a meeting is an extremely useful provision and could help to solve potential problems. For a management meeting to serve its purpose, it will need to be held immediately the need arises so the Engineer should under Sub-Clause 3.2 [*Delegation by the Engineer*] delegate the duty of attending any such meeting to a resident engineer or other assistant.

The most likely reason for the Contractor to wish to discuss the arrangements for future work is that he is aware of some potential problem and wants to discuss the options available to avoid or minimise delay or additional cost. For example, if the Contractor gives notice under Sub-Clause 8.3 [*Programme*] of '*specific probable future events or circumstances which may adversely affect the work, increase the Contract Price or delay the execution of the Works*', then it is likely that to avoid or reduce the effect of the problem will require action from the Engineer as well as from the Contractor. Similarly, many of the situations which lead to a reference to the Engineer, under Sub-Clause 3.5 [*Determinations*], might be avoided or the consequences reduced by a meeting held when the problem was first reported.

The final sentence of this additional FIDIC Sub-Clause, referring to responsibilities for any actions to be taken, is confusing and difficult to understand, and could be omitted. Sub-Clause 3.1 [*Engineer's Duties and Authority*] is clear that the Engineer has no authority to amend the Contract so it is difficult to see how the minutes of a meeting could impose responsibilities that are not in accordance with the Contract, unless the Engineer has obtained a relevant approval from the Employer.

FIDIC Users' Guide
ISBN 978-0-7277-5856-9

ICE Publishing: All rights reserved
http://dx.doi.org/10.1680/fug.58569.125

Chapter 10
Clause 4: The Contractor

In the RED Book the Contract Agreement confirms that the Contractor will '*execute and complete the Works and remedy any defects therein, in conformity with the provisions of the Contract*'. In order to meet this primary obligation the Contractor accepts a large number of secondary obligations. Clause 4 defines and confirms details of many of these obligations. However, this is not the only Clause that imposes detailed obligations on the Contractor and it must be read in conjunction with the other Clauses in the Conditions of Contract. When unexpected problems and costs occur during the construction of a project the Contractor may submit claims in order to recover Costs and/or lost time. The detailed requirements in Clause 4 are often used to support or respond to these claims.

Clause 4 is the longest and one of the most important clauses in the Contract. Clause 4 Sub-Clauses cover a wide range of subjects and frequently include topics that would not be anticipated from the headings.

4.1 Contractor's General Obligations
The Contractor shall design (to the extent specified in the Contract), execute and complete the Works in accordance with the Contract and with the Engineer's instructions, and shall remedy any defects in the Works.

The Contractor shall provide the Plant and Contractor's Documents specified in the Contract, and all Contractor's Personnel, Goods, consumables and other things and services, whether of a temporary or permanent nature, required in and for this design, execution, completion and remedying of defects.

The Contractor shall be responsible for the adequacy, stability and safety of all Site operations and of all methods of construction. Except to the extent specified in the Contract, the Contractor (i) shall be responsible for all Contractor's Documents, Temporary Works, and such design of each item of Plant and Materials as is required for the item to be in accordance with the Contract, and (ii) shall not otherwise be responsible for the design or specification of the Permanent Works.

The Contractor shall, whenever required by the Engineer, submit details of the arrangements and methods which the Contractor proposes to adopt for the execution of the Works. No significant alteration to these arrangements and methods shall be made without this having previously been notified to the Engineer.

If the Contract specifies that the Contractor shall design any part of the Permanent Works, then unless otherwise stated in the Particular Conditions:

(a) *the Contractor shall submit to the Engineer the Contractor's Documents for this part in accordance with the procedures specified in the Contract;*

(b) *these Contractor's Documents shall be in accordance with the Specification and Drawings, shall be written in the language for communications defined in Sub-Clause 1.4 [Law and Language], and shall include additional information required by the Engineer to add to the Drawings for co-ordination of each Party's designs;*

(c) *the Contractor shall be responsible for this part and it shall, when the Works are completed, be fit for such purposes for which the part is intended as are specified in the Contract; and*

(d) *prior to the commencement of the Tests on Completion, the Contractor shall submit to the Engineer the 'as-built' documents and operation and maintenance manuals in accordance with the Specification and in sufficient detail for the Employer to operate, maintain, dismantle, reassemble, adjust and repair this part of the Works. Such part shall not be considered to be completed for the purposes of taking-over under Sub-Clause 10.1 [Taking Over of the Works and Sections] until these documents and manuals have been submitted to the Engineer.*

The FIDIC *Conditions of Contract for Construction* are intended to be used for projects with the design provided by the Employer. The Contractor's obligation is to '*execute and complete the Works*' and '*remedy any defects*'. These overall obligations must be read in conjunction with the requirements of other Clauses, such as *inter alia* to '*proceed with the Works with due expedition and without delay*' at Sub-Clause 8.1 [*Commencement of Works*] and '*take full responsibility for the care of the Works*' at Sub-Clause 17.2 [*Contractor's Care of the Works*].

The phrase '*execute and complete*' may seem to be repetitive, but it draws attention to the importance of the procedures at Completion, such as those given at Clauses 8, 9 and 10. The requirement to execute and complete can also give an obligation to complete any item of work which is necessary for total completion of the Works, but which may not have been shown in detail on the Drawings. However, this is an obligation to carry out and complete the Works and the question of whether payment is included in the Accepted Contract Amount is a separate issue.

If the Employer requires the Contractor to carry out the design of part of the Permanent Works then the requirement must be specified in the Contract. The obligations and procedures given at subparagraphs (a) to (d) will apply and must be read in conjunction with other Clauses which refer to the same subjects.

Care must be taken to co-ordinate the various documents that make up the Contract. Problems often arise when different documents are prepared by different consultants and information on the requirements for the Contractor's design is scattered throughout different technical specifications and other documents.

The standard of design performance is stated at subparagraph (c) as '*be fit for such purposes for which the part is intended as are specified in the Contract*'. The details of this requirement must be clearly stated and compliance will often be checked by the Tests on Completion, which are carried out under Clause 9, or Tests after Completion, which are specified separately. This is a much more rigorous standard than the '*reasonable skill, care and diligence*' obligation which governs most Consultants' Contracts. This difference in design standards would lead to serious problems in the event of a dispute as to whether a failure was caused by defective design by either the Employer's designer or the Contractor.

4.2 Performance Security

The Contractor shall obtain (at his cost) a Performance Security for proper performance, in the amount and currencies stated in the Appendix to Tender. If an amount is not stated in the Appendix to Tender, this Sub-Clause shall not apply.

The Contractor shall deliver the Performance Security to the Employer within 28 days after receiving the Letter of Acceptance, and shall send a copy to the Engineer. The Performance Security shall be issued by an entity and from within a country (or other jurisdiction) approved by the Employer, and shall be in the form annexed to the Particular Conditions or in another form approved by the Employer.

The Contractor shall ensure that the Performance Security is valid and enforceable until the Contractor has executed and completed the Works and remedied any defects. If the terms of the Performance Security specify its expiry date, and the Contractor has not become entitled to receive the Performance Certificate by the date 28 days prior to the expiry date, the Contractor shall extend the validity of the Performance Security until the Works have been completed and any defects have been remedied.

The Employer shall not make a claim under the Performance Security, except for amounts to which the Employer is entitled under the Contract in the event of:

(a) failure by the Contractor to extend the validity of the Performance Security as described in the preceding paragraph, in which event the Employer may claim the full amount of the Performance Security;
(b) failure by the Contractor to pay the Employer an amount due, as either agreed by the Contractor or determined under Sub-Clause 2.5 [Employer's Claims] or Clause 20 [Claims, Disputes and Arbitration], within 42 days after this agreement or determination;
(c) failure by the Contractor to remedy a default within 42 days after receiving the Employer's notice requiring the default to be remedied; or
(d) circumstances which entitle the Employer to termination under Sub-Clause 15.2 [Termination by Employer], irrespective of whether notice of termination has been given.

The Employer shall indemnify and hold the Contractor harmless against and from all damages, losses and expenses (including legal fees and expenses) resulting from a claim under the Performance Security to the extent to which the Employer was not entitled to make the claim.

The Employer shall return the Performance Security to the Contractor within 21 days after receiving a copy of the Performance Certificate.

The requirements for the Performance Security will only apply when the amount of the Security is stated in the Appendix to Tender.

The FIDIC Annexes to the *Guidance for the Preparation of Particular Conditions* include example forms C and D for Performance Securities as a Demand Guarantee or as a Surety Bond. Reference is made to guides published by the International Chamber of Commerce (38 cours Albert ler, 75008 Paris, France). The form that is to be used should be included in the Particular Conditions.

The Performance Security must be delivered to the Employer within 28 days from receipt of the Letter of Acceptance and must be approved by the Employer. The receipt and approval of the Performance Security is critical for the start of the project. As Sub-Clause 2.1 [*Right of Access to the Site*] allows the Employer to refuse access to the Site if the Performance Security has not been received, it is desirable that this be completed before the issue of the notice of commencement under Sub-Clause 8.1 [*Commencement of Works*]. A problem may occur if the Commencement Date is earlier than 28 days after the Letter of Acceptance. In this case the Contractor must submit the required Performance Security early if he requires access to the Site in accordance with Sub-Clause 2.1. For practical purposes this is not always possible since banks or other financial institutions that usually underwrite the Performance Securities need a certain amount of time to issue the required documents. There is thus arguably the possibility for the Contractor to be delayed by the Employer without any recourse under the Contract. A remedy may be possible under the Law, and needs to be reviewed on a case by case basis.

Also Sub-Clause 14.6 [*Issue of Interim Payment Certificates*] states that the Engineer will not issue an Interim Payment Certificate, for the advance payment or any other payment, until the Employer has received and approved the Performance Security.

Any claim by the Employer under the Performance Security must follow the procedures of Sub-Clause 2.5 [*Employer's Claims*]. The Employer should proceed with caution before making a claim because Sub-Clause 4.2 [*Performance Security*] includes an indemnity to the Contractor if the Employer was not entitled to make the claim. The Employer must also be aware that the 42-day time periods at (b) and (c) are very short by comparison with other time periods for actions by the Contractor. It is possible that the Contractor might have already started a procedure to rectify these situations, in which case the Performance Security should not be invoked.

4.3 Contractor's Representative
The Contractor shall appoint the Contractor's Representative and shall give him all authority necessary to act on the Contractor's behalf under the Contract.

Unless the Contractor's Representative is named in the Contract, the Contractor shall, prior to the Commencement Date, submit to the Engineer for consent the name and particulars of the person the Contractor proposes to appoint as Contractor's Representative. If consent is withheld or subsequently revoked, or if the appointed person fails to act as Contractor's Representative, the Contractor shall similarly submit the name and particulars of another suitable person for such appointment.

The Contractor shall not, without the prior consent of the Engineer, revoke the appointment of the Contractor's Representative or appoint a replacement.

The whole time of the Contractor's Representative shall be given to directing the Contractor's performance of the Contract. If the Contractor's Representative is to be temporarily absent from the Site during the execution of the Works, a suitable replacement person shall be appointed, subject to the Engineer's prior consent, and the Engineer shall be notified accordingly.

The Contractor's Representative shall, on behalf of the Contractor, receive instructions under Sub-Clause 3.3 [Instructions of the Engineer].

The Contractor's Representative may delegate any powers, functions and authority to any competent person, and may at any time revoke the delegation. Any delegation or revocation shall not take effect until the Engineer has received prior notice signed by the Contractor's Representative, naming the person and specifying the powers, functions and authority being delegated or revoked.

The Contractor's Representative and all these persons shall be fluent in the language for communications defined in Sub-Clause 1.4 [Law and Language].

The Contract gives onerous requirements for the Contractor's Representative.

- He must have either been named in the Contract or had his name and particulars submitted to the Engineer before the Commencement Date.
- He must have received the consent of the Engineer, which can subsequently be revoked.
- He must not be removed or replaced without the prior consent of the Engineer.
- He must have the authority to act on the Contractor's behalf under the Contract.
- He must spend the whole of his time directing the Contractor's performance of the Contract.
- He must be on the Site whenever work is in progress, or be replaced by an approved substitute.
- He must be fluent in the language for communications stated in the Appendix to Tender.
- He must ensure that any delegation is to a competent person, fluent in the language for communications and that the Engineer is notified of any delegation.

The FIDIC *Guidance for the Preparation of Particular Conditions* includes additional paragraphs for the situation when the Contractor's Representative is required to be fluent in more than one language, or when it would be acceptable to use an interpreter.

The Contractor's Representative is, in effect, the Contractor's equivalent to the Engineer. Some Tender documents specify the required qualifications and experience and that the Contractor's Representative must be named in the Tender. This causes problems when the named person is no longer available at the Commencement Date. However, Sub-Clause 4.3 [*Contractor's Representative*] allows for a named person to be replaced, subject to the Engineer's consent to the replacement.

4.4 Subcontractors

The Contractor shall not subcontract the whole of the Works.

The Contractor shall be responsible for the acts or defaults of any Subcontractor, his agents or employees, as if they were the acts or defaults of the Contractor. Unless otherwise stated in the Particular Conditions:

(a) *the Contractor shall not be required to obtain consent to suppliers of Materials, or to a subcontract for which the Subcontractor is named in the Contract;*

(b) *the prior consent of the Engineer shall be obtained to other proposed Subcontractors;*

(c) *the Contractor shall give the Engineer not less than 28 days' notice of the intended date of the commencement of each Subcontractor's work, and of the commencement of such work on the Site; and*

(d) *each subcontract shall include provisions which would entitle the Employer to require the subcontract to be assigned to the Employer under Sub-Clause 4.5 [Assignment of Benefit of Subcontract] (if or when applicable) or in the event of termination under Sub-Clause 15.2 [Termination by Employer].*

This Sub-Clause leaves the Contractor free in his choice of suppliers for materials but requires him to obtain prior consent of the Engineer for all subcontractors that are not named in the Contract. This begs the question, what if a supplier that also installs his material/equipment is or becomes a subcontractor. The definition under Sub-Clause 1.1.2.8 is not helpful and thus to avoid potential disputes any prudent Contractor should raise the issue with the Engineer before placing any order. Alternatively the Parties may decide to include a definition of a Supplier for '*Goods and Materials only*' under Sub-Clause 1.1.2.

The requirement under subparagraph (c) stipulates that the Contractor shall give notice of not less than 28 days to the Engineer whenever a subcontractor intends to commence work and of the commencement of such work on the Site. This is a double notification requirement if the work involved, for example, is first commenced off-site in a factory and subsequently transported to site for installation.

Lastly the provision at subparagraph (d) may cause problems under the governing law and may be difficult to apply.

The involvement of subcontractors on an international project is almost unavoidable, and the need for a compatible (to the RED) subcontract form was identified. As such FIDIC decided to rewrite the *Conditions of Subcontract for Works of Civil Engineering Construction* (first edition, 1994) specifically for use in conjunction with RED. This was achieved through the publication of *Conditions of Subcontract for Construction for Building and Engineering Works designed by the Employer* (first edition, 2011) also referred to in this book as the Subcontract Book. It is further detailed under Part 4.

4.5 Assignment of Benefit of Subcontract

If a Subcontractor's obligations extend beyond the expiry date of the relevant Defects Notification Period and the Engineer, prior to this date, instructs the Contractor to assign the benefit of such obligations to the Employer, then the Contractor shall do so. Unless otherwise stated in the assignment, the Contractor shall have no liability to the Employer for the work carried out by the Subcontractor after the assignment takes effect.

It is essential to check the provisions of the applicable law concerning the rights and obligations of Subcontractors. Under some legal systems the Subcontractor may have the right of direct access to the Employer in order to obtain payment of sums that have been withheld by the Contractor.

4.6 Co-operation

The Contractor shall, as specified in the Contract or as instructed by the Engineer, allow appropriate opportunities for carrying out work to:

(a) the Employer's Personnel,

(b) any other contractors employed by the Employer, and

(c) the personnel of any legally constituted public authorities, who may be employed in the execution on or near the Site of any work not included in the Contract.

Any such instruction shall constitute a Variation if and to the extent that it causes the Contractor to incur Unforeseeable Cost. Services for these personnel and other contractors may include the use of Contractor's Equipment, Temporary Works or access arrangements which are the responsibility of the Contractor.

If, under the Contract, the Employer is required to give to the Contractor possession of any foundation, structure, plant or means of access in accordance with Contractor's Documents, the Contractor shall submit such documents to the Engineer in the time and manner stated in the Specification.

The heading '*Co-operation*' implies that the Sub-Clause deals with mutual co-operation between the Employer and the Contractor. In fact, the co-operation is one sided and refers to the provision of facilities by the Contractor, and by Subcontractors, to anyone who has been instructed by the Employer to carry out work on the Site. The reciprocal requirement, at Sub-Clause 2.3(a) [*Employer's Personnel*], is that the Employer's Personnel and other contractors on the Site shall co-operate with the Contractor's efforts.

The Sub-Clause requires the Contractor to provide the use of his Equipment, Temporary Works and access arrangements, but allows for all Unforeseeable Costs to be claimed and paid as a Variation. Any delays as a consequence of the application of this Sub-Clause would qualify for an extension of time under Sub-Clause 8.4(e) [*Extension of Time for*

Completion]. Alternatively, if the other Contractor is a legally constituted public authority then the delay might be covered by Sub-Clause 8.5 [*Delays Caused by Authorities*]. This compensation would be important in the event of misuse or damage caused by the other Contractor.

The final paragraph of Sub-Clause 4.6 [*Co-operation*] refers to a more complex situation. A '*foundation, structure, plant or means of access*' is occupied by the Employer and the Contractor requires possession in order to execute his work. The Contractor is required to submit the relevant Contractor's Documents in the time and manner stated in the Specification. The reference to Contractor's Documents implies that the situation relates to work that was designed by the Contractor. If not then the Employer's obligation is given at the first paragraph of Sub-Clause 2.1 [*Right of Access to the Site*].

The duties of the Contractor under this Sub-Clause are not to be confused with '*Coordination*' with other contractors or persons on the Site. This is always the responsibility of the Employer, unless otherwise agreed between the Parties.

4.7 Setting Out

The Contractor shall set out the Works in relation to original points, lines and levels of reference specified in the Contract or notified by the Engineer. The Contractor shall be responsible for the correct positioning of all parts of the Works, and shall rectify any error in the positions, levels, dimensions or alignment of the Works.

The Employer shall be responsible for any errors in these specified or notified items of reference, but the Contractor shall use reasonable efforts to verify their accuracy before they are used.

If the Contractor suffers delay and/or incurs Cost from executing work which was necessitated by an error in these items of reference, and an experienced contractor could not reasonably have discovered such error and avoided this delay and/or Cost, the Contractor shall give notice to the Engineer and shall be entitled, subject to Sub-Clause 20.1 [Contractor's Claims] to:

(a) *an extension of time for any such delay, if completion is or will be delayed, under Sub-Clause 8.4 [Extension of Time for Completion]; and*
(b) *payment of any such Cost plus reasonable profit, which shall be included in the Contract Price.*

After receiving this notice, the Engineer shall proceed in accordance with Sub-Clause 3.5 [Determinations] to agree or determine (i) whether and (if so) to what extent the error could not reasonably have been discovered, and (ii) the matters described in subparagraphs (a) and (b) above related to this extent.

The Employer and the Contractor are each responsible for their own work. If setting out information from the Employer, or the Engineer on his behalf, is not accurate then the Contractor can give notice and follow the usual claims procedure for a determination by the Engineer.

However, the Contractor is obliged to use reasonable efforts to check the accuracy of the reference items. To what extent this is feasible will depend on the circumstances. If the Contractor finds an error then he would be required to give notice under the final paragraph of Sub-Clause 1.8 [*Care and Supply of Documents*].

4.8 Safety Procedures
The Contractor shall:

(a) *comply with all applicable safety regulations;*
(b) *take care for the safety of all persons entitled to be on the Site;*
(c) *use reasonable efforts to keep the Site and Works clear of unnecessary obstruction so as to avoid danger to these persons;*
(d) *provide fencing, lighting, guarding and watching of the Works until completion and taking over under Clause 10 [Employer's Taking Over]; and*
(e) *provide any Temporary Works (including roadways, footways, guards and fences) which may be necessary, because of the execution of the Works, for the use and protection of the public and of owners and occupiers of adjacent land.*

Most countries have their own health and safety regulations although the details vary considerably. Paragraph (a) of Sub-Clause 4.8 [*Safety Procedures*] requires the Contractor to comply with the local regulations. The Sub-Clause includes safety requirements that may be less onerous, or more onerous, than the local regulations. The requirements for safety may be expanded in the Particular Conditions.

If the Contractor does not have exclusive possession of the Site, as permitted by Sub-Clause 2.1 [*Right of Access to the Site*], then the Particular Conditions should clarify the responsibility for the Contractor to comply with this Sub-Clause.

4.9 Quality Assurance
The Contractor shall institute a quality assurance system to demonstrate compliance with the requirements of the Contract. The system shall be in accordance with the details stated in the Contract. The Engineer shall be entitled to audit any aspect of the system.

Details of all procedures and compliance documents shall be submitted to the Engineer for information before each design and execution stage is commenced. When any document of a technical nature is issued to the Engineer, evidence of the prior approval by the Contractor himself shall be apparent on the document itself.

Compliance with the quality assurance system shall not relieve the Contractor of any of his duties, obligations or responsibilities under the Contract.

The requirements for a quality assurance system depend on the details of the system which is stated in the Contract. If the Employer does not intend to include such details then the Sub-Clause should be deleted. If the Sub-Clause is retained and no details are provided within the Contract then there is ambiguity between the first two sentences in the first paragraph leading to a potential dispute. To resolve this potential conflict the Specification should include details of a suitable quality assurance system.

When considering whether to include a requirement for a Contractor's quality assurance system the Employer must consider whether the benefits will outweigh the cost. If the tenderers are experienced in the operation of their own quality control then the additional cost will probably be justified. However, with a Contractor who is not accustomed to operating his own quality control systems, the Employer may prefer to rely on the quality control exercised by the Engineer in accordance with the Contract. A good quality assurance system, operated by experienced personnel from both Parties and the Engineer, should avoid quality problems, benefit the project and justify the additional cost.

4.10 Site Data

The Employer shall have made available to the Contractor for his information, prior to the Base Date, all relevant data in the Employer's possession on sub-surface and hydrological conditions at the Site, including environmental aspects. The Employer shall similarly make available to the Contractor all such data which come into the Employer's possession after the Base Date. The Contractor shall be responsible for interpreting all such data.

To the extent which was practicable (taking account of cost and time), the Contractor shall be deemed to have obtained all necessary information as to risks, contingencies and other circumstances which may influence or affect the Tender or Works. To the same extent, the Contractor shall be deemed to have inspected and examined the Site, its surroundings, the above data and other available information, and to have been satisfied before submitting the Tender as to all relevant matters, including (without limitation):

(a) the form and nature of the Site, including sub-surface conditions;

(b) the hydrological and climatic conditions;

(c) the extent and nature of the work and Goods necessary for the execution and completion of the Works and the remedying of any defects;

(d) the Laws, procedures and labour practices of the Country; and

(e) the Contractor's requirements for access, accommodation, facilities, personnel, power, transport, water and other services.

The heading and first sentence of this Sub-Clause do not give any indication of the wide scope of the Sub-Clause. Even the list of items about which the Contractor must have satisfied himself does not limit the scope of these matters.

For the Employer to *'make available to the Contractor for his information'* the details of sub-surface and other conditions does not mean that copies of the data will have been provided in the Tender documents. The Employer may just have indicated that certain information is available for inspection. The phrase *'all relevant data in the Employer's possession'* is a very wide requirement and must include all the information that the Employer has acquired concerning the Site. Although the Employer is only required to make available *'relevant'* data, it is advisable for the Employer to disclose all data relating to the Site. Data that may seem not to be relevant at the Tender stage may become relevant if problems occur during construction. Information which appears to be inconsistent may be a clue to sub-surface problems and is important for prospective Contractors. The information must be made available at least 28 days before the latest date for the submission of Tenders (refer Sub-Clause 1.1.3.1 [*Base Date*]), in order to give tenderers time to make their own studies and enquiries. If the Employer obtains further information after the Base Date this must also be made available to the Contractor.

The Contractor is responsible for interpreting the Site data and making his own enquiries in order to satisfy himself that his Tender is adequate, as required by Sub-Clause 4.11 [*Sufficiency of the Accepted Contract Amount*]. The extent and detail of the Contractor's enquiries can be restricted as stated at the second paragraph of Sub-Clause 4.10 [*Site Data*]. The interpretation of the phrase '*which was practicable (taking account of cost and time)*' is subjective. If the Contractor submits a claim, perhaps under Sub-Clause 4.12 [*Unforeseeable Physical Conditions*], the Engineer may respond that the Contractor should have made further enquiries and investigations before submitting his Tender.

The second paragraph of Sub-Clause 4.10 requires tenderers to have examined the surroundings to the Site, as well as the Site itself. The extent and detail of this examination is not clear. If the project involves deep excavation close to the Site boundary then the tenderer will have to consider the support necessary to avoid causing problems for the adjoining site. However, if there is a potential for a problem to originate from the adjoining or nearby area, at no fault of the Contractor, then it would be unreasonable to expect a tenderer to allow for the cost of protecting his Works against that problem. If a tenderer is expected to include any such allowance in his Tender then the potential problem should have been identified in the Tender documents so that all Tenders are prepared on the same basis. This is an important consideration because claims against the Employer that result from activities or events on an adjoining area will be difficult to substantiate. They may be a matter of Employer's Risks as Sub-Clause 17.3(h) [*Employer's Risks*], Insurance as Clause 18 [*Insurance*], Force Majeure as Sub-Clause 19.1 [*Definition of Force Majeure*], or a claim against a third party.

The Contractor should keep records of any studies and investigations that he has made in compliance with this Sub-Clause. Records to show that he has taken reasonable action in all the circumstances may be useful to support any claim for unforeseeable physical conditions, in accordance with the final paragraph of Sub-Clause 4.12 [*Unforeseeable Physical Conditions*].

The second part of Sub-Clause 4.10 [*Site Data*] covers a much wider range of subjects than is indicated by the heading '*Site Data*'. Subparagraph (c) requires the Contractor to have studied his proposed method of construction and considered his requirements and the availability of the Materials, Equipment and Temporary Works that will be required. Subparagraph (d) requires a review of '*the Laws, procedures and labour practices of the Country*'. This is a very wide subject and includes many subjects mentioned elsewhere in the Contract, such as health and safety regulations, normally accepted practices for local labour, restrictions on imported labour and equipment and the local tax regulations. A lack of knowledge on any of these subjects will not be accepted as justification for a claim.

4.11 Sufficiency of the Accepted Contract Amount
The Contractor shall be deemed to:

(a) *have satisfied himself as to the correctness and sufficiency of the Accepted Contract Amount; and*

(b) have based the Accepted Contract Amount on the data, interpretations, necessary information, inspections, examinations and satisfaction as to all relevant matters referred to in Sub-Clause 4.10 [Site Data].

Unless otherwise stated in the Contract, the Accepted Contract Amount covers all the Contractor's obligations under the Contract (including those under Provisional Sums, if any) and all things necessary for the proper execution and completion of the Works and the remedying of any defects.

This is another sub-clause that refers to actions by the Contractor during the preparation of his Tender. The Contractor is deemed to have based his Tender submission on the requirements of Sub-Clause 4.10 [*Site Data*]. Hence, if there is an error in any of the information that was provided by the Employer under Sub-Clause 4.10 then the Contractor may have the basis for a claim for additional payment.

If the Contractor discovered any errors or omissions in the information that was provided in the Tender documents, then he should have raised the problem before becoming committed to the Accepted Contract Amount.

4.12 Unforeseeable Physical Conditions

In this Sub-Clause, 'physical conditions' means natural physical conditions and man-made and other physical obstructions and pollutants, which the Contractor encounters at the Site when executing the Works, including sub-surface and hydrological conditions but excluding climatic conditions.

If the Contractor encounters adverse physical conditions which he considers to have been Unforeseeable, the Contractor shall give notice to the Engineer as soon as practicable.

This notice shall describe the physical conditions, so that they can be inspected by the Engineer, and shall set out the reasons why the Contractor considers them to be Unforeseeable. The Contractor shall continue executing the Works, using such proper and reasonable measures as are appropriate for the physical conditions, and shall comply with any instructions which the Engineer may give. If an instruction constitutes a Variation, Clause 13 [Variations and Adjustments] shall apply.

If and to the extent that the Contractor encounters physical conditions which are Unforeseeable, gives such a notice, and suffers delay and/or incurs Cost due to these conditions, the Contractor shall be entitled subject to Sub-Clause 20.1 [Contractor's Claims] to:

(a) an extension of time for any such delay, if completion is or will be delayed, under Sub-Clause 8.4 [Extension of Time for Completion]; and
(b) payment of any such Cost, which shall be included in the Contract Price.

After receiving such notice and inspecting and/or investigating these physical conditions, the Engineer shall proceed in accordance with Sub-Clause 3.5 [Determinations] to agree or determine (i) whether and (if so) to what extent these physical conditions were

Unforeseeable, and (ii) the matters described in subparagraphs (a) and (b) above related to this extent.

However, before additional Cost is finally agreed or determined under subparagraph (ii), the Engineer may also review whether other physical conditions in similar parts of the Works (if any) were more favourable than could reasonably have been foreseen when the Contractor submitted the Tender. If and to the extent that these more favourable conditions were encountered, the Engineer may proceed in accordance with Sub-Clause 3.5 [Determinations] to agree or determine the reductions in Cost which were due to these conditions, which may be included (as deductions) in the Contract Price and Payment Certificates. However, the net effect of all adjustments under sub-paragraph (b) and all these reductions, for all the physical conditions encountered in similar parts of the Works, shall not result in a net reduction in the Contract Price.

The Engineer may take account of any evidence of the physical conditions foreseen by the Contractor when submitting the Tender, which may be made available by the Contractor, but shall not be bound by any such evidence.

Unforeseeable physical conditions are probably the most common source of claims and disputes in construction projects. The definition of *'Physical Conditions'* is wider than might be expected and will include many of the unexpected situations that the Contractor *'encounters at the Site when executing the Works'*. It was the Employer who decided to construct the project on this particular Site and designed the project to suit the Site; in principle, the Employer should take responsibility for the consequences of any problems present on his Site. The difficult question is whether a particular problem should have been anticipated and allowed for in the Accepted Contract Amount, or whether an experienced Contractor could not have foreseen the situation. The restraints of cost and time referred in Sub-Clause 14.10 [*Statement at Completion*] must be considered in making this assessment.

When a problem arises the Contractor must give a notice *'as soon as practicable'* describing the physical conditions. This initial notice may result in the Engineer deciding to issue an instruction for a Variation under Clause 13 so the delay and additional costs would be covered by the variation procedures. However, the Contractor may also need to consider notices under other clauses, such as the final two paragraphs of Sub-Clause 8.3 [*Programme*], Sub-Clause 8.4 [*Extension of Time for Completion*] for an extension of time, Sub-Clause 13.2 [*Value Engineering*] if he has a Value Engineering proposal, Sub-Clause 19.2 [*Notice of Force Majeure*] for Force Majeure, and the notice under Sub-Clause 20.1 [*Contractor's Claims*] which is essential for any claims.

The Contractor will need to comply with the appropriate time periods for any notices. The Contractor may also have a claim under an unforeseeable circumstances provision in the governing law.

Physical conditions situations frequently require a change to the design or method of working. The Contractor is obliged to continue the Works, so an immediate notice followed by prompt action from the Engineer is essential. The Contractor can take *'such proper and reasonable measures as are appropriate for the physical conditions*' but, if these measures constitute a Variation, an instruction is required from the Engineer. Clearly there is a potential for misunderstanding and claims, particularly if the Engineer fails to give instructions and the Contractor feels obliged to make decisions as to what action to take, although in all these situations the Contractor would be obliged to also comply with the requirements of Sub-Clause 1.9 [*Delayed Drawings or Instructions*].

The situations that may be covered by this Sub-Clause give a clear example of the need for a meeting between the Engineer and the Contractor in order to discuss the problem and agree on the best technical solution. While an immediate meeting would be normal practice on most projects it is unfortunate that the Contract fails to put a clear obligation on both the Contractor and the Engineer to call and attend such a meeting. This is another reason for introducing a provision for management meetings, as discussed at the proposed Sub-Clause 3.6.

Having received the Contractor's notices the Engineer will investigate the situation and proceed in accordance with the usual claims procedures at Sub-Clauses 20.1, 8.4 and 3.5.

Sub-Clause 4.12 [*Unforeseeable Physical Conditions*] also includes a provision to enable the Engineer to take into account physical conditions at other parts of the Site that were more favourable than could reasonably have been foreseen. This is controversial for a variety of reasons. While it may seem reasonable that better than foreseeable should balance worse than foreseeable, in practice the situation is rarely straightforward. The Contractor has made assumptions, brought on Site the appropriate materials and equipment and made the appropriate assumptions in his programme. Any change to these initial assumptions, whether more or less favourable, can result in delay and additional cost. Analysis of actual time and cost, compared to the time and cost assumed in the Accepted Contract Amount, will be a lengthy and complex procedure. Evidence of the Contractor's Tender assumptions, as the final paragraph of Sub-Clause 4.12, could be crucial in the discussions, but which the Contractor is not contractually obliged to make available. The Engineer is to determine his valuation of any deductions to the claimed amount under the Sub-Clause 3.5 [*Determinations*] procedure, which requires him to be fair in accordance with the Contract and to take into account all relevant circumstances. But in the absence of any hard facts as to what exactly the Contractor has foreseen and allowed for in his tender calculation, it is very difficult to make such a determination beyond doubt. Under the PNK, where the wording is changed from '*may*' to '*shall*' placing an obligation on the Contractor to provide this information, the Engineer is able to make a determination. However, there is a further complication that, under some legal jurisdictions, the deduction as such might be construed as a contractual entitlement to the Employer, thereby potentially putting the onerous obligation of the burden of proof onto the Employer, making it even more difficult to implement the subparagraph.

This provision is worded as a plus and minus review and calculation by the Engineer in his valuation of the Contractor's claim, rather than as a separate claim by the Employer under Sub-Clause 2.5. It is again a question of law as to the validity of this provision, as any deduction under this subparagraph would in essence be a contractual benefit to the Employer arbitrarily exercised by the Engineer.

The FIDIC Particular Conditions include a suggested provision for the risk of sub-surface conditions being shared between the Parties. The percentage of the cost that is to be borne by the Contractor would be stated in the Particular Conditions. For this provision to be both fair and workable in practice the Contractor must be given the time and opportunity to carry out investigations at Tender stage.

The potential for technical problems, claims and disputes due to unforeseeable physical conditions, enhanced by any uncertainty at Tender stage, demonstrates the importance of the Employer carrying out a proper investigation before calling Tenders. The value of site investigation is not just in order to prepare the design, but also to enable the Contractor to prepare a realistic Tender.

4.13 Rights of Way and Facilities
The Contractor shall bear all costs and charges for special and/or temporary rights-of-way which he may require, including those for access to the Site. The Contractor shall also obtain, at his risk and cost, any additional facilities outside the Site which he may require for the purposes of the Works.

This Sub-Clause recognises that the Contractor may require additional facilities outside the Site. The location and boundaries of the Site are defined in the Contract. The use of these additional facilities would require the Engineer's agreement under Sub-Clause 4.23 [*Contractor's Operations on Site*].

Under Sub-Clause 7.3 [*Inspection*] the Contractor must provide access and facilities for inspection of work in these additional areas, as required by the Employer's Personnel. Matters of insurance and payment for work done must also be considered for all off-site facilities.

4.14 Avoidance of Interference
The Contractor shall not interfere unnecessarily or improperly with:

(a) the convenience of the public; or
(b) the access to and use and occupation of all roads and footpaths, irrespective of whether they are public or in the possession of the Employer or of others.

The Contractor shall indemnify and hold the Employer harmless against and from all damages, losses and expenses (including legal fees and expenses) resulting from any such unnecessary or improper interference.

The indemnity under this Sub-Clause only covers cost resulting from interference that is '*unnecessary or improper*'. This does not imply that there is some form of necessary or proper interference that is not the Contractor's responsibility. Presumably the necessary and proper interference was anticipated and allowed for in the Contract, for example, in relation to Sub-Clause 4.15 [*Access Route*]. However, the indemnity could have wider implications, possibly including a breach of the Sub-Clause 4.23 [*Contractor's operations on site*] requirement to keep Contractor's Equipment and Personnel within the Site and off adjacent land.

4.15 Access Route

The Contractor shall be deemed to have been satisfied as to the suitability and availability of access routes to the Site. The Contractor shall use reasonable efforts to prevent any road or bridge from being damaged by the Contractor's traffic or by the Contractor's Personnel. These efforts shall include the proper use of appropriate vehicles and routes. Except as otherwise stated in these Conditions:

(a) *the Contractor shall (as between the Parties) be responsible for any maintenance which may be required for his use of access routes;*

(b) *the Contractor shall provide all necessary signs or directions along access routes, and shall obtain any permission which may be required from the relevant authorities for his use of routes, signs and directions;*

(c) *the Employer shall not be responsible for any claims which may arise from the use or otherwise of any access route;*

(d) *the Employer does not guarantee the suitability or availability of particular access routes; and*

(e) *Costs due to non-suitability or non-availability, for the use required by the Contractor, of access routes shall be borne by the Contractor.*

The first sentence of this Sub-Clause repeats the requirement of Sub-Clause 4.10(e) [*Site Data*] with regard to access routes to the Site. If access for heavy Contractor's Equipment or Plant is impossible because of the lack of suitable roads or bridges then the problem should have been raised by the Contractor before he submitted his Tender.

The obligations and responsibilities imposed by this Sub-Clause are between the Contractor and the Employer. If the Contractor causes damage to any roads or bridges, or fails to comply with any regulation, then he will also have a problem with the relevant authority and a liability under the governing law.

4.16 Transport of Goods

Unless otherwise stated in the Particular Conditions:

(a) *the Contractor shall give the Engineer not less than 21 days' notice of the date on which any Plant or a major item of other Goods will be delivered to the Site;*

(b) *the Contractor shall be responsible for packing, loading, transporting, receiving, unloading, storing and protecting all Goods and other things required for the Works; and*

(c) *the Contractor shall indemnify and hold the Employer harmless against and from all damages, losses and expenses (including legal fees and expenses) resulting from the transport of Goods, and shall negotiate and pay all claims arising from their transport.*

The definition of Goods covers virtually everything that will be delivered to the Site and this Sub-Clause repeats the Contractor's general obligations under the Contract. The requirement for 21 days' notice of delivery only applies to Plant or major items or other Goods. It seems strange that the notice shall be for all Plant (big or small) and not for all Goods. The Employer could specify items of Plant and Goods for which this Sub-Clause shall apply within the tender documents. Failing this, the Contractor could raise the issue during tender negotiations resulting in a mutually agreed amendment to the Particular Conditions. The reason for giving a notice under this Sub-Clause is not stated. The arrangements for access, delivery and storage are the Contractor's responsibility and so the Engineer will only be interested from his usual concerns for progress, payment and supervision. The notice should confirm the dates that have already been given in the Contractor's programme and monthly progress reports.

The indemnity for costs arising from the transport of Goods includes an additional requirement that the Contractor shall negotiate any claims from third persons that rise from the transport of Goods.

4.17 Contractor's Equipment
The Contractor shall be responsible for all Contractor's Equipment. When brought on to the Site, Contractor's Equipment shall be deemed to be exclusively intended for the execution of the Works. The Contractor shall not remove from the Site any major items of Contractor's Equipment without the consent of the Engineer. However, consent shall not be required for vehicles transporting Goods or Contractor's Personnel off Site.

By definition, Contractor's Equipment includes any workshop or manufacturing equipment that is used to manufacture items for the Works. Any such Equipment that is installed on the Site cannot be used to manufacture items for sale to other projects.

When the Contractor has completed the work for which a particular item of Contractor's Equipment was brought to the Site then that item will obviously be removed. If the Engineer, for whatever reason, does not give prompt consent, then the Contractor will probably claim for the cost of keeping the equipment on the Site.

The FIDIC *Guidance for the Preparation of Particular Conditions* includes an additional paragraph with provision that the Contractor's Equipment would be deemed to be the property of the Employer when it is on the Site. This provision could lead to other problems and may not be recognised by the governing law.

4.18 Protection of the Environment
The Contractor shall take all reasonable steps to protect the environment (both on and off the Site) and to limit damage and nuisance to people and property resulting from pollution, noise and other results of his operations. The Contractor shall ensure that emissions, surface discharges and effluent from the Contractor's activities shall not exceed the values indicated in the Specification, and shall not exceed the values prescribed by applicable Laws.

Under Sub-Clause 4.10(d) the Contractor is deemed to have obtained information on the Laws and procedures of the Country and, under Sub-Clause 1.13 [*Compliance with Laws*], is required to comply with applicable Laws.

Sub-Clause 4.18 [*Protection of the Environment*] also states that maximum values for emissions, surface discharges and effluent from the Contractor's activities will be given in the Specification. These values may be higher but not lower than the figures in local regulations.

Sub-Clause 4.18 does not require an indemnity from the Contractor if anyone should claim against the Employer, but the general indemnity at Sub-Clause 17.1 [*Indemnities*] would probably cover breaches of this Sub-Clause.

The requirement for the Contractor to protect the environment both on and off the Site could be potentially problematic. There is no definition for what constitutes '*off Site*', although any damage to off Site properties or persons would ordinarily be dealt with under the applicable laws of the country that the Contractor has to comply with in any case as per Sub-Clause 1.4 [*Law and Language*] and Sub-Clause 1.13 [*Compliance with Laws*].

4.19 Electricity, Water and Gas
The Contractor shall, except as stated below, be responsible for the provision of all power, water and other services he may require.

The Contractor shall be entitled to use for the purposes of the Works such supplies of electricity, water, gas and other services as may be available on the Site and of which details and prices are given in the Specification. The Contractor shall, at his risk and cost, provide any apparatus necessary for his use of these services and for measuring the quantities consumed.

The quantities consumed and the amounts due (at these prices) for such services shall be agreed or determined by the Engineer in accordance with Sub-Clause 2.5 [Employer's Claims] and Sub-Clause 3.5 [Determinations]. The Contractor shall pay these amounts to the Employer.

Sub-Clause 4.19 [*Electricity, Water and Gas*] is reviewed together with Sub-Clause 4.20 [*Employer's Equipment and Free-Issue Material*].

4.20 Employer's Equipment and Free-Issue Material
The Employer shall make the Employer's Equipment (if any) available for the use of the Contractor in the execution of the Works in accordance with the details, arrangements and prices stated in the Specification. Unless otherwise stated in the Specification:

(a) the Employer shall be responsible for the Employer's Equipment, except that
(b) the Contractor shall be responsible for each item of Employer's Equipment while any of the Contractor's Personnel is operating it, driving it, directing it or in possession or control of it.

The appropriate quantities and the amounts due (at such stated prices) for the use of Employer's Equipment shall be agreed or determined by the Engineer in accordance with Sub-Clause 2.5 [Employer's Claims] and Sub-Clause 3.5 [Determinations]. The Contractor shall pay these amounts to the Employer.

The Employer shall supply, free of charge, the 'free-issue materials' (if any) in accordance with the details stated in the Specification. The Employer shall, at his risk and cost, provide these materials at the time and place specified in the Contract. The Contractor shall then visually inspect them, and shall promptly give notice to the Engineer of any shortage, defect or default in these materials. Unless otherwise agreed by both Parties, the Employer shall immediately rectify the notified shortage, defect or default.

After this visual inspection, the free-issue materials shall come under the care, custody and control of the Contractor. The Contractor's obligations of inspection, care, custody and control shall not relieve the Employer of liability for any shortage, defect or default not apparent from a visual inspection.

If the Contractor wishes to use services that are available on the Site, or equipment that is made available by the Employer, he must pay at the prices stated in the Specification or Employer's Requirements. Under Sub-Clause 2.5 [*Employer's Claims*] the Employer does not have to give notice of a claim for this payment. However, the quantity or time for each item must be recorded and agreed. If there is any disagreement then the Engineer will act under Sub-Clauses 2.5 and 3.5. Payment can then be made as a deduction in the Contractor's monthly statement under Sub-Clause 14.3(f) [*Application for Interim Payment Certificates*].

While it may be convenient for the Employer to provide free-issue material or the use of equipment, the procedure can cause problems. If, for example, the Contractor is relying on this supply and the materials do not comply with the Specification then the Employer must find an alternative. Any such division of responsibility that changes the Contractor's general obligations is difficult to control. The costs to the Employer of any problems will be more than might be expected.

4.21 Progress Reports

Unless otherwise stated in the Particular Conditions, monthly progress reports shall be prepared by the Contractor and submitted to the Engineer in six copies. The first report shall cover the period up to the end of the first calendar month following the Commencement Date. Reports shall be submitted monthly thereafter, each within 7 days after the last day of the period to which it relates.

Reporting shall continue until the Contractor has completed all work which is known to be outstanding at the completion date stated in the Taking-Over Certificate for the Works.

Each report shall include:

(a) charts and detailed descriptions of progress, including each stage of design (if any), Contractor's Documents, procurement, manufacture, delivery to Site, construction, erection and testing; and including these stages for work by each nominated Subcontractor (as defined in Clause 5 [Nominated Subcontractors]);

(b) photographs showing the status of manufacture and of progress on the Site;

143

(c) for the manufacture of each main item of Plant and Materials, the name of the manufacturer, manufacture location, percentage progress, and the actual or expected dates of:

    (i)   commencement of manufacture,

    (ii) Contractor's inspections,

    (iii) tests, and

    (iv) shipment and arrival at the Site;

(d) the details described in Sub-Clause 6.10 [Records of Contractor's Personnel and Equipment];

(e) copies of quality assurance documents, test results and certificates of Materials;

(f) list of notices given under Sub-Clause 2.5 [Employer's Claims] and notices given under Sub-Clause 20.1 [Contractor's Claims];

(g) safety statistics, including details of any hazardous incidents and activities relating to environmental aspects and public relations; and

(h) comparisons of actual and planned progress, with details of any events or circumstances which may jeopardise the completion in accordance with the Contract, and the measures being (or to be) adopted to overcome delays.

The requirement for the Contractor to provide monthly progress reports is an important task that will enable both Parties to efficiently plan and execute their respective obligations under the Contract. The detailed requirements are onerous and are not just a matter of reporting progress on the Site and elsewhere but include safety statistics, lists of claims and other matters. The Particular Conditions may include further requirements, such as submitting the progress report for discussion at a meeting or replacing the six copies by an electronic submission. A useful addition that has been included as a standard requirement under the YLW Book is a list of Variations. On the other hand, some of the requirements may not be relevant to a particular project and the format of the report should be agreed at the start of the project.

The report will be a substantial document and must be submitted within seven days from the last day of the relevant month, which means five working days. To meet this requirement the Contractor must record and collect the information during the month, including information from Subcontractors. The Contractor will probably have recorded most of the information as part of his own internal procedures and records to monitor the project so the aim must be to keep records in a form that will meet the requirements of Sub-Clause 4.21 [Progress Reports] as well as to satisfy the Contractor's internal procedures. The collection and collation of information from Subcontractors may cause problems for the Contractor.

Under subparagraph (a) the Progress Report shall include charts and detailed descriptions of progress. It is unclear in what format this shall be done. Common sense would dictate this to be in the same format in which the Sub-Clause 8.3 [Programme] Programme or any update thereto is submitted.

Some Employers and the building laws of some countries also require Contractors to keep progress and other records; these records should be combined wherever possible. The format of the monthly progress report does not have to be agreed with the Engineer, but some discussion is desirable in order to minimise the work and co-ordinate the format and avoid duplication with other submissions such as the Clause 20.1 claims records.

The progress report is submitted to the Engineer, but does not have to be approved or agreed by the Engineer. However, the information in the progress reports will undoubtedly be used in support of claims. For example, if figures have not been disputed at the time it may be difficult for the Employer to reject them at a later date. A regular progress meeting to discuss the report would enable the Engineer to raise any points that he wishes to query.

Under Sub-Clause 14.3 [*Application for Interim Payment Certificates*] a copy of the progress report must be included with the Contractor's Statement for interim payments, and as per Sub-Clause 14.6 [*Issue of Interim Payment Certificates*] the Engineer is obligated to issue a corresponding interim payment certificate within 28 days of receiving the application including supporting particulars which would include such monthly progress report.

4.22 Security of the Site
Unless otherwise stated in the Particular Conditions:

(a) *the Contractor shall be responsible for keeping unauthorised persons off the Site; and*
(b) *authorised persons shall be limited to the Contractor's Personnel and the Employer's Personnel; and to any other personnel notified to the Contractor, by the Employer or the Engineer, as authorised personnel of the Employer's other contractors on the Site.*

The procedure that the Contractor will adopt to control access to the Site will depend on the circumstances. Any special requirements should be stated in the Particular Conditions and detailed in the Specification.

The FIDIC *Guidance for the Preparation of Particular Conditions* draws attention to the potential problems that may arise when the Employer has more than one Contractor working on the Site. The Contractor can identify his own employees and the employees of his Subcontractors. To identify and check the Employer's Personnel may be more difficult but only requires co-operation from the Engineer. To identify and check the employees of another Contractor will be extremely difficult and requires a rigorous identification procedure. However, the Contractor is responsible for Site security and for keeping unauthorised persons out of the Site.

The Contractor's responsibility for Site security must be considered in relation to other Sub-Clauses, including Sub-Clause 4.6 [*Co-operation*] concerning co-operation with other Contractors, Sub-Clause 4.8 [*Safety Procedures*] concerning Site fencing and other temporary Works, Sub-Clause 17.2 [*Contractor's Care of the Works*] concerning the Contractor's responsibility for the care of the Works and the Clause 18 insurance requirements. If the Contractor does not have total control over Site access there will be implications for his other responsibilities.

Where more than one Contractor is working on the Site it is important that Site security procedures are described in detail in the Particular Conditions and Specifications. Sub-

Clause 4.6 [*Co-operation*] would enable the Contractor to claim any additional costs that he incurs as a consequence of the work involved in controlling Site access by other Contractors.

If the Site is part of a larger Site or existing operation, or the project involves work that the Employer considers to be particularly sensitive or confidential, then the Particular Conditions should include additional requirements for controlling access to the Site.

4.23 Contractor's Operations on Site

The Contractor shall confine his operations to the Site, and to any additional areas which may be obtained by the Contractor and agreed by the Engineer as working areas. The Contractor shall take all necessary precautions to keep Contractor's Equipment and Contractor's Personnel within the Site and these additional areas, and to keep them off adjacent land.

During the execution of the Works, the Contractor shall keep the Site free from all unnecessary obstruction, and shall store or dispose of any Contractor's Equipment or surplus materials. The Contractor shall clear away and remove from the Site any wreckage, rubbish and Temporary Works which are no longer required.

Upon the issue of a Taking-Over Certificate, the Contractor shall clear away and remove, from that part of the Site and Works to which the Taking-Over Certificate refers, all Contractor's Equipment, surplus material, wreckage, rubbish and Temporary Works. The Contractor shall leave that part of the Site and the Works in a clean and safe condition. However, the Contractor may retain on Site, during the Defects Notification Period, such Goods as are required for the Contractor to fulfil obligations under the Contract.

The requirements of Sub-Clause 4.23 [*Contractor's Operations on Site*] are related to the requirements of other Sub-Clauses. The requirement that work can only be carried out on the Site, or locations '*which have been agreed by the Engineer as working areas*', will apply to Subcontractors and staff or consultants who are carrying out any design work. While total compliance may be impractical, the requirement is important in relation to the Engineer's right to inspect materials, workmanship and work in progress, as per Sub-Clause 7.3 [*Inspection*]. The requirement is presumably not intended to apply to the manufacture of mechanical and electrical items that will then be incorporated into Plant for the Works. However, the Parties are advised to clarify this issue to avoid future conflicts and disputes. The extent of off-site work and the requirements for notification and inspection should be discussed between the Contractor and the Engineer.

The requirement to keep the Contractor's Personnel off adjacent land relates to the Sub-Clause 4.14 [*Avoidance of Interference*] requirement not to interfere with the convenience of the public and the Sub-Clause 4.8(e) [*Safety Procedures*] requirement to provide Temporary Works for the protection of owners and occupiers of adjacent land. The requirement to keep the Site free from rubbish and other obstructions also relates to the safety provisions at Sub-Clause 4.8.

The requirements for clearing the Site on completion must be read in conjunction with Clause 10 [*Employer's Taking Over*] and Clause 11 [*Defects Liability*]. Sub-Clause 11.7

[*Right of Access*] gives the Contractor the right of access during the Defects Notification Period and the final sentence of Sub-Clause 4.23 [*Contractor's Operations on Site*] gives him the right to retain on the Site such Goods, that is Equipment, Materials and Temporary Works, as are required to fulfil his obligations. An area of the Site will need to be agreed for his use.

4.24 Fossils

All fossils, coins, articles of value or antiquity, and structures and other remains or items of geological or archaeological interest found on the Site shall be placed under the care and authority of the Employer. The Contractor shall take reasonable precautions to prevent Contractor's Personnel or other persons from removing or damaging any of these findings.

The Contractor shall, upon discovery of any such finding, promptly give notice to the Engineer, who shall issue instructions for dealing with it. If the Contractor suffers delay and/or incurs Cost from complying with the instructions, the Contractor shall give a further notice to the Engineer and shall be entitled subject to Sub-Clause 20.1 [Contractor's Claims] to:

(a) an extension of time for any such delay, if completion is or will be delayed, under Sub-Clause 8.4 [Extension of Time for Completion]; and

(b) payment of any such Cost, which shall be included in the Contract Price.

After receiving this further notice, the Engineer shall proceed in accordance with Sub-Clause 3.5 [Determinations] to agree or determine these matters.

The procedure when fossils or similar objects are found on the Site follows the typical notice and claims procedures and Sub-Clause 20.1 [*Contractor's Claims*]. However, the Contractor is required to give an initial notice when the fossils are discovered, in order that the Engineer can examine the situation and issue any appropriate instructions. This is one of the various Sub-Clauses that require a so-called double notification (refer to Appendix 1, 2 and 3). These instructions may constitute a Variation under Clause 13. The Contractor will also be obliged to comply with the governing law and any local regulations. It may also have been necessary to permit archaeological investigation of the Site before work commenced, which may be covered by Sub-Clause 1.13 [*Compliance with Laws*] and included as Site data under Sub-Clause 4.10 [*Site Data*].

FIDIC Users' Guide
ISBN 978-0-7277-5856-9

ICE Publishing: All rights reserved
http://dx.doi.org/10.1680/fug.58569.149

Chapter 11
Clause 5: Nominated Subcontractors

Clause 5 gives a procedure whereby the Employer can impose a particular Subcontractor on the Contractor, either by the nominated Subcontractor being stated as such in the Contract, or by an instruction from the Engineer as a Variation under Clause 13. The provisions of Sub-Clause 4.4 [*Subcontractors*] must be seen in addition to this Clause 5, and care must be taken to ensure compliance with both. While there may be good reasons why the Employer would like a particular Subcontractor to be used for certain work, there are considerable potential problems created when imposing a Subcontractor on the Contractor. The procedure should be used with caution and a full appreciation of the potential problems is necessary, including consideration of any indemnities that may be required by the Contractor.

Attempts to impose a particular Subcontractor but, at the same time, trying to avoid the nomination route as per Clause 5, perhaps by a Plant specification which can only be achieved by one supplier, could result in the Subcontractor being deemed to have been nominated. Although such a 'deemed provision' is not included in the text of Clause 5, mandatory governance of the Contract by the Law as per Sub-Clause 1.4 [*Law and Language*] may impose such an effect on the Parties.

Under these procedures, Sub-Clause 4.4(a) makes it clear that the Contractor does not need to obtain the Engineer's consent to the nominated Subcontractor. Sub-Clause 4.4 requires that the Contractor '*shall be responsible for the acts or defaults of any Subcontractor, his agents or employees, as if they were the acts or defaults of the Contractor*'. The use of the phrase '*any Subcontractor*' must include a nominated Subcontractor. Sub-Clause 5.2(b) and (c) [*Objection to Nomination*] enable the Contractor to object to the nomination if the subcontract does not indemnify the Contractor against failures by the Subcontractor, but indemnities do not lessen the Contractor's obligations to the Employer. For example, the extension of time provisions at Sub-Clause 8.4 [*Extension of Time for Completion*] do not necessarily provide for an extension in the event of delay caused by a nominated Subcontractor. As it is reasonable to say that the Contractor has a duty to supervise and control the nominated Subcontractor on a day-to-day basis, it is extremely difficult to ascertain with certainty to what extent the Contractor suffered undue damage in the case of non-performance by such a nominated Subcontractor.

The Contract is silent on the procedures for the replacement of a nominated Subcontractor, in the event of default by that Subcontractor. The Engineer might argue that he is entitled to issue a Variation to nominate a different Subcontractor. Alternatively, the Engineer might argue that, under Sub-Clause 4.4 [*Subcontractors*], it is the Contractor's problem and, under Sub-Clause 8.1 [*Commencement of Works*], the Contractor must '*proceed with the Works with due expedition and without delay*'. The consequences in cost and delay could be considerable and the Contractor would no doubt wish to claim for reimbursement of his loss, possibly under a provision of the local law. Alternatively, the Contractor may have sought an indemnity by the Employer against such an eventuality under the provisions of Sub-Clause 5.2 [*Objection to Nomination*].

5.1 Definition of 'nominated Subcontractor'
In the Contract, 'nominated Subcontractor' means a Subcontractor:

(a) *who is stated in the Contract as being a nominated Subcontractor; or*
(b) *whom the Engineer, under Clause 13 [Variations and Adjustments], instructs the Contractor to employ as a Subcontractor.*

Previous FIDIC Contracts had a definition of nominated Subcontractors that included anyone who supplied Goods for which a Provisional Sum had been included in the Contract, as well as those who executed work. Sub-Clause 5.1 [*Definition of 'nominated Subcontractor'*] now refers only to Subcontractors who are either named in the Contract or are the subject of a Variation under Clause 13. Subcontractors are defined at Sub-Clause 1.1.2.8 to include a person who is appointed as a Subcontractor for a part of the Works and '*Materials*' are defined at Sub-Clause 1.1.5.3 as forming part of the Permanent Works. Hence, under FIDIC, a material supplier who is nominated would appear to be a nominated Subcontractor and be covered by the provisions of Clause 5.

Instructions for the purchase of Plant, Materials or services are also covered at Sub-Clause 13.5 [*Provisional Sums*] concerning Provisional Sums. Paragraph (b) refers to Plant, Materials or services to be purchased from a nominated Subcontractor or otherwise.

5.2 Objection to Nomination
The Contractor shall not be under any obligation to employ a nominated Subcontractor against whom the Contractor raises reasonable objection by notice to the Engineer as soon as practicable, with supporting particulars. An objection shall be deemed reasonable if it arises from (among other things) any of the following matters, unless the Employer agrees to indemnify the Contractor against and from the consequences of the matter:

(a) *there are reasons to believe that the Subcontractor does not have sufficient competence, resources or financial strength;*
(b) *the subcontract does not specify that the nominated Subcontractor shall indemnify the Contractor against and from any negligence or misuse of Goods by the nominated Subcontractor, his agents and employees; or*
(c) *the subcontract does not specify that, for the subcontracted work (including design, if any), the nominated Subcontractor shall:*
   (i) *undertake to the Contractor such obligations and liabilities as will enable the Contractor to discharge his obligations and liabilities under the Contract, and*

(ii) *indemnify the Contractor against and from all obligations and liabilities arising under or in connection with the Contract and from the consequences of any failure by the Subcontractor to perform these obligations or to fulfil these liabilities.*

If the Contractor does not wish to employ a particular nominated Subcontractor he must raise an objection '*as soon as practicable*'. If the nominated Subcontractor was named in the Tender documents then the Contractor had the opportunity to object before signing the Contract. The inclusion of the words '*among other things*' means that the causes for an objection are not limited to the matters stated at paragraphs (a) to (c). However, they must be '*reasonable*'. Any dispute about whether an objection is reasonable could result in considerable problems and delays. If the Contractor does not wish to employ a particular Subcontractor, then for the Employer to insist could result in problems later if there is a query concerning the Subcontractor's performance. Some queries as to whether a Subcontractor should be imposed on a Contractor might be overcome by an indemnity from the Employer as provided for at Sub-Clause 5.2 [*Objection to Nomination*]. But this cannot be a blanket indemnity as the Contractor has an implied responsibility to supervise and control the Subcontractor.

The provisions whereby either the Employer or the Subcontractor may be required to indemnify the Contractor against the consequences of failure by the Subcontractor are indicative of the potential problems that can arise from the use of nominated Subcontractors.

5.3 Payments to nominated Subcontractors
The Contractor shall pay to the nominated Subcontractor the amounts which the Engineer certifies to be due in accordance with the subcontract. These amounts plus other charges shall be included in the Contract Price in accordance with subparagraph (b) of Sub-Clause 13.5 [Provisional Sums], except as stated in Sub-Clause 5.4 [Evidence of Payments].

Payments to a nominated Subcontractor are certified by the Engineer as being due under the particular subcontract. These payments are certified by the Engineer in the Clause 14 Payment Certificates and are included in the Contract Price as Provisional Sums under Sub-Clause 13.5(b) [*Provisional Sums*]. The Contractor is also paid for overheads and profit at the percentage stated either in an appropriate Schedule or in the Appendix to Tender.

5.4 Evidence of Payments
Before issuing a Payment Certificate which includes an amount payable to a nominated Subcontractor, the Engineer may request the Contractor to supply reasonable evidence that the nominated Subcontractor has received all amounts due in accordance with previous Payment Certificates, less applicable deductions for retention or otherwise. Unless the Contractor:

(a) submits this reasonable evidence to the Engineer; or
(b) (i) satisfies the Engineer in writing that the Contractor is reasonably entitled to withhold or refuse to pay these amounts, and
(ii) submits to the Engineer reasonable evidence that the nominated Subcontractor has been notified of the Contractor's entitlement;

then the Employer may (at his sole discretion) pay, direct to the nominated Subcontractor, part or all of such amounts previously certified (less applicable deductions) as are due to the nominated Subcontractor and for which the Contractor has failed to submit the evidence described in sub-paragraphs (a) or (b) above. The Contractor shall then repay, to the Employer, the amount which the nominated Subcontractor was directly paid by the Employer.

The Engineer is entitled to check that the Contractor has passed on the payments to the nominated Subcontractor and, in certain circumstances, to make payments direct to the Subcontractor. Any payment that has been made to the nominated Subcontractor, after having been paid to the Contractor, will be recovered from the Contractor.

Before making any direct payment, the Engineer will need to check why the money has not been paid to the nominated Subcontractor, following the procedure of paragraph (b) (i). Any subsequent deduction from the Contractor will need to follow the procedures of Sub-Clause 2.5 [*Employer's Claims*]. Any action for direct payments to a nominated Subcontractor must also be considered in conjunction with any relevant provisions in the governing law.

FIDIC Users' Guide
ISBN 978-0-7277-5856-9

ICE Publishing: All rights reserved
http://dx.doi.org/10.1680/fug.58569.153

Chapter 12
Clause 6: Staff and Labour

Under Sub-Clause 4.1 [*Contractor's General Obligations*] the Contractor is required to provide '*all Contractor's Personnel*'. Clause 6 covers requirements for the recruitment, employment and behaviour of the Contractor's staff and labour, together with specific requirements for superintendence and records.

The description '*staff and labour*' appears to refer to personnel who are employed by the Contractor, whereas '*Contractor's Personnel*', as defined at Sub-Clause 1.1.2.7, includes the employees of Subcontractors and '*any other personnel assisting the Contractor in the execution of the Works*'. Many of the requirements of Clause 6 include Subcontractors and the conditions of subcontracts should include similar provisions.

Additional requirements to suit local conditions may be included in the Particular Conditions. Example clauses are included in the FIDIC *Guidance for the Preparation of Particular Conditions* for Foreign Staff and Labour; Measures against Insect and Pest Nuisance; Alcohol Liquor or Drugs; Arms and Ammunition; Festivals and Religious Customs. The PNK Book also has additional Sub-Clauses which may be considered for inclusion within the Particular Conditions covering Health and Safety (additional paragraphs on HIV/AIDS), Foreign Personnel, Supply of Foodstuffs, Supply of Water, Measures against Insect and Pest Nuisance, Alcoholic Liquor or Drugs, Arms and Ammunition, Festivals and Religious Customs, Funeral Arrangements, Forced Labour, Child Labour, Employment Records of Workers, Workers Organisations and Non-Discrimination and Equal Opportunity (refer to Chapter 35). When it is considered necessary to include these matters they are likely to be the subject of local Laws or regulations.

Although the Clause is headed Staff and Labour, the records required under Sub-Clause 6.10 [*Records of Contractor's Personnel and Equipment*] include Contractor's Equipment.

6.1 Engagement of Staff and Labour
Except as otherwise stated in the Specification, the Contractor shall make arrangements for the engagement of all staff and labour, local or otherwise, and for their payment, housing, feeding and transport.

Sub-Clause 6.1 requires the Contractor to make all the necessary arrangements for his staff and labour, but allows for this to be overruled by the Specification. The Specification could, for example, state that the Employer would provide some aspect of recruitment, payment, housing, feeding or transport, either directly or through some other agency. This

153

exception is important because, without a specific exception, if there is any difference between the Contract Clause and the Specification the requirement of the Contract Clause would take priority under Sub-Clause 1.5 [*Priority of Documents*].

6.2 Rates of Wages and Conditions of Labour

The Contractor shall pay rates of wages, and observe conditions of labour, which are not lower than those established for the trade or industry where the work is carried out. If no established rates or conditions are applicable, the Contractor shall pay rates of wages and observe conditions which are not lower than the general level of wages and conditions observed locally by employers whose trade or industry is similar to that of the Contractor.

The Sub-Clause refers to conditions of labour established for the trade or industry where the work is carried out. The wording is not specific on what constitutes a proper established condition for a trade or industry: for example, The Law or a Union negotiated package. The conditions shall furthermore not be lower than the general level observed locally by similar contractors. This again is open to interpretation.

The Particular Conditions may include additional requirements to conform to local require-ments. The second paragraph contained within PNK may also be considered for inclusion in the Particular Conditions. The Contractor will also need to comply with any local Laws and regulations that apply to wages and working conditions, in accordance with Sub-Clause 1.13 [*Compliance with Laws*] and Sub-Clause 6.4 [*Labour Laws*].

Before submitting a binding tender any potential contractor is advised to ascertain in detail as to what is expected from him under this Sub-Clause.

6.3 Persons in the Service of Employer

The Contractor shall not recruit, or attempt to recruit, staff and labour from amongst the Employer's Personnel.

Employees of the Employer will have knowledge of local conditions that may be valuable for the Contractor. However, the Employer will not want to lose the services of valuable staff. At pre-qualification stage the Contractor may have been asked how he proposes to recruit staff for the project, either by local recruitment or by bringing in expatriate staff and labour.

Sub-Clause 6.3 attempts to prevent the Contractor from recruiting, or attempting to recruit, staff and labour from amongst the Employer's Personnel. The subtitle refers to '*Persons in the Service of the Employer's Personnel*' and the Sub-Clause, by referring to '*Employer's Personnel*', covers a wider range than people who are directly employed by the Employer, as defined at Sub-Clause 1.1.2.6.

While the intention of this Sub-Clause is clear, experience of similar Clauses in practice suggests its application may be more difficult. If the person concerned decided that he wanted to change his employment, then it is difficult for the Employer to prevent his depar-ture. If he has already left the employment of the Employer before the Contractor recruits him, then the requirement may not be applicable. However, it would be a breach of Contract for the Contractor, or a Subcontractor, to take the initiative and persuade someone to leave the employment of the Employer.

Interestingly the Sub-Clause does not apply in the situation where the Employer decides to employ or recruit his staff from amongst the Contractor's Personnel (ref [Sub-Clause 1.1.2.7]).

6.4 Labour Laws
The Contractor shall comply with all the relevant labour Laws applicable to the Contractor's Personnel, including Laws relating to their employment, health, safety, welfare, immigration and emigration, and shall allow them all their legal rights.

The Contractor shall require his employees to obey all applicable Laws, including those concerning safety at work.

Sub-Clause 6.4 is reviewed together with Sub-Clause 6.5.

6.5 Working Hours
No work shall be carried out on the Site on locally recognised days of rest, or outside the normal working hours stated in the Appendix to Tender, unless:

(a) otherwise stated in the Contract;
(b) the Engineer gives consent; or
(c) the work is unavoidable, or necessary for the protection of life or property or for the safety of the Works, in which case the Contractor shall immediately advise the Engineer.

The normal working hours may be restricted by the Employer, due to security or other local requirements, or may be left open for the Appendix to Tender to be completed by the Contractor.

The Contractor's normal working hours will also depend on whether he is bringing workers from outside the Country. Expatriate workers will often expect to work long hours, with fewer than normal days for weekends. Workers from the Country, or nearby countries, may have a tradition of long weekends after pay day. The Contractor will be deemed to have made enquiries and taken local custom into account when preparing his Tender, as required by Sub-Clause 4.10 [*Site Data*] and Sub-Clause 4.11 [*Sufficiency of the Accepted Contract Amount*].

Work outside the normal working hours requires the consent of the Engineer. The additional work will presumably benefit the project so consent should not normally be refused. When it is obvious that the Works will include significant operations that generally require work outside normal working hours, such as large concrete pours that cannot be interrupted, or those operations that have to be executed at specific times of the day outside the normal working hours due to climatic conditions, it may be wise to take this into account before submitting a tender.

Exceptions to the normal working hours may also be necessary for particular operations, when the Contractor wishes to overcome delays, or when the Engineer has invoked the procedures of Sub-Clause 8.6 [*Rate of Progress*] and required the Contractor to adopt measures to expedite progress. Sub-Clause 8.6 enables the Employer to claim from the Contractor any additional supervision Costs and similar provisions may be instructed as a condition of the Engineer's consent to work out of normal hours.

6.6 Facilities for Staff and Labour

Except as otherwise stated in the Specification, the Contractor shall provide and maintain all necessary accommodation and welfare facilities for the Contractor's Personnel. The Contractor shall also provide facilities for the Employer's Personnel as stated in the Specification.

The Contractor shall not permit any of the Contractor's Personnel to maintain any temporary or permanent living quarters within the structures forming part of the Permanent Works.

Sub-Clause 6.6 repeats the same requirement and exclusion as Sub-Clause 6.1 [*Engagement of Staff and Labour*], specifically for the provision of accommodation and welfare facilities. The Contractor is also required to provide the facilities for the Employer's Personnel, including the Engineer's Site staff as defined in Sub-Clause 1.1.2.6, which are stated in the Specification.

The Employer's Personnel is defined as those persons having been notified as such to the Contractor. There is no time limit on this notification, thus the list can be expanded as the project progresses. If the Contractor incurs extra costs under Sub-Clause 6.6 as a result of an addition to the list of Employer's Personnel, then the Contractor may have a claim subject to the requirements of Sub-Clause 20.1 [*Contractor's Claims*].

6.7 Health and Safety

The Contractor shall at all times take all reasonable precautions to maintain the health and safety of the Contractor's Personnel. In collaboration with local health authorities, the Contractor shall ensure that medical staff, first aid facilities, sick bay and ambulance service are available at all times at the Site and at any accommodation for Contractor's and Employer's Personnel, and that suitable arrangements are made for all necessary welfare and hygiene requirements and for the prevention of epidemics.

The Contractor shall appoint an accident prevention officer at the Site, responsible for maintaining safety and protection against accidents. This person shall be qualified for this responsibility, and shall have the authority to issue instructions and take protective measures to prevent accidents. Throughout the execution of the Works, the Contractor shall provide whatever is required by this person to exercise this responsibility and authority.

The Contractor shall send, to the Engineer, details of any accident as soon as practicable after its occurrence. The Contractor shall maintain records and make reports concerning health, safety and welfare of persons, and damage to property, as the Engineer may reasonably require.

This Sub-Clause must be considered together with the health and safety regulations that apply under the applicable Laws. The Contractor is also required to appoint an accident prevention officer at the Site, who is suitably qualified and is responsible for ensuring that these requirements are followed. It is not clear what constitutes a suitably qualified person, and it is proposed that the Parties agree on this before executing the Contract. The Site is defined under Sub-Clause 1.1.6.7 and may include any area so specified in the Contract. If this definition includes multiple geographical areas (e.g. on overhead power lines or projects where significant substructures are manufactured elsewhere), then it should be considered if all those areas require a separate officer. This officer shall have the authority to issue instructions, but it is not clear as to whom these instructions shall be given. In modern day construction the provision of an accident prevention officer is considered important (compare the UK HSE provisions), and his/her role should be clearly defined and advertised to all involved in the project so as to maximise efficiency and effectiveness.

The requirement to ensure that medical facilities are available at the Site could be an onerous requirement for a project that is located a long way away from any centre of population. If the project is intended to encourage further development in the area, the Employer should consider including the provision of a permanent medical facility as a part of the project.

The Contractor must keep records of safety matters as required by the Engineer and these records will form a basis for the safety statistics which are necessary for the monthly progress reports under Sub-Clause 4.21(g) [*Progress Reports*].

6.8 Contractor's Superintendence

Throughout the execution of the Works, and as long thereafter as is necessary to fulfil the Contractor's obligations, the Contractor shall provide all necessary superintendence to plan, arrange, direct, manage, inspect and test the work.

Superintendence shall be given by a sufficient number of persons having adequate knowledge of the language for communications (defined in Sub-Clause 1.4 [Law and Language]) and of the operations to be carried out (including the methods and techniques required, the hazards likely to be encountered and methods of preventing accidents), for the satisfactory and safe execution of the Works.

Sub-Clause 6.8 imposes an all-embracing obligation on the Contractor to provide '*all necessary superintendence*', which is described as the people who will '*plan, arrange, direct, manage, inspect and test the work*' and to comply with the detailed requirements. This requirement may seem to be obvious, but it gives the Engineer the opportunity to raise the matter, as a breach of this Sub-Clause, if he considers that the progress or quality of the work has suffered due to the lack of a suitable person to fulfil one of these functions.

6.9 Contractor's Personnel

The Contractor's Personnel shall be appropriately qualified, skilled and experienced in their respective trades or occupations. The Engineer may require the Contractor to remove (or cause to be removed) any person employed on the Site or Works, including the Contractor's Representative if applicable, who:

(a) persists in any misconduct or lack of care;
(b) carries out duties incompetently or negligently;
(c) fails to conform with any provisions of the Contract; or
(d) persists in any conduct which is prejudicial to safety, health, or the protection of the environment.

If appropriate, the Contractor shall then appoint (or cause to be appointed) a suitable replacement person.

This provision applies, by the definition at Sub-Clause 1.1.2.7, to personnel employed by all Subcontractors (including nominated if any) as well the Contractor's own personnel. If the person concerned is in a supervisory position then Sub-Clause 6.8 could also be relevant.

6.10 Records of Contractor's Personnel and Equipment

The Contractor shall submit, to the Engineer, details showing the number of each class of Contractor's Personnel and of each type of Contractor's Equipment on the Site. Details shall be submitted each calendar month, in a form approved by the Engineer, until the Contractor has completed all work which is known to be outstanding at the completion date stated in the Taking-Over Certificate for the Works.

The records and monthly reports under Sub-Clause 6.10 must include Contractor's Equipment and Personnel, as well as '*staff and labour*'. The Contractor's Equipment and Personnel includes those provided by all Subcontractors. The report is included in the monthly progress report under Sub-Clause 4.21(d) [*Progress Reports*]. For a large or complex project it may be useful for the Particular Conditions to require that the numbers are allocated to particular parts of the project. It is also essential that any contemporary records, which are submitted under Sub-Clause 20.1 [*Contractor's Claims*], are consistent with these records.

It may be advisable to define the various classes of Contractor's Personnel and types of Equipment in the Particular Conditions to enable effective record keeping.

6.11 Disorderly Conduct

The Contractor shall at all times take all reasonable precautions to prevent any unlawful, riotous or disorderly conduct by or amongst the Contractor's Personnel, and to preserve peace and protection of persons and property on and near the Site.

Failure to comply with these requirements could result in a claim under the insurance provided under Sub-Clause 18.3 [*Insurance against Injury to Persons and Damage to Property*] or the indemnity under Sub-Clause 17.1 [*Indemnities*], as well as under the provisions of Sub-Clause 4.14 [*Avoidance of Interference*]. It could also affect any liability of the Employer under other clauses, excluding Force Majeure under Sub-Clause 19.1(iii) [*Definition of Force Majeure*] and liability as an Employer's risk under Sub-Clause 17.3(c) [*Employer's Risks*].

FIDIC Users' Guide
ISBN 978-0-7277-5856-9

ICE Publishing: All rights reserved
http://dx.doi.org/10.1680/fug.58569.159

Chapter 13
Clause 7: Plant, Materials and Workmanship

Clause 7 covers the Contractor's obligations concerning the quality of his work and the procedures to be followed for tests and in the event that an item of work fails the test. The matter of the time when an item of Plant or Materials becomes the property of the Employer is covered at Sub-Clause 7.7 [*Ownership of Plant and Materials*] and Royalties are dealt with at Sub-Clause 7.8 [*Royalties*].

If the Contract includes a requirement for a quality assurance system, as Sub-Clause 4.9 [*Quality Assurance*], then the application of Clause 7 will need to be considered in relation to that system.

The FIDIC *Guidance for the Preparation of Particular Conditions* includes an additional Sub-Clause for use when the financing institution imposes restrictions on the use of its funds. Interestingly, the PNK Book includes an eligibility statement for equipment, materials and services in Sub-Clause 4.1 [*Contractor's General Obligations*] rather than under this Clause.

7.1 Manner of Execution
The Contractor shall carry out the manufacture of Plant, the production and manufacture of Materials, and all other execution of the Works:

(a) in the manner (if any) specified in the Contract;
(b) in a proper workmanlike and careful manner, in accordance with recognised good practice; and
(c) with properly equipped facilities and non-hazardous Materials, except as otherwise specified in the Contract.

The quality of Materials and standard of workmanship will be specified elsewhere in the Contract documents, and will normally refer to the national standard specifications of the Country of the project. Phrases such as '*proper workmanlike and careful manner*', '*recognised good practice*' and '*properly equipped facilities*', which are used in this Sub-Clause are not precise. These requirements will be interpreted by the Engineer in relation to the actual Goods supplied and the work that is executed by the Contractor.

7.2 Samples
The Contractor shall submit the following samples of Materials, and relevant information, to the Engineer for consent prior to using the Materials in or for the Works:

(a) *manufacturer's standard samples of Materials and samples specified in the Contract, all at the Contractor's cost; and*

(b) *additional samples instructed by the Engineer as a Variation.*

Each sample shall be labelled as to origin and intended use in the Works.

When samples are specified or instructed then the requirements and details should be clear. The requirement for '*manufacturer's standard samples*' is not always clear. However, the Sub-Clause refers only to samples of Materials that, by definition, excludes Plant. The samples are provided to the Engineer for consent which, under Sub-Clause 1.3 [*Communications*], shall not be unreasonably withheld or delayed.

7.3 Inspection
The Employer's Personnel shall at all reasonable times:

(a) *have full access to all parts of the Site and to all places from which natural Materials are being obtained; and*

(b) *during production, manufacture and construction (at the Site and elsewhere), be entitled to examine, inspect, measure and test the materials and workmanship, and to check the progress of manufacture of Plant and production and manufacture of Materials.*

The Contractor shall give the Employer's Personnel full opportunity to carry out these activities, including providing access, facilities, permissions and safety equipment. No such activity shall relieve the Contractor from any obligation or responsibility.

The Contractor shall give notice to the Engineer whenever any work is ready and before it is covered up, put out of sight, or packaged for storage or transport. The Engineer shall then either carry out the examination, inspection, measurement or testing without unreasonable delay, or promptly give notice to the Contractor that the Engineer does not require to do so. If the Contractor fails to give the notice, he shall, if and when required by the Engineer, uncover the work and thereafter reinstate and make good, all at the Contractor's cost.

Sub-Clause 7.3 requires the Contractor to allow the Employer's Personnel to enter the Site and any factories, quarries or other places where work is being carried out for the Works at all reasonable times. The Contractor must provide '*access, facilities, permissions and safety equipment*' to enable the person concerned to inspect and check the Materials, workmanship and progress. This must be seen in relation to the Sub-Clause 6.7 [*Health and Safety*] obligation to employ an accident prevention officer who has the authority to issue instructions, for example, to not enter a specific area until it is considered safe to do so.

Clearly, the Contractor can insist that anyone carrying out an inspection shall have proper authority to enter the Site and inspect the work. This is covered by the definition of '*Employer's Personnel*' at Sub-Clause 1.1.2.6, which requires that all Employers' Personnel have been notified to the Contractor by the Employer or Engineer. The person concerned can inspect, but cannot give any instructions unless they have been given delegated authority under Sub-Clause 3.3 [*Instructions of the Engineer*]. The Contractor will also insist that anyone entering the Site or other premises shall follow the appropriate safety regulations.

The Contractor is required to give notice when any work is ready for inspection and before it is concealed in any way. The Specification will often include a more detailed procedure and standard Notice for Inspection, or the Contractor may introduce his own standard form. The notice should also state when the Contractor wishes to proceed with the work in order to comply with his programme. Standard notice forms often have space for the Inspector to sign that the work has passed or failed the inspection. Any such signature must have been authorised in accordance with Sub-Clause 3.3 and the Engineer may decide only to authorise an 'I have inspected' signature.

The Engineer will either inspect '*without unreasonable delay*' or indicate that he does not need to do so. If the Contractor fails to give notice that work is ready then he can be required to uncover the work and reinstate at his own cost. This will be at the Contractor's cost, regardless of whether the work is found to be faulty or not.

If the Contractor gives notice but, for whatever reason, the work is covered before it is inspected, then the Contractor would appear to be in breach of Sub-Clause 7.3 [*Inspection*], unless the Engineer failed to inspect within a reasonable time.

The provisions of this Sub-Clause are typical of the provisions for the supervision of civil engineering work. If the Works include mechanical and electrical work then the application of the provisions may be difficult. Plant may include items of standard manufacture that the Subcontractor purchases when they are required. The extent of detail which is required for compliance with paragraph (b) and with paragraph (c) of the monthly progress report under Sub-Clause 4.21 [*Progress Reports*], will need to be agreed to suit the actual procedures that are normally adopted. Items which are specified by performance and subject to Tests on, or after, Completion would not normally be subject to the same detailed inspection during manufacture as is required for civil engineering work.

7.4 Testing
This Sub-Clause shall apply to all tests specified in the Contract, other than the Tests after Completion (if any).

The Contractor shall provide all apparatus, assistance, documents and other information, electricity, equipment, fuel, consumables, instruments, labour, materials, and suitably qualified and experienced staff, as are necessary to carry out the specified tests efficiently. The Contractor shall agree, with the Engineer, the time and place for the specified testing of any Plant, Materials and other parts of the Works.

The Engineer may, under Clause 13 [Variations and Adjustments], vary the location or details of specified tests, or instruct the Contractor to carry out additional tests. If these varied or additional tests show that the tested Plant, Materials or workmanship is not in accordance with the Contract, the cost of carrying out this Variation shall be borne by the Contractor, notwithstanding other provisions of the Contract.

161

The Engineer shall give the Contractor not less than 24 hours' notice of the Engineer's intention to attend the tests. If the Engineer does not attend at the time and place agreed, the Contractor may proceed with the tests, unless otherwise instructed by the Engineer, and the tests shall then be deemed to have been made in the Engineer's presence.

If the Contractor suffers delay and/or incurs Cost from complying with these instructions or as a result of a delay for which the Employer is responsible, the Contractor shall give notice to the Engineer and shall be entitled subject to Sub-Clause 20.1 [Contractor's Claims] to:

(a) an extension of time for any such delay, if completion is or will be delayed, under Sub-Clause 8.4 [Extension of Time for Completion]; and
(b) payment of any such Cost plus reasonable profit, which shall be included in the Contract Price.

After receiving this notice, the Engineer shall proceed in accordance with Sub-Clause 3.5 [Determinations] to agree or determine these matters.

The Contractor shall promptly forward to the Engineer duly certified reports of the tests. When the specified tests have been passed, the Engineer shall endorse the Contractor's test certificate, or issue a certificate to him, to that effect. If the Engineer has not attended the tests, he shall be deemed to have accepted the readings as accurate.

Sub-Clause 7.4 gives the procedures for tests specified in the Contract and additional tests instructed under Clause 13. Tests on Completion are covered at Clause 9, which refers back to Sub-Clause 7.4. Tests after Completion, as the YLW Book Clause 12, require different procedures because, after completion, the project is occupied and controlled by the Employer. For a RED Book Contract, Tests after Completion must be required in the Particular Conditions, as referred to in Sub-Clause 1.1.3.6.

The Contractor will have given notice under Sub-Clause 7.3 [*Inspection*] that the item is ready to be tested. The Engineer will then give the Contractor not less than 24 hours' notice of his intention to attend the tests and the time and place will be agreed between Engineer and Contractor. The Contractor will provide all the facilities necessary for the test. If the Engineer wants to change any of the specified details then he must issue a Variation under Clause 13. If the Engineer fails to attend the test, without issuing an appropriate instruction, then the Contractor can proceed with the test. The tests are deemed to have been made in the Engineer's presence and he will be deemed to have accepted the readings as accurate.

After the tests, the Contractor will send to the Engineer a certified report and, when the test has been successful, the Engineer will issue a Certificate. If an item fails a test then the Engineer can reject that item under Sub-Clause 7.5 [*Rejection*], but must give reasons for the rejection. The Contractor must then make good the item and ensure that it complies with the Contract. Alternatively the Employer may agree to accept the item under the procedures of Sub-Clause 9.4 [*Failure to Pass Tests on Completion*].

7.5 Rejection

If, as a result of an examination, inspection, measurement or testing, any Plant, Materials or workmanship is found to be defective or otherwise not in accordance with the Contract, the Engineer may reject the Plant, Materials or workmanship by giving notice to the Contractor, with reasons. The Contractor shall then promptly make good the defect and ensure that the rejected item complies with the Contract.

If the Engineer requires this Plant, Materials or workmanship to be retested, the tests shall be repeated under the same terms and conditions. If the rejection and retesting cause the Employer to incur additional costs, the Contractor shall subject to Sub-Clause 2.5 [Employer's Claims] pay these costs to the Employer.

The cost of tests is dealt with at Sub-Clause 7.4 [*Testing*] and Sub-Clause 7.5 [*Rejection*].

- The Sub-Clause refers to the result of 'a' test, indicating that the provisions are valid for specified as well as instructed tests under Clause 13.
- Tests that are specified are clearly included in the Accepted Contract Amount. If quantities are stated in the Bill of Quantities they would be re-measured in accordance with Clause 12. If the item fails the test then the Contractor will have to promptly make good the defect and ensure compliance with the Contract. That would certainly require a re-test, and it is reasonable to assume that such a re-test would not be re-measured under Clause 12 for the purpose of payment under Clause 14. However, if the Engineer requires the test to be repeated, presumably either before or after remedial work and the Employer incurs additional costs, then these can be claimed from the Contractor using the procedures at Sub-Clause 2.5 [*Employer's Claims*].
- The cost of tests that are instructed by the Engineer under Clause 13 should be paid for by the Employer. However, if such a test is unsuccessful, than the Contractor is obligated to ensure compliance with the Contract, and any resulting required re-test would be at the cost of the Contractor.

7.6 Remedial Work

Notwithstanding any previous test or certification, the Engineer may instruct the Contractor to:

(a) *remove from the Site and replace any Plant or Materials which is not in accordance with the Contract;*

(b) *remove and re-execute any other work which is not in accordance with the Contract; and*

(c) *execute any work which is urgently required for the safety of the Works, whether because of an accident, unforeseeable event or otherwise.*

The Contractor shall comply with the instruction within a reasonable time, which shall be the time (if any) specified in the instruction, or immediately if urgency is specified under sub-paragraph (c).

If the Contractor fails to comply with the instruction, the Employer shall be entitled to employ and pay other persons to carry out the work. Except to the extent that the Contractor would have been entitled to payment for the work, the Contractor shall subject to Sub-Clause 2.5 [Employer's Claims] pay to the Employer all costs arising from this failure.

Sub-Clause 7.6 gives powers to the Engineer to deal with items that, in his opinion, are not in accordance with the Contract. The Engineer may instruct that Plant, Materials or any other work is removed from the Site and the item replaced. The Contractor must comply with the instruction within a reasonable time, which may be stated in the instruction. This could be problematic under certain legal jurisdictions where the supplier, in terms of his guarantee provisions, has a fundamental right to repair before replacing.

Under Sub-Clause 7.6(c) the Engineer may also issue instructions for any work which is required for the safety of the Works, which must be carried out immediately. If the Contractor fails, or is unable, to comply with this instruction then the Employer may employ others to carry out the work and claim the Costs arising from the failure to the Contractor under Sub-Clause 2.5 *[Employer's Claims]*.

7.7 Ownership of Plant and Materials

Each item of Plant and Materials shall, to the extent consistent with the Laws of the Country, become the property of the Employer at whichever is the earlier of the following times, free from liens and other encumbrances:

(a) when it is delivered to the Site;
(b) when the Contractor is entitled to payment of the value of the Plant and Materials under Sub-Clause 8.10 [Payment for Plant and Materials in Event of Suspension].

Sub-Clause 7.7 states that any item of Plant and Materials that has been delivered to the Site becomes the property of the Employer. The definition of '*Site*' at Sub-Clause 1.1.6.7 includes places specified in the Contract as forming part of the Site. The Plant and Materials that become the property of the Employer should have been included in Interim Payment Certificates under Sub-Clause 14.5 *[Plant and Materials intended for the Works]*, if the Contractor has provided all the paperwork required by this Sub-Clause. This provision could be unfair on the Contractor in the event that the Employer went into liquidation before the payment had actually been made. This issue is changed in PNK where the phrase '*delivered to*' is changed to '*incorporated in*'.

Another complication may develop where the Contractor has not paid for the delivered Plant or Material himself yet, and the terms and Conditions of those purchases dictate ownership to remain with the original supplier until such payment has been made. This situation is common; it happens frequently in many modern day projects where contractors buy their materials only shortly before they are needed, and where they have negotiated 30–60[+] day payment terms.

The reference to the Laws of the Country is extremely important and should also refer to the governing law as stated in the Appendix to Tender. Any conflict between the law and

Sub-Clause 7.7 should have been checked by the Employer and clarified in the Particular Conditions.

7.8 Royalties
Unless otherwise stated in the Specification, the Contractor shall pay all royalties, rents and other payments for:

(a) natural Materials obtained from outside the Site; and
(b) the disposal of material from demolitions and excavations and of other surplus material (whether natural or man-made), except to the extent that disposal areas within the Site are specified in the Contract.

The cost of royalties for Materials and for the disposal of material can be anticipated and should be allowed in the Contractor's Tender.

FIDIC Users' Guide
ISBN 978-0-7277-5856-9

ICE Publishing: All rights reserved
http://dx.doi.org/10.1680/fug.58569.167

Chapter 14
Clause 8: Commencement, Delays and Suspension

Clause 8 covers three very important subjects, all of which are related to the period during which the Contractor will construct the Works.

- The start and duration of the construction period, at Sub-Clauses 8.1–8.3.
- Programme, delays and extension of time, at Sub-Clauses 8.4–8.7.
- Suspension of work by the Engineer, at Sub-Clauses 8.8–8.12.

The procedures for the completion of the Works are given at Clause 10.

Although Clause 8 deals primarily with matters that are a function of time, the requirement for a notice under Sub-Clause 8.3 [*Programme*] also applies to events or circumstances that may adversely affect the work or increase the Contract Price.

8.1 Commencement of Work
The Engineer shall give the Contractor not less than 7 days' notice of the Commencement Date. Unless otherwise stated in the Particular Conditions, the Commencement Date shall be within 42 days after the Contractor receives the Letter of Acceptance.

The Contractor shall commence the execution of the Works as soon as is reasonably practicable after the Commencement Date, and shall then proceed with the Works with due expedition and without delay.

The Contract starts with the Letter of Acceptance or, if required by the governing law or other regulation, the Contract Agreement as defined at Sub-Clause 1.1.1.2. Notification by the Letter of Acceptance constitutes the formation of the Contract, so any negotiations and discussions on bond or insurance arrangements must have been completed before it is issued. If the Employer decides, for whatever reason, to issue a letter that he intends to enter into a Contract, it is important that the letter of intent is worded so as to be clear that it is not a Letter of Acceptance. Even if the governing law requires a Contract Agreement in order to commence the formal contractual relationship, it is important that any letter from the Employer to the Contractor before the Contract Agreement is signed should be very carefully worded. The issue of the Letter of Acceptance creates obligations and starts a sequence of events and time periods which require the Employer to hand over the Site and the Contractor to start work on the project. The Employer is then liable to pay the Contractor for Costs incurred in order to comply with the Letter of Acceptance.

The '*Commencement Date*' is the start of the '*Time for Completion*' that is the period within which the Contractor has agreed to construct the Works. When the Commencement Date has been determined the Engineer should calculate the calendar date for completion. Potential arguments can be avoided by agreeing the calendar date at the start of the construction period. The number of days in the Time for Completion is given in the Appendix to Tender and may refer to the whole of the Works, or a designated Section of the Works. '*day*' is defined as a calendar day and not a working day and so the number of days includes weekends and holidays.

The Commencement Date is fixed by the Engineer, subject to the requirements of Sub-Clause 8.1 [*Commencement of Works*].

- The Commencement Date shall be within 42 days after the Contractor receives the Letter of Acceptance, unless a different period is stated in the Particular Conditions.
- The Engineer shall give the Contractor not less than 7 days' notice of the Commencement Date.

The 7-day notice period is short, particularly for a design-build Contract that extends over a period of several years, and the Contractor must commence his preparation immediately he receives the Letter of Acceptance. Sub-Clause 2.1 [*Right of Access to the Site*] then requires the Employer to give the Contractor access to and possession of the Site within the number of days from the Commencement Date stated in the Appendix to Tender.

If, for any reason, it is not possible for the Engineer to meet these dates, or for the Employer to give possession of the Site, then a change to the Conditions of Contract must be agreed between the Contractor and the Employer. The Engineer does not have the authority to issue an instruction to change these requirements.

The Contractor is then required to start the execution of the Works '*as soon as is reasonably practicable*' and proceed '*with due expedition and without delay*'. The precise interpretation of '*due expedition*' and '*without delay*' will depend on the circumstances but the general requirement imposes an overall obligation on the Contractor to continue working, even when some problem has arisen.

8.2 Time for Completion
The Contractor shall complete the whole of the Works, and each Section (if any), within the Time for Completion for the Works or Section (as the case may be), including:

(a) *achieving the passing of the Tests on Completion; and*

(b) *completing all work which is stated in the Contract as being required for the Works or Section to be considered to be completed for the purposes of taking-over under Sub-Clause 10.1 [Taking Over of the Works and Sections].*

Under Sub-Clause 8.2 the Contractor is obliged to complete all the work which is required for taking over under Sub-Clause 10.1 [*Taking-Over of the Works and Sections*], including passing the Tests on Completion as Clause 9, before expiry of the Time for Completion.

If Sections of the Works are required to be completed before the overall Time for Completion then the Sections must be described in the Appendix to Tender, together with the Time for Completion and delay damages for each Section.

Sub-Clause 10.1 just refers back to Sub-Clause 8.2, which is not very helpful. The Particular Conditions could include a more detailed list of the work that must be completed, in addition to the actual construction and tests.

8.3 Programme

The Contractor shall submit a detailed time programme to the Engineer within 28 days after receiving the notice under Sub-Clause 8.1[Commencement of Works]. The Contractor shall also submit a revised programme whenever the previous programme is inconsistent with actual progress or with the Contractor's obligations. Each programme shall include:

(a) the order in which the Contractor intends to carry out the Works, including the anticipated timing of each stage of design (if any), Contractor's Documents, procurement, manufacture of Plant, delivery to Site, construction, erection and testing;

(b) each of these stages for work by each nominated Subcontractor (as defined in Clause 5 [Nominated Subcontractors]);

(c) the sequence and timing of inspections and tests specified in the Contract; and

(d) a supporting report which includes:

    (i) a general description of the methods which the Contractor intends to adopt, and of the major stages, in the execution of the Works, and

    (ii) details showing the Contractor's reasonable estimate of the number of each class of Contractor's Personnel and of each type of Contractor's Equipment, required on the Site for each major stage.

Unless the Engineer, within 21 days after receiving a programme, gives notice to the Contractor stating the extent to which it does not comply with the Contract, the Contractor shall proceed in accordance with the programme, subject to his other obligations under the Contract. The Employer's Personnel shall be entitled to rely upon the programme when planning their activities.

The Contractor shall promptly give notice to the Engineer of specific probable future events or circumstances which may adversely affect the work, increase the Contract Price or delay the execution of the Works. The Engineer may require the Contractor to submit an estimate of the anticipated effect of the future event or circumstances, and/or a proposal under Sub-Clause 13.3 [Variation Procedure].

If, at any time, the Engineer gives notice to the Contractor that a programme fails (to the extent stated) to comply with the Contract or to be consistent with actual progress and the

Contractor's stated intentions, the Contractor shall submit a revised programme to the Engineer in accordance with this Sub-Clause.

Sub-Clause 8.3 requires the Contractor to submit a detailed programme to the Engineer within 28 days after receiving the notice of the Commencement Date. The Contractor will obviously have started to prepare his programme at Tender stage and continued to plan his work during any negotiations and when he has received the Letter of Acceptance or Contract Agreement. However, he cannot prepare his final programme until he learns the actual calendar dates for the construction period, which is when he receives the notice of the Commencement Date. The detailed programme may also include reduced working during certain periods of the year.

This Sub-Clause gives detailed requirements for the information to be included in this programme and these requirements may be amplified or extended in the Particular Conditions or Specifications. When deciding on the form of the programme and the detail to be included, the Contractor should remember that the programme will be used to demonstrate whether any delay situation will cause a delay to completion. In general terms the programme must include the following.

- The order in which the Contractor intends to carry out the Works.
- The anticipated timing of each stage from design, through procurement, manufacture, construction and testing; including the stages of the work of nominated Subcontractors.
- A report with a general description and details of the methods the Contractor intends to adopt and estimates of the numbers of personnel and type of equipment that will be required on Site.

The information required is extensive and requires the Contractor to have planned the work in detail. If the Engineer has any further requirements for the layout of the information, or particular computer software which is to be used, then these requirements must be given in the Particular Conditions or Specification in order that the Contractor can allow for the costs in his Tender.

Commissioning is not a defined term, so will presumably have its usual meaning of '*bringing into operation*'. This is a task that will be related to the trial operation. There is a sequence of post-installation events including testing, training, submission of documents, commissioning and trial operation that should be included in the programme. These tasks will sometimes need to be included in the Particular Conditions for RED Book Contracts. The RED Book allows for the provision of Plant as part of the Permanent Works and any item of machinery or other Plant will need to be commissioned after installation and tests. The trial operation would normally refer to the operation of the completed project and so is applicable when the Contractor has designed the complete Works. However, there are circumstances when trial operation may be appropriate for Plant that has been designed by the Contractor under a RED Book Contract. In a RED Book contract the requirements for commissioning and trial operation must be amplified in the Specification, read in conjunction with Sub-Clauses 4.21 [*Progress Reports*] and 9.1 [*Contractor's Obligations*] and the Particular Conditions may also include requirements from the YLW Book Clause 5 [*Design*].

The Engineer is not expected or required to approve, or even to consent to, the programme. The Engineer has 21 days within which he can state that the programme does not comply with the Contract. In such case the Contractor shall submit a revised programme to the Engineer. When there are no (more) comments from the Engineer the Contractor is then required to work in accordance with the latest submitted programme – subject to his other obligations under the Contract. An intention or requirement of the programme cannot change any obligation under the Contract. It is often erroneously thought that a delayed completion date shown in an updated Sub-Clause 8.3 programme, and which is not further commented on by the Engineer constitutes an acceptance of that later completion date as an extension to the Time for Completion.

It is significant that the information required is stated as '*the Contractor intends*' and '*the anticipated timing*'. If circumstances cause a change in the Contractor's intentions or anticipated timing then he is presumably allowed to change the programme, provided of course that the revised programme meets the Time for Completion and further complies with the Contract.

The information that is required in the programme and supporting report is also listed in the requirements for the Contractor's monthly progress report under Sub-Clause 4.21 [*Progress Reports*]. The programme and the various reports will need to be prepared using the same format or computer software.

If the actual progress falls behind the programme, is in any way different to the programme, or the programme fails to comply with the Contract, the Contractor may be required to take action, such as, either

- to submit a revised programme under the first or last paragraph of Sub-Clause 8.3, or
- to submit a revised programme and supporting report under Sub-Clause 8.6 [*Rate of Progress*] showing how he proposes to expedite progress.

This Sub-Clause also states that the Employer's Personnel are entitled to rely on the programme when planning their activities so any change must be made known immediately to enable the Engineer and his staff to adjust their own plans to suit the change. If the Contractor works in advance of his programme, in order to allow for any future delays, then the Engineer may object on the grounds that he is not ready to issue additional drawings or to provide the appropriate supervision. The use of an 'early start' and 'late start' type of programme can be useful, or work in advance of the programme should be agreed with the Engineer. Any such advance working might also be highlighted in the monthly progress report.

The Contractor is required to give notice to the Engineer of any events or circumstances that may adversely affect the work, increase the Contract Price or delay the execution of the Works. This requirement for a notice from the Contractor covers virtually anything that has an adverse effect on the activities of the Contractor and not just matters that affect the programme or for which the Contractor intends to submit a claim. It is in addition to the requirement for claims notices under Sub-Clause 20.1 [*Contractor's Claims*] although a delay or price situation will frequently result in notices under both clauses. The Sub-Clause

8.3 notice must be given 'promptly', which is much quicker than the 'not later than 28 days' requirement for the Sub-Clause 20.1 notice. However, this notice appears to serve the same purpose as the potential delay notices under other Clauses, such as the delayed drawing or instruction notice under Sub-Clause 1.9 [*Delayed Drawings or Instructions*].

The Sub-Clause 8.3 notice is in effect an 'early warning notice' and gives the Engineer the opportunity to take action to overcome the problem before the Contractor incurs delay or additional cost. There is no requirement for the Contractor to meet with the Engineer to discuss the problems and possible solutions, but the Sub-Clause does enable the Engineer to require the Contractor to submit estimates and proposals. For good management the Engineer would normally meet with the Contractor to discuss the potential problem and the best way to overcome the problem.

8.4 Extension of Time for Completion

The Contractor shall be entitled subject to Sub-Clause 20.1 [Contractor's Claims] to an extension of the Time for Completion if and to the extent that completion for the purposes of Sub-Clause 10.1 [Taking Over of the Works and Sections] is or will be delayed by any of the following causes:

(a) *a Variation (unless an adjustment to the Time for Completion has been agreed under Sub-Clause 13.3 [Variation Procedure]) or other substantial change in the quantity of an item of work included in the Contract;*

(b) *a cause of delay giving an entitlement to extension of time under a Sub-Clause of these Conditions;*

(c) *exceptionally adverse climatic conditions;*

(d) *Unforeseeable shortages in the availability of personnel or Goods caused by epidemic or governmental actions; or*

(e) *any delay, impediment or prevention caused by or attributable to the Employer, the Employer's Personnel, or the Employer's other contractors on the Site.*

If the Contractor considers himself to be entitled to an extension of the Time for Completion, the Contractor shall give notice to the Engineer in accordance with Sub-Clause 20.1 [Contractor's Claims]. When determining each extension of time under Sub-Clause 20.1, the Engineer shall review previous determinations and may increase, but shall not decrease, the total extension of time.

Sub-Clause 8.4 lists the situations that may entitle the Contractor to an extension of the Time for Completion. It is not sufficient for the event to cause delay or disruption to the Contractor's work; the Contractor must demonstrate that it will actually delay completion under Sub-Clause 10.1 [*Taking-Over of the Works and Sections*]. The Contractor must comply with the notice and other requirements of Sub-Clause 20.1 [*Contractor's Claims*], which means that the Engineer will then follow the procedures of Sub-Clause 3.5 [*Determinations*] in order to determine any extension of time.

Sub-Clause 3.5 does not give the Engineer a time period for making this determination although it must not be '*unreasonably withheld or delayed*' under Sub-Clause 1.3

[*Communications*]. Any determination of an extension of time should be made quickly in order that the Contractor can revise his programme to suit any revision to the Date for Completion or can accelerate to meet the previous Date for Completion. If the determination is delayed then the Contractor may have a justifiable claim for unnecessary aceleration or additional Costs due to the late determination.

Sub-Clause 8.4 only entitles the Contractor to an extension of time, which brings relief from delay damages under Sub-Clause 8.7 [*Delay Damages*], but not to additional payment. In principle under FIDIC contracts, claims for an extension of time and claims for additional payment should be submitted as separate issues, each under their respective Sub-Clause. In practical terms the Contractor will generally submit his claim under the sub-clause that also allows for additional payment, in addition to the relevant paragraph of this Sub-Clause.

The matters listed are as follows.

(*a*) *Variations.* Clause 13 gives the Engineer the power to issue instructions to vary the Works and the Engineer may ask for a proposal from the Contractor before issuing the instruction. If the Engineer has not asked for such a proposal then the Contractor must give notices promptly under Sub-Clause 8.3 [*Programme*] and within 28 days under Sub-Clause 20.1 [*Contractor's Claims*] if he considers that the Variation may delay completion. The Contractor may also claim an extension of time if there is a substantial change in the quantity of an item of work, which would presumably occur as a consequence of the measurement procedure under Clause 12.

(*b*) *Other Sub-Clauses.* The other sub-clauses that entitle the Contractor to an extension of time are discussed under the relevant Sub-Clause:

    1.9   [*Delayed Drawings or Instructions*]
    2.1   [*Right of Access to the Site*]
    4.7   [*Setting Out*]
    4.12 [*Unforeseeable Physical Conditions*]
    4.24 [*Fossils*]
    7.4   [*Testing*]
    8.5   [*Delays Caused by Authorities*]
    8.9   [*Consequences of Suspension*]
    9.2   [*Delayed Tests*]
    10.3 [*Interference with Tests on Completion*]
    13.1 [*Right to Vary*]
    13.5 [*Provisional Sums*]
    13.7 [*Adjustments for Changes in Legislation*]
    16.1 [*Contractor's Entitlement to Suspend Work*]
    17.4 [*Consequences of Employer's Risks*]
    19.4 [*Consequences of Force Majeure*].

(*c*) *Climatic conditions.* To justify an extension of time the Contractor must demonstrate that the climatic conditions were exceptionally adverse and actually delayed completion. It will be necessary to submit records for the normal weather over a period of, say, five years. The Employer may already have such records and made them available at Tender stage under Sub-Clause 4.10 [*Site Data*] or the Contractor should have obtained all available information under Sub-Clause 4.10(b). In order to record the actual conditions, it will be necessary for the Contractor to have the necessary equipment in place from the start of the project, such that rainfall or other conditions are recorded automatically when they occur, even though such conditions may occur outside normal working hours. Subject to a review of the Laws of the Country, claims for climatic conditions will only result in additional time, but not money, and they are specifically excluded from the unforeseeable physical conditions situations under Sub-Clause 4.12 [*Unforeseeable Physical Conditions*]. Under Sub-Clause 17.3(h) [*Employer's Risks*] there would only be an entitlement for additional costs to the extent that the inclement weather conditions actually resulted in a damage or loss to the Works or Goods needing repair or making good.

(d) *Shortages of personnel or Goods.* The shortage must have been Unforeseeable by an experienced Contractor, as defined at Sub-Clause 1.1.6.8 and caused by epidemic or governmental action. Governmental action is not restricted to the government of the Country of the project.

(e) *Employer causes.* If the Employer causes a delay then the Contractor, as the other Party to the Contract, should be entitled to compensation. Part (e) gives an overall right to an extension of time but many of the events that are covered will also be covered under Employer's Risks as Sub-Clause 17.3 [*Employer's Risks*] and Sub-Clause 17.4 [*Consequences of Employer's Risks*]. The reference to the Employer's other Contractors on the Site must be read in conjunction with Sub-Clause 4.6 [*Co-operation*], which refers to the reimbursement of cost, but not delays.

The provision at Sub-Clause 20.1 [*Contractor's Claims*] for claims '*under any Clause of these Conditions or otherwise in connection with the Contract*', together with the development of paragraphs (a) to (e) and the provision at Sub-Clause 8.5 [*Delays Caused by Authorities*], should be sufficiently general to cover any situation for which a Contractor should reasonably be entitled to an extension of time.

8.5 Delays Caused by Authorities
If the following conditions apply, namely:

(a) *the Contractor has diligently followed the procedures laid down by the relevant legally constituted public authorities in the Country;*
(b) *these authorities delay or disrupt the Contractor's work; and*
(c) *the delay or disruption was Unforeseeable;*

then this delay or disruption will be considered as a cause of delay under subparagraph (b) of Sub-Clause 8.4 [Extension of Time for Completion].

This Sub-Clause gives an additional entitlement to the grounds for an extension of time under Sub-Clause 8.4(b) [*Extension of Time for Completion*]. The Sub-Clause seems to indicate that under given circumstances an entitlement to an extension of time is automatic, however this does not override the provisions of Sub-Clause 20.1 [*Contractor's Claims*], specifically with regards to the required notices. The Sub-Clause also does not require the Engineer to follow the procedures of Sub-Clause 3.5 [*Determinations*] in order to determine the extension but this will again be required under the provisions of Sub-Clause 20.1.

The precise meaning of '*legally constituted public authorities in the Country*' under the governing law is a potential source of dispute. The trend to privatise public bodies may reduce the scope of this Sub-Clause.

8.6 Rate of Progress
If, at any time:

(a) *actual progress is too slow to complete within the Time for Completion; and/or*
(b) *progress has fallen (or will fall) behind the current programme under Sub-Clause 8.3 [Programme];*

other than as a result of a cause listed in Sub-Clause 8.4 [Extension of Time for Completion], then the Engineer may instruct the Contractor to submit, under Sub-Clause 8.3 [Programme], a revised programme and supporting report describing the revised methods which the Contractor proposes to adopt in order to expedite progress and complete within the Time for Completion.

Unless the Engineer notifies otherwise, the Contractor shall adopt these revised methods, which may require increases in the working hours and/or in the numbers of Contractor's Personnel and/or Goods, at the risk and cost of the Contractor. If these revised methods cause the Employer to incur additional costs, the Contractor shall subject to Sub-Clause 2.5 [Employer's Claims] pay these costs to the Employer, in addition to delay damages (if any) under Sub-Clause 8.7 below.

Sub-Clause 8.6 entitles the Engineer to instruct the Contractor to submit proposals to revise his programme and to accelerate the work in order to achieve completion by the due date. If the acceleration measures cause the Employer to incur additional Costs then the Employer may submit a claim under Sub-Clause 2.5 [*Employer's Claims*] and the Engineer will make a determination under the procedures of Sub-Clause 3.5 [*Determinations*].

Sub-Clause 8.6 states that the provision only applies when the reason for the delay is not covered by Sub-Clause 8.4 [*Extension of Time for Completion*], such as, situations where the delay would entitle the Contractor to an extension of time. Whether or not the delay was caused by a Sub-Clause 8.4 event depends on an Engineer's determination under Sub-Clause 3.5 [*Determinations*], or a subsequent decision by the Dispute Adjudication Board (DAB). If the Contractor decides to refer the matter to the DAB under Sub-Clause 20.4 [*Obtaining Dispute Adjudication Board's Decision*] then he should first give notices under Clause 20.1 and any other relevant clauses to claim reimbursement of his acceleration Costs together with any Employer's Costs and delay damages.

If the Contractor incurs acceleration Costs under this Sub-Clause and the DAB later decides that the Contractor is entitled to an extension of time then there will be a potential claim situation.

8.7 Delay Damages

If the Contractor fails to comply with Sub-Clause 8.2 [Time for Completion], the Contractor shall subject to Sub-Clause 2.5 [Employer's Claims] pay delay damages to the Employer for this default. These delay damages shall be the sum stated in the Appendix to Tender, which shall be paid for every day which shall elapse between the relevant Time for Completion and the date stated in the Taking-Over Certificate.

However, the total amount due under this Sub-Clause shall not exceed the maximum amount of delay damages (if any) stated in the Appendix to Tender.

These delay damages shall be the only damages due from the Contractor for such default, other than in the event of termination under Sub-Clause 15.2 [Termination by Employer] prior to completion of the Works. These damages shall not relieve the Contractor from his obligation to complete the Works, or from any other duties, obligations or responsibilities which he may have under the Contract.

The Appendix to Tender must state the daily sum and the maximum total amount of the damages due from the Contractor to the Employer if the Works are not completed by the due date. These figures are normally required to be given as percentages of the final Contract Price. That is the final sum, as Sub-Clause 14.1 [*The Contract Price*], after taking into account all adjustments for re-measurement, Variations and otherwise under the Contract. Hence the sum for delay damages cannot be determined finally until after agreement of the Final Statement.

By the normal definition of the word '*damages*', the figures for delay damages should be a reasonable estimate of the actual losses that will be incurred by the Employer. If the governing law entitles the Employer to deduct '*penalties*' then the Sub-Clause and the Appendix to Tender should be amended by the Particular Conditions.

8.8 Suspension of Work

The Engineer may at any time instruct the Contractor to suspend progress of part or all of the Works. During such suspension, the Contractor shall protect, store and secure such part or the Works against any deterioration, loss or damage.

The Engineer may also notify the cause for the suspension. If and to the extent that the cause is notified and is the responsibility of the Contractor, the following Sub-Clauses 8.9, 8.10 and 8.11 shall not apply.

Sub-Clause 8.8 entitles the Engineer to instruct the Contractor to suspend progress of part or all of the Works. Surprisingly, the Engineer is not obliged to give the reason for the suspension but 'may' notify the cause. Clearly the reasonable Engineer should tell the Contractor the reason and likely extent of the suspension in order that the Contractor can decide how to meet his obligation to 'protect, store and secure' that part of the Works.

8.9 Consequences of Suspension

If the Contractor suffers delay and/or incurs Cost from complying with the Engineer's instructions under Sub-Clause 8.8 [Suspension of Work] and/or from resuming the work, the Contractor shall give notice to the Engineer and shall be entitled subject to Sub-Clause 20.1 [Contractor's Claims] to:

(a) an extension of time for any such delay, if completion is or will be delayed, under Sub-Clause 8.4 [Extension of Time for Completion]; and
(b) payment of any such Cost, which shall be included in the Contract Price.

After receiving this notice, the Engineer shall proceed in accordance with Sub-Clause 3.5 [Determinations] to agree or determine these matters.

The Contractor shall not be entitled to an extension of time for, or to payment of the Cost incurred in, making good the consequences of the Contractor's faulty design, workmanship or materials, or of the Contractor's failure to protect, store or secure in accordance with Sub-Clause 8.8 [Suspension of Work].

The Contractor must give the usual notices under Clause 20.1 in order to claim any Costs and extension of time. This Sub-Clause also covers the Costs of resuming work, such as any re-mobilisation Costs.

8.10 Payment for Plant and Materials in Event of Suspension
The Contractor shall be entitled to payment of the value (as at the date of suspension) of Plant and/or Materials which have not been delivered to Site, if:

(a) the work on Plant or delivery of Plant and/or Materials has been suspended for more than 28 days; and
(b) the Contractor has marked the Plant and/or Materials as the Employer's property in accordance with the Engineer's instructions.

Any request for payment for Plant and/or Materials that have not been delivered to Site would presumably require notice from the Contractor and confirmation from the Engineer, either as an item in the monthly Statement under Sub-Clause 14.3 [Application for Interim Payment Certificates] or as a claim under Sub-Clause 20.1 [Contractor's Claims].

8.11 Prolonged Suspension
If the suspension under Sub-Clause 8.8 [Suspension of Work] has continued for more than 84 days, the Contractor may request the Engineer's permission to proceed. If the Engineer does not give permission within 28 days after being requested to do so, the Contractor may, by giving notice to the Engineer, treat the suspension as an omission under Clause 13 [Variations and Adjustments] of the affected part of the Works. If the suspension affects the whole of the Works, the Contractor may give notice of termination under Sub-Clause 16.2 [Termination by Contractor].

If the Contractor makes his request at the end of the 84th day and the Engineer waits for the full 28 days before replying to the request, then the suspension may have lasted 112 days, by which time the Contractor may have incurred substantial costs and disruption to planning his allocation of resources to different projects. This would presumably be the time to cancel the suspended work.

If the suspension is not lifted and the Contractor chooses to treat suspension of part of the Works as an omission under Sub-Clause 13.1(d) [*Right to Vary*] then the omitted work will either be valued by agreement, or under Sub-Clause 12.4 [*Omissions*]. Following omission under Sub-Clause 13.1(d) the omitted work cannot be carried out by others. If the whole of the Works was suspended and the Contractor decides to give notice of termination under Sub-Clause 16.2(f) [*Termination by Contractor*] then the payment will be made under the provisions of Sub-Clause 19.6 [*Optional Termination, Payment and Release*], as for Force Majeure termination, plus loss of profit and other losses or damage under Sub-Clause 16.4(c) [*Payment on Termination*]. Clearly this would be expensive for the Employer, so sustained suspension should be avoided and should not be used for ulterior motives or as a weapon against the Contractor.

8.12 Resumption of Work
After the permission or instruction to proceed is given, the Contractor and the Engineer shall jointly examine the Works and the Plant and Materials affected by the suspension. The Contractor shall make good any deterioration or defect in or loss of the Works or Plant or Materials, which has occurred during the suspension.

If the Contractor and Engineer fail to agree on the results of their joint inspection then the Engineer would presumably issue instructions and the Contractor would have the option of giving notice under Sub-Clause 20.1 [*Contractor's Claims*] and any other relevant clause. The same procedure would be followed if there were a dispute about who pays the Costs of making good. The Contract insurances would have remained in place during the suspension and the insurer would have been notified of the suspension in accordance with Sub-Clause 18.1 [*General Requirements for Insurances*].

FIDIC Users' Guide
ISBN 978-0-7277-5856-9

Chapter 15
Clause 9: Tests on Completion

Tests on Completion are the tests that are carried out after the Works have been completed and before the Engineer will issue the Taking-Over Certificate. The responsibilities and procedures for the '*Tests on Completion*' are given at Clause 9, which also requires compliance with the testing procedures at Sub-Clause 7.4 [*Testing*] and the submission of Contractor's Documents at Sub-Clause 4.1(d) [*Contractor's General Obligations*]. The technical details of the required tests will be given in the Specification.

Under a RED Book Contract the Materials and other aspects of the Works are normally tested during construction and further tests before the issue of the Taking-Over Certificate should only be specified if testing at an earlier stage was not possible. If the Contract includes items of Plant, or other work that is designed by the Contractor to a performance specification, then the YLW Book procedures will be appropriate.

9.1 Contractor's Obligations
The Contractor shall carry out the Tests on Completion in accordance with this Clause and Sub-Clause 7.4 [Testing], after providing the documents in accordance with sub-paragraph (d) of Sub-Clause 4.1 [Contractor's General Obligations].

The Contractor shall give to the Engineer not less than 21 days' notice of the date after which the Contractor will be ready to carry out each of the Tests on Completion. Unless otherwise agreed, Tests on Completion shall be carried out within 14 days after this date, on such day or days as the Engineer shall instruct.

In considering the results of the Tests on Completion, the Engineer shall make allowances for the effect of any use of the Works by the Employer on the performance or other characteristics of the Works. As soon as the Works, or a Section, have passed any Tests on Completion, the Contractor shall submit a certified report of the results of these Tests to the Engineer.

Sub-Clause 8.2 [*Time for Completion*] requires that the Works have passed any Tests on Completion before issue of the Taking-Over Certificate under Clause 10. It is not stated whether the Tests on Completion can be carried out immediately the particular item is complete, or whether all Tests on Completion must be carried out immediately prior to the issue of the Taking-Over Certificate. In practice this will depend on the circumstances and any requirements should be detailed in the Specification.

Sub-Clause 9.1 states that the Tests on Completion cannot be carried out until the Contractor has provided the Engineer with the documents listed at Sub-Clause 4.1(d) [*Contractor's*

General Obligations]. This refers to the as-built drawings and the operation and maintenance manuals for any part of the Permanent Works that had been designed by the Contractor. The items required to be tested after they are complete will often be items of Plant that have been designed by the Contractor to meet a performance specification.

The Contract requires the documents to have been submitted to the Engineer, but does not require them to have been approved by the Engineer. However, Sub-Clause 4.1(d) requires these documents to be '*in accordance with the Specification and in sufficient detail for the Employer to operate, maintain, dismantle, reassemble, adjust and repair this part of the Works*'. Hence, the Employer will expect the Engineer to confirm that the submitted documents are acceptable.

The Contractor must give 21 days' notice of the date when he will be ready to carry out the Tests on Completion, which gives time for the Engineer to arrange for any specialist engineers to attend and for the Employer to make any necessary arrangements, particularly in parts of the Works which may already have been taken over by the Employer. The Employer may require that his staff that will maintain the Works should observe the tests. The tests must be carried out within 14 days after this date, on a date instructed by the Engineer.

The additional requirements in the YLW Book are essential when the Works include Plant, as distinct from Materials. Where appropriate they should also be included in the Particular Conditions for a RED Book Contract. Further details of the tests and the criteria for acceptance must be given in the Specification.

9.2 Delayed Tests

If the Tests on Completion are being unduly delayed by the Employer, Sub-Clause 7.4 [Testing] (fifth paragraph) and/or Sub-Clause 10.3 [Interference with Tests on Completion] shall be applicable.

If the Tests on Completion are being unduly delayed by the Contractor, the Engineer may by notice require the Contractor to carry out the Tests within 21 days after receiving the notice. The Contractor shall carry out the Tests on such day or days within that period as the Contractor may fix and of which he shall give notice to the Engineer.

If the Contractor fails to carry out the Tests on Completion within the period of 21 days, the Employer's Personnel may proceed with the Tests at the risk and cost of the Contractor. The Tests on Completion shall then be deemed to have been carried out in the presence of the Contractor and the results of the Tests shall be accepted as accurate.

This Sub-Clause is specific in that if the Tests on Completion are delayed by the Employer then the provisions of Sub-Clause 9.2 will apply. If the tests are so delayed by the Employer then the Contractor can give notice to the Engineer under Sub-Clause 7.4 [*Testing*] and follow the Sub-Clause 20.1 [*Contractor's Claims*] procedure to claim for an extension of time and additional payment. These tests may also be delayed by others, for example, the Engineer. It is therefore recommended to amend this Sub-Clause through the Particular Conditions by adding the words 'and/or Employer's Personnel' after the word 'Employer' in the first sentence of the first paragraph.

If the delay lasts for more than 14 days then, under Sub-Clause 10.3 [*Interference with Tests on Completion*], the Employer is deemed to have taken over the Works or Section on the date when the Tests on Completion would otherwise have been completed. The tests must then be carried out as soon as practicable before the expiry date of the Defects Notification Period. The procedures for the Engineer to issue a Taking-Over Certificate and for any claims are given at Sub-Clause 10.3.

If the tests are unduly delayed by the Contractor then the Engineer may give notice for the Contractor to carry out the tests within 21 days, or the Employer's Personnel may proceed with the tests, at the Contractor's risk and Cost.

9.3 Retesting
If the Works, or a Section, fail to pass the Tests on Completion, Sub-Clause 7.5 [Rejection] shall apply, and the Engineer or the Contractor may require the failed Tests, and Tests on Completion on any related work, to be repeated under the same terms and conditions.

Sub-Clause 9.3 requires that Sub-Clause 7.5 [*Rejection*] shall apply to '*the Works, or a Section*', which has failed to pass the Tests on Completion. Sub-Clause 7.5 provides that the Engineer may reject the '*Plant, Materials or workmanship*' and the Contractor must promptly make good the defect. The Engineer can require a repeat test.

9.4 Failure to Pass Tests on Completion
If the Works, or a Section, fail to pass the Tests on Completion repeated under Sub-Clause 9.3 [Retesting], the Engineer shall be entitled to:

(a) order further repetition of Tests on Completion under Sub-Clause 9.3;
(b) if the failure deprives the Employer of substantially the whole benefit of the Works or Section, reject the Works or Section (as the case may be), in which event the Employer shall have the same remedies as are provided in subparagraph (c) of Sub-Clause 11.4 [Failure to Remedy Defects]; or
(c) issue a Taking-Over Certificate, if the Employer so requests.

In the event of subparagraph (c), the Contractor shall proceed in accordance with all other obligations under the Contract, and the Contract Price shall be reduced by such amount as shall be appropriate to cover the reduced value to the Employer as a result of this failure. Unless the relevant reduction for this failure is stated (or its method of calculation is defined) in the Contract, the Employer may require the reduction to be (i) agreed by both Parties (in full satisfaction of this failure only) and paid before this Taking-Over Certificate is issued, or (ii) determined and paid under Sub-Clause 2.5 [Employer's Claims] and Sub-Clause 3.5 [Determinations].

Sub-Clause 9.4 refers to failure by '*the Works, or a Section*'. If a particular part of the Works, or a single item of Plant fails a test then the whole of the Works or Section has failed. If the

Engineer then decides to order a repetition of the tests, as option (a), then his decision is a technical matter and may, or may not, resolve the problem. If the item being tested again fails the test, despite the making good as Sub-Clause 7.5 [*Rejection*], then the Engineer will consider other technical alternatives before going to options (b) or (c).

Sub-Clause 7.6 [*Remedial Work*] enables the Engineer to instruct the Contractor to remove and replace Plant or Materials or remove and re-execute other work. If the Contractor fails to carry out this instruction then the Employer is entitled to arrange for other persons to carry out the work and claim the Costs under Sub-Clause 2.5 [*Employer's Claims*].

Option (b) refers to the procedure at subparagraph (c) of Sub-Clause 11.4 [*Failure to Remedy Defects*]. The Employer can terminate the Contract, in whole or in part, and recover all his Costs as defined at subparagraph (c). It is difficult to envisage circumstances where this could be appropriate under a RED Book Contract and before invoking this procedure the Engineer should also consider alternative actions available under Sub-Clause 11.4.

Any acceptance of work that has failed a test must be agreed by the Employer and Contractor and must be confirmed as a Variation, signed by both Parties, stating the terms and conditions under which the work has been accepted.

FIDIC Users' Guide
ISBN 978-0-7277-5856-9

ICE Publishing: All rights reserved
http://dx.doi.org/10.1680/fug.58569.183

Chapter 16
Clause 10: Employer's Taking Over

This Clause provides procedures to follow when the Works are taken over by the Employer. The procedure can apply to the Works as a whole, or to any Section of the Works that has been defined in the Appendix to Tender.

10.1 Taking-Over of the Works and Sections
Except as stated in Sub-Clause 9.4 [Failure to Pass Tests on Completion], the Works shall be taken over by the Employer when (i) the Works have been completed in accordance with the Contract, including the matters described in Sub-Clause 8.2 [Time for Completion] and except as allowed in subparagraph (a) below, and (ii) a Taking-Over Certificate for the Works has been issued, or is deemed to have been issued in accordance with this Sub-Clause.

The Contractor may apply by notice to the Engineer for a Taking-Over Certificate not earlier than 14 days before the Works will, in the Contractor's opinion, be complete and ready for taking over. If the Works are divided into Sections, the Contractor may similarly apply for a Taking-Over Certificate for each Section.

The Engineer shall, within 28 days after receiving the Contractor's application:

(a) issue the Taking-Over Certificate to the Contractor, stating the date on which the Works or Section were completed in accordance with the Contract, except for any minor outstanding work and defects which will not substantially affect the use of the Works or Section for their intended purpose (either until or while this work is completed and these defects are remedied); or

(b) reject the application, giving reasons and specifying the work required to be done by the Contractor to enable the Taking-Over Certificate to be issued. The Contractor shall then complete this work before issuing a further notice under this Sub-Clause.

If the Engineer fails either to issue the Taking-Over Certificate or to reject the Contractor's application within the period of 28 days, and if the Works or Section (as the case may be) are substantially in accordance with the Contract, the Taking-Over Certificate shall be deemed to have been issued on the last day of that period.

Sub-Clause 10.1 requires that the Works will be taken over when completed in accordance with the Contract. This requirement specifically includes the matters described at Sub-Clause 8.2 [*Time for Completion*], which are

- passing the Tests on Completion, as Clause 9; and
- completing all the work as required by the Contract.

However, Sub-Clause 10.1(a) states that it is not necessary to complete '*any minor outstanding work and defects which will not substantially affect the use of the Works or Section for their intended purpose (either until or while this work is completed and these defects are remedied)*'. This specific exclusion is a considerable improvement on the wording of previous Conditions, which referred to the Works being '*substantially completed*'. It is now clear that some outstanding work or defects will not delay the issue of the Taking-Over Certificate, provided that they do not substantially affect the use of the Works for their intended purpose. The question of whether a particular item will '*substantially affect*' such use will be a matter for the Engineer to exercise his judgement, but the emphasis on the use of the Works should help towards a clearer definition of when the Works are ready to be taken over by the Employer.

The minor outstanding work must be completed during the Defects Notification Period as instructed by the Engineer under Sub-Clause 11.1(a) [*Completion of Outstanding Work and Remedying Defects*].

The procedure for the Employer to take over the Works is given at Sub-Clause 10.1 as follows.

(*a*) When the Contractor decides that the Works are within 14 days of being ready to be taken over, he issues a notice to apply to the Engineer for a Taking-Over Certificate;
(*b*) Within 28 days of receiving the Contractor's application, the Engineer must either:
    (*i*) issue the Taking-Over Certificate stating the date when the Works were completed in accordance with the Contract; or
    (*ii*) reject the application.

If the Engineer rejects the application he must give his reasons and specify the work that must be done by the Contractor to enable the Taking-Over Certificate to be issued. The Contractor must then complete this work and issue another notice.

If the Engineer fails either to issue the Taking-Over Certificate or to reject the application within the 28-day period then the Contract states that the Taking-Over Certificate shall be deemed to have been issued on the last day of the 28-day period, provided that the Works are substantially completed in accordance with the Contract. As per subparagraph (b)(i) above a Taking-Over Certificate shall state the date on which the Works were completed in accordance with the Contract! In the event of the Engineer failing to issue or reject the Works or Section the 'deemed' provision merely states that the Taking-Over Certificate is issued on the last day of the 28-day period. There is no provision for the factual date for completion of the Works or Section. However, for example, for the proper functioning of Sub-Clause 8.7 [*Delay Damages*], such a date must be stated on the Certificate. It is advised to amend the last paragraph of Sub-Clause 10.1 through the Particular Conditions by adding at the end of the last sentence: *and the date on which the Works or Section were completed in accordance with the Contract will be the date applied for by the Contractor in accordance with this Sub-Clause.*

In addition the procedure to establish whether or not the Works are substantially completed in accordance with the Contract, in the absence of a statement by the Engineer, is not clear and could cause problems. Presumably the Contractor would first refer the matter direct to the Employer and then, if necessary, to the DAB or the Arbitrator.

The issue of the Taking-Over Certificate for the Works (and not the stated date on which the Works or Section were completed in accordance with the Contract) is the start of the 84-day period for the Contractor to issue his Statement at completion, which includes an estimate of any further claims, in accordance with Sub-Clause 14.10 [*Statement at Completion*].

10.2 Taking-Over of Parts of the Works

The Engineer may, at the sole discretion of the Employer, issue a Taking-Over Certificate for any part of the Permanent Works.

The Employer shall not use any part of the Works (other than as a temporary measure which is either specified in the Contract or agreed by both Parties) unless and until the Engineer has issued a Taking-Over Certificate for this part. However, if the Employer does use any part of the Works before the Taking-Over Certificate is issued:

(a) the part which is used shall be deemed to have been taken over as from the date on which it is used;

(b) the Contractor shall cease to be liable for the care of such part as from this date; when responsibility shall pass to the Employer; and

(c) if requested by the Contractor, the Engineer shall issue a Taking-Over Certificate for this part.

After the Engineer has issued a Taking-Over Certificate for a part of the Works, the Contractor shall be given the earliest opportunity to take such steps as may be necessary to carry out any outstanding Tests on Completion. The Contractor shall carry out these Tests on Completion as soon as practicable before the expiry date of the relevant Defects Notification Period.

If the Contractor incurs Cost as a result of the Employer taking over and/or using a part of the Works, other than such use as is specified in the Contract or agreed by the Contractor, the Contractor shall (i) give notice to the Engineer and (ii) be entitled subject to Sub-Clause 20.1 [Contractor's Claims] to payment of any such Cost plus reasonable profit, which shall be included in the Contract Price. After receiving this notice, the Engineer shall proceed in accordance with Sub-Clause 3.5 [Determinations] to agree or determine this Cost and profit.

If a Taking-Over Certificate has been issued for a part of the Works (other than a Section), the delay damages thereafter for completion of the remainder of the Works shall be reduced. Similarly, the delay damages for the remainder of the Section (if any) in which this part is included shall also be reduced. For any period of delay after the date stated in this Taking-Over Certificate, the proportional reduction in these delay damages shall be calculated as the proportion which the value of the part so certified bears to the value of the

Works or Section (as the case may be) as a whole. The Engineer shall proceed in accordance with Sub-Clause 3.5 [Determinations] to agree or determine these proportions. The provisions of this paragraph shall only apply to the daily rate of delay damages under Sub-Clause 8.7 [Delay Damages], and shall not affect the maximum amount of these damages.

The provisions of Sub-Clause 10.1 [*Taking-Over of the Works and Sections*] and Sub-Clause 10.3 [*Interference with Tests on Completion*] refer to the whole of the Works, or Sections of the Works that are designated in the Appendix to Tender. Sub-Clause 10.2 relates to parts of the Works that have not been designated as Sections.

Only the Employer has the right to decide that a certain part of the Works will be taken over before the remainder of the Works. This discretion may be exercised when the Employer wishes to use a part of the Works before the whole of the Works is complete. The Employer can require the Engineer to issue a Taking-Over Certificate and the Employer would then be responsible for the care of that part.

If the Employer's use of part of the Works is a temporary measure, which is specified in the Contract or agreed by both Parties, then a Taking-Over Certificate is not required. If the Employer uses a part of the Works without a Taking-Over Certificate then Sub-Clause 10.2 requires the following.

(*a*) The part that is used, shall be deemed to have been taken over, as from the date on which it is used.
(*b*) The Contractor shall cease to be liable for the care of such part as from this date, when responsibility shall pass to the Employer.
(*c*) If requested by the Contractor, the Engineer shall issue a Taking-Over Certificate for this part.

Under this Sub-Clause the taking over of a part of the Works was not envisaged in the Contract and is likely to cause problems for any Tests on Completion. Any such tests must be carried out as soon as practicable, but may have to be coordinated with tests in other parts of the Works.

Sub-Clause 10.2 provides that if the Contractor incurs Cost as a result of the Employer using and/or taking over a part of the Works then the Contractor will give notice and proceed as Sub-Clause 20.1 [*Contractor's Claims*] and Sub-Clause 3.5 [*Determinations*]. The Contractor could then be entitled to his Cost plus reasonable profit.

This Sub-Clause does not refer to any entitlement for an extension of time but, if completion of another part of the Works has been delayed by this situation, the Contractor will presumably claim under a different clause, possibly Sub-Clause 8.4(e) [*Extension of Time for Completion*] as a delay attributable to the Employer.

The Sub-Clause also provides that the daily rate for any delay damages will be reduced in proportion to the value of any part of the Works for which a Taking-Over Certificate has been issued. The figure will be determined by the Engineer under Sub-Clause 3.5 [*Determinations*].

10.3 Interference with Tests on Completion

If the Contractor is prevented, for more than 14 days, from carrying out the Tests on Completion by a cause for which the Employer is responsible, the Employer shall be deemed to have taken over the Works or Section (as the case may be) on the date when the Tests on Completion would otherwise have been completed.

The Engineer shall then issue a Taking-Over Certificate accordingly, and the Contractor shall carry out the Tests on Completion as soon as practicable, before the expiry date of the Defects Notification Period. The Engineer shall require the Tests on Completion to be carried out by giving 14 days' notice and in accordance with the relevant provisions of the Contract.

If the Contractor suffers delay and/or incurs Cost as a result of this delay in carrying out the Tests on Completion, the Contractor shall give notice to the Engineer and shall be entitled subject to Sub-Clause 20.1 [Contractor's Claims] to:

(a) an extension of time for any such delay, if completion is or will be delayed, under Sub-Clause 8.4 [Extension of Time for Completion]; and

(b) payment of any such Cost plus reasonable profit, which shall be included in the Contract Price.

After receiving this notice, the Engineer shall proceed in accordance with Sub-Clause 3.5 [Determinations] to agree or determine these matters.

Sub-Clause 10.3 states that the Works shall be deemed to have been taken over by the Employer if the Contractor is prevented, for more than 14 days, from carrying out the Tests on Completion by a cause for which the Employer is responsible. The Engineer is required to issue a Taking-Over Certificate for the date on which the tests would have been completed if they had not been delayed by this cause. This Sub-Clause is similarly restricted as Sub-Clause 9.2 [*Delayed Tests*] in that it only refers to a delay caused by the Employer. A similar amendment through the Particular Conditions is recommended.

If the Contractor suffers delay and/or incurs Costs as a result of this delay to the Tests on Completion he can give notice under Sub-Clause 10.3 and follow the procedures of Sub-Clause 20.1 [*Contractor's Claims*] and Sub-Clause 3.5 [*Determinations*].

The delay to the tests may be due to more than one cause, or the critical cause may be disputed. The Contractor's claim would include a statement that a certain cause was critical and establish the responsibility for the critical cause.

When the cause that prevented the Tests on Completion from being carried out has been removed, the Engineer can require them to be carried out by giving 14 days' notice. The Contractor is then obliged to carry out the tests as soon as practicable and before the expiry date of the Defects Notification Period. In the unlikely event that this cause continues to the end of the Defects Notification Period then the Engineer will need to consider what action to take, by agreement with the Employer and Contractor. The situation would be outside the provisions of the Contract and therefore outside the authority of the Engineer.

10.4 Surfaces Requiring Reinstatement

Except as otherwise stated in a Taking-Over Certificate, a certificate for a Section or part of the Works shall not be deemed to certify completion of any ground or other surfaces requiring reinstatement.

Sub-Clause 10.4 covers the situation when the ground or other surface needs to be reinstated after the Works are complete. Final reinstatement may not be physically possible until after the Contractor has completed the Works, or even until the end of the Defects Notification Period. This Sub-Clause states that a Taking-Over Certificate shall not be deemed to cover such reinstatement unless it is stated in the Certificate. The completion of reinstatement would become '*minor outstanding work*' to be carried out during the Defects Notification Period, or as an unfulfilled obligation under Sub-Clause 11.10 [*Unfulfilled Obligations*].

FIDIC Users' Guide
ISBN 978-0-7277-5856-9

Chapter 17
Clause 11: Defects Liability

Clause 11 deals with the procedures during the Defects Notification Period, immediately after the Works have been taken over by the Employer. During this period the Contractor is responsible for correcting any defects. If, for any reason, the Employer does not fully occupy and use the project immediately after taking it over then the Employer will lose the benefits of the full Defects Notification Period.

The duration of the Defects Notification Period is stated in the Appendix to Tender. Most of the items in the Appendix to Tender have been left blank in the FIDIC form for details to be inserted by the Employer before calling Tenders, but the figure of 365 days has been printed for the Defects Notification Period. This period may need to be changed by the Employer in the Tender documents. While a period of one year will generally be suitable for civil engineering projects, a longer period may be required for electrical, mechanical or building services work. For example, performance tests on air conditioning plant must be carried out during hot weather, may be specified as Tests after Completion and are often followed by balancing and adjustment of the plant. For the Defects Notification Period to include a full hot-weather season after the completion of the balancing will require a two-year or 730-day period.

The procedures under Clause 11 generally require notifications and actions by the Employer, whereas the Engineer would have undertaken similar actions before Completion. This is a logical change because the Employer has occupied the Works, will be aware of any defects or other problems and the Contractor will need to liaise with the Employer in order to carry out repairs. The Employer will need to make the appropriate arrangements to identify any defects and must designate a representative to liaise with the Contractor. In practice, as the Engineer is now defined as Employer's Personnel, it may be convenient for these tasks to be carried out by the Engineer.

During this period the Engineer has certain powers and responsibilities, as stated in the Sub-Clauses, but no longer has the power to issue instructions for Variations. Under Sub-Clause 13.1 [*Right to Vary*] Variations can only be issued *prior* to the issue of the Taking-Over Certificate.

If any claims or disputes arise during the Disputes Notification Period the provisions of Clause 20 will still apply. The Dispute Adjudication Board is still operative during this period.

11.1 Completion of Outstanding Work and Remedying Defects

In order that the Works and Contractor's Documents, and each Section, shall be in the condition required by the Contract (fair wear and tear excepted) by the expiry date of the relevant Defects Notification Period or as soon as practicable thereafter, the Contractor shall:

(a) *complete any work which is outstanding on the date stated in a Taking-Over Certificate, within such reasonable time as is instructed by the Engineer; and*

(b) *execute all work required to remedy defects or damage, as may be notified by (or on behalf) of the Employer on or before the expiry date of the Defects Notification Period for the Works or Section (as the case may be).*

If a defect appears or damage occurs, the Contractor shall be notified accordingly, by (or on behalf of) the Employer.

Sub-Clause 11.1 gives the overall requirement and procedures for the Defects Notification Period. Normally during the Defects Notification Period, the Works have been occupied and are being used by the Employer. At the end of the period, the Works must be in the condition required by the Contract, with the exception of 'fair wear and tear'. Routine maintenance and problems caused by the Employer's use of the Works are the responsibility of the Employer. This means that

■ any work which was outstanding at the date of the Taking-Over Certificate will have been completed; and

■ any defects or damage which have been notified to the Contractor by the Employer in accordance with this Sub-Clause will have been repaired.

Any defect will be notified to the Contractor by (or on behalf of) the Employer and any such notification should ideally be copied to the Engineer, in order that he can take any necessary action under other Sub-Clauses.

In earlier Conditions of Contracts this period was known as the '*Maintenance Period*' or the '*Defects Liability Period*'. The change to the name '*Defects Notification Period*' emphasises that the Contractor's liability is to repair any defects that have been notified during the period.

11.2 Cost of Remedying Defects

All work referred to in sub-paragraph (b) of Sub-Clause 11.1 [Completion of Outstanding Work and Remedying Defects] shall be executed at the risk and cost of the Contractor, if and to the extent that the work is attributable to:

(a) *any design for which the Contractor is responsible;*

(b) *Plant, Materials or workmanship not being in accordance with the Contract; or*

(c) *failure by the Contractor to comply with any other obligation.*

If and to the extent that such work is attributable to any other cause, the Contractor shall be notified promptly by (or on behalf of) the Employer, and Sub-Clause 13.3 [Variation Procedure] shall apply.

Sub-Clause 11.2 explains the procedure by which the liability for the Cost of the repair is established. If the defect or damage has been notified by the Employer under Sub-Clause 11.1(b) [*Completion of Outstanding Work and Remedying Defects*], and the work is attributable to the Contractor as per Sub-Clause 11.2(a), (b) and (c), then it shall be done at the risk and cost of the Contractor. If the work is not the Contractor's liability then the Contractor must be notified promptly and the Cost to be paid to the Contractor is decided by the Engineer as a Variation, using the procedures at Sub-Clause 13.3 [*Variation Procedure*].

The Defects Notification Period does not give the Employer the right to use the Contractor to carry out routine maintenance and repairs to damage caused by the Employer's Personnel. Paragraphs (a) to (c) in the RED Book list the types of defects that must be repaired at the Contractor's Cost, any other work must be paid for by the Employer. While category (c) may seem to cover a wide range of problems, it is still necessary for the Employer to state and prove with which obligation the Contractor has failed to comply.

11.3 Extension of Defects Notification Period
The Employer shall be entitled subject to Sub-Clause 2.5 [Employer's Claims] to an extension of the Defects Notification Period for the Works or a Section if and to the extent that the Works, Section or a major item of Plant (as the case may be, and after taking over) cannot be used for the purposes for which they are intended by reason of a defect or damage. However, a Defects Notification Period shall not be extended by more than two years.

If delivery and/or erection of Plant and/or Materials was suspended under Sub-Clause 8.8 [Suspension of Work] or Sub-Clause 16.1 [Contractor's Entitlement to Suspend Work], the Contractor's obligations under this Clause shall not apply to any defects or damage occurring more than two years after the Defects Notification Period for the Plant and/or Materials would otherwise have expired.

The duration of the Defects Notification Period is stated in the Appendix to Tender but may be extended under Sub-Clause 11.3. If the Employer considers that he is entitled to an extension of this period, then either the Employer or the Engineer must give notice to the Contractor. The notice must be given, under Sub-Clause 2.5 [*Employer's Claims*], as soon as practicable after the Employer became aware of the circumstances and before the expiry of the period. The Engineer will then make a determination under Sub-Clause 3.5 [*Determinations*].

Potentially controversial is the fact that the wording of the Sub-Clause does not explicitly state that the defect or damage must be attributable to the Contractor as contemplated under Sub-Clause 11.2 [*Cost of Remedying Defects*]. Under certain legal jurisdictions this may make the Sub-Clause void or unenforceable. It is therefore recommended to amend this wording similar to the PNK Book through the Particular Conditions by changing the latter part of the first sentence in the first paragraph into '…by reason of a defect or by reason of damage attributable to the Contractor.'

To establish an entitlement to an extension, the Employer must thus prove that the whole or a Section of the Works or a major item of Plant could not be used for the purpose for which it was intended due to a defect or a damage attributable to the Contractor. The Defects Notification Period for the whole, or the Section of the Works, or the major item of Plant, could then be extended for an appropriate period, which must not exceed two years.

If the delivery and/or erection of Plant and/or Materials were suspended under Sub-Clause 8.8 [*Suspension of Work*] or Sub-Clause 16.1 [*Contractor's Entitlement to Suspend Work*] then the calendar dates of the Defects Notification Period will be delayed. However, under Sub-Clause 11.3, the Contractor's obligation to repair defects or damage shall not apply to any defect or damage that occurs more than two years after the Defects Notification Period for that Plant and/or Materials would have expired. This limitation presumably takes precedence over the other requirements in this Clause and refers specifically to problems with Plant and/or Materials.

11.4 Failure to Remedy Defects

If the Contractor fails to remedy any defect or damage within a reasonable time, a date may be fixed by (or on behalf of) the Employer, on or by which the defect or damage is to be remedied. The Contractor shall be given reasonable notice of this date.

If the Contractor fails to remedy the defect or damage by this notified date and this remedial work was to be executed at the cost of the Contractor under Sub-Clause 11.2 [Cost of Remedying Defects], the Employer may (at his option):

(a) *carry out the work himself or by others, in a reasonable manner and at the Contractor's cost, but the Contractor shall have no responsibility for this work; and the Contractor shall subject to Sub-Clause 2.5 [Employer's Claims] pay to the Employer the costs reasonably incurred by the Employer in remedying the defect or damage;*

(b) *require the Engineer to agree or determine a reasonable reduction in the Contract Price in accordance with Sub-Clause 3.5 [Determinations]; or*

(c) *if the defect or damage deprives the Employer of substantially the whole benefit of the Works or any major part of the Works, terminate the Contract as a whole, or in respect of such major part which cannot be put to the intended use. Without prejudice to any other rights, under the Contract or otherwise, the Employer shall then be entitled to recover all sums paid for the Works or for such part (as the case may be), plus financing costs and the cost of dismantling the same, clearing the Site and returning Plant and Materials to the Contractor.*

Before taking any action the Employer's preference must be discussed with the Engineer and Contractor, in an attempt to reach agreement on the proposed action, and any appropriate notices given to the Contractor. The Employer may do the following.

■ Make other arrangements to carry out the work and claim the Cost against the Contractor, under Sub-Clause 2.5 [*Employer's Claims*].

■ Accept the work including the defect and reduce the Contract price, under the procedures of Sub-Clause 3.5 [*Determinations*] (this is a practical procedure and can be used when the remedial work would cause substantial inconvenience or damage and the Employer would prefer to accept the out of specification work).

■ Terminate the Contract as a whole or in respect of the relevant part of the Works. This would obviously be a very serious action and would result in the Employer claiming substantial sums of money from the Contractor. The termination procedures at Clause 15 would apply, with the additional requirements of Sub-Clause 11.4(c).

As with the similar provision at Sub-Clause 9.4 [*Failure to Pass Tests on Completion*], it is difficult to envisage circumstances where the provisions at subparagraph (c) would be appropriate under a RED Book Contract.

11.5 Removal of Defective Work
If the defect or damage cannot be remedied expeditiously on the Site and the Employer gives consent, the Contractor may remove from the Site for the purposes of repair such items of Plant as are defective or damaged.

This consent may require the Contractor to increase the amount of the Performance Security by the full replacement cost of these items, or to provide other appropriate security.

The Employer must give his consent before any item of Plant is removed from the Site for repair. Before giving consent the Employer will want to know how long the repair will take and what action is proposed to enable the Works to be used during its absence. The situation envisaged at Sub-Clause 11.5 could result in an extension to the Defects Notification Period under Sub-Clause 11.3 [*Extension of Defects Notification Period*].

11.6 Further Tests
If the work of remedying of any defect or damage may affect the performance of the Works, the Engineer may require the repetition of any of the tests described in the Contract. The requirement shall be made by notice within 28 days after the defect or damage is remedied.

These tests shall be carried out in accordance with the terms applicable to the previous tests, except that they shall be carried out at the risk and cost of the Party liable, under Sub-Clause 11.2 [Cost of Remedying Defects], for the cost of the remedial work.

It is logical that if work that has previously been subject to tests has been repaired because of a defect, then the original tests should be repeated. However, the details of the tests may need to be modified to suit the different circumstances now that the Works have been occupied by the Employer. Any modification would need to be agreed by the Parties.

11.7 Right of Access
Until the Performance Certificate has been issued, the Contractor shall have such right of access to the Works as is reasonably required in order to comply with this Clause, except as may be inconsistent with the Employer's reasonable security restrictions.

Obviously the Contractor needs access to the Site in order to carry out his obligations. Times and arrangements must be agreed, to suit both the Employer and the Contractor. The Contractor may also need office and storage facilities on Site in order to meet his obligations during the Defects Notification Period. The details must be agreed between the Contractor and the Engineer. The removal of these facilities and any reinstatement would be covered by Sub-Clause 10.4 [*Surfaces Requiring Reinstatement*] and Sub-Clause 11.11 [*Clearance of Site*] concerning surfaces requiring reinstatement and clearance of the Site.

11.8 Contractor to Search

The Contractor shall, if required by the Engineer, search for the cause of any defect, under the direction of the Engineer. Unless the defect is to be remedied at the cost of the Contractor under Sub-Clause 11.2 [Cost of Remedying Defects], the Cost of the search plus reasonable profit shall be agreed or determined by the Engineer in accordance with Sub-Clause 3.5 [Determinations] and shall be included in the Contract Price.

After the Employer has notified the Contractor of the defect, the Engineer will make a technical assessment of the problem. Sub-Clause 11.8 enables the Engineer to give instructions to the Contractor to investigate the problem. This Sub-Clause entitles the Contractor to additional payment, but does not refer to giving notice under Sub-Clause 20.1 [*Contractor's Claims*]. However, any such payment could be disputed and so the procedures of Sub-Clause 20.1 should be followed.

11.9 Performance Certificate

Performance of the Contractor's obligations shall not be considered to have been completed until the Engineer has issued the Performance Certificate to the Contractor, stating the date on which the Contractor completed his obligations under the Contract.

The Engineer shall issue the Performance Certificate within 28 days after the latest of the expiry dates of the Defects Notification Periods, or as soon thereafter as the Contractor has supplied all the Contractor's Documents and completed and tested all the Works, including remedying any defects. A copy of the Performance Certificate shall be issued to the Employer.

Only the Performance Certificate shall be deemed to constitute acceptance of the Works.

The Performance Certificate is issued when the Engineer is satisfied that the Contractor has fulfilled his obligations during the Defects Notification Period. The delay of 28 days from the end of the Defects Notification Period allows time for a joint inspection of the Works and for the Contractor to complete any outstanding work.

The final sentence of Sub-Clause 11.9 confirms the Contract requirements, such as Sub-Clause 3.2(a) [*Delegation of the Engineer*], which stipulate that any failure to notice and report, at the time, any item of work that does not comply with the Specifications does not mean that the work has been accepted.

The receipt of the Performance Certificate starts the 56-day period for the Contractor to submit a draft final account that includes all outstanding claims.

11.10 Unfulfilled Obligations

After the Performance Certificate has been issued, each Party shall remain liable for the fulfilment of any obligation which remains unperformed at that time. For the purposes of determining the nature and extent of unperformed obligations, the Contract shall be deemed to remain in force.

Sub-Clause 11.9 requires that the Performance Certificate state the date on which the Contractor completed his obligations under the Contract.

This is clearly inconsistent with the reference at Sub-Clause 11.10 to '*Unfulfilled Obligations*'. However, there are some obligations that cannot be carried out until after the issue of the Performance Certificate.

- Clearance of the Site under Sub-Clause 11.11 and reinstatement under Sub-Clause 10.4 [*Surfaces Requiring Reinstatement*].
- The application for Final Payment Certificate under Sub-Clause 14.11 [*Application for Final Payment Certificate*].
- Investigation and correction of any latent defects which are not noticed until after the issue of the Performance Certificate, but for which the Contractor may have a liability under the applicable law.

11.11 Clearance of Site
Upon receiving the Performance Certificate, the Contractor shall remove any remaining Contractor's Equipment, surplus material, wreckage, rubbish and Temporary Works from the Site.

If all these items have not been removed within 28 days after the Employer receives a copy of the Performance Certificate, the Employer may sell or otherwise dispose of any remaining items. The Employer shall be entitled to be paid the costs incurred in connection with, or attributable to, such sale or disposal and restoring the Site.

Any balance of the moneys from the sale shall be paid to the Contractor. If these moneys are less than the Employer's costs, the Contractor shall pay the outstanding balance to the Employer.

The time periods for actions under Sub-Clause 11.11 start from the receipt of the Performance Certificate by the Employer. This is changed in the PNK Book to receipt by the Contractor which is more logical. The Engineer must record the dates of receipt. The requirements for the Contractor's final Site clearance may also be the subject of regulations under the governing law.

FIDIC Users' Guide
ISBN 978-0-7277-5856-9

ICE Publishing: All rights reserved
http://dx.doi.org/10.1680/fug.58569.197

Chapter 18
Clause 12: Measurement and Evaluation

Clause 12 in the RED Book is written for a re-measurement contract in which the Accepted Contract Amount is based on estimated quantities but the Contractor is obliged to carry out all the work which is required by the Specification and Drawings and is paid for the actual quantities of work which he has executed. The Clause covers the procedures for the measurement and evaluation of the Works that have been executed, or have been omitted by a Variation.

It is clear that the quantities in the Bill of Quantities are only estimates. Hence, any changes due to re-measurement are not Variations, but are part of the original obligations of both Parties. If the Employer, for his own internal accounting purposes, requires the additional Cost due to changes in quantities to be confirmed by a Variation order then this is just an administrative procedure and does not indicate that the change is a Variation.

If the Contract is based on a lump sum, or is on a cost-plus or other basis, then Clause 12 must be omitted from the General Conditions and alternative arrangements included in the Particular Conditions. Sub-Clause 14.1 [*The Contract Price*] of the FIDIC *Guidance for the Preparation of Particular Conditions* includes recommendations for Contracts on a cost-plus or lump sum basis. If the RED Book Contract includes some lump sum items then these must be paid from a Schedule of Payments as Sub-Clause 14.4 [*Schedule of Payments*].

12.1 Works to be Measured
The Works shall be measured, and valued for payment, in accordance with this Clause.

Whenever the Engineer requires any part of the Works to be measured, reasonable notice shall be given to the Contractor's Representative, who shall:

(a) promptly either attend or send another qualified representative to assist the Engineer in making the measurement; and
(b) supply any particulars requested by the Engineer.

If the Contractor fails to attend or send a representative, the measurement made by (or on behalf of) the Engineer shall be accepted as accurate.

Except as otherwise stated in the Contract, wherever any Permanent Works are to be measured from records, these shall be prepared by the Engineer. The Contractor shall, as and when requested, attend to examine and agree the records with the Engineer, and shall sign the same when agreed. If the Contractor does not attend, the records shall be accepted as accurate.

If the Contractor examines and disagrees the records, and/or does not sign them as agreed, then the Contractor shall give notice to the Engineer of the respects in which the records are asserted to be inaccurate. After receiving this notice, the Engineer shall review the records and either confirm or vary them. If the Contractor does not so give notice to the Engineer within 14 days after being requested to examine the records, they shall be accepted as accurate.

The procedures for measurement of the Works are given at Sub-Clause 12.1 as follows.

- The Engineer decides that he requires a part of the Works to be measured and notifies the Contractor.
- The Contractor attends and assists the Engineer in making the measurement.
- Alternatively, the Contractor fails to attend and the Engineer's measurements are accepted as accurate.

If a part of the Works is to be measured from records of its construction then the details should be specified in the Tender documents. A similar procedure applies in that the Engineer prepares the records and the Contractor examines and agrees or disagrees with the records. For the latter, the wording is potentially contentious in that the Contractor shall examine and agree these records 'as and when requested'. This can be interpreted such that the Engineer has an option but not an obligation to request the Contractor to examine and agree the records prepared by him. It is recommended to amend the wording through the Particular Conditions by omitting the words 'as and when requested, attend to' from the second sentence in the fourth paragraph.

If the Engineer requires the Contractor to carry out the work for either measurement or the preparation of records then this should be stated in the Specification. In practice the Contractor often does play a greater part in this work than is required by the Contract. The Contractor is more likely to have the staff and equipment available and some Contractors prefer to make the measurements themselves, rather than assist the Engineer. The Engineer will then check and confirm the Contractor's measurement.

The Contract does not stipulate any fixed periods or timing for the measurement and this is a matter for the Engineer. The timing of measurement notifications will depend on the progress of the Works and the completion of convenient parts or items in the Bill of Quantities, although any work that is to be buried or covered up must be measured before it is buried or covered. The Contractor will have notified the Engineer under Sub-Clause 7.3 [*Inspection*] before the Work is buried or covered up.

The measurement must be completed in time for the Contractor to prepare and submit his Statement at Completion, as Sub-Clause 14.10 [*Statement at Completion*], but does not have to be related to interim payments. The provision for interim payments, at Sub-Clause 14.3 [*Application for Interim Payment Certificates*], is based on '*the estimated value of the*

Works executed' and not on a final valuation. The estimated value may be based on an interim or approximate measurement and will be adjusted when the final measurement figures have been agreed. Nevertheless it is good practice for the Contractor and the Engineer to agree on some form of measurement for each Interim Certificate, especially as under Sub-Clause 14.6 [*Issue of Interim Payment Certificates*] the Engineer is required to make a fair determination of the amounts due for such interim payment application.

12.2 Method of Measurement

Except as otherwise stated in the Contract and notwithstanding local practice:

(a) *measurement shall be made of the net actual quantity of each item of the Permanent Works; and*

(b) *the method of measurement shall be in accordance with the Bill of Quantities or other applicable Schedules.*

The procedure for the actual measurement of the different work items can be standardised for different projects and for consistency within a project by the use of a published standard method of measurement. The FIDIC Conditions of Contract do not require the use of a standard method of measurement but Sub-Clause 12.2 states that the method of measurement will be in accordance with the Bill of Quantities '*or other applicable Schedules*'. If a standard method of measurement, such as the *Civil Engineering Standard Method of Measurement* published by The Institution of Civil Engineers, is to be used then this requirement should be stated in the Particular Conditions.

Alternatively, the Bill of Quantities should include a detailed explanation of the method of measurement that will be used. For example, the phrase '*net actual quantity*' needs to be explained. It will not mean the actual quantity that has been executed by the Contractor. Particularly for work below ground level, the Contractor may have provided additional concrete, or carried out additional excavation, to suit his method of working. Net actual quantity would then mean the minimum quantities that would be required in order to provide the Permanent Works.

12.3 Evaluation

Except as otherwise stated in the Contract, the Engineer shall proceed in accordance with Sub-Clause 3.5 [Determinations] to agree or determine the Contract Price by evaluating each item of work, applying the measurement agreed or determined in accordance with the above Sub-Clauses 12.1 and 12.2 and the appropriate rate or price for the item.

For each item of work, the appropriate rate or price for the item shall be the rate or price specified for such item in the Contractor, if there is no such item, specified for similar work. However, a new rate or price shall be appropriate for an item of work if:

(a) (i) *the measured quantity of the item is changed by more than 10% from the quantity of this item in the Bill of Quantities or other Schedule,*
(ii) *this change in quantity multiplied by such specified rate for this item exceeds 0.01% of the Accepted Contract Amount,*

(iii) this change in quantity directly changes the Cost per unit quantity of this item by more than 1%, and

(iv) this item is not specified in the Contract as a 'fixed rate item'; or

(b) (i) the work is instructed under Clause 13 [Variations and Adjustments],

(ii) no rate or price is specified in the Contract for this item, and

(iii) no specified rate or price is appropriate because the item of work is not of similar character, or is not executed under similar conditions, as any item in the Contract.

Each new rate or price shall be derived from any relevant rates or prices in the Contract, with reasonable adjustments to take account of the matters described in sub-paragraph (a) and/or (b), as applicable. If no rates or prices are relevant for the derivation of a new rate or price, it shall be derived from the reasonable Cost of executing the work, together with reasonable profit, taking account of any other relevant matters.

Until such time as an appropriate rate or price is agreed or determined, the Engineer shall determine a provisional rate or price for the purposes of interim Payment Certificates.

Sub-Clause 12.3 requires the Engineer to agree or determine the Contract Price by applying the measurement and the appropriate rate or price for each item. The appropriate rate or price is stated to be the Contract rate or price for the item, or for similar work, unless

■ the measured quantity has changed by more than the amounts stated at Sub-Clause 12.3(a) and the item is not specified as a 'fixed rate item', or

■ the work is a Variation under Clause 13 and there is no appropriate rate or price in the Contract.

The percentage figures given at subparagraphs (a) (i) and (ii) are probably too low and will result in a requirement to discuss an excessive number of new rates, which may be virtually unchanged from the original rates. The PNK Book, which is discussed later in this book, has used 25% and 0.25%. Item (a) (iii) refers to changes of Cost, which is defined as expenditure reasonably incurred by the Contractor for the particular item. This is highly contentious and has been the source of many disputes over the years. As the Contractor is the only Party that knows his costs accurately there is an incentive to be subjective. However, any evaluation or new rate or price will be determined by the Engineer in accordance with the procedures in Sub-Clause 3.5 [*Determinations*]. That is, the Engineer will consult with both Parties and try to reach agreement. Failing agreement the Engineer will make a fair determination. Either Party has the right to invoke the Clause 20 dispute procedures if it is not satisfied with the Engineer's determination.

12.4 Omissions
Whenever the omission of any work forms part (or all) of a Variation, the value of which has not been agreed, if:

(a) the Contractor will incur (or has incurred) cost which, if the work had not been omitted, would have been deemed to be covered by a sum forming part of the Accepted Contract Amount;

(b) the omission of the work will result (or has resulted) in this sum not forming part of the Contract Price; and

(c) this cost is not deemed to be included in the evaluation of any substituted work;

then the Contractor shall give notice to the Engineer accordingly, with supporting particulars. Upon receiving this notice, the Engineer shall proceed in accordance with Sub-Clause 3.5 [Determinations] to agree or determine this cost, which shall be included in the Contract Price.

When any work is omitted it is likely that the Contractor will already have incurred some expenditure in the preparation, provision of Goods and ordering Materials, if not in the actual execution of work on the Site. This Sub-Clause requires the Engineer to make a fair assessment of any such Cost.

Interestingly the notice provision is not subject to Sub-Clause 20.1 [Contractor's Claims] thus arguably removing the time-bar. The Contractor is seemingly only obligated to submit the notice in writing (refer Sub-Clause 1.3 [Communications]). However, as this omission forms part of a Variation, its evaluation must be in accordance with the last paragraph of Sub-Clause 13.3 [Variation Procedure] that refers to Clause 12 (see Figure 5.13 in Chapter 5).

FIDIC Users' Guide
ISBN 978-0-7277-5856-9

ICE Publishing: All rights reserved
http://dx.doi.org/10.1680/fug.58569.203

Chapter 19
Clause 13: Variations and Adjustments

Clause 13 covers the procedures for work being added, omitted, or changed from the original contract work, together with other matters that may increase or decrease the Contract Price. Under Sub-Clause 13.3 [*Variation Procedure*] the Contractor must acknowledge receipt of any Engineer's instruction for a Variation.

In any construction project there will be a need to change the initial requirements as the construction proceeds on the Site. This may be a matter of the Employer changing his mind about some requirement, or the Engineer may need to issue further information which involves changes to the initial requirements, or it may be necessary to correct a mistake in the information issued to the Contractor. An instruction can be given to change an item of the Works that has been completed and the Cost of making the change will be included in the valuation of the Variation. However, instructions for changes cannot be given after the issue of the Taking-Over Certificate for the Works.

The FIDIC Conditions allow the Engineer, but not the Employer, to issue an instruction to change the Works. The Contractor is not permitted to change the Permanent Works unless the Engineer has instructed or approved the Variation. If the Employer wants to make any changes he must request that the Engineer issues an instruction. If the Employer gives an instruction direct to the Contractor then the Contractor must obtain the Engineer's confirmation and instruction before he executes the change. Strict adherence to these requirements is essential for the Engineer to maintain technical and financial control over the project.

Sub-Clause 3.3 gives the Engineer the power to issue instructions that may, or may not, constitute a Variation under Clause 13. If the Contractor considers that an instruction constitutes a Variation then he should confirm the instruction as a Variation, by acknowledging receipt and stating that it is as required by Sub-Clause 13.3. If the Engineer disagrees then the Contractor must still comply with the instruction but can follow the appropriate claim procedures under Sub-Clause 20.1 [*Contractor's Claims*].

13.1 Right to Vary
Variations may be initiated by the Engineer at any time prior to issuing the Taking-Over Certificate for the Works, either by an instruction or by a request for the Contractor to submit a proposal.

The Contractor shall execute and be bound by each Variation, unless the Contractor promptly gives notice to the Engineer stating (with supporting particulars) that the

Contractor cannot readily obtain the Goods required for the Variation. Upon receiving this notice, the Engineer shall cancel, confirm or vary the instruction.

Each Variation may include:

(a) changes to the quantities of any item of work included in the Contract (however, such changes do not necessarily constitute a Variation);

(b) changes to the quality and other characteristics of any item of work;

(c) changes to the levels, positions and/or dimensions of any part of the Works;

(d) omission of any work unless it is to be carried out by others;

(e) any additional work, Plant, Materials or services necessary for the Permanent Works, including any associated Tests on Completion, boreholes and other testing and exploratory work; or

(f) changes to the sequence or timing of the execution of the Works.

The Contractor shall not make any alteration and/or modification of the Permanent Works, unless and until the Engineer instructs or approves a Variation.

Under the RED Book, the Engineer can issue instructions to change a wide range of matters concerning the Works, as listed at Sub-Clause 13.1(a) to (f). However, these are all matters that concern the Permanent Works that have been defined in the Contract. The Engineer does not have authority to change the Contract. Nevertheless, if an Engineer gives an instructed Variation of any kind, even one that is perceived to change the Contract, the Contractor must comply due to the provision of the second paragraph of Sub-Clause 13.1. In this case, the Contractor must assume that the Engineer has the authority to make such an instruction in accordance with the fourth paragraph of Sub-Clause 3.1 [*Engineer's Duties and Authority*]. If the Contractor acts on any such instruction that is later found to be given without authority then the Contractor may claim under Sub-Clause 20.1 [*Contractor's Claims*].

The Engineer cannot issue instructions under this Clause for additional work unless it is '*necessary for the Permanent Works*', as Sub-Clause 13.1(e). However, the words necessary for the Permanent Works are highly contentious. If the Employer wants the Contractor to carry out work that changes the scope of the Works then it may be necessary to negotiate a change to the Contract.

The Contractor can object that he cannot readily obtain the Goods: Equipment, Plant, Materials or Temporary Works; and the Engineer can either cancel, confirm or vary the instruction. However, the Contractor cannot object to the instruction because of difficulties in obtaining management staff or labour.

A significant change to previous FIDIC Conditions of Contracts is that the Engineer can initiate the Variation either by issuing an instruction or '*by a request for the Contractor to submit a proposal*' (refer Sub-Clause 13.3 [*Variation Procedure*]). A request for a proposal is not just a matter of submitting a price for a proposed Variation. The Engineer may ask for a detailed technical proposal, together with details of the impact on the programme and on other aspects of the Works.

Where, under a RED Book Contract, the Contractor is responsible for a significant part of the design, consideration should be given to restricting the Engineer's power to vary that part of the Works, as stipulated in the YLW Book.

The Engineer can then either cancel, confirm or vary the instruction. If the Engineer confirms an instruction following objections by the Contractor then the Employer is taking responsibility for the potential problems that have been notified by the Contractor. If the Variation has design or safety implications then it could also bring additional responsibilities to both the Employer and the Engineer under the applicable law.

If, for any reason the Contractor is of the opinion that an instruction for a Variation is required, but the Engineer is not doing so, he should give notice of this fact to the Engineer under Sub-Clause 1.9 [*Delayed Drawings or Instructions*], which in effect will require the Engineer to act in accordance with Sub-Clause 3.5 [*Determinations*].

13.2 Value Engineering

The Contractor may, at any time, submit to the Engineer a written proposal which (in the Contractor's opinion) will, if adopted, (i) accelerate completion, (ii) reduce the cost to the Employer of executing, maintaining or operating the Works, (iii) improve the efficiency or value to the Employer of the completed Works, or (iv) otherwise be of benefit to the Employer.

The proposal shall be prepared at the cost of the Contractor and shall include the items listed in Sub-Clause 13.3 [Variation Procedure].

If a proposal, which is approved by the Engineer, includes a change in the design of part of the Permanent Works, then unless otherwise agreed by both Parties:

(a) the Contractor shall design this part;
(b) subparagraphs (a) to (d) of Sub-Clause 4.1 [Contractor's General Obligations] shall apply; and
(c) if this change results in a reduction in the contract value of this part, the Engineer shall proceed in accordance with Sub-Clause 3.5 [Determinations] to agree or determine a fee, which shall be included in the Contract Price. This fee shall be half (50%) of the difference between the following amounts:
  (i) such reduction in contract value, resulting from the change, excluding adjustments under Sub-Clause 13.7 [Adjustments for Changes in Legislation] and Sub-Clause 13.8 [Adjustments for Changes in Cost], and
  (ii) the reduction (if any) in the value to the Employer of the varied works, taking account of any reductions in quality, anticipated life or operational efficiencies.

However, if amount (i) is less than amount (ii), there shall not be a fee.

The provision for value engineering is another significant addition to previous FIDIC Contracts. This provision enables the Employer to benefit from the Contractor's experience and proposals. However, the procedures that have been introduced may not encourage Contractors to put forward proposals.

If the Contractor wishes to submit a proposal that will benefit the Employer in any of the ways listed in the first paragraph of Sub-Clause 13.2 then he must prepare, at his own cost, a proposal that includes the information which is required for a Variation proposal under Sub-Clause 13.3 [*Variation Procedure*].

The third paragraph of Sub-Clause 13.2 draws a distinction between proposals that include a change in the design of part of the Permanent Works and those that do not include a design change. If the proposal includes a design change then, unless the Parties agree otherwise, the Contractor shall design the changed part of the Works and receive a fee, provided that the reduction in Contract value is greater than any reduction in the value of the Works to the Employer. The fee is calculated by the Engineer, following the Sub-Clause 3.5 [*Determinations*] procedure. If the proposal does not include a design change then the Contractor will have to rely on any potential benefit from savings in his own costs.

The preparation of a value engineering proposal could involve the Contractor in substantial costs which may, or may not, be recovered thus significantly reducing the incentive for going through all the trouble. Any proposal should be discussed and agreed in principle before the Contractor incurs costs that may not be recoverable.

13.3 Variation Procedure
If the Engineer requests a proposal, prior to instructing a Variation, the Contractor shall respond in writing as soon as practicable, either by giving reasons why he cannot comply (if this is the case) or by submitting:

(a) a description of the proposed work to be performed and a programme for its execution;
(b) the Contractor's proposal for any necessary modifications to the programme according to Sub-Clause 8.3 [Programme] and to the Time for Completion; and
(c) the Contractor's proposal for evaluation of the Variation.

The Engineer shall, as soon as practicable after receiving such proposal (under Sub-Clause 13.2 [Value Engineering] or otherwise), respond with approval, disapproval or comments. The Contractor shall not delay any work while awaiting a response.

Each instruction to execute a Variation, with any requirements for the recording of Costs, shall be issued by the Engineer to the Contractor, who shall acknowledge receipt.

Each Variation shall be evaluated in accordance with Clause 12 [Measurement and Evaluation], unless the Engineer instructs or approves otherwise in accordance with this Clause.

When the Engineer is considering whether to issue instructions for a Variation, he has the option of first asking the Contractor for a proposal. The Contractor can either give the reasons why he cannot comply or provide '*a description of the proposed work to be performed*', modifications to the programme and date for Completion and his proposal for the evaluation of the Variation. There is no time limit on the provision of this proposal, only that it must be submitted '*as soon as practicable*'.

The purpose of this Sub-Clause is apparently to reduce potential disputes by encouraging agreement of the consequences of Variations before the instruction is issued. If the Contractor considers that to prepare a proposal is likely to be time consuming, or the time and cost implications are uncertain, then he may either find a reason not to submit a proposal, or give high estimates for the programme modification and evaluation. The proposal could then be subject to discussion before the Engineer decides to issue the Variation.

The Sub-Clause 13.3 proposal is not a firm offer to carry out the Variation for a fixed price and time extension. Payment will be made by measurement in accordance with Clause 12, unless the Engineer decides otherwise. The Engineer can ask for records of Costs. Any extension of time would be determined under the procedures at Sub-Clause 8.4 [*Extension of Time for Completion*].

Sub-Clause 13.3, unlike Sub-Clause 13.2 [*Value Engineering*], does not state that the proposal shall be made at the Contractor's Cost, so Costs could presumably be claimed, or would be covered in the valuation of the Variation.

If a Variation affects the work which is currently being executed by the Contractor then the Engineer would need to issue the instructions immediately, without waiting for a proposal. The Contractor must not delay any work during this process, which could result in a need for the Variation to include the removal of completed work. The Cost of any disruption caused by the Variation must also be considered, possibly requiring a separate claim under Sub-Clause 20.1 [*Contractor's Claims*].

Variations are to be valued in accordance with Clause 12. Omissions will be valued in accordance with Sub-Clause 12.4 [*Omissions*].

13.4 Payment in Applicable Currencies
If the Contract provides for payment of the Contract Price in more than one currency, then whenever an adjustment is agreed, approved or determined as stated above, the amount payable in each of the applicable currencies shall be specified. For this purpose, reference shall be made to the actual or expected currency proportions of the Cost of the varied work, and to the proportions of various currencies specified for payment of the Contract Price.

This Sub-Clause requires the currency provisions in the Contract to be carried through into the evaluation of a Variation. It does not address the problem of changes in exchange rates between the Base Date and the Variation date, but it would presumably give the opportunity for the Contractor to raise any problems for discussion, or if he deems it necessary to submit a claim subject to Sub-Clause 20.1 [*Contractor's Claims*].

13.5 Provisional Sums
Each Provisional Sum shall only be used, in whole or in part, in accordance with the Engineer's instructions, and the Contract Price shall be adjusted accordingly. The total sum paid to the Contractor shall include only such amounts, for the work, supplies or services to which the Provisional Sum relates, as the Engineer shall have instructed. For each Provisional Sum, the Engineer may instruct:

(a) work to be executed (including Plant, Materials or services to be supplied) by the Contractor and valued under Sub-Clause 13.3 [Variation Procedure]; and/or

(b) Plant, Materials or services to be purchased by the Contractor, from a nominated Subcontractor (as defined in Clause 5 [Nominated Subcontractors]) or otherwise; and for which there shall be included in the Contract Price:

    (i) the actual amounts paid (or due to be paid) by the Contractor, and

    (ii) a sum for overhead charges and profit, calculated as a percentage of these actual amounts by applying the relevant percentage rate (if any) stated in the appropriate Schedule. If there is no such rate, the percentage rate stated in the Appendix to Tender shall be applied.

The Contractor shall, when required by the Engineer, produce quotations, invoices, vouchers and accounts or receipts in substantiation.

A Provisional Sum is defined as a sum of money, which is included in the Contract Price, but has been allocated for a particular part of the Works, or the supply of Plant, Materials or services, as specified in the Contract. The money can only be used when instructed by the Engineer. The instruction may take the form of a Variation for additional work, which is then valued in accordance with Sub-Clause 13.3 [*Variation Procedure*]. Alternatively, it may be an instruction for the purchase of Plant, Materials or services. This would be valued as the actual amount paid, based on invoices, plus a percentage for overheads and profit. The percentage must be stated in the appropriate schedule or alternatively the Appendix to Tender. The Engineer has the option of stating the source to be used, which becomes a nominated Subcontractor. The Provisional Sum can only be used for the specified purpose. Any money surplus to these requirements cannot be used for other work and will result in a reduction in the Contract Price. If a sum of money that is included as a Provisional Sum is actually intended to be a contingency sum then this must be clearly defined in the Particular Conditions.

13.6 Daywork

For work of a minor or incidental nature, the Engineer may instruct that a Variation shall be executed on a daywork basis. The work shall then be valued in accordance with the Daywork Schedule included in the Contract, and the following procedure shall apply. If a Daywork Schedule is not included in the Contract, this Sub-Clause shall not apply.

Before ordering Goods for the work, the Contractor shall submit quotations to the Engineer. When applying for payment, the Contractor shall submit invoices, vouchers and accounts or receipts for any Goods.

Except for any items for which the Daywork Schedule specifies that payment is not due, the Contractor shall deliver each day to the Engineer accurate statements in duplicate which shall include the following details of the resources used in executing the previous day's work:

(a) the names, occupations and time of Contractor's Personnel;

(b) the identification, type and time of Contractor's Equipment and Temporary Works; and

(c) the quantities and types of Plant and Materials used.

One copy of each statement will, if correct, or when agreed, be signed by the Engineer and returned to the Contractor. The Contractor shall then submit priced statements of these resources to the Engineer, prior to their inclusion in the next Statement under Sub-Clause 14.3 [Application for Interim Payment Certificates].

Payment on a daywork basis means that the Contractor submits daily records of the resources used in the execution of the previous day's work, including the information that is listed at Sub-Clause 13.6, and is paid at the rates stated in the Daywork Schedule. If there is no Daywork Schedule in the Contract, then in principle, payment cannot be made on a daywork basis, unless the Parties agree on such a schedule retrospectively.

It is therefore recommended that a Daywork Schedule is included in the Tender documents, completed by the Contractor and included with the Letter of Tender and Letter of Acceptance so as to form part of the Contract.

Sub-Clause 13.6 restricts the use of Daywork to work '*of a minor or incidental nature*'. Minor or incidental work generally creates more disruption than is normal for the total Cost of the work, so daywork rates are usually high.

If work is executed on a daywork basis the Contractor should expect the Engineer to give instructions on the quantities and types of Plant and Materials to be used and to report accordingly. If no such instruction is received the Contractor must expect the Engineer to request justification of the quantities and types of Plant and Material used.

13.7 Adjustments for Changes in Legislation
The Contract Price shall be adjusted to take account of any increase or decrease in Cost resulting from a change in the Laws of the Country (including the introduction of new Laws and the repeal or modification of existing Laws) or in the judicial or official governmental interpretation of such Laws, made after the Base Date, which affect the Contractor in the performance of obligations under the Contract.

If the Contractor suffers (or will suffer) delay and/or incurs (or will incur) additional Cost as a result of these changes in the Laws or in such interpretations, made after the Base Date, the Contractor shall give notice to the Engineer and shall be entitled subject to Sub-Clause 20.1 [Contractor's Claims] to:

(a) an extension of time for any such delay, if completion is or will be delayed, under Sub-Clause 8.4 [Extension of Time for Completion]; and
(b) payment of any such Cost, which shall be included in the Contract Price.

After receiving this notice, the Engineer shall proceed in accordance with Sub-Clause 3.5 [Determinations] to agree or determine these matters.

Sub-Clause 13.7 refers to increases or decreases in Cost due to changes in legislation. The Sub-Clause only refers to changes in the Laws of the Country. If the governing law is different to the Laws of the Country, or the Contractor is carrying out work in a different country, as discussed in the commentary on Sub-Clause 1.4 [*Law and Language*], then the provisions of this Sub-Clause would not apply. If the Contractor suffers delay or additional Cost then he submits a claim under the procedures of Sub-Clause 20.1 [*Contractor's Claims*]. Similarly, if there were a possible decrease in Cost then the Employer would have to submit a claim under Sub-Clause 2.5 [*Employer's Claims*].

Changes in legislation will normally mean the introduction of new laws or regulations. The possibility of a change in the interpretation of the law introduces a more difficult subject with subjective opinions. Also, it is not unknown for a country's legal system to change, for new countries to be created or for country borders to move and for a site to effectively move into a different country. Any additional Costs would be claimed under Sub-Clause 20.1 and the legal basis of the claim discussed in relation to the particular circumstances.

Changes in legislation may affect several items of work in different ways and it is important to ensure that the additional payment does not duplicate other payments.

13.8 Adjustments for Changes in Cost

In this Sub-Clause, 'table of adjustment data' means the completed table of adjustment data included in the Appendix to Tender. If there is no such table of adjustment data, this Sub-Clause shall not apply.

If this Sub-Clause applies, the amounts payable to the Contractor shall be adjusted for rises or falls in the cost of labour, Goods and other inputs to the Works, by the addition or deduction of the amounts determined by the formulae prescribed in this Sub-Clause. To the extent that full compensation for any rise or fall in Costs is not covered by the provisions of this or other Clauses, the Accepted Contract Amount shall be deemed to have included amounts to cover the contingency of other rises and falls in costs.

The adjustment to be applied to the amount otherwise payable to the Contractor, as valued in accordance with the appropriate Schedule and certified in Payment Certificates, shall be determined from formulae for each of the currencies in which the Contract Price is payable. No adjustment is to be applied to work valued on the basis of Cost or current prices. The formulae shall be of the following general type:

$$Pn = a + b\,\frac{Ln}{Lo} + c\,\frac{En}{Eo} + d\,\frac{Mn}{Mo} + \cdots$$

where:

'Pn' is the adjustment multiplier to be applied to the estimated contract value in the relevant currency of the work carried out in period 'n', this period being a month unless otherwise stated in the Appendix to Tender;

'a' is a fixed coefficient, stated in the relevant table of adjustment data, representing the non-adjustable portion in contractual payments; 'b', 'c', 'd', ... are coefficients representing the estimated proportion of each cost element related to the execution of the Works, as stated

in the relevant table of adjustment data; such tabulated cost elements may be indicative of resources such as labour, equipment and materials;

'Ln', 'En', 'Mn', ... are the current cost indices or reference prices for period 'n', expressed in the relevant currency of payment, each of which is applicable to the relevant tabulated cost element on the date 49 days prior to the last day of the period (to which the particular Payment Certificate relates); and 'Lo', 'Eo', 'Mo', ... are the base cost indices or reference prices, expressed in the relevant currency of payment, each of which is applicable to the relevant tabulated cost element on the Base Date.

The cost indices or reference prices stated in the table of adjustment data shall be used. If their source is in doubt, it shall be determined by the Engineer. For this purpose, reference shall be made to the values of the indices at stated dates (quoted in the fourth and fifth columns respectively of the table) for the purposes of clarification of the source; although these dates (and thus these values) may not correspond to the base cost indices.

In cases where the 'currency of index' (stated in the table) is not the relevant currency of payment, each index shall be converted into the relevant currency of payment at the selling rate, established by the central bank of the Country, of this relevant currency on the above date for which the index is required to be applicable.

Until such time as each current cost index is available, the Engineer shall determine a provisional index for the issue of Interim Payment Certificates. When a current cost index is available, the adjustment shall be recalculated accordingly.

If the Contractor fails to complete the Works within the Time for Completion, adjustment of prices thereafter shall be made using either (i) each index or price applicable on the date 49 days prior to the expiry of the Time for Completion of the Works, or (ii) the current index or price: whichever is more favourable to the Employer.

The weightings (coefficients) for each of the factors of cost stated in the table(s) of adjustment data shall only be adjusted if they have been rendered unreasonable, unbalanced or inapplicable, as a result of Variations.

If the Employer wishes to include provision to reimburse the Contractor for changes in the Cost of labour, equipment, Materials, or other items then the table in the Appendix to Tender must be completed in accordance with the provisions of Sub-Clause 13.8.

To decide on the various coefficients to be included in the adjustment formula, together with the other information that is required, is not a simple matter and will depend on the information that is available for the Country, as well as on the details of the project. Similarly, the administration and evaluation of Cost adjustments will be time consuming for both the Contractor and the Engineer and could lead to disputes as to the correct evaluation.

Employers who currently use different procedures for evaluation of inflation costs may prefer to continue to use their own procedures and make the appropriate provisions in the Particular Conditions.

International circumstances may result in a significant increase in the price of a single material, which is not reflected in the general price increases. This situation is not covered in the FIDIC Contracts, but may be covered in an '*unforeseen or exceptional circumstances*' provision in the governing law, or by a government regulation.

FIDIC Users' Guide
ISBN 978-0-7277-5856-9

ICE Publishing: All rights reserved
http://dx.doi.org/10.1680/fug.58569.213

Chapter 20
Clause 14: Contract Price and Payment

In the RED Book the Contract Price is defined at Sub-Clause 1.1.4.2 as '*the price defined in Sub-Clause 14.1*', which includes any adjustments which are provided for in the Contract. The Contract Price must be distinguished from the Accepted Contract Amount, which is defined at Sub-Clause 1.1.4.1 as '*the amount accepted in the Letter of Acceptance*'. The Accepted Contract Amount is fixed, but the Contract Price can change and will probably increase, due to the measurement of actual quantities, Variations and other adjustments. The Contract Agreement states that the Employer will pay the Contractor '*the Contract Price at the times and in the manner prescribed by the Contract*'.

The FIDIC *Guidance for the Preparation of Particular Conditions* includes the statement:

> '*When writing the Particular Conditions, consideration should be given to the amount and timing of payment(s) to the Contractor. A positive cash flow is clearly of benefit to the Contractor, and tenderers will take account of the interim payment procedures when preparing their tenders.*'

This statement is extremely important. The structure of the FIDIC payment and bonding procedures is such that the Contractor is unlikely to achieve a positive cash flow throughout the project, but the Employer's interests are covered by the bonds and the work that has been completed by the time he makes an interim payment. Hence, all tenderers will study the provisions for payment, including the advance payment, and may include an allowance for the Contractor having to pay interest on any negative cash flow. It is therefore important that consideration is given to the cash flow situation in both the preparation of the Particular Conditions and the administration of the Contract. The time periods for the measurement of work on Site, the submission of the Contractor's statement, issue of the Engineer's certificate and the payment to the Contractor are such that the work on Site will normally be several months further advanced by the time the Employer actually makes the payment. This fact should be taken into account when preparing the Contract documents, considering the percentage and limit of retention as Sub-Clause 14.3(c) [*Application for Interim Payment Certificates*] and calculating an interim payment.

14.1 The Contract Price
Unless otherwise stated in the Particular Conditions:

(a) *the Contract Price shall be agreed or determined under Sub-Clause 12.3 [Evaluation] and be subject to adjustments in accordance with the Contract;*

(b) the Contractor shall pay all taxes, duties and fees required to be paid by him under the Contract, and the Contract Price shall not be adjusted for any of these costs except as stated in Sub-Clause 13.7 [Adjustments for Changes in Legislation];

(c) any quantities which may be set out in the Bill of Quantities or other Schedule are estimated quantities and are not to be taken as the actual and correct quantities:

(i) of the Works which the Contractor is required to execute, or

(ii) for the purposes of Clause 12 [Measurement and Evaluation]; and

(d) the Contractor shall submit to the Engineer, within 28 days after the Commencement Date, a proposed breakdown of each lump sum price in the Schedules. The Engineer may take account of the breakdown when preparing Payment Certificates, but shall not be bound by it.

Sub-Clause 14.1 gives the following four general requirements, which relate to other Sub-Clauses.

- The Contract Price, which is stated in the Contract Agreement to be the amount which the Employer will pay to the Contractor, is agreed or determined by the Engineer. The Engineer will determine the Contract Price in accordance with Sub-Clause 12.3 [*Evaluation*], as the sum of the valuations of each item of work, including any Variations, together with other adjustments.
  If payment is to be made on a lump sum basis then the *Guidance for the Preparation of Particular Conditions* gives an alternative wording:
  '*(a) the Contract Price shall be the lump sum Accepted Contract Amount and be subject to adjustments in accordance with the Contract.*'
  For a lump sum contract it is essential that full details of the Employer's requirements are given in the Tender documents to enable the Contractor to submit a realistic Tender. Any uncertainties in the Tender documents, or additional information issued by the Engineer, are likely to result in claims.
  The Guidance suggests that Clause 12 and the last sentence of Sub-Clause 13.3 [*Variation Procedure*] should be deleted for a lump sum Contract. Variations should be valued by the Engineer using the procedures of Sub-Clause 3.5 [*Determinations*]. The Sub-Clause 13.3 procedure for the Contractor to submit a proposal before a Variation is issued would seem to be essential if disputes are to be avoided. Sub-Clause 14.1(b), (c) and (d) may also need to be revised for a lump sum Contract.
- Sub-Clause 1.13(b) [*Compliance with Laws*] requires the Contractor to pay all taxes, duties and fees. The Employer only reimburses any of these that are the result of changes in legislation as per Sub-Clause 13.7 [*Adjustments for Changes in Legislation*]. If the Employer intends to reimburse any other charges they must be defined in the Particular Conditions. The *Guidance for the Preparation of Particular Conditions* includes suggested Sub-Clauses if Sub-Clause 14.1(b) is not to apply. These Sub-Clauses, in the Particular Conditions, as per Sub-Clause 1.5 [*Priority of Documents*] would have priority over the relevant part of Sub-Clause 1.13(b) [*Compliance with Laws*].
- Sub-Clause 14.1(c) must be revised if the quantities in the Bill of Quantities or other Schedules are to be taken as the actual and correct quantities.
- Sub-Clause 14.1(d) imposes another requirement on the Contractor. Within 28 days from the Commencement Date the Contractor must provide a proposed breakdown of

every lump sum price in the Schedules. These breakdowns are only proposals, for the information of the Engineer, but should be consistent with other financial submissions. The Engineer is not bound to follow the Contractor's breakdown but should be prepared to discuss any queries or changes to the breakdown.

14.2 Advance Payment

The Employer shall make an advance payment, as an interest-free loan for mobilisation, when the Contractor submits a guarantee in accordance with this Sub-Clause. The total advance payment, the number and timing of instalments (if more than one), and the applicable currencies and proportions, shall be as stated in the Appendix to Tender.

Unless and until the Employer receives this guarantee, or if the total advance payment is not stated in the Appendix to Tender, this Sub-Clause shall not apply.

The Engineer shall issue an Interim Payment Certificate for the first instalment after receiving a Statement (under Sub-Clause 14.3 [Application for Interim Payment Certificates]) and after the Employer receives (i) the Performance Security in accordance with Sub-Clause 4.2 [Performance Security], and (ii) a guarantee in amounts and currencies equal to the advance payment. This guarantee shall be issued by an entity and from within a country (or other jurisdiction) approved by the Employer, and shall be in the form annexed to the Particular Conditions or in another form approved by the Employer.

The Contractor shall ensure that the guarantee is valid and enforceable until the advance payment has been repaid, but its amount may be progressively reduced by the amount repaid by the Contractor as indicated in the Payment Certificates. If the terms of the guarantee specify its expiry date, and the advance payment has not been repaid by the date 28 days prior to the expiry date, the Contractor shall extend the validity of the guarantee until the advance payment has been repaid.

The advance payment shall be repaid through percentage deductions in Payment Certificates. Unless other percentages are stated in the Appendix to Tender:

(a) *deductions shall commence in the Payment Certificate in which the total of all certified interim payments (excluding the advance payment and deductions and repayments of retention) exceeds ten per cent (10%) of the Accepted Contract Amount less Provisional Sums; and*

(b) *deductions shall be made at the amortisation rate of one quarter (25%) of the amount of each Payment Certificate (excluding the advance payment and deductions and repayments of retention) in the currencies and proportions of the advance payment, until such time as the advance payment has been repaid.*

If the advance payment has not been repaid prior to the issue of the Taking-Over Certificate for the Works or prior to termination under Clause 15 [Termination by Employer], Clause 16 [Suspension and Termination by Contractor] or Clause 19 [Force Majeure] (as the case may be), the whole of the balance then outstanding shall immediately become due and payable by the Contractor to the Employer.

If the Employer intends to make an advance payment then the relevant information must be provided in the Appendix to Tender. The figure for the advance payment is given as a percentage of the Accepted Contract Amount and the currency in which it is to be paid must be stated. If the advance payment is to be paid in instalments then the number and timing of instalments must be stated.

In accordance with Sub-Clause 14.7(a) [*Payment*], the first instalment of the advance payment must be paid by the later of

- 42 days from issuing the Letter of Acceptance; or
- 21 days from receiving the Performance Security under Sub-Clause 4.2 [*Performance Security*] and the Advance Payment Guarantee and other documents under Sub-Clause 14.2.

The Contractor should ensure that the Performance Security and Advance Payment Guarantee are submitted to the Employer within 21 days from the Letter of Acceptance. The forms that are acceptable to the Employer should have been annexed to the Particular Conditions, preferably as the FIDIC Annex C or D and E.

The procedure for payment, given at Sub-Clause 14.2, is as follows.

- The Contractor submits a Statement for the first instalment under Sub-Clause 14.3 [*Application for Interim Payment Certificates*]. This Statement should be submitted immediately after the Letter of Acceptance, so that the various time periods can be concurrent, rather than consecutive.
- The Engineer must issue an Interim Payment Certificate immediately after receiving the Statement. The Engineer does not have the usual 28-day period, as Sub-Clause 14.6 [*Issue of Interim Payment Certificates*], to issue his Certificate because this would prevent the Employer from making payment by the due date. If the Engineer fails to issue the Certificate then the provisions of Sub-Clause 16.1 [*Contractor's Entitlement to Suspend Work*], for the Contractor first to give notice to suspend or reduce the rate of work and eventually Sub-Clause 16.2(b) [*Termination by Contractor*], for the Contractor to terminate the Contract, would apply.
- The Employer must have received the Performance Security and Advance Payment Guarantee. If these documents have not been received then the Engineer can delay issuing the Interim Payment Certificate. Some Employers may also require that the insurance documents have also been received.
- The Employer pays within 21 days from receiving the Performance Security and Advance Payment Guarantee, or within 42 days from issuing the Letter of Acceptance if this is later. This 42-day period coincides with the 42-day period for the Commencement Date, although the advance payment period starts from the Employer issuing the Letter of Acceptance and the Commencement Date period from its receipt by the Contractor. If the Employer fails to pay within these periods then the provisions of Sub-Clauses 16.1, for the Contractor to give notice to suspend or reduce the rate of work, and Sub-Clause 16.2(c), for the Contractor to terminate the Contract, will apply.

The Contractor will also be entitled to finance charges as Sub-Clause 14.8 [*Delayed Payment*].

The advance payment is stated at the first sentence of Sub-Clause 14.2 to be an '*interest-free loan for mobilisation*', so any delay in payment could result in the Contractor claiming that his mobilisation will be delayed. Similarly, if payment is made but mobilisation is slow then the Employer could ask whether the payment has been used for its proper purpose.

The payment procedures are appropriate for the first instalment, which must be paid as soon as possible after the Letter of Acceptance. If the Particular Conditions state that the advance payment will be made in several instalments then the procedures for the further instalments must also be stated. For the payment to be an advance for mobilisation then the total advance payment must be paid quickly and be related to the timing of the Contractor's mobilisation costs. If the timing of instalments is intended to relate to further mobilisation costs that will arise during the project, then the timing of the date for payment must relate to these costs. Sub-Clause 14.3(d) [*Application for Interim Payment Certificates*] allows for advance payments to be included in the Contractor's statements for Interim Payment Certificates. However, this would not comply with the requirements for an advance payment due to the lengthy time period between the Contractor's statement and the receipt of payment.

Sub-Clause 14.2 requires that repayment of the advance payment will commence when the certified interim payments exceed 10% of the Accepted Contract Amount and will be calculated as 25% of the amount of each Payment Certificate. The *Guidance for the Preparation of Particular Conditions* states that these figures were calculated on the assumption that the total advance payment is less than 22% of the Accepted Contract Amount. If the total advance payment is more than 22% of the Accepted Contract Amount, which seems unlikely, then the repayments should be increased. If it is substantially less than 22% the repayments could be reduced. Any outstanding balance must be repaid immediately on the issue of the Taking-Over Certificate for the Works or prior to termination under Clauses 15, 16 or 19.

14.3 Application for Interim Payment Certificates
The Contractor shall submit a Statement in six copies to the Engineer after the end of each month, in a form approved by the Engineer, showing in detail the amounts to which the Contractor considers himself to be entitled, together with supporting documents which shall include the report on the progress during this month in accordance with Sub-Clause 4.21 [Progress Reports].

The Statement shall include the following items, as applicable, which shall be expressed in the various currencies in which the Contract Price is payable, in the sequence listed:

(a) *the estimated contract value of the Works executed and the Contractor's Documents produced up to the end of the month (including Variations but excluding items described in sub-paragraphs (b) to (g) below);*

(b) *any amounts to be added and deducted for changes in legislation and changes in cost, in accordance with Sub-Clause 13.7 [Adjustments for Changes in Legislation] and Sub-Clause 13.8 [Adjustments for Changes in Cost];*

217

(c) any amount to be deducted for retention, calculated by applying the percentage of retention stated in the Appendix to Tender to the total of the above amounts, until the amount so retained by the Employer reaches the limit of Retention Money (if any) stated in the Appendix to Tender;

(d) any amounts to be added and deducted for the advance payment and repayments in accordance with Sub-Clause 14.2 [Advance Payment];

(e) any amounts to be added and deducted for Plant and Materials in accordance with Sub-Clause 14.5 [Plant and Materials intended for the Works];

(f) any other additions or deductions which may have become due under the Contract or otherwise, including those under Clause 20 [Claims, Disputes and Arbitration]; and

(g) the deduction of amounts certified in all previous Payment Certificates.

The procedure for the Contractor to receive payments from the Employer starts with the Contractor's application for payment. The Contractor submits a Statement to the Engineer each month giving the amounts that he is claiming, together with supporting documents. The Statement is submitted '*after the end of each month*' and the sooner the Contractor collects the necessary information and submits the Statement, the sooner he will be paid. The Statement must be '*in a form approved by the Engineer*' and include the items listed as (a) to (g), in the order in which they are listed. Early agreement on the form and layout of the Statement is important to avoid any later delaying issues, and if the Engineer has any particular requirements they should be stated in the Particular Conditions.

Item (a) refers to '*the estimated contract value of the Works executed*'. This will include any Works that have been measured in accordance with Clause 12, together with an estimate of the value of work that has not yet been measured by the Engineer.

Item (f) refers to amounts due as claims under Clause 20. This will include the claims and interim claims described at Sub-Clause 20.1 [*Contractor's Claims*]. The reference to Clause 20 emphasises the importance of claims being submitted under Clause 20 in addition to the notices under other clauses, such as Sub-Clause 1.9 [*Delayed Drawings or Instructions*].

Item (g) refers to the deduction of amounts previously certified, but does not require a Statement of any amounts which have been certified by the Engineer but were not paid by the Employer. The Contractor should normally confirm the Certificates that have been paid and the due dates for unpaid Certificates. The Contractor is entitled to finance charges for delayed payments '*without formal notice or certification*', as Sub-Clause 14.8 [*Delayed Payment*], but it would obviously be prudent to include a calculation of any finance charges under item (f).

When submitting the first Statement the Contractor should demonstrate that he has received, or state when he expects to receive, the Employer's approval to the Performance Security. Under Sub-Clause 14.6 [*Issue of Interim Payment Certificates*] the Engineer will not certify any payment until the Employer has approved the Performance Security. Under Sub-Clause 1.3 [*Communications*], the approval must not be '*unreasonably withheld or delayed*'. If the Contract includes provision for an advance payment then the Performance Security will already have been approved before payment of the advance payment. The timing of the first monthly Statement for work executed will depend on the progress of the Works being such that the net amount due to be paid exceeds the minimum amount of Interim Payment Certificates stated in the Appendix to Tender.

The Contractor's Statement should be sent by a method that requires a receipt. Sub-Clause 14.7 [*Payment*] requires the Employer to make payment of the sum that is certified by the Engineer within 56 days from the date the Engineer receives the Contractor's Statement. This puts an obligation on the Engineer to review the application and issue a certificate without delay, so as to enable the Employer to make timely payments. Any delay in making payment would entitle the Contractor to receive interest as per Sub-Clause 14.8 [*Delayed Payment*].

The supporting documents with the Contractor's Statement must be complete, as required by Sub-Clause 14.3, and must include the Contractor's monthly progress report under Sub-Clause 4.21 [*Progress Reports*].

14.4 Schedule of Payments
If the Contract includes a schedule of payments specifying the instalments in which the Contract Price will be paid, then unless otherwise stated in this schedule:

(a) *the instalments quoted in this schedule of payments shall be the estimated contract values for the purposes of sub-paragraph (a) of Sub-Clause 14.3 [Application for Interim Payment Certificates];*

(b) *Sub-Clause 14.5 [Plant and Materials intended for the Works] shall not apply; and*

(c) *if these instalments are not defined by reference to the actual progress achieved in executing the Works, and if actual progress is found to be less than that on which this schedule of payments was based, then the Engineer may proceed in accordance with Sub-Clause 3.5 [Determinations] to agree or determine revised instalments, which shall take account of the extent to which progress is less than that on which the instalments were previously based.*

If the Contract does not include a schedule of payments, the Contractor shall submit non-binding estimates of the payments which he expects to become due during each quarterly period. The first estimate shall be submitted within 42 days after the Commencement Date. Revised estimates shall be submitted at quarterly intervals, until the Taking-Over Certificate has been issued for the Works.

If the Contract Price is to be paid in instalments then the Contract must include a Schedule of Payments, giving the timing and details of the instalments, subject to the provisions of Sub-Clause 14.4. The schedule must be clear and unambiguous and relate to the Contractor's programme and the actual progress that has been recorded on the Site. Subparagraph (c) allows the Engineer to reduce the payment if actual progress is less than had been assumed. Logically the Engineer should also increase the payment if actual progress is greater than had been assumed.

If the Contract does not include a Schedule of Payments then the Contractor must submit esti-mates of the amounts he expects will become due during each quarterly period. Even though the wording of this Sub-Clause is not entirely clear, but has similarity to Sub-Clause 14.3 [*Application for Interim Payment Certificates*] where the Contractor shall submit monthly statements, it may be assumed that these estimates shall be the monthly amounts the Contractor expects to become due, updated quarterly. The first estimate must be submitted within 42 days after the Commencement Date, that is, during the first half of that quarter. Further estimates are required '*at quarterly intervals*', but the next estimate is not necessarily required exactly three months after the first estimate. The dates for submission should be agreed between the Contractor and the Engineer. This requirement replaces the provision for a cash flow estimate in earlier FIDIC Conditions of Contracts. If the Employer requires the estimate to include particular details, then the requirements must be stated in the Particular Conditions. The estimates should also relate to the Contractor's programme.

14.5 Plant and Materials intended for the Works

If this Sub-Clause applies, Interim Payment Certificates shall include, under subparagraph (e) of Sub-Clause 14.3, (i) an amount for Plant and Materials which have been sent to the Site for incorporation in the Permanent Works, and (ii) a reduction when the contract value of such Plant and Materials is included as part of the Permanent Works under subparagraph (a) of Sub-Clause 14.3 [Application for Interim Payment Certificates].

If the lists referred to in subparagraphs (b)(i) or (c)(i) below are not included in the Appendix to Tender, this Sub-Clause shall not apply.

The Engineer shall determine and certify each addition if the following conditions are satisfied:

(a) the Contractor has:
    (i)   kept satisfactory records (including the orders, receipts, Costs and use of Plant and Materials) which are available for inspection, and
    (ii) submitted a statement of the Cost of acquiring and delivering the Plant and Materials to the Site, supported by satisfactory evidence;
and either:
(b) the relevant Plant and Materials:
    (i)   are those listed in the Appendix to Tender for payment when shipped,
    (ii) have been shipped to the Country, en route to the Site, in accordance with the Contract; and
    (iii) are described in a clean shipped bill of lading or other evidence of shipment, which has been submitted to the Engineer together with evidence of payment of freight and insurance, any other documents reasonably required, and a bank guarantee in a form and issued by an entity approved by the Employer in amounts and currencies equal to the amount due under this Sub-Clause: this guarantee may be in a similar form to the form referred to in Sub-Clause 14.2 [Advance Payment] and shall be valid until the Plant and Materials are properly stored on Site and protected against loss, damage or deterioration;
or
(c) the relevant Plant and Materials:
    (i)   are those listed in the Appendix to Tender for payment when delivered to the Site, and

(ii) *have been delivered to and are properly stored on the Site, are protected against loss, damage or deterioration, and appear to be in accordance with the Contract.*

The additional amount to be certified shall be the equivalent of eighty percent of the Engineer's determination of the cost of the Plant and Materials (including delivery to Site), taking account of the documents mentioned in this Sub-Clause and of the contract value of the Plant and Materials.

The currencies for this additional amount shall be the same as those in which payment will become due when the contract value is included under sub-paragraph (a) of Sub-Clause 14.3 [Application for Interim Payment Certificates]. At that time, the Payment Certificate shall include the applicable reduction which shall be equivalent to, and in the same currencies and proportions as, this additional amount for the relevant Plant and Materials.

It is normal practice for the Employer to make interim payments for Plant and Materials that have been allocated to the Works but which are not yet incorporated into the Works. The Appendix to Tender has provision for the Employer to list the Plant and Materials for which payment will be made, either when they have been shipped to the Country, or when they have been delivered to the Site.

If payment is to be made for Plant and Materials that have been shipped to the Country then the Contractor must provide a bank guarantee for the amount due. The guarantee must be similar to the Advance Payment Guarantee and approved by the Engineer. The amounts to be paid and the necessary supporting documents are described at Sub-Clause 14.5.

14.6 Issue of Interim Payment Certificates

No amount will be certified or paid until the Employer has received and approved the Performance Security. Thereafter, the Engineer shall, within 28 days after receiving a Statement and supporting documents, issue to the Employer an Interim Payment Certificate which shall state the amount which the Engineer fairly determines to be due, with supporting particulars.

However, prior to issuing the Taking-Over Certificate for the Works, the Engineer shall not be bound to issue an Interim Payment Certificate in an amount which would (after retention and other deductions) be less than the minimum amount of Interim Payment Certificates (if any) stated in the Appendix to Tender. In this event, the Engineer shall give notice to the Contractor accordingly.

An Interim Payment Certificate shall not be withheld for any other reason, although:

(a) *if any thing supplied or work done by the Contractor is not in accordance with the Contract, the cost of rectification or replacement may be withheld until rectification or replacement has been completed; and/or*

(b) *if the Contractor was or is failing to perform any work or obligation in accordance with the Contract, and had been so notified by the Engineer, the value of this work or obligation may be withheld until the work or obligation has been performed.*

The Engineer may in any Payment Certificate make any correction or modification that should properly be made to any previous Payment Certificate. A Payment Certificate shall not be deemed to indicate the Engineer's acceptance, approval, consent or satisfaction.

The Engineer must decide the amount which is due to be paid to the Contractor and issue a certificate, including supporting particulars, within 28 days after receiving the Contractor's Statement. The calculation of the sum due must be made strictly in accordance with the provisions of the Contract, including the provisions at Sub-Clause 14.6(a) and (b). Under Sub-Clause 14.6 the Engineer *'fairly determines'* the amount. If the Contractor is not satisfied with the figure he may inform the Engineer and declare the issue is a dispute and refer the matter to the DAB.

If the Engineer fails to issue the Interim Payment Certificate the Contractor may give 21 days' notice and then suspend or reduce the rate of work, under Sub-Clause 16.1 [*Contractor's Entitlement to Suspend Work*]. If the Engineer has still failed to issue the Interim Payment Certificate after 56 days from receipt of the Contractor's Statement and supporting documents then the Contractor may terminate the Contract under Sub-Clause 16.2(b) [*Termination by Contractor*]. These provisions emphasise the importance of regular interim payments to enable the Contractor to carry out his obligations. If the problem is caused by the Engineer, rather than the Employer, then the Employer could offer to make an interim payment, of an agreed amount, while the problem is rectified.

14.7 Payment
The Employer shall pay to the Contractor:

(a) *the first instalment of the advance payment within 42 days after issuing the Letter of Acceptance or within 21 days after receiving the documents in accordance with Sub-Clause 4.2 [Performance Security] and Sub-Clause 14.2 [Advance Payment], whichever is later;*

(b) *the amount certified in each Interim Payment Certificate within 56 days after the Engineer receives the Statement and supporting documents; and*

(c) *the amount certified in the Final Payment Certificate within 56 days after the Employer receives this Payment Certificate.*

Payment of the amount due in each currency shall be made into the bank account, nominated by the Contractor, in the payment country (for this currency) specified in the Contract.

Interim payments must be made within 56 days after the Engineer receives the Contractor's Statement and supporting documents. For the Final Payment, the 56-day period for payment starts when the Employer receives the Payment Certificate. The Employer has the benefit of the cash flow estimates provided under Sub-Clause 14.4 [*Schedule of Payments*], together with advice from the Engineer, so it should be possible for these periods to be reduced in the Particular Conditions. Anything that can be done to improve the Contractor's cash flow should result in lower tenders and savings in the overall project cost.

The Employer's obligation is to pay the sum that is certified by the Engineer, without any deductions. If the Employer is entitled to any deduction or payment from the Contractor then the amount will have been claimed under Sub-Clause 2.5 [*Employer's Claims*], subject to the exceptions listed. The total of any deductions will then be listed in the Engineer's Certificate and should have been included in the Contractor's Statement under Sub-Clause 14.3(f) [*Application for Interim Payment Certificates*]. If the Employer has a query concerning the Engineer's Interim Payment Certificate then the payment must still be made and any correction included in the next month's certificate.

If the Employer fails to comply with these requirements then the Contractor may give 21 days' notice and then suspend or reduce the rate of work under Sub-Clause 16.1 [*Contractor's Entitlement to Suspend Work*]. The Contractor may terminate the Contract if payment is not received within 42 days of the due date under Sub-Clause 16.2(c) [*Termination by Contractor*].

14.8 Delayed Payment

If the Contractor does not receive payment in accordance with Sub-Clause 14.7 [Payment], the Contractor shall be entitled to receive financing charges compounded monthly on the amount unpaid during the period of delay. This period shall be deemed to commence on the date for payment specified in Sub-Clause 14.7 [Payment], irrespective (in the case of its sub-paragraph (b)) of the date on which any Interim Payment Certificate is issued.

Unless otherwise stated in the Particular Conditions, these financing charges shall be calculated at the annual rate of three percentage points above the discount rate of the central bank in the country of the currency of payment, and shall be paid in such currency.

The Contractor shall be entitled to this payment without formal notice or certification, and without prejudice to any other right or remedy.

Sub-Clause 14.8 gives the procedures for calculating the financing charges that are due to the Contractor if a payment is not made by the due date.

The financing charges will be included in the Contractor's Statements and the Payment Certificates under Sub-Clause 14.3(f) [*Application for Interim Payment Certificates*] and so are carried forward if the next Certificate is not paid. Financing charges are damages – the reimbursement of money which has been deemed to be spent as a consequence of a breach of Contract – but should still be checked against the provisions of the applicable law.

14.9 Payment of Retention Money

When the Taking-Over Certificate has been issued for the Works, the first half of the Retention Money shall be certified by the Engineer for payment to the Contractor. If a Taking-Over Certificate is issued for a Section or part of the Works, a proportion of the Retention Money shall be certified and paid. This proportion shall be two-fifths (40%) of the proportion calculated by dividing the estimated contract value of the Section or part, by the estimated final Contract Price.

Promptly after the latest of the expiry dates of the Defects Notification Periods, the outstanding balance of the Retention Money shall be certified by the Engineer for payment to the Contractor. If a Taking-Over Certificate was issued for a Section, a proportion of the second half of the Retention Money shall be certified and paid promptly after the expiry date of the Defects Notification Period for the Section. This proportion shall be two-fifths (40%) of the proportion calculated by dividing the estimated contract value of the Section by the estimated final Contract Price.

However, if any work remains to be executed under Clause 11 [Defects Liability], the Engineer shall be entitled to withhold certification of the estimated cost of this work until it has been executed.

When calculating these proportions, no account shall be taken of any adjustments under Sub-Clause 13.7 [Adjustments for Changes in Legislation] and Sub-Clause 13.8 [Adjustments for Changes in Cost].

Sub-Clause 14.9 gives the procedures for the Employer to pay Retention Money, which has been deducted from Interim Payment Certificates as Sub-Clause 14.3(c) [*Application for Interim Payment Certificates*]. The first half of the Retention Money is to be certified by the Engineer when the Taking-Over Certificate has been issued for the Works. For a Section or part of the Works, 40% of the proportion by value will be certified. No time periods are stated for the certification and payment and Sub-Clause 14.3(c) concerning the deduction of retention does not refer to the payment of retention. The Contractor should ensure that the retention is included in a Statement as soon as possible, not waiting for the Statement at Completion as Sub-Clause 14.10 [*Statement at Completion*]. It should then be possible for the Certificate and payment to be issued well within the periods of 28 and 56 days from the Statement.

The outstanding balance of the Retention Money is to be certified '*promptly*' after the latest of the expiry dates of the Defects Notification Period, subject to a possible deduction for the estimated cost of any defects work under Clause 11 which has not been completed. The use of the word '*promptly*' could cause problems but it surely implies a matter of a few days, rather than weeks. Again, it would be prudent for the Contractor to issue a request for payment, asking that the certificate will be issued and payment made well within the 28- and 56-day periods.

The FIDIC *Guidance for the Preparation of Particular Conditions* includes an example Sub-Clause for use if the Employer is prepared to agree to early release of all or part of the Retention Money against a bank guarantee. When the Retention Money reaches 60% of the limit in the Appendix to Tender the Employer makes a payment of 50% of the limit, against the bank guarantee. The 50% of the limit that is due when the Taking-Over Certificate is issued, is paid at that time and the Guarantee continues in force until any defects have been corrected, as specified for the Performance Certificate.

When considering whether to accept a bank guarantee, or to reduce or even omit the requirement for Retention Money, the Employer should consider the cash flow benefits to the Contractor and the positive value of work on Site that has been executed but not yet paid for.

14.10 Statement at Completion

Within 84 days after receiving the Taking-Over Certificate for the Works, the Contractor shall submit to the Engineer six copies of a Statement at completion with supporting documents, in accordance with Sub-Clause 14.3 [Application for Interim Payment Certificates], showing:

(a) *the value of all work done in accordance with the Contract up to the date stated in the Taking-Over Certificate for the Works;*

(b) *any further sums which the Contractor considers to be due; and*

(c) *an estimate of any other amounts which the Contractor considers will become due to him under the Contract. Estimated amounts shall be shown separately in this Statement at completion.*

The Engineer shall then certify in accordance with Sub-Clause 14.6 [Issue of Interim Payment Certificates].

The Statement at completion can only be issued following the issue of the Taking-Over Certificate for all of the Works. Before the issue of this certificate, the Contractor has been able to submit Statements for Interim Payment Certificates. It is in the Contractor's interest to prepare the Statement in advance for issue well within the 84-day period.

The Statement at completion must include amounts or estimates for every payment for which the Contractor considers the Employer has a liability, including all claims and potential claims, or the Employer may reject liability under Sub-Clause 14.14(b) [*Cessation of Employer's Liability*]. However, if the Contractor has followed the notices and procedures under Sub-Clause 20.1 [*Contractor's Claims*] this further submission would be a reminder and record rather than a new submission.

14.11 Application for Final Payment Certificate

Within 56 days after receiving the Performance Certificate, the Contractor shall submit, to the Engineer, six copies of a draft final statement with supporting documents showing in detail in a form approved by the Engineer:

(a) *the value of all work done in accordance with the Contract; and*

(b) *any further sums which the Contractor considers to be due to him under the Contract or otherwise.*

If the Engineer disagrees with or cannot verify any part of the draft final statement, the Contractor shall submit such further information as the Engineer may reasonably require and shall make such changes in the draft as may be agreed between them. The Contractor shall then prepare and submit to the Engineer the final statement as agreed. This agreed statement is referred to in these Conditions as the 'Final Statement'.

However if, following discussions between the Engineer and the Contractor and any changes to the draft final statement which are agreed, it becomes evident that a dispute exists, the Engineer shall deliver to the Employer (with a copy to the Contractor) an Interim Payment Certificate for the agreed parts of the draft final statement. Thereafter, if the dispute is finally resolved under Sub-Clause 20.4 [Obtaining Dispute Adjudication Board's Decision] or Sub-Clause 20.5 [Amicable Settlement], the Contractor shall then prepare and submit to the Employer (with a copy to the Engineer) a Final Statement.

Sub-Clauses 14.10 and 14.11 give the procedures for the Contractor to submit Statements and the Engineer to certify interim payments, following the issue of the Taking-Over Certificate and the Performance Certificate. The Employer must make the interim payments within 56 days from the Contractor's Statement in accordance with Sub-Clause 14.7(b) [*Payment*].

The Final Statement, following the Performance Certificate, is first submitted as a draft, which is discussed between the Contractor and the Engineer. If the draft statement is agreed then the Contractor submits a Final Statement. However, if there are matters that cannot be agreed, they are considered under the Clause 20 disputes procedures. Any sums that have been agreed must be paid on an Interim Payment Certificate and the draft Final Statement remains open until the disputes are eventually resolved. By the time the agreed matters have been finalised it may be too late for the Employer to pay within 56 days from when the Engineer received the draft Final Statement. The Sub-Clause does not require the Contractor to submit a further interim Statement before the Engineer prepares this Interim Payment Certificate so the Contractor could be entitled to financing charges under Sub-Clause 14.8 [*Delayed Payment*].

If the dispute is resolved by the DAB or by amicable settlement, then the Contractor must prepare a Final Statement to start the payment sequence. However, if the dispute is resolved by arbitration then any payment must be made on the Arbitration Tribunal's Award and additional Contract documentation is not necessary.

14.12 Discharge
When submitting the Final Statement, the Contractor shall submit a written discharge which confirms that the total of the Final Statement represents full and final settlement of all moneys due to the Contractor under or in connection with the Contract. This discharge may state that it becomes effective when the Contractor has received the Performance Security and the outstanding balance of this total, in which event the discharge shall be effective on such date.

Sub-Clause 14.12 requires the Contractor to include with the Final Statement confirmation that it covers all moneys due to him under or in connection with the Contract. The Final Statement must cover any claims that have been, or should have been, submitted. However, Sub-Clause 11.10 [*Unfulfilled Obligations*] refers to the continuing liability for any obligations that have not been fulfilled at the issue of the Performance Certificate. Sub-Clause 14.14 [*Cessation of Employer's Liability*] also refers to the Employer's continuing liability under certain Clauses.

It is only if parts of the draft Final Statement are eventually settled in arbitration that the Contractor does not have to submit a Final Statement and hence would not submit a discharge. The submissions to the arbitration tribunal would normally cover any amounts that the Contractor considers to be due.

14.13 Issue of Final Payment Certificate

Within 28 days after receiving the Final Statement and written discharge in accordance with Sub-Clause 14.11 [Application for Final Payment Certificate] and Sub-Clause 14.12 [Discharge], the Engineer shall issue, to the Employer, the Final Payment Certificate which shall state:

(a) the amount which is finally due; and

(b) after giving credit to the Employer for all amounts previously paid by the Employer and for all sums to which the Employer is entitled, the balance (if any) due from the Employer to the Contractor or from the Contractor to the Employer, as the case may be.

If the Contractor has not applied for a Final Payment Certificate in accordance with Sub-Clause 14.11 [Application for Final Payment Certificate] and Sub-Clause 14.12 [Discharge], the Engineer shall request the Contractor to do so. If the Contractor fails to submit an application within a period of 28 days, the Engineer shall issue the Final Payment Certificate for such amount as he fairly determines to be due.

When the Final Statement has eventually been agreed between the Contractor and the Engineer, it is submitted to the Engineer together with the Sub-Clause 14.12 [*Discharge*].

Sub-Clause 14.13 gives the procedures for the Engineer to issue the Final Payment Certificate. If the Contractor does not make the proper application for the Final Payment Certificate then the Engineer shall request the Contractor to do so. If the Contractor still fails to do so within a period of 28 days, than the Engineer shall issue a Certificate for the amount he '*fairly determines to be due*'. Under this procedure the Contractor has not submitted a written discharge and the Engineer should demonstrate that all claims that have been received have been considered and dealt with in accordance with the Contract.

The Employer must pay the amount in the Final Payment Certificate within 56 days from receipt, in accordance with Sub-Clause 14.7(c) [*Payment*].

14.14 Cessation of Employer's Liability

The Employer shall not be liable to the Contractor for any matter or thing under or in connection with the Contract or execution of the Works, except to the extent that the Contractor shall have included an amount expressly for it:

(a) in the Final Statement and; also

(b) (except for matters or things arising after the issue of the Taking-Over Certificate for the Works) in the Statement at completion described in Sub-Clause 14.10 [Statement at Completion].

However, this Sub-Clause shall not limit the Employer's liability under his indemnification obligations, or the Employer's liability in any case of fraud, deliberate default or reckless misconduct by the Employer.

Sub-Clause 14.14 refers to the Employer's liability *'for any matter or thing under or in connection with the Contract or execution of the Works'*. If the Contractor is not satisfied with any valuation, the response to any claim, or anything else whatsoever related to the Contract, then he must ensure that the matter is mentioned in the Final Statement, together with an amount of money in compensation. The Employer can disclaim liability for any matter that is not mentioned and valued in the Final Statement.

The nearest equivalent Clause for cessation of the Contractor's liability is Sub-Clause 11.9 [*Performance Certificate*] for the issue of the Performance Certificate. However, after the Performance Certificate has been issued, both Parties still remain liable for unfulfilled obligations, as Sub-Clause 11.10 [*Unfulfilled Obligations*]. For the Employer, Sub-Clause 11.10 appears to be superseded and the general liability presumably ceases with the Final Statement, except for the exceptions noted in Sub-Clause 14.14. Sub-Clause 14.14 must be considered in conjunction with any provisions in the governing law.

14.15 Currencies of Payment
The Contract Price shall be paid in the currency or currencies named in the Appendix to Tender. Unless otherwise stated in the Particular Conditions, if more than one currency is so named, payments shall be made as follows:

(a) *if the Accepted Contract Amount was expressed in Local Currency only:*
    (i) *the proportions or amounts of the Local and Foreign Currencies, and the fixed rates of exchange to be used for calculating the payments, shall be as stated in the Appendix to Tender, except as otherwise agreed by both Parties;*
    (ii) *payments and deductions under Sub-Clause 13.5 [Provisional Sums] and Sub-Clause 13.7 [Adjustments for Changes in Legislation] shall be made in the applicable currencies and proportions; and*
    (iii) *other payments and deductions under sub-paragraphs (a) to (d) of Sub-Clause 14.3 [Application for Interim Payment Certificates] shall be made in the currencies and proportions specified in sub-paragraph (a)(i) above;*
(b) *payment of the damages specified in the Appendix to Tender shall be made in the currencies and proportions specified in the Appendix to Tender;*
(c) *other payments to the Employer by the Contractor shall be made in the currency in which the sum was expended by the Employer, or in such currency as may be agreed by both Parties;*
(d) *if any amount payable by the Contractor to the Employer in a particular currency exceeds the sum payable by the Employer to the Contractor in that currency, the Employer may recover the balance of this amount from the sums otherwise payable to the Contractor in other currencies; and*
(e) *if no rates of exchange are stated in the Appendix to Tender, they shall be those prevailing on the Base Date and determined by the central bank of the Country.*

Sub-Clause 14.15 must be read together with the Appendix to Tender and the FIDIC form for the Letter of Tender, which are printed at the end of the Conditions of Contract and are reproduced in Appendix 8 of this book. These documents allow for different provisions for the currency of payment and the documents must be clear as to the precise provisions that will apply.

The FIDIC form for the Letter of Tender includes provision for the Contractor to insert the currency or currencies in which the Contract Price will be paid. This conforms to the first option in the Appendix to Tender. Alternatively, the Employer can stipulate the currency of payment in the Appendix to Tender, or the percentages of the Contract Price which will be paid in different currencies, together with the rates of exchange.

If only one currency is named in the Letter of Tender or the Appendix to Tender, then all payments will be made in that currency. However, if the Letter of Acceptance gives the Accepted Contract Amount in the Local Currency, but the Letter of Tender or the Appendix to Tender allow for different currencies then the provisions of Sub-Clause 14.15(a) will apply. Sub-Clauses 14.15(b) to (e) will apply whenever the Letter of Tender or Appendix to Tender allow for more than one currency.

The Letter of Acceptance is generally written following some negotiations and any queries or problems concerning currencies for payment should be clarified and agreed at that stage.

If the parts of the Contract Price that are to be paid in different currencies are to be based on the cost of different items, such as items of Plant, Materials or labour, then suitable provision, in sufficient detail to avoid misunderstanding, must be made in the Particular Conditions.

The provisions for the currencies of payment do not include any provision for changes in the rates of exchange. If rates of exchange are to be adjusted to allow for currency fluctuations then suitable provision must be included in the Particular Conditions.

The FIDIC *Guidance for the Preparation of Particular Conditions* includes an example Sub-Clause for a single-currency Contract, with all payments in Local Currency. The Local Currency is assumed to be fully convertible, with the Employer being liable for any additional costs if the situation should change.

FIDIC Users' Guide
ISBN 978-0-7277-5856-9

ICE Publishing: All rights reserved
http://dx.doi.org/10.1680/fug.58569.231

Chapter 21
Clause 15: Termination by Employer

Clause 15 describes the circumstances in which the Employer is entitled to terminate the Contract, the procedures that must be followed and the financial arrangements that will apply.

Termination is an extremely serious step, which inevitably causes hardship to both Parties and should not be invoked without considerable discussion and attempts to overcome any problems and rectify the situation. If the Contractor objects to the termination, and a DAB or Arbitrator later decides that the Employer was not entitled to terminate, then the financial consequences to the Employer would be substantial.

If the Employer does decide to terminate the Contract it is essential that he follow the correct procedures. The governing law may also have requirements in addition to, or which may supersede, the Contract procedures.

15.1 Notice to Correct
If the Contractor fails to carry out any obligation under the Contract, the Engineer may by notice require the Contractor to make good the failure and to remedy it within a specified reasonable time.

The Sub-Clause refers to a situation where the Contractor fails to comply with an obligation under the Contract. If the Engineer is planning to issue a notice under this Sub-Clause it is imperative that he ensures beyond doubt that such non-compliance is factually correct and is under the control of the Contractor. Any premature or incorrect notice will inevitably lead to less effective relationships within the project.

When the Engineer does decide to instruct the Contractor requiring him to make good a failure to carry out some obligation, it may refer to a relatively minor matter or it may be an obligation that is crucial to the success of the project. However, it could be the first step towards termination of the Contract.

A '*Notice to Correct*' under Sub-Clause 15.1 [*Notice to Correct*] is the starting point of one of the routes towards termination of the Contract by the Employer. To avoid possible disputes as to whether the termination procedure was followed correctly, any Notice to Correct should refer specifically to Sub-Clause 15.1. The issue of such a notice may be a sensible additional step in some of the alternative routes to termination, in order to emphasise the seriousness of the situation as perceived by the Engineer and to give the Contractor a final warning as to the consequences of failure to comply with the particular obligation.

15.2 Termination by Employer

The Employer shall be entitled to terminate the Contract if the Contractor:

(a) *fails to comply with Sub-Clause 4.2 [Performance Security] or with a notice under Sub-Clause 15.1 [Notice to Correct];*

(b) *abandons the Works or otherwise plainly demonstrates the intention not to continue performance of his obligations under the Contract;*

(c) *without reasonable excuse fails:*

   (i) *to proceed with the Works in accordance with Clause 8 [Commencement, Delays and Suspension], or*

   (ii) *to comply with a notice issued under Sub-Clause 7.5 [Rejection] or Sub-Clause 7.6 [Remedial Work], within 28 days after receiving it;*

(d) *subcontracts the whole of the Works or assigns the Contract without the required agreement;*

(e) *becomes bankrupt or insolvent, goes into liquidation, has a receiving or administration order made against him, compounds with his creditors, or carries on business under a receiver, trustee or manager for the benefit of his creditors, or if any act is done or event occurs which (under applicable Laws) has a similar effect to any of these acts or events; or*

(f) *gives or offers to give (directly or indirectly) to any person any bribe, gift, gratuity, commission or other thing of value, as an inducement or reward:*

   (i) *for doing or forbearing to do any action in relation to the Contract, or*

   (ii) *for showing or forbearing to show favour or disfavour to any person in relation to the Contract;*

or if any of the Contractor's Personnel, agents or Subcontractors gives or offers to give (directly or indirectly) to any person any such inducement or reward as is described in this sub-paragraph (f). However, lawful inducements and rewards to Contractor's Personnel shall not entitle termination.

In any of these events or circumstances, the Employer may, upon giving 14 days' notice to the Contractor, terminate the Contract and expel the Contractor from the Site. However, in the case of sub-paragraph (e) or (f), the Employer may by notice terminate the Contract immediately.

The Employer's election to terminate the Contract shall not prejudice any other rights of the Employer, under the Contract or otherwise.

The Contractor shall then leave the Site and deliver any required Goods, all Contractor's Documents, and other design documents made by or for him, to the Engineer. However, the Contractor shall use his best efforts to comply immediately with any reasonable instructions included in the notice (i) for the assignment of any subcontract, and (ii) for the protection of life or property or for the safety of the Works.

After termination, the Employer may complete the Works and/or arrange for any other entities to do so. The Employer and these entities may then use any Goods, Contractor's Documents and other design documents made by or on behalf of the Contractor.

The Employer shall then give notice that the Contractor's Equipment and Temporary Works will be released to the Contractor at or near the Site. The Contractor shall promptly arrange their removal, at the risk and cost of the Contractor. However, if by this time the Contractor has failed to make a payment due to the Employer, these items may be sold by the Employer in order to recover this payment. Any balance of the proceeds shall then be paid to the Contractor.

The Employer is entitled to terminate the Contract either because of some default by the Contractor, as listed in this Sub-Clause, or for convenience, under Sub-Clause 15.5 [*Employer's Entitlement to Termination*]. Sub-Clause 19.6 [*Optional Termination, Payment and Release*] also includes provision for termination, by either Party, if the execution of substantially all of the Works in progress is prevented by Force Majeure for a continuous period of 84 days. The provisions are very wide and include actions by a Subcontractor who may have acted without the knowledge or consent of the Contractor. In these circumstances a provision to require the Contractor to terminate the subcontract would seem to be more appropriate than termination of the Contract.

The circumstances described at subparagraphs (a) to (d) refer to failures to carry out his obligations in accordance with specific Clauses of the Contract. These could include failure to carry out some minor obligation that would seem to be an excessively harsh remedy for a minor default. Although in a situation where the Employer actually wants to terminate in any case, the provisions under this Sub-Clause give him opportunity to do so. Sub-Clauses, such as 11.4(c), for failure to remedy defects, also entitle the Employer to terminate all or part of the Contract and in these circumstances the Employer should also follow the Clause 15 procedures, together with any requirements of the particular Sub-Clause.

These procedures require the Employer, as distinct from the Engineer, to give 14 days' notice of the termination. This gives a final opportunity for the Contractor to comply with the relevant obligation or to discuss the matter with the Engineer or directly with the Employer.

Subparagraphs (e) and (f) refer to insolvency and bribery, which entitle the Employer to give notice terminating the Contract immediately. If the Employer intends to terminate the Contract under any of these provisions he must be careful to ensure that there is legally acceptable proof of the circumstances. While the insolvency or similar situation under subparagraph (e) should be capable of being proved, the bribery circumstances under subparagraph (f) could be very difficult to prove unless there has been a criminal prosecution. Even an admission of guilt may not be sufficient, if it is denied by the Contractor and not supported by independent evidence which would be acceptable in a Court of Law.

The termination notice is required to include any instructions concerning safety and the assignment of subcontracts and should also include any other instructions concerning the Contractor's departure from the Site and the valuation of the Works. Any of the termination provisions should be checked against the governing law before being invoked.

15.3 Valuation at Date of Termination
As soon as practicable after a notice of termination under Sub-Clause 15.2 [Termination by Employer] has taken effect, the Engineer shall proceed in accordance with Sub-Clause 3.5

[Determinations] to agree or determine the value of the Works, Goods and Contractor's Documents, and any other sums due to the Contractor for work executed in accordance with the Contract.

The Engineer will agree or determine the sums due to the Contractor '*as soon as practicable*' after the notice of termination have taken effect. While the agreement of the value of some items may require lengthy negotiations, any measurement of work executed or agreement of Materials on Site must be carried out immediately, preferably during the period between the notice and the Contractor's departure from the Site.

The Parties are advised to check the legal definition under the applicable Law in accordance with Sub-Clause 1.4 *[Law and Language]* of when a notice has '*taken effect*'. However, in most jurisdictions this would probably be the date the notice was sent.

15.4 Payment after Termination

After a notice of termination under Sub-Clause 15.2 [Termination by Employer] has taken effect, the Employer may:

(a) *proceed in accordance with Sub-Clause 2.5 [Employer's Claims];*
(b) *withhold further payments to the Contractor until the costs of execution, completion and remedying of any defects, damages for delay in completion (if any), and all other costs incurred by the Employer, have been established; and/or*
(c) *recover from the Contractor any losses and damages incurred by the Employer and any extra costs of completing the Works, after allowing for any sum due to the Contractor under Sub-Clause 15.3 [Valuation at Date of Termination]. After recovering any such losses, damages and extra costs, the Employer shall pay any balance to the Contractor.*

Subparagraphs (a) to (c) refer to three procedures whereby the Employer may recover money from the Contractor. However, the word '*may*' in the opening sentence of this Sub-Clause could lead to confusion. The choice of (a), (b) and/or (c) will depend on the circumstances and the Employer must follow the Contract procedures.

Subparagraphs (a) and (b) refer to procedures for the correction of problems caused by the Contractor whose Contract has been terminated. Subparagraph (a) requires that claims which would be covered by Sub-Clause 2.5 *[Employer's Claims]* are to follow the procedure of Sub-Clause 2.5. Similarly claims under (b) arose from actions by this Contractor and should, as far as is practical, follow the same procedure.

The procedure for subparagraph (c) is more difficult. If the termination occurs early in the project, or there is a delay before a new Contractor can start on Site, the extra Costs incurred by the Employer could be substantial and may be disputed by the Contractor. Any such claim by the Employer would also seem to be covered by Sub-Clause 2.5. The Employer must ensure that all the necessary records are kept to substantiate any such claim.

Any dispute that arises under the termination procedure could be referred to the DAB. The final paragraph of Sub-Clause 20.2 *[Appointment of the Dispute Adjudication Board]* states that the appointment of the full term DAB expires when it has given its decision on the last dispute which has been referred to it or when the discharge as Sub-Clause 14.12 [*Discharge*] has become effective. The Sub-Clause 14.12 discharge depends on the issue of the Contractor's Final Statement. It is possible that this will not be agreed until after the Works have been completed by the replacement Contractor and the Employer has established any extra Costs that he has incurred.

15.5 Employer's Entitlement to Termination

The Employer shall be entitled to terminate the Contract, at any time for the Employer's convenience, by giving notice of such termination to the Contractor. The termination shall take effect 28 days after the later of the dates on which the Contractor receives this notice or the Employer returns the Performance Security. The Employer shall not terminate the Contract under this Sub-Clause in order to execute the Works himself or to arrange for the Works to be executed by another contractor.

After this termination, the Contractor shall proceed in accordance with Sub-Clause 16.3 [Cessation of Work and Removal of Contractor's Equipment] and shall be paid in accordance with Sub-Clause 19.6 [Optional Termination, Payment and Release].

The Employer is entitled to terminate the Contract at any time for his own convenience, that is, without needing to show any default by the Contractor or other justification. However, the consequences of such action may be against the interests of the Employer in that the Employer is not permitted to complete the Works by using another contractor or by executing the remaining works himself. However, there can be circumstances when the Employer no longer requires the project and so termination is necessary.

For this termination for convenience, the financial cost to the Employer follows the Force Majeure termination procedure at Sub-Clause 19.6 [*Optional Termination, Payment and Release*], which is more beneficial to the Contractor than Sub-Clause 15.4 [*Payment after Termination*]. However, it does not include compensation for loss of profit, which seems unreasonable when the Employer has decided on termination for his own convenience and not due to any default by the Contractor. This should however be checked against the Law, as in many jurisdictions this issue is governed by specific legislation that may override the provision of the Sub-Clause.

FIDIC Users' Guide
ISBN 978-0-7277-5856-9

ICE Publishing: All rights reserved
http://dx.doi.org/10.1680/fug.58569.237

Chapter 22
Clause 16: Suspension and Termination by Contractor

Payment by the Employer to the Contractor, in accordance with the provisions of the Contract, is an essential requirement of any construction contract. This obligation on the Employer is clearly stated in the FIDIC Contract Agreement and the procedures for payment are given at Clause 14. When a Contractor prepares his Tender he will base his calculations on the assumption that the money he pays out, for example, for labour and Materials, will be reimbursed in accordance with the provisions of the Contract. If this does not happen then the Contractor may have no alternative but to stop work. Clause 16 enables the Contractor to reduce the rate of work, suspend all work or terminate the Contract if the Employer fails to comply with his obligations for payment or to provide the information concerning his financial arrangements as required by Sub-Clause 2.4 [*Employer's Financial Arrangements*].

16.1 Contractor's Entitlement to Suspend Work

If the Engineer fails to certify in accordance with Sub-Clause 14.6 [Issue of Interim Payment Certificates] or the Employer fails to comply with Sub-Clause 2.4 [Employer's Financial Arrangements] or Sub-Clause 14.7 [Payment], the Contractor may, after giving not less than 21 days' notice to the Employer, suspend work (or reduce the rate of work) unless and until the Contractor has received the Payment Certificate, reasonable evidence or payment, as the case may be and as described in the notice.

The Contractor's action shall not prejudice his entitlements to financing charges under Sub-Clause 14.8 [Delayed Payment] and to termination under Sub-Clause 16.2 [Termination by Contractor].

If the Contractor subsequently receives such Payment Certificate, evidence or payment (as described in the relevant Sub-Clause and in the above notice) before giving a notice of termination, the Contractor shall resume normal working as soon as is reasonably practicable.

If the Contractor suffers delay and/or incurs Cost as a result of suspending work (or reducing the rate of work) in accordance with this Sub-Clause, the Contractor shall give notice to the Engineer and shall be entitled subject to Sub-Clause 20.1 [Contractor's Claims] to:

(a) an extension of time for any such delay, if completion is or will be delayed, under Sub-Clause 8.4 [Extension of Time for Completion]; and

(b) payment of any such Cost plus reasonable profit, which shall be included in the Contract Price.

After receiving this notice, the Engineer shall proceed in accordance with Sub-Clause 3.5 [Determinations] to agree or determine these matters.

The Contractor is entitled to reduce the rate of work or suspend work for the following reasons.

- The Employer fails to provide information concerning his financial arrangements as Sub-Clause 2.4 [Employer's Financial Arrangements].
- The Engineer fails to issue an Interim Payment Certificate as Sub-Clause 14.6 [Issue of Interim Payment Certificates].
- The Employer fails to pay the Contractor the sum due in compliance with Sub-Clause 14.7 [Payment].

If the Contractor wishes to invoke the provisions of this Sub-Clause he must give at least 21 days' notice and describe the information, Certificate or payment that he has not received. If the information, Certificate or payment is not received during the notice period then the Contractor can reduce or suspend work, but must resume normal working 'as soon as is reasonably practicable' if the information, Certificate or payment is received before he gives a notice of termination under Sub-Clause 16.2 [Termination by Contractor].

The Contractor should take extreme care to ensure that the allegations on which he is basing his decision to reduce the rate of work, or even suspend the work are absolutely correct, failing which the Employer would be entitled to terminate himself under Sub-Clause 15.2(b) or (c) [Termination by Employer].

When the Contractor takes action to reduce or suspend work he will inevitably incur additional Costs and delay, which may delay completion. The Contractor should then give notice under Sub-Clause 16.1 and follow the procedures of Sub-Clause 8.4 [Extension of Time for Completion] and Sub-Clause 20.1 [Contractor's Claims]. The Engineer will endeavour to reach agreement or make a determination in accordance with Sub-Clause 3.5 [Determinations]. The Contractor's Costs would include the Cost of resuming work and he is also entitled to profit on those Costs.

16.2 Termination by Contractor
The Contractor shall be entitled to terminate the Contract if:

(a) *the Contractor does not receive the reasonable evidence within 42 days after giving notice under Sub-Clause 16.1 [Contractor's Entitlement to Suspend Work] in respect of a failure to comply with Sub-Clause 2.4 [Employer's Financial Arrangements];*

(b) *the Engineer fails, within 56 days after receiving a Statement and supporting documents, to issue the relevant Payment Certificate;*

(c) *the Contractor does not receive the amount due under an Interim Payment Certificate within 42 days after the expiry of the time stated in Sub-Clause 14.7 [Payment] within which payment is to be made (except for deductions in accordance with Sub-Clause 2.5 [Employer's Claims]);*

(d) *the Employer substantially fails to perform his obligations under the Contract;*

(e) *the Employer fails to comply with Sub-Clause 1.6 [Contract Agreement] or Sub-Clause 1.7 [Assignment];*

(f) *a prolonged suspension affects the whole of the Works as described in Sub-Clause 8.11 [Prolonged Suspension]; or*

(g) *the Employer becomes bankrupt or insolvent, goes into liquidation, has a receiving or administration order made against him, compounds with his creditors, or carries on business under a receiver, trustee or manager for the benefit of his creditors, or if any act is done or event occurs which (under applicable Laws) has a similar effect to any of these acts or events.*

In any of these events or circumstances, the Contractor may, upon giving 14 days' notice to the Employer, terminate the Contract. However, in the case of subparagraph (f) or (g), the Contractor may by notice terminate the Contract immediately.

The Contractor's election to terminate the Contract shall not prejudice any other rights of the Contractor, under the Contract or otherwise.

Sub-Clause 16.2 lists the reasons that would entitle the Contractor to terminate the Contract, including failure to comply with the requirements of Sub-Clause 16.1 [*Contractor's Entitlement to Suspend Work*]. The Contractor must give 14 days' notice before he terminates the Contract except that he may terminate immediately in case of prolonged suspension as Sub-Clause 8.11 [*Prolonged Suspension*] or the Employer becoming bankrupt or having any of the problems listed at paragraph (g) of this Sub-Clause. The Contractor may also terminate the Contract in the event of prolonged Force Majeure as Sub-Clause 19.6 [*Optional Termination, Payment and Release*].

Again and before taking action under this Sub-Clause the Contractor must ensure that he has legally acceptable proof that the Employer has failed to meet the relevant obligation. In the examples such as failure to make payment, this should be easy to establish. However, for a provision such as (d) '*the Employer substantially fails to perform his obligations under the Contract*' it will be more difficult to establish proof that would satisfy the Dispute Adjudication Board or Arbitration Tribunal if the Employer should dispute the termination. Any significant failure on the part of the Employer would presumably have been the subject of correspondence and probably have already resulted in claims under other Clauses of the Contract.

16.3 Cessation of Work and Removal of Contractor's Equipment
After a notice of termination under Sub-Clause 15.5 [Employer's Entitlement to Termination], Sub-Clause 16.2 [Termination by Contractor] or Sub-Clause 19.6 [Optional Termination, Payment and Release] has taken effect, the Contractor shall promptly:

(a) *cease all further work, except for such work as may have been instructed by the Engineer for the protection of life or property or for the safety of the Works;*

(b) hand over Contractor's Documents, Plant, Materials and other work, for which the Contractor has received payment; and

(c) remove all other Goods from the Site, except as necessary for safety, and leave the Site.

This Sub-Clause gives the requirements for the Contractor to leave the Site and applies to Sub-Clause 15.5 [*Employer's Entitlement to Termination*], Sub-Clause 19.6 [*Optional Termination, Payment and Release*] and Sub-Clause 16.2 [*Termination by Contractor*].

16.4 Payment on Termination

After a notice of termination under Sub-Clause 16.2 [Termination by Contractor] has taken effect, the Employer shall promptly:

(a) *return the Performance Security to the Contractor;*

(b) *pay the Contractor in accordance with Sub-Clause 19.6 [Optional Termination, Payment and Release]; and*

(c) *pay to the Contractor the amount of any loss of profit or other loss or damage sustained by the Contractor as a result of this termination.*

Under Sub-Clause 16.2 [*Termination by Contractor*] the termination is caused by failure by the Employer, so the Contractor is entitled to receive back his Performance Security and receive loss of profit and other losses or damage in addition to payment as Sub-Clause 19.6 [*Optional Termination, Payment and Release*].

FIDIC Users' Guide
ISBN 978-0-7277-5856-9

ICE Publishing: All rights reserved
http://dx.doi.org/10.1680/fug.58569.241

Chapter 23
Clause 17: Risk and Responsibility

The FIDIC form for the Contract Agreement states clearly that the Contractor is responsible for the execution and completion of the Works and will remedy any defects in the Works. In return, the Employer will pay the Contract Price to the Contractor. Clause 17 refers to risks and responsibilities for which one Party indemnifies the other Party against losses and some additional risks for which the Employer accepts responsibility for the Cost of repairing any damage to the Works. These are not all-inclusive lists and must be considered in conjunction with the risks and responsibilities stated or implied in other clauses of the Contract.

Clause 17 also contains an important Sub-Clause that limits the liability of the Parties for consequential loss and also limits the total liability of the Contractor.

The legal meaning of the phrase '*indemnify and hold harmless*' should be checked under the governing law.

17.1 Indemnities
The Contractor shall indemnify and hold harmless the Employer, the Employer's Personnel, and their respective agents, against and from all claims, damages, losses and expenses (including legal fees and expenses) in respect of:

(a) *bodily injury, sickness, disease or death, of any person whatsoever arising out of or in the course of or by reason of the Contractor's design (if any), the execution and completion of the Works and the remedying of any defects, unless attributable to any negligence, wilful act or breach of the Contract by the Employer, the Employer's Personnel, or any of their respective agents; and*

(b) *damage to or loss of any property, real or personal (other than the Works), to the extent that such damage or loss:*
    (i) *arises out of or in the course of or by reason of the Contractor's design (if any), the execution and completion of the Works and the remedying of any defects, and*
    (ii) *is attributable to any negligence, wilful act or breach of the Contract by the Contractor, the Contractor's Personnel, their respective agents, or anyone directly or indirectly employed by any of them.*

The Employer shall indemnify and hold harmless the Contractor, the Contractor's Personnel, and their respective agents, against and from all claims, damages, losses and expenses (including legal fees and expenses) in respect of (1) bodily injury, sickness, disease or death, which is attributable to any negligence, wilful act or breach of the Contract by the Employer, the Employer's Personnel, or any of their respective agents, and (2) the matters for which liability may be excluded from insurance cover, as described in sub-paragraphs (d)(i), (ii) and (iii) of Sub-Clause 18.3 [Insurance Against Injury to Persons and Damage to Property].

Sub-Clause 17.1 includes separate indemnities from the Contractor and the Employer. The indemnities cover similar losses, which are different due to the different roles of the Parties. Both indemnities cover the other Party's personnel as well as that Party itself. This is a considerable extension of the normal contractual obligations to the other Party to the Contract. Employer's Personnel includes anyone who has been notified to the Contractor. It is important that any specialist advisers or other visitors to Site are notified to the Contractor so as to be covered by the indemnity. Contractor's Personnel includes employees of Subcontractors and *'any other personnel assisting the Contractor in the execution of the Works'*, however, unlike Employer's Personnel, the names of the individuals do not have to be notified to the Employer or the Engineer.

These indemnities must be considered in conjunction with the insurance provisions at Clause 18 and cover losses that may be excluded from the insurance cover. The indemnities must also be considered together with the indemnities at other Sub-Clauses listed below.

1.13   Failure to give notices required by Law and comply with regulations.
4.2   Claims under Performance Security.
4.14   Interference with the convenience of the public.
4.16   Claims from transport of Goods.
5.2   Objection to nominated Subcontractor.
17.5   Intellectual and industrial property rights.

17.2 Contractor's Care of the Works

The Contractor shall take full responsibility for the care of the Works and Goods from the Commencement Date until the Taking-Over Certificate is issued (or is deemed to be issued under Sub-Clause 10.1 [Taking Over of the Works and Sections]) for the Works, when responsibility for the care of the Works shall pass to the Employer. If a Taking-Over Certificate is issued (or is so deemed to be issued) for any Section or part of the Works, responsibility for the care of the Section or part shall then pass to the Employer.

After responsibility has accordingly passed to the Employer, the Contractor shall take responsibility for the care of any work which is outstanding on the date stated in a Taking-Over Certificate, until this outstanding work has been completed.

If any loss or damage happens to the Works, Goods or Contractor's Documents during the period when the Contractor is responsible for their care, from any cause not listed in Sub-Clause 17.3 [Employer's Risks], the Contractor shall rectify the loss or damage at the Contractor's risk and cost, so that the Works, Goods and Contractor's Documents conform with the Contract.

The Contractor shall be liable for any loss or damage caused by any actions performed by the Contractor after a Taking-Over Certificate has been issued. The Contractor shall also be liable for any loss or damage which occurs after a Taking-Over Certificate has been issued and which arose from a previous event for which the Contractor was liable.

The Contractor is responsible for the care of the Works until the Taking-Over Certificate is issued. After this time his responsibility for the care of the Works is limited to any outstanding work listed in the Taking-Over Certificate and to any damage to the Works caused by his own actions.

17.3 Employer's Risks

The risks referred to in Sub-Clause 17.4 below are:

(a) war, hostilities (whether war be declared or not), invasion, act of foreign enemies;

(b) rebellion, terrorism, revolution, insurrection, military or usurped power, or civil war, within the Country;

(c) riot, commotion or disorder within the Country by persons other than the Contractor's Personnel and other employees of the Contractor and Subcontractors;

(d) munitions of war, explosive materials, ionising radiation or contamination by radio-activity, within the Country, except as may be attributable to the Contractor's use of such munitions, explosives, radiation or radio-activity;

(e) pressure waves caused by aircraft or other aerial devices travelling at sonic or supersonic speeds;

(f) use or occupation by the Employer of any part of the Permanent Works, except as may be specified in the Contract;

(g) design of any part of the Works by the Employer's Personnel or by others for whom the Employer is responsible; and

(h) any operation of the forces of nature which is Unforeseeable or against which an experienced contractor could not reasonably have been expected to have taken adequate preventative precautions.

The list of Employer's risks includes the items listed at Sub-Clause 19.1 [*Definition of Force Majeure*] as constituting Force Majeure with the following changes and additions.

(*a*) No change.

(*b*) The Employer's risks are restricted to actions within the Country.

(*c*) The Employer's risks are restricted to actions within the Country and do not include strikes or lockouts.

(*d*) The Employer's risks are restricted to actions within the Country.

(*e*) The Employer's risks are not included under Force Majeure.

(*f*) The Employer's risks are not included under Force Majeure.

(*g*) The Employer's risks are not included under Force Majeure.

(*h*) Covers similar situations to the Force Majeure natural catastrophes, but is both more general and more restrictive.

Items (f) and (g) are situations for which the Employer must take responsibility, hence, payment of profit is applicable under Sub-Clause 17.4 [*Consequences of Employer's Risks*]. For some other items a shared responsibility might be considered more appropriate.

However, if the Employer is local and the Contractor is from overseas then it might be appropriate for the Employer to accept the risk. The provision of insurance must also be considered.

17.4 Consequences of Employer's Risks

If and to the extent that any of the risks listed in Sub-Clause 17.3 above results in loss or damage to the Works, Goods or Contractor's Documents, the Contractor shall promptly give notice to the Engineer and shall rectify this loss or damage to the extent required by the Engineer.

If the Contractor suffers delay and/or incurs Cost from rectifying this loss or damage, the Contractor shall give a further notice to the Engineer and shall be entitled subject to Sub-Clause 20.1 [Contractor's Claims] to:

(a) an extension of time for any such delay, if completion is or will be delayed, under Sub-Clause 8.4 [Extension of Time for Completion]; and

(b) payment of any such Cost, which shall be included in the Contract Price. In the case of sub-paragraphs (f) and (g) of Sub-Clause 17.3 [Employer's Risks], reasonable profit on the Cost shall also be included.

After receiving this further notice, the Engineer shall proceed in accordance with Sub-Clause 3.5 [Determinations] to agree or determine these matters.

Some of the items listed at Sub-Clause 17.3 are similar to events that are noted elsewhere in the Contract as giving justification for a claim by the Contractor. Any claims should always be considered together with the provisions for Force Majeure at Clause 19. The distinction is that the Employer's risk is only for rectifying the loss or damage which has occurred to the Works, Goods or Contractor's Documents. It does not cover any other Costs that may have been incurred by the Contractor. When damage occurs the Contractor must give the usual notices to the Engineer but is only required to rectify the loss or damage to the extent that is required by the Engineer.

17.5 Intellectual and Industrial Property Rights

In this Sub-Clause, 'infringement' means an infringement (or alleged infringement) of any patent, registered design, copyright, trade mark, trade name, trade secret or other intellectual or industrial property right relating to the Works; and 'claim' means a claim (or proceedings pursuing a claim) alleging an infringement.

Whenever a Party does not give notice to the other Party of any claim within 28 days of receiving the claim, the first Party shall be deemed to have waived any right to indemnity under this Sub-Clause.

The Employer shall indemnify and hold the Contractor harmless against and from any claim alleging an infringement which is or was:

(a) an unavoidable result of the Contractor's compliance with the Contract; or

(b) *a result of any Works being used by the Employer:*

    (i) *for a purpose other than that indicated by, or reasonably to be inferred from, the Contract, or*

    (ii) *in conjunction with any thing not supplied by the Contractor,*

unless such use was disclosed to the Contractor prior to the Base Date or is stated in the Contract.

The Contractor shall indemnify and hold the Employer harmless against and from any other claim which arises out of or in relation to (i) the manufacture, use, sale or import of any Goods, or (ii) any design for which the Contractor is responsible.

If a Party is entitled to be indemnified under this Sub-Clause, the indemnifying Party may (at its cost) conduct negotiations for the settlement of the claim, and any litigation or arbitration which may arise from it. The other Party shall, at the request and cost of the indemnifying Party, assist in contesting the claim. This other Party (and its Personnel) shall not make any admission which might be prejudicial to the indemnifying Party, unless the indemnifying Party failed to take over the conduct of any negotiations, litigation or arbitration upon being requested to do so by such other Party.

Sub-Clause 17.5 includes separate indemnities from the Employer and the Contractor. The indemnities cover claims that are made against the other Party by someone who alleges that their rights have been infringed. Unlike Sub-Clause 17.1 [*Indemnities*], these indemnities do not include legal fees and expenses of the innocent Party. The details differ in that the roles of the Parties, and hence the details of possible claims, are different.

In order to exercise the right to this indemnity, notice must be given within 28 days of receiving the claim alleging an infringement under this Sub-Clause (as opposed to, and not to be confused with, claims as interpreted under Sub-Clause 2.5 [*Employer's Claims*] and/ or Sub-Clause 20.1 [*Contractor's Claims*]) from the person who alleges infringement. Notice should be given at the earliest possible date and failure to give prompt notice of the claim might restrict the ability of the indemnifying Party to defend the claim.

17.6 Limitation of Liability

Neither Party shall be liable to the other Party for loss of use of any Works, loss of profit, loss of any contract or for any indirect or consequential loss or damage which may be suffered by the other Party in connection with the Contract, other than under Sub-Clause 16.4 [Payment on Termination] and Sub-Clause 17.1 [Indemnities].

The total liability of the Contractor to the Employer, under or in connection with the Contract other than under Sub-Clause 4.19 [Electricity, Water and Gas], Sub-Clause 4.20 [Employer's Equipment and Free-Issue Material], Sub-Clause 17.1 [Indemnities] and Sub-Clause 17.5 [Intellectual and Industrial Property Rights], shall not exceed the sum stated in the Particular Conditions or (if a sum is not so stated) the Accepted Contract Amount.

This Sub-Clause shall not limit liability in any case of fraud, deliberate default or reckless misconduct by the defaulting Party.

Sub-Clause 17.6 removes any liability of either Party to the other for losses other than for direct Costs, except for indemnities under Sub-Clause 17.1 [*Indemnities*] and the Contractor's losses as Sub-Clause 16.4(c) [*Payment on Termination*] for losses on termination following default by the Employer.

The Contractor's total liability to the Employer is restricted to the Accepted Contract Amount or the sum stated in the Particular Conditions, except under the Sub-Clauses listed. There is no similar limitation on the Employer's liability.

Both Parties' liability for fraud and certain other causes remains in accordance with the applicable law and is not limited by this Sub-Clause. The meaning and enforcement of this Sub-Clause could be controversial and will depend on any relevant provisions in the applicable law.

FIDIC Users' Guide
ISBN 978-0-7277-5856-9

ICE Publishing: All rights reserved
http://dx.doi.org/10.1680/fug.58569.247

Chapter 24
Clause 18: Insurance

Clause 18 covers the requirements for construction insurances for the Works and Contractor's Equipment, injury to persons, damage to property and for the Contractor's Personnel.

The FIDIC Contracts Guide acknowledges that all the insurance cover described in this Clause may not be available. The available cover will depend on the type of Works and the Country in which the cover is being sought. It is therefore essential that the Employer checks on the available cover when preparing the Particular Conditions. Specialist advice from an insurance expert may be necessary.

When part of the design is being carried out by the Contractor then Professional Indemnity Insurance to cover the design liability may also be required. This may cause a problem because of the fitness for purpose requirement at Sub-Clause 4.1 [*Contractor's General Obligations*].

The Multilateral Development Banks Harmonised Edition (the MDB Edition or Pink Book or PNK) is reviewed in Part 2. In the Introduction to the MDB Edition, FIDIC suggest that the Employer should consider a dedicated single policy for the project which would cover all the insurance requirements. Details of the policy should be included in the invitation to tender.

18.1 General Requirements for Insurances
In this Clause, 'insuring Party' means, for each type of insurance, the Party responsible for effecting and maintaining the insurance specified in the relevant Sub-Clause.

Wherever the Contractor is the insuring Party, each insurance shall be effected with insurers and in terms approved by the Employer. These terms shall be consistent with any terms agreed by both Parties before the date of the Letter of Acceptance. This agreement of terms shall take precedence over the provisions of this Clause.

Wherever the Employer is the insuring Party, each insurance shall be effected with insurers and in terms consistent with the details annexed to the Particular Conditions.

If a policy is required to indemnify joint insured, the cover shall apply separately to each insured as though a separate policy had been issued for each of the joint insured. If a policy indemnifies additional joint insured, namely in addition to the insured specified in this Clause, (i) the Contractor shall act under the policy on behalf of these additional joint insured except that the Employer shall act for Employer's Personnel, (ii) additional joint insured shall not be entitled to receive payments directly from the insurer or to have any other direct dealings with the insurer, and (iii) the insuring Party shall require all additional joint insured to comply with the conditions stipulated in the policy.

Each policy insuring against loss or damage shall provide for payments to be made in the currencies required to rectify the loss or damage. Payments received from insurers shall be used for the rectification of the loss or damage.

The relevant insuring Party shall, within the respective periods stated in the Appendix to Tender (calculated from the Commencement Date), submit to the other Party:

(a) evidence that the insurances described in this Clause have been effected; and

(b) copies of the policies for the insurances described in Sub-Clause 18.2 [Insurance for Works and Contractor's Equipment] and Sub-Clause 18.3 [Insurance against Injury to Persons and Damage to Property].

When each premium is paid, the insuring Party shall submit evidence of payment to the other Party. Whenever evidence or policies are submitted, the insuring Party shall also give notice to the Engineer.

Each Party shall comply with the conditions stipulated in each of the insurance policies. The insuring Party shall keep the insurers informed of any relevant changes to the execution of the Works and ensure that insurance is maintained in accordance with this Clause.

Neither Party shall make any material alteration to the terms of any insurance without the prior approval of the other Party. If an insurer makes (or attempts to make) any alteration, the Party first notified by the insurer shall promptly give notice to the other Party.

If the insuring Party fails to effect and keep in force any of the insurances it is required to effect and maintain under the Contract, or fails to provide satisfactory evidence and copies of policies in accordance with this Sub-Clause, the other Party may (at its option and without prejudice to any other right or remedy) effect insurance for the relevant coverage and pay the premiums due. The insuring Party shall pay the amount of these premiums to the other Party, and the Contract Price shall be adjusted accordingly.

Nothing in this Clause limits the obligations, liabilities or responsibilities of the Contractor or the Employer, under the other terms of the Contractor otherwise. Any amounts not insured or not recovered from the insurers shall be borne by the Contractor and/or the Employer in accordance with these obligations, liabilities or responsibilities. However, if the insuring Party fails to effect and keep in force an insurance which is available and which it is required to effect and maintain under the Contract, and the other Party neither approves the omission nor effects insurance for the coverage relevant to this default, any moneys which should have been recoverable under this insurance shall be paid by the insuring Party.

Payments by one Party to the other Party shall be subject to Sub-Clause 2.5 [Employer's Claims] or Sub-Clause 20.1 [Contractor's Claims], as applicable.

The second paragraph of this Sub-Clause implies and envisages a meeting between the Employer and the Contractor, before the issue of the Letter of Acceptance, at which the terms of the insurance will be agreed. FIDIC recognise that there may be difficulties in obtaining the insurance as required by Clause 18 and it is certainly desirable that the terms should be agreed before the Letter of Acceptance is issued or the Contract Agreement is signed. However, it is preferable that any problems are resolved before the Base Date, in order that all Tenders are submitted on the same basis. If changes are agreed after the submission of the Tender then they must be recorded and attached to the Letter of Acceptance or Contract Agreement.

The Contractor is required to effect and maintain the insurances as required by Sub-Clauses 18.2, 18.3 and 18.4 unless it is stated in the Particular Conditions that the Employer will effect these insurances. It may be preferable for the Contractor to effect the insurances which cover his Equipment and Materials as well as matters for which he has obligations to the Employer. The Contractor may be able to take advantage of favourable terms from insurance companies who know his past record. However, in some circumstances the Employer may be able to obtain favourable terms for certain insurance cover and decide to effect parts of the insurances himself. When a project involves several contracts, the Employer may prefer to take an overall insurance covering the separate contracts. Any such arrangements may require specialist insurance advice. The FIDIC *Guide for the Preparation of Particular Conditions* includes an example Clause, but warns that the Employer may not be aware of the details of the Contractor's Equipment. Any additional requirements, such as the use of insurance companies who are established in the Country, must be stated in the Particular Conditions.

The Appendix to Tender must give the number of days from the Commencement Date for the submission of evidence that the insurance has been effected and for submission of copies of the relevant policies. These periods will be determined to give the Contractor time to effect the insurances but the insurance should normally be arranged before the Commencement Date. The Sub-Clause recognises that the insurance terms and details will probably have been discussed before the issue of the Letter of Acceptance. Any changes to the provisions of Clause 18 must be confirmed by the Contractor and accepted by the Employer in the Letter of Acceptance. The instructions to tenderers may have required insurance proposals to be included with the Contractor's Tender.

Sub-Clause 18.1 emphasises that the insurance requirements do not limit the obligations, liabilities or responsibilities for the Parties under the other terms of the Contract or otherwise. Either Party can claim from the other Party for any Cost that is not recoverable from the insurer. Any such claims must be submitted under Sub-Clause 2.5 [*Employer's Claims*] for claims by the Employer or Sub-Clause 20.1 [*Contractor's Claims*] for claims by the Contractor, and will be determined by the Engineer in accordance with the provisions of the Contract.

18.2 Insurance for Works and Contractor's Equipment
The insuring Party shall insure the Works, Plant, Materials and Contractor's Documents for not less than the full reinstatement cost including the costs of demolition, removal of

debris and professional fees and profit. This insurance shall be effective from the date by which the evidence is to be submitted under sub-paragraph (a) of Sub-Clause 18.1 [General Requirements for Insurances], until the date of issue of the Taking-Over Certificate for the Works.

The insuring Party shall maintain this insurance to provide cover until the date of issue of the Performance Certificate, for loss or damage for which the Contractor is liable arising from a cause occurring prior to the issue of the Taking-Over Certificate, and for loss or damage caused by the Contractor in the course of any other operations (including those under Clause 11 [Defects Liability]).

The insuring Party shall insure the Contractor's Equipment for not less than the full replacement value, including delivery to Site. For each item of Contractor's Equipment, the insurance shall be effective while it is being transported to the Site and until it is no longer required as Contractor's Equipment.

Unless otherwise stated in the Particular Conditions, insurances under this Sub-Clause:

(a) shall be effected and maintained by the Contractor as insuring Party;

(b) shall be in the joint names of the Parties, who shall be jointly entitled to receive payments from the insurers, payments being held or allocated between the Parties for the sole purpose of rectifying the loss or damage;

(c) shall cover all loss and damage from any cause not listed in Sub-Clause 17.3 [Employer's Risks];

(d) shall also cover loss or damage to a part of the Works which is attributable to the use or occupation by the Employer of another part of the Works, and loss or damage from the risks listed in sub-paragraphs (c), (g) and (h) of Sub-Clause 17.3 [Employer's Risks], excluding (in each case) risks which are not insurable at commercially reasonable terms, with deductibles per occurrence of not more than the amount stated in the Appendix to Tender (if an amount is not so stated, this sub-paragraph (d) shall not apply); and

(e) may however exclude loss of, damage to, and reinstatement of:

(i)   a part of the Works which is in a defective condition due to a defect in its design, materials or workmanship (but cover shall include any other parts which are lost or damaged as a direct result of this defective condition and not as described in sub-paragraph (ii) below),

(ii)   a part of the Works which is lost or damaged in order to reinstate any other part of the Works if this other part is in a defective condition due to a defect in its design, materials or workmanship,

(iii) a part of the Works which has been taken over by the Employer, except to the extent that the Contractor is liable for the loss or damage, and

(iv)   Goods while they are not in the Country, subject to Sub-Clause 14.5 [Plant and Materials intended for the Works].

If, more than one year after the Base Date, the cover described in sub-paragraph (d) above ceases to be available at commercially reasonable terms, the Contractor shall (as insuring

Party) give notice to the Employer, with supporting particulars. The Employer shall then (i) be entitled subject to Sub-Clause 2.5 [Employer's Claims] to payment of an amount equivalent to such commercially reasonable terms as the Contractor should have expected to have paid for such cover, and (ii) be deemed, unless he obtains the cover at commercially reasonable terms, to have approved the omission under Sub-Clause 18.1 [General Requirements for Insurances].

Sub-Clause 18.2 requires the insurance for Works and Contractor's Equipment to be effective from the date by which the evidence is to be submitted, as stated in the Particular Conditions. The Contractor will need to ensure that he is properly insured for his own requirements after this date. The insuring Party is also required to submit evidence of payment to the other Party and to the Engineer. For the initial premium this would be part of the evidence that the insurance has been effected, because insurance is not effective until the premium has been paid. However, the initial payment will probably be for one year only and it is important that evidence is provided of payment of the renewal premiums.

The insurance must cover the full reinstatement cost, as defined in the first sentence of the Sub-Clause. The insurance must commence by the date by which the evidence of insurance must have been submitted as stated in the Appendix to Tender and Sub-Clause 18.1 [*General Requirements for Insurances*] subparagraph (a) and continue until the date of issue of the

- Taking-Over Certificate for the Works, for all the cover as required by this Sub-Clause
- Performance Certificate for loss or damage for which the Contractor is liable as defined at the second paragraph of this Sub-Clause.

If the Contract includes some design by the Contractor and requires Tests after Completion then the second paragraph should be amended in the Particular Conditions to follow the YLW Book requirement. The insuring Party must notify the insurer of any changes or delays to either of these dates.

The requirements for this insurance are given at paragraphs (a) to (e), which may have been modified in the Particular Conditions. The insurance covers all causes not listed as Employer's risks at Sub-Clause 17.3 [*Employer's Risks*], but the provisions of paragraph (d) are complex. The paragraph states that certain risks may be excluded if cover cannot be obtained '*at commercially reasonable terms*' with deductibles per occurrence of not more than the amount stated in the Appendix to Tender. The risks listed, items (c), (g) and (h) of Sub-Clause 17.3 and certain loss or damage following use or occupation by the Employer of another part of the Works are all risks for which the Employer would be liable. The matter of whether cover can be obtained at commercially reasonable terms should be discussed and agreed, preferably before issue of the Letter of Acceptance.

The final paragraph of Sub-Clause 18.2 introduces a further complication to paragraph (d). If cover is no longer available at commercially reasonable rates more than one year from the Base Date, then the cover may be agreed to be omitted and the Employer can claim the sum which the Contractor would have allowed as premium, subject to the provisions of

Sub-Clause 2.5 [*Employer's Claims*]. However, if the Employer is prepared to dispense with paragraph (d) insurance because of an increase in the premium, it might be queried whether it is necessary in the first place, or could be omitted from the Sub-Clause.

The exclusions at subparagraph (e) are the same in the RED Book and the YLW Book, including the reference to design defects, although design may be the responsibility of either the Employer or the Contractor. This requires further study when preparing the Particular Conditions in order to establish the Employer's requirements for this insurance.

18.3 Insurance against Injury to Persons and Damage to Property

The insuring Party shall insure against each Party's liability for any loss, damage, death or bodily injury which may occur to any physical property (except things insured under Sub-Clause 18.2 [Insurance for Works and Contractor's Equipment]) or to any person (except persons insured under Sub-Clause 18.4 [Insurance for Contractor's Personnel]), which may arise out of the Contractor's performance of the Contract and occurring before the issue of the Performance Certificate.

This insurance shall be for a limit per occurrence of not less than the amount stated in the Appendix to Tender, with no limit on the number of occurrences. If an amount is not stated in the Appendix to Tender, this Sub-Clause shall not apply.

Unless otherwise stated in the Particular Conditions, the insurances specified in this Sub-Clause:

(a) shall be effected and maintained by the Contractor as insuring Party;
(b) shall be in the joint names of the Parties;
(c) shall be extended to cover liability for all loss and damage to the Employer's property (except things insured under Sub-Clause 18.2) arising out of the Contractor's performance of the Contract; and
(d) may however exclude liability to the extent that it arises from:
   (i) the Employer's right to have the Permanent Works executed on, over, under, in or through any land, and to occupy this land for the Permanent Works,
   (ii) damage which is an unavoidable result of the Contractor's obligations to execute the Works and remedy any defects, and
   (iii) a cause listed in Sub-Clause 17.3 [Employer's Risks], except to the extent that cover is available at commercially reasonable terms.

The minimum amount of third party insurance required by this Sub-Clause must be stated in the Appendix to Tender. If no amount is stated then the Sub-Clause will not apply. The Contractor will then need to ensure that he has adequate insurance to cover his liabilities to third parties.

No date is given for the cover to commence but the insurance covers events arising out of the Contractor's performance of the Contract and so must commence before the Contractor takes any action in connection with the Contract. The insurance must cover any such event up to the issue of the Performance Certificate.

The requirements for the insurance are given at paragraphs (a) to (d). Paragraph (d) refers to exclusion of circumstances where the Employer has some responsibility, but the exclusion of the Employer's risks listed at Sub-Clause 17.3 [*Employer's Risks*] does not apply if cover is available at commercially reasonable terms. The meaning of '*commercially reasonable*' could be open to debate and the matter should be discussed and agreed, as noted under Sub-Clause 18.2 [*Insurance for Works and Contractor's Equipment*], preferably before the issue of the Letter of Acceptance.

18.4 Insurance for Contractor's Personnel

The Contractor shall effect and maintain insurance against liability for claims, damages, losses and expenses (including legal fees and expenses) arising from injury, sickness, disease or death of any person employed by the Contractor or any other of the Contractor's Personnel.

The Employer and the Engineer shall also be indemnified under the policy of insurance, except that this insurance may exclude losses and claims to the extent that they arise from any act or neglect of the Employer or of the Employer's Personnel.

The insurance shall be maintained in full force and effect during the whole time that these personnel are assisting in the execution of the Works. For a Subcontractor's employees, the insurance may be effected by the Subcontractor, but the Contractor shall be responsible for compliance with this Clause.

The insurance to cover liabilities for claims concerning Contractor's Personnel must be maintained for the whole time that such persons are assisting with the execution of the Works.

Sub-Clause 18.2 [*Insurance for Works and Contractor's Equipment*] and Sub-Clause 18.3 [*Insurance against Injury to Persons and Damage to Property*] refer to '*the insuring Party*', which is stated to be the Contractor but could be the Employer if this is required by the Particular Conditions. However, Sub-Clause 18.4 does not refer to the insuring Party but states that the insurance shall be effected by the Contractor.

FIDIC Users' Guide
ISBN 978-0-7277-5856-9

ICE Publishing: All rights reserved
http://dx.doi.org/10.1680/fug.58569.255

Chapter 25
Clause 19: Force Majeure

Clause 19 includes a definition of 'Force Majeure' and gives the procedures that must be followed and the consequences if a Force Majeure event occurs. Force Majeure is a new concept for FIDIC Civil Engineering Contracts (1999 onwards), but is covered by the law of many countries. Any provisions of the governing law that cover Force Majeure or refer to situations such as *'unexpected circumstances'* must be checked.

For an event to be classed as a Force Majeure event the Clause includes a triple test.

- It should be an exceptional event or circumstance.
- It should meet the criteria contained within Sub-Clause 19.1(a) to (d) [*Definition of Force Majeure*].
- It should be of the kind or similar, listed under Sub-Clause 19.1(i) to (v).

Sub-Clause 19.7 [*Release from Performance under the Law*] refers to events or circumstances outside the control of the Parties and is not limited to the definition of Force Majeure. The procedures for payment on termination under Sub-Clause 19.6 [*Optional Termination, Payment and Release*] and 19.7 do not include provision for the resolution of any disputes, but the procedures of Sub-Clause 2.5 [*Employer's Claims*] and Sub-Clause 20.1 [*Contractor's Claims*] would apply.

The definition of Force Majeure at Sub-Clause 1.1.6.4 just refers back to Clause 19.

19.1 Definition of Force Majeure
In this Clause, 'Force Majeure' means an exceptional event or circumstance:

(a) which is beyond a Party's control;
(b) which such Party could not reasonably have provided against before entering into the Contract;
(c) which, having arisen, such Party could not reasonably have avoided or overcome; and
(d) which is not substantially attributable to the other Party.

Force Majeure may include, but is not limited to, exceptional events or circumstances of the kind listed below, so long as conditions (a) to (d) above are satisfied:

(i)   war, hostilities (whether war be declared or not), invasion, act of foreign enemies,
(ii) rebellion, terrorism, revolution, insurrection, military or usurped power, or civil war,

(iii) *riot, commotion, disorder, strike or lockout by persons other than the Contractor's Personnel and other employees of the Contractor and Subcontractors,*

(iv) *munitions of war, explosive materials, ionising radiation or contamination by radio-activity, except as may be attributable to the Contractor's use of such munitions, explosives, radiation or radio-activity, and*

(v) *natural catastrophes such as earthquake, hurricane, typhoon or volcanic activity.*

Sub-Clause 19.1 gives the four conditions that must all be met for an event or circumstance to qualify as Force Majeure. The event or circumstance does not have to be Unforeseeable, but it must prevent the Party from performing one or more of its obligations, and more importantly, which such Party could not reasonably have avoided or overcome [Sub-Clause 19.1(c)].

These conditions are similar to the conditions for '*unexpected circumstances*' that are included in the Civil Codes of some countries. A comparison with the conditions and conse-quences as given in the governing law, as stated in the Appendix to Tender, will be necessary. If the governing law is more advantageous to the Contractor then the claim may be submitted under these provisions, or as '*otherwise in connection with the Contract*' under Sub-Clause 20.1 [*Contractor's Claims*].

A non-exclusive list of five examples is given, all of which are events outside the control of either Party. The list is significant as the Contractor's right to claim for Cost as well as delay depends on the category of the Force Majeure, as provided at Sub-Clause 19.4 [*Consequences of Force Majeure*].

The Force Majeure list includes similar events that are also listed at Sub-Clause 17.3 [*Employer's Risks*] (refer to Chapter 23 for comparison). The distinction is that the Employer's risk relates to loss or damage to the Works, Goods or Contractor's Documents. Force Majeure refers to the situation when either Party is prevented from performing any of its obligations under the Contract. The relevant forces of nature are described in greater detail under Force Majeure as '*earthquake, hurricane, lightning, typhoon or volcanic activity*'. However, they are not limited to these examples unlike Employer's Risks where the list is finite. Any claim under Force Majeure should always be considered together with the provisions for Employer's Risks.

For example, under 'Employer's risks' a Contractor may include profit in a claim for additional cost where it relates to use or occupation or design by Employer's Personnel. Similar claims under Force Majeure may not include profit subject to the provisions of the Law in accordance with Sub-Clause 1.4 [*Law and Language*].

19.2 Notice of Force Majeure
If a Party is or will be prevented from performing any of its obligations under the Contract by Force Majeure, then it shall give notice to the other Party of the event or circumstances constituting the Force Majeure and shall specify the obligations, the performance of which is or will be prevented. The notice shall be given within 14 days after the Party became aware (or should have become aware), of the relevant event or circumstance constituting Force Majeure.

The Party shall, having given notice, be excused performance of such obligations for so long as such Force Majeure prevents it from performing them.

Notwithstanding any other provision of this Clause, Force Majeure shall not apply to obligations of either Party to make payments to the other Party under the Contract.

Sub-Clause 19.2 is reviewed together with Sub-Clause 19.3.

19.3 Duty to Minimise Delay

Each Party shall at all times use all reasonable endeavours to minimise any delay in the performance of the Contract as a result of Force Majeure.

A Party shall give notice to the other Party when it ceases to be affected by the Force Majeure.

Sub-Clauses 19.2 and 19.3 refer to the notices that must be given within 14 days of the Party becoming aware of the relevant event or circumstance, or when the Party should have become aware of it. Another notice must be given when the Party is no longer affected by the Force Majeure. The notice under Sub-Clause 19.2 may duplicate a notice given under another Sub-Clause.

In particular the last paragraph of Sub-Clause 20.1 [*Contractor's Claims*] states that for any claim by the Contractor the Sub-Clause requirements are in addition to any other Sub-Clause. This would indicate that a claim for the consequences of a Force Majeure event would be prejudiced if the appropriate notifications under Sub-Clause 19.2 and 19.3 have not been given properly (i.e. triple notification!)

Therefore and to avoid potential confusion and problems it is important that each notice refers to every Sub-Clause under which it is given and includes exactly the information that is stipulated in each Sub-Clause. If the Force Majeure prevents the performance of obligations by both Parties then they must each give the due notices to the other. For Force Majeure to apply the Party must be prevented from performing an obligation. It is not sufficient that performance may be more difficult or more expensive.

Sub-Clause 19.2 also includes the effect of the Force Majeure that:

'*The Party shall, having given notice, be excused performance of such obligations for so long as such Force Majeure prevents it from performing them.*'

For a Party to be excused from performing some obligation, it must be prevented from performing that particular obligation by the Force Majeure. All other obligations, which are not prevented by the Force Majeure, must be performed as required by the Contract. In particular, the final paragraph of Sub-Clause 19.2 states that the obligation to make payments is not affected by the provisions for Force Majeure. If the Force Majeure has serious consequences for the Works as a whole then the Engineer may consider suspension under Sub-Clause 8.8 [*Suspension of Work*].

There is apparently no need for the Engineer or other Party to confirm or agree that the event constitutes Force Majeure or the performance of the obligation has been prevented. However, it is clearly advisable for the Contractor to ensure that the Engineer agrees that the event constitutes Force Majeure before he assumes that he can be excused from some obligation and stops work on the relevant item of work. Under Sub-Clause 19.3 both Parties must use '*all reasonable endeavours*' to minimise any delay in the performance of the Contract, which repeats the usual legal obligation to mitigate damage due to an event outside the Party's control. The application of the phrase '*all reasonable endeavours*' is a subjective interpretation which may be argued and would be determined initially by the Engineer and then, if necessary, by the Dispute Adjudication Board.

19.4 Consequences of Force Majeure

If the Contractor is prevented from performing any of his obligations under the Contract by Force Majeure of which notice has been given under Sub-Clause 19.2 [Notice of Force Majeure], and suffers delay and/or incurs Cost by reason of such Force Majeure, the Contractor shall be entitled subject to Sub-Clause 20.1 [Contractor's Claims] to:

(a) *an extension of time for any such delay, if completion is or will be delayed, under Sub-Clause 8.4 [Extension of Time for Completion]; and*
(b) *if the event or circumstance is of the kind described in subparagraphs (i) to (iv) of Sub-Clause 19.1 [Definition of Force Majeure] and, in the case of subparagraphs (ii) to (iv), occurs in the Country, payment of any such Cost.*

After receiving this notice, the Engineer shall proceed in accordance with Sub-Clause 3.5 [Determinations] to agree or determine these matters.

If the Contractor suffers delay or additional Cost due to the Force Majeure then the rights to claim depend on which category of the list at Sub-Clause 19.1 [*Definition of Force Majeure*] is appropriate. An extension of time can be claimed, as Sub-Clause 8.4(b) [*Extension of Time for Completion*], if completion is, or will be, delayed. However, Cost can only be claimed following the warlike events at subparagraphs (i) to (iv) of Sub-Clause 19.1 and not as a consequence of natural catastrophes. If the event is a war, hostilities, invasion or act of foreign enemies, as paragraph (i), then the event may occur anywhere in the world, although it would be necessary to prove the effect on the Contract. Paragraphs (ii) to (iv) refer to more localized situations and only events in the Country are covered.

19.5 Force Majeure Affecting Subcontractor

If any Subcontractor is entitled under any contract or agreement relating to the Works to relief from force majeure on terms additional to or broader than those specified in this Clause, such additional or broader force majeure events or circumstances shall not excuse the Contractor's non-performance or entitle him to relief under this Clause.

Also refer to the appropriate section for the SUB Book for more information as to the procedures involving Force Majeure in a Main Contract versus Subcontract scenario.

19.6 Optional Termination, Payment and Release

If the execution of substantially all the Works in progress is prevented for a continuous period of 84 days by reason of Force Majeure of which notice has been given under Sub-Clause 19.2 [Notice of Force Majeure], or for multiple periods which total more than 140 days due to the same notified Force Majeure, then either Party may give to the other Party a notice of termination of the Contract. In this event, the termination shall take effect 7 days after the notice is given, and the Contractor shall proceed in accordance with Sub-Clause 16.3 [Cessation of Work and Removal of Contractor's Equipment].

Upon such termination, the Engineer shall determine the value of the work done and issue a Payment Certificate which shall include:

(a) the amounts payable for any work carried out for which a price is stated in the Contract;

(b) the Cost of Plant and Materials ordered for the Works which have been delivered to the Contractor, or of which the Contractor is liable to accept delivery; this Plant and Materials shall become the property of (and be at the risk of) the Employer when paid for by the Employer, and the Contractor shall place the same at the Employer's disposal;

(c) any other Cost or liability which in the circumstances was reasonably incurred by the Contractor in the expectation of completing the Works;

(d) the Cost of removal of Temporary Works and Contractor's Equipment from the Site and the return of these items to the Contractor's works in his country (or to any other destination at no greater cost); and

(e) the Cost of repatriation of the Contractor's staff and labour employed wholly in connection with the Works at the date of termination.

Under Sub-Clause 19.6, either Party may give a notice of termination of the Contract. The procedures are a mixture of the provisions for termination of Clauses 15 and 16, which reflects the fact that Force Majeure is not caused by any default by either Party.

19.7 Release from Performance Under the Law

Notwithstanding any other provision of this Clause, if any event or circumstance outside the control of the Parties (including, but not limited to, Force Majeure) arises which makes it impossible or unlawful for either or both Parties to fulfil its or their contractual obligations or which, under the law governing the Contract, entitles the Parties to be released from further performance of the Contract, then upon notice by either Party to the other Party of such event or circumstance:

(a) the Parties shall be discharged from further performance, without prejudice to the rights of either Party in respect of any previous breach of the Contract; and

(b) the sum payable by the Employer to the Contractor shall be the same as would have been payable under Sub-Clause 19.6 [Optional Termination, Payment and Release] if the Contract had been terminated under Sub-Clause 19.6.

259

Sub-Clause 19.7 is much wider than the Force Majeure provisions of Sub-Clause 19.1 [*Definition of Force Majeure*] and acknowledges that situations can arise when it is impossible for the Contract to continue. This is a very serious action and should only be taken after legal advice and consideration of the applicable law.

FIDIC Users' Guide
ISBN 978-0-7277-5856-9

Chapter 26
Clause 20: Claims, Disputes and Arbitration

This is arguably the most important Clause of the Contract in that it is the principle tool for the Contractor (compare Sub-Clause 2.5 [*Employer's Claims*] for the Employer) to state and submit his contractual entitlements. The Clause deals with strict procedures for the timely submission of notices and subsequent fully particularised claims (Sub-Clause 20.1 [*Contractor's Claims*]) as well as any follow-on actions through the DAB (Sub-Clauses 20.2 to 20.4), further amicable settlement (Sub-Clause 20.5) and arbitration (Sub-Clause 20.6).

In the introduction to the PNK Book, FIDIC states that the changes to the dispute provisions, at Sub-Clauses 20.2 to 20.8 (dispute resolution procedures) and the Appendix, are an improvement on the RED Book wording and could be introduced through the Particular Conditions. Any such change requires a check on other related Clauses and, in particular, Rule 2 of the Annex: Procedural Rules.

Previous FIDIC Conditions included a requirement that any dispute must first be referred to the Engineer. Only after the Engineer had made a decision on the dispute under Clause 67 could it be referred to an outside dispute resolver for amicable settlement or arbitration. The 1996 Supplement to the old Red Book introduced an option for a Dispute Adjudication Board (DAB) and in the 1999 RED Book the DAB is the standard procedure.

The FIDIC *Guidance for the Preparation of Particular Conditions* includes an alternative paragraph for Sub-Clause 20.4 [*Obtaining Dispute Adjudication Board's Decision*] that enables the Engineer to be appointed as the DAB, but this cannot be recommended. In practice, the Engineer tends to be regarded as an instrument of the Employer and in the 1999 Conditions he is defined as Employer's Personnel, which can only enhance this belief. Generally the Engineer will already have given a determination under Sub-Clause 3.5 [*Determinations*] and a decision in principle under Sub-Clause 20.1 [*Contractor's Claims*] before a dispute is referred to the DAB. This makes it virtually impossible for him to act as an independent DAB. The DAB, provided it is properly constituted, is the only way to obtain a fast and truly independent decision on a dispute.

Sub-Clauses 20.2 to 20.4 (DAB procedures) must be read together with the Appendix: General Conditions of Dispute Adjudication Agreement and the Annex: Procedural Rules, as well as the Appendix to Tender. The arrangement of information in these different documents is not always logical and all the documents must be read together in order to understand and correctly apply the DAB procedures.

In the introduction to the PNK Book, FIDIC states that the changes to the dispute procedures, at Sub-Clauses 20.2 to 20.8 (dispute resolution procedures) and the Appendix, are an improvement on the earlier wording and could be introduced into the RED Book. Any such change requires a check on other related Clauses and Rule 2 of the Annex: Procedural Rules should also be changed as in the MDB Edition.

20.1 Contractor's Claims

If the Contractor considers himself to be entitled to any extension of the Time for Completion and/or any additional payment, under any Clause of these Conditions or otherwise in connection with the Contract, the Contractor shall give notice to the Engineer, describing the event or circumstance giving rise to the claim. The notice shall be given as soon as practicable, and not later than 28 days after the Contractor became aware, or should have become aware, of the event or circumstance.

If the Contractor fails to give notice of a claim within such period of 28 days, the Time for Completion shall not be extended, the Contractor shall not be entitled to additional payment, and the Employer shall be discharged from all liability in connection with the claim. Otherwise, the following provisions of this Sub-Clause shall apply.

The Contractor shall also submit any other notices which are required by the Contract, and supporting particulars for the claim, all as relevant to such event or circumstance.

The Contractor shall keep such contemporary records as may be necessary to substantiate any claim, either on the Site or at another location acceptable to the Engineer. Without admitting the Employer's liability, the Engineer may, after receiving any notice under this Sub-Clause, monitor the record-keeping and/or instruct the Contractor to keep further contemporary records. The Contractor shall permit the Engineer to inspect all these records, and shall (if instructed) submit copies to the Engineer.

Within 42 days after the Contractor became aware (or should have become aware) of the event or circumstance giving rise to the claim, or within such other period as may be proposed by the Contractor and approved by the Engineer, the Contractor shall send to the Engineer a fully detailed claim which includes full supporting particulars of the basis of the claim and of the extension of time and/or additional payment claimed. If the event or circumstance giving rise to the claim has a continuing effect:

(a) *this fully detailed claim shall be considered as interim;*
(b) *the Contractor shall send further interim claims at monthly intervals, giving the accumulated delay and/or amount claimed, and such further particulars as the Engineer may reasonably require; and*
(c) *the Contractor shall send a final claim within 28 days after the end of the effects resulting from the event or circumstance, or within such other period as may be proposed by the Contractor and approved by the Engineer.*

Within 42 days after receiving a claim or any further particulars supporting a previous claim, or within such other period as may be proposed by the Engineer and approved by the Contractor, the Engineer shall respond with approval, or with disapproval and detailed comments. He may also request any necessary further particulars, but shall nevertheless give his response on the principles of the claim within such time.

Each Payment Certificate shall include such amounts for any claim as have been reasonably substantiated as due under the relevant provision of the Contract. Unless and until the particulars supplied are sufficient to substantiate the whole of the claim, the Contractor shall only be entitled to payment for such part of the claim as he has been able to substantiate.

The Engineer shall proceed in accordance with Sub-Clause 3.5 [Determinations] to agree or determine (i) the extension (if any) of the Time for Completion (before or after its expiry) in accordance with Sub-Clause 8.4 [Extension of Time for Completion], and/or (ii) the additional payment (if any) to which the Contractor is entitled under the Contract.

The requirements of this Sub-Clause are in addition to those of any other Sub-Clause which may apply to a claim. If the Contractor fails to comply with this or another Sub-Clause in relation to any claim, any extension of time and/or additional payment shall take account of the extent (if any) to which the failure has prevented or prejudiced proper investigation of the claim, unless the claim is excluded under the second paragraph of this Sub-Clause.

Sub-Clause 20.1 [*Contractor's Claims*] includes the overall requirement that the Contractor shall give notice to the Engineer of any claim, either for time or money, for whatever reason, '*as soon as practicable, and not later than 28 days after the Contractor became aware, or should have become aware, of the event or circumstance*'. If the Contractor fails to give this notice then he loses any entitlement to additional time or money and '*the Employer shall be discharged from all liability in connection with the claim*'. This is an extremely serious sanction for failing to submit a notice on time and it must be queried whether the Courts will enforce the sanction. In particular many civil jurisdictions are constructed so as to render this so-called time-bar unenforceable. Also the well-known 'prevention-principle' as known in some common law jurisdictions may provide relief to the Contractor. It is particularly harsh when the claim may have been caused by a default on the part of the Employer, or when the Employer will enjoy the benefit of some additional work, or when the Engineer initially recognised the validity of the claim but then withdrew his initial acceptance. If the Engineer rejects a claim on the basis of this paragraph then he must be prepared for the Contractor to pursue his claim and continue to follow the procedures under Clause 20 [*Claims, Disputes and Arbitration*]. If the immediate decision is overturned in a later decision by a DAB or arbitration tribunal then the Employer will be at a considerable disadvantage if the Engineer has not checked, monitored and maintained the appropriate records. Nevertheless, any Contractor is well advised to implement adequate procedures within his own organisation that would ensure compliance with the Sub-Clause 20.1 time-bar at all times.

In order to avoid losing the right to claim many Contractors use a standard form to submit notices for every imaginable event, many of which will not be followed by a detailed claim. Unconfirmed statistics put the number of actually pursued claims versus submitted notifications as low as 10–15%. The contractors' reasoning is simple: rather submit a notice and decide later whether to claim or not to submit and never be able to claim. The result is a drastic increase in administrative management of the projects with a commensurate increase in project overhead costs.

It is also necessary to consider the meaning of the phrase *'event or circumstance'*. These are not defined terms but, according to the dictionary, an event is the fact of something that happened. A circumstance seems to be a wider situation such as *'external conditions affecting or that might affect action'*. Clearly there is scope for debate on what exactly was the *'event or circumstance giving rise to the claim'*. There is also the possibility that the Contractor might not have been immediately aware that the event was going to entitle him to additional time and/or money. This sanction is bound to be controversial and Employers should consider whether it might be preferable to delete this paragraph and rely on the final paragraph of Sub-Clause 20.1. This paragraph results in a reduction in a claim if the Engineer was not given the opportunity to investigate the claim, but does not affect the entitlement in principle.

Whether a failure to give a notice or provide information can remove a legal entitlement is a matter that may depend on the applicable law. In any event, in order to avoid potential problems, this notice is obviously essential for the Contractor to establish his legal entitlements and must be given in writing, to the appropriate address, as required by Sub-Clause 1.3 [*Communications*].

Sub-Clause 20.1 only refers to claims by the Contractor. The equivalent provision for claims by the Employer is at Sub-Clause 2.5 [*Employer's Claims*], which is much less stringent in its requirements. Notably the requirement to issue a notice when the Employer *'should have been aware'* and any time-bar are missing. A list of notices under Sub-Clause 2.5 [*Employer's Claims*] and Sub-Clause 20.1 [*Contractor's Claims*], is to be included in the Contractor's monthly progress report under Sub-Clause 4.21 [*Progress Reports*].

The Sub-Clause 20.1 [*Contractor's Claims*] notice is in addition to the other requirements for similar notices, such as the Sub-Clauses that refer matters to the Engineer for a determination under Sub-Clause 3.5 [*Determinations*], the sequence of notices at the RED Book Sub-Clause 1.9 [*Delayed Drawings or Instructions*] and the notice hidden in Sub-Clause 8.3 [*Programme*]. Clearly these notices can be combined, but the supporting information that is required and the further actions by the Engineer and the Contractor vary for each notice.

These notices are important

- to enable the Engineer to make his own observations and records of the problem
- to enable the Engineer to consider possible actions to overcome the problem
- to put the problem on record and make it possible for the Contractor to receive a prompt decision on his entitlements.

For a quick overview as to what Sub-Clause requires a notice (or even a double or triple notice) refer to Appendix 1, 2 and 3.

Sub-Clause 20.1 [*Contractor's Claims*] includes detailed requirements for the Contractor to submit contemporary records to substantiate the claim. Contemporary records are essential to substantiate any claim, regardless of whether a resident engineer is considering it sympathetically, or if it will eventually be determined by a DAB or ultimately by an Arbitration Tribunal. The extent of records that are appropriate are the Contractor's responsibility, because it is the Contractor who will eventually have to prove his entitlement to additional time or money. It is good advice for any contractor to discuss and agree with the Engineer at the start of a claim procedure exactly what records shall be kept and in what format. This will most certainly result in a more efficient claim review procedure with probably less potential for dispute.

The Engineer is entitled but not obliged to inspect the contemporary records and to request copies. If the Engineer has any doubts about the accuracy of the records then he should raise any queries and give the Contractor the opportunity to discuss and justify the records. The Engineer may also keep his own records and is not obliged to show these records to the Contractor. However, if the Engineer fails to query the Contractor's records at the time it will be difficult for him to argue later that they are not accurate.

Having submitted all the appropriate notices and contemporary records the Contractor must now prepare his detailed claim, for submission within 42 days of when he became aware of the event. Note that this period runs concurrent with the notification time-bar. This may be just an interim evaluation of the claim but it is important for the Contractor to submit the principle that justifies the claim as soon as possible. The Engineer must then respond with approval or disapproval on the principles of the claim within a further 42 days. The Engineer must give detailed comments with his approval or disapproval. The Contract includes provisions for extending these periods but it is to everyone's advantage for the principle to be established as quickly as possible. If the principle is agreed, the Contractor will receive interim payments and can revise his programme to incorporate the decision. The Employer will know that he has an additional liability and that completion will be delayed.

Sub-Clause 20.1 [*Contractor's Claims*] includes a requirement for the Engineer to carry out Sub-Clause 3.5 [*Determinations*] determinations. However, the paragraph with this requirement is positioned after he has made his decision on the principles of the claim. The other claims clauses, such as Sub-Clause 2.1 [*Employer's Claims*], simply state that the Contractor gives notice and '*after receiving this notice the Engineer shall proceed in accordance with Sub-Clause 3.5*'. From a claims management point of view, the Sub-Clause 3.5 consultations should proceed as soon as possible after both sides have established their positions on the merits of the claim. In the opinion of the authors, the Contract should state that the Engineer may proceed with his Sub-Clause 3.5 determination when he considers that the time is right, or at the request of either Party. The determination might cover just the principle, or also the details, as appropriate. It does not seem right for consultation and determination to wait several months after the claim event and the Contractor's expenditure of money. Also, if the claim results in a dispute then it should be referred to the DAB as soon as possible.

20.2 Appointment of the Dispute Adjudication Board

Disputes shall be adjudicated by a DAB in accordance with Sub-Clause 20.4 [Obtaining Dispute Adjudication Board's Decision]. The Parties shall jointly appoint a DAB by the date stated in the Appendix to Tender.

The DAB shall comprise, as stated in the Appendix to Tender, either one or three suitably qualified persons ('the members'). If the number is not so stated and the Parties do not agree otherwise, the DAB shall comprise three persons.

If the DAB is to comprise three persons, each Party shall nominate one member for the approval of the other Party. The Parties shall consult both these members and shall agree upon the third member, who shall be appointed to act as chairman.

However, if a list of potential members is included in the Contract, the members shall be selected from those on the list, other than anyone who is unable or unwilling to accept appointment to the DAB.

The agreement between the Parties and either the sole member ('adjudicator') or each of the three members shall incorporate by reference the General Conditions of Dispute Adjudication Agreement contained in the Appendix to these General Conditions, with such amendments as are agreed between them.

The terms of the remuneration of either the sole member or each of the three members, including the remuneration of any expert whom the DAB consults, shall be mutually agreed upon by the Parties when agreeing the terms of appointment. Each Party shall be responsible for paying one-half of this remuneration.

If at any time the Parties so agree, they may jointly refer a matter to the DAB for it to give its opinion. Neither Party shall consult the DAB on any matter without the agreement of the other Party.

If at any time the Parties so agree, they may appoint a suitably qualified person or persons to replace (or to be available to replace) any one or more members of the DAB. Unless the Parties agree otherwise, the appointment will come into effect if a member declines to act or is unable to act as a result of death, disability, resignation or termination of appointment.

If any of these circumstances occurs and no such replacement is available, a replacement shall be appointed in the same manner as the replaced person was required to have been nominated or agreed upon, as described in this Sub-Clause.

The appointment of any member may be terminated by mutual agreement of both Parties, but not by the Employer or the Contractor acting alone. Unless otherwise agreed by both Parties, the appointment of the DAB (including each member) shall expire when the discharge referred to in Sub-Clause 14.12 [Discharge] shall have become effective.

The RED Book DAB procedure requires the appointment of the DAB at the start of the project and this is referred to as a 'full term' or 'standing' DAB, as opposed to an 'ad-hoc'

DAB under the YLW Book. The DAB is kept informed about the project, makes regular visits to the Site and makes decision on any disputes arising.

The DAB is an independent panel of three people, or a single person, who give decisions on any disputes. The DAB decision must be implemented but if either Party is not satisfied with the DAB decision it can refer the original dispute to arbitration. Sub-Clause 20.4 [*Obtaining Dispute Adjudication Board's Decision*] states specifically that the DAB does not act as arbitrators and so are not covered by any arbitration provisions in the governing law. It is necessary for the Contract to state the procedures, powers and authority of the DAB and the actions that can or must be taken by the Parties after the DAB has given its decision on a dispute. These procedures, powers and authority are given at Sub-Clauses 20.2 to 20.4 (dispute resolution procedures), together with the Appendix and Annex that follow Clause 20 [*Claims, Disputes and Arbitration*] and the DAB Agreement printed at the end of the FIDIC document.

For the DAB to operate successfully it is essential that the following occurs.

- The procedure for the selection of members of the DAB allows for people suggested separately by each Party and the people appointed are completely independent of both Parties.
- The DAB is appointed at the time stated in the Contract and Dispute Adjudication Agreements are signed by both parties and each member of the DAB.
- Any decision by the DAB is implemented immediately (refer Sub-Clause 20.4 [*Obtaining Dispute Adjudication Board's Decision*])

Sub-Clause 20.2 [*Appointment of the Dispute Adjudication Board*] requires the DAB to be appointed at the start of the Contract, within 28 days of the Commencement Date, as stated in the Appendix to Tender.

For a three-person DAB the Sub-Clause requires that each Party shall nominate one member, for the approval of the other Party. The Chairman is then agreed between the Parties, after consultation with the two members, or appointed by the appointing authority as Sub-Clause 20.3 [*Failure to Agree Dispute Adjudication Board*]. This is sometimes referred to as the 'bottom-up' approach. Recent experience has shown that it may be advantageous to use the 'top-down' approach, that is, the Parties agree on the Chairperson, who in turn has the authority to select his two board members. One advantage is the fact that the Chairperson can choose his wingmen with the hindsight of the technical and/or legal/commercial requirements of the dispute.

A one-person DAB is chosen by agreement or appointed by the appointing authority. Different procedures for the selection of the person to be nominated are envisaged in the Contract, as follows.

- Each Party could nominate one member for the other Party's agreement during the period after the submission of the Tender and before the Letter of Acceptance. The agreed names would be confirmed in the Letter of Acceptance.

- Sub-Clause 20.2 [*Appointment of the Dispute Adjudication Board*] refers to the possibility that a list of potential members is included in the Contract. The FIDIC Guid*ance for the Preparation of Particular Conditions* suggests that such a list could be useful if the DAB is not to be appointed at the commencement of the Contract, but emphasises that '*it is essential that candidates for this position (as DAB member) are not imposed by either Party on the other Party*'. Hence, any such list would need to be prepared after the Employer has entered negotiations with the preferred tenderer. The list must not be prepared just by the Employer, but must include people suggested by the Contractor.
- One problem when preparing a list of potential DAB members is that people on the list may no longer be available when the appointment is actually made. It will also be necessary to check for any conflict of interest when the successful tenderer is known.
- The FIDIC example for the Letter of Tender allows the Contractor to accept or reject names proposed by the Employer and to include the Contractor's suggestions for his nominee. If this procedure is used it is essential that the tenderer does not feel under any pressure to accept the Employer's suggestions but feels free to add his own suggestions.
- Nominations can be submitted and agreed during the period after the Commencement Date. This is a last minute procedure which has the advantage, or possibly the disadvantage, that the individuals who will be working on the project can be involved in the selection procedure.
- If the Parties fail to agree then the FIDIC Appendix to Tender designates the President of FIDIC as the appointing authority. The Appendix to Tender may have been amended to designate a different appointing authority.

The FIDIC *Guidance for the Preparation of Particular Conditions* emphasises the importance of each Party being free to nominate one member and the benefits of appointing the DAB *before* the Letter of Acceptance.

It is important that any person appointed to a DAB has appropriate construction experience, including experience of claims and dispute resolution, knowledge of contract interpretation and knowledge of the DAB procedures. The DAB acts as a team, not as representatives of the Parties, so ideally there should be a balance of experience and professional expertise within the team. While this may be difficult to achieve for the members nominated by the separate Parties, it should be considered in the choice of the Chairman. It is always useful for the Parties to discuss their prospective nominees before making a final decision.

Training courses for DAB members have already been established and no doubt will develop further as the 1999 FIDIC Contracts are used more extensively. Lists of suitable people have been prepared by several organisations, including FIDIC (refer to the President's List of Approved Dispute Adjudicators), The Institution of Civil Engineers (ICE) in London, The International Chamber of Commerce (ICC) in Paris, the Dispute Board Federation (DBF) in Geneva and other organisations in different countries.

The FIDIC Appendix to Tender names the President of FIDIC as the appointing authority but other organisations can be substituted in the Appendix to Tender for the particular Contract or by agreement between the Parties.

If a dispute occurs and there is no DAB in place it is unlikely that the Parties will agree on the composition of a DAB. In such an event Sub-Clause 20.3 [*Failure to Agree Dispute Adjudication Board*] does not really help unless it is a one-man board. In this case Sub-Clause 20.3(a) applies even if it may result in an ex-parte situation with a bipartite Dispute Adjudication Agreement.

In the case of a three-man board the problem is more complex. Sub-Clause 20.2 [*Appointment of the Dispute Adjudication Board*] requires the agreement of both Parties to the respective nominations of DAB members. However, failing such an agreement there is no provision under Sub-Clause 20.3 [*Failure to Agree Dispute Adjudication Board*] to prevent one Party consistently refusing to accept the other Parties nominee thereby extending the process indefinitely (the Sub-Clause 20.2 trap). A further complication is that one Party may insist that Sub-Clause 20.8 [*Expiry of Dispute Adjudication Board's Appointment*] applies attempting to force the issue to arbitration. He would do this at a time when the Parties are still formally attempting to form a DAB. If this attempt at arbitration were successful, it would imply that Sub-Clause 20.8 could be invoked even before an attempt at establishing a DAB has been made. This could never have been the intention of the drafting committee.

The solution is to ensure the timely establishment of a DAB in accordance with Sub-Clause 20.2 [*Appointment of the Dispute Adjudication Board*]. Alternatively the Parties may consider an amendment to Sub-Clause 20.3 [*Failure to Agree Dispute Adjudication Board*] to circumvent the Sub-Clause 20.2 trap. Having accepted the appointment, each member of the DAB signs an Agreement with both Parties. Standard forms for the Dispute Adjudication Agreement for a one-person and three-person DAB are included with the FIDIC Contract. The Agreement refers to the '*General Conditions of Dispute Adjudication Agreement*' and the '*Procedural Rules*', which are annexed to Clause 20. The Conditions and Rules give detailed procedures for the working of the DAB and emphasise the importance of the DAB being impartial.

The General Conditions Dispute of Adjudication Agreement state in Sub-Clause 2 [*General Provisions*] '*When the Dispute Adjudication Agreement has taken effect, the Employer and the Contractor shall each give notice to the Member accordingly*'. It is unclear why such a notice is required and if not followed may well lead to problems. It is interesting to note that the PNK Book has removed this requirement from its General Conditions of Dispute Board Agreement. Similar changes should be considered when making appointments under the RED and YLW Books.

In accordance with the final paragraph of Sub-Clause 20.2 [*Appointment of the Dispute Adjudication Board*] a full term DAB's appointment will expire when the Contractor's discharge under Sub-Clause 14.12 [*Discharge*] becomes effective. This will normally be when the Contractor has received back his Performance Security, together with the amount due under the Final Payment Certificate. However, if the Engineer and the Contractor cannot agree on the Final Payment Certificate then the resulting dispute can be referred to the DAB, amicable settlement or arbitration. The discharge does not then become effective and the DAB would remain in place. The Parties might then agree to terminate the appointment of the DAB.

Sub-Clause 20.2 [*Appointment of the Dispute Adjudication Board*] includes a provision that the Parties may refer any matter to the DAB for an opinion at any time. The matter does not have to be a dispute, as defined at Sub-Clause 20.4 [*Obtaining Dispute Adjudication Board's Decision*], but may be a claim or a difference of opinion. This provision is important because an opinion from the DAB, or just the fact of preparing presentations to the DAB, may enable the staff on the Site to resolve their differences. Such a reference to the DAB must be with the agreement of both Parties. However, if one Party wants to refer some matter to the DAB for an opinion, the other Party should normally agree. If they do not agree to refer any matter to the DAB then the Party who wants an opinion can turn the problem into a dispute and ask the DAB to give a decision. This would mean a longer, more time consuming and hence more expensive procedure. If the DAB considers that matters are being referred unnecessarily, or frivolously, they can comment on this in their opinion. The Sub-Clause does not require the DAB opinion to be in writing, but Procedural Rule 3 requires the DAB to prepare a report on its activities before leaving the Site at the conclusion of each visit.

20.3 Failure to Agree Dispute Adjudication Board
If any of the following conditions apply, namely:

(a) *the Parties fail to agree upon the appointment of the sole member of the DAB by the date stated in the first paragraph of Sub-Clause 20.2;*

(b) *either Party fails to nominate a member (for approval by the other Party) of a DAB of three persons by such date;*

(c) *the Parties fail to agree upon the appointment of the third member (to act as chairman) of the DAB by such date; or*

(d) *the Parties fail to agree upon the appointment of a replacement person within 42 days after the date on which the sole member or one of the three members declines to act or is unable to act as a result of death, disability, resignation or termination of appointment;*

then the appointing entity or official named in the Particular Conditions shall, upon the request of either or both of the Parties and after due consultation with both Parties, appoint this member of the DAB. This appointment shall be final and conclusive. Each Party shall be responsible for paying one-half of the remuneration of the appointing entity or official.

The Sub-Clause refers to the entity or official named in the Particular Conditions. The Appendix to Tender names the President of FIDIC, which should be confirmed in the Particular Conditions. If the Employer wishes to substitute the name of a different organis-ation then care must be taken to ensure that it is an organisation with a list of potential DAB members who are suitably qualified, trained and monitored.

The Employer may also wish to investigate the extent of consultation with the Parties that the appointing entity would normally carry out, and the time it normally takes to make an appointment. An enquiry as to any particular requirements by either Party, or whether people on the appointing entity's list have already been considered and rejected by either

Party, would be reasonable. However, it is important that a reluctant Party deliberately prolonging the consultation process does not delay the appointment. This is an appointment, not a nomination or suggestion, so the Parties have to accept whomever the appointing entity decides to appoint. The Sub-Clause does not give any time limit for making this appointment, but it should be made within a matter of days, rather than weeks.

The fact that the appointing entity will make a final and conclusive appointment is in conflict with the opening paragraph of the Dispute Adjudication Agreement (DAA) as attached to the FIDIC that states that the Parties *'desire jointly to appoint the Member'*. The member has already been selected and appointed by the appointing entity so there is no real opportunity to negotiate fees or other details. However, it is generally accepted that in such a situation the appointed DAB member(s) shall be realistic in their demands. The person(s) appointed must also have studied and accepted the Procedural Rules, including warranties.

The Sub-Clause states that one half of the remuneration of the appointing entity is paid by each Party. However, as per Clause 6 of the *'General Conditions of Dispute Adjudication Agreement'* the DAB will send their invoices in the first instance to the Contractor, who shall pay them in full and in turn apply to the Employer through regular Clause 14 [*Contract Price and Payment*] statements for reimbursement of 50% of the invoice total.

20.4 Obtaining Dispute Adjudication Board's Decision

If a dispute (of any kind whatsoever) arises between the Parties in connection with, or arising out of, the Contract or the execution of the Works, including any dispute as to any certificate, determination, instruction, opinion or valuation of the Engineer, either Party may refer the dispute in writing to the DAB for its decision, with copies to the other Party and the Engineer. Such reference shall state that it is given under this Sub-Clause.

For a DAB of three persons, the DAB shall be deemed to have received such reference on the date when it is received by the chairman of the DAB.

Both Parties shall promptly make available to the DAB all such additional information, further access to the Site, and appropriate facilities, as the DAB may require for the purposes of making a decision on such dispute. The DAB shall be deemed to be not acting as arbitrator(s).

Within 84 days after receiving such reference, or within such other period as may be proposed by the DAB and approved by both Parties, the DAB shall give its decision, which shall be reasoned and shall state that it is given under this Sub-Clause. The decision shall be binding on both Parties, who shall promptly give effect to it unless and until it shall be revised in an amicable settlement or an arbitral award as described below. Unless the Contract has already been abandoned, repudiated or terminated, the Contractor shall continue to proceed with the Works in accordance with the Contract.

If either Party is dissatisfied with the DAB's decision, then either Party may, within 28 days after receiving the decision, give notice to the other Party of its dissatisfaction. If the DAB fails to give its decision within the period of 84 days (or as otherwise approved) after receiving such reference, then either Party may, within 28 days after this period has expired, give notice to the other Party of its dissatisfaction.

In either event, this notice of dissatisfaction shall state that it is given under this Sub-Clause, and shall set out the matter in dispute and the reason (s) for dissatisfaction. Except as stated in Sub-Clause 20.7 [Failure to Comply with Dispute Adjudication Board's Decision] and Sub-Clause 20.8 [Expiry of Dispute Adjudication Board's Appointment], neither Party shall be entitled to commence arbitration of a dispute unless a notice of dissatisfaction has been given in accordance with this Sub-Clause.

If the DAB has given its decision as to a matter in dispute to both Parties, and no notice of dissatisfaction has been given by either Party within 28 days after it received the DAB's decision, then the decision shall become final and binding upon both Parties.

FIDIC does not define what is meant by the word '*dispute*'. Presumably the word will have its normal meaning, that is, any statement, complaint, request, allegation or claim which has been rejected and the rejection is not acceptable to the person who made the original statement or complaint, also referred to as a 'rejection of a rejection'. It is clearly not necessary for a complaint to have been considered by the Engineer under the Sub-Clause 2.5 [*Employer's Claims*] or Sub-Clause 3.5 [*Determinations*] procedure in order to create a dispute, unless it refers to a subject about which notice must be given under the Contract. However, the DAB may suggest to the Parties that they should follow the Contract procedures before asking for a formal decision. However, for the DAB to have the proper jurisdiction to factually make a decision on the matter, it is imperative that there is a dispute.

The strength of the DAB procedure is that whenever there is any problem between the people on Site – whether it is caused by a difference of opinion on a technical matter, a problem of interpretation of the Contract, communication or simply a misunderstanding – it can be referred quickly to an independent tribunal. The problem can then be resolved, with the assistance of the DAB, whether this requires an explanation, opinion or binding decision.

Sub-Clause 20.4 [*Obtaining Dispute Adjudication Board's Decision*] lays down the procedure for the DAB and further details are given in the FIDIC Annex: Procedural Rules, including the following.

- Both Parties make available to the DAB any information or facilities which it requires. The DAB may decide to conduct a hearing to consider submissions on the dispute from the Employer and the Contractor.
- The DAB gives its decision within 84 days or other agreed period, from the date the chairman receives the dispute reference. The decision will be in writing, with reasons and shall state that '*it is made pursuant to Sub-Clause 20.4*';
- Both Parties will comply immediately with the decision, which is binding until revised by agreement or arbitration.
- If either Party is dissatisfied with the decision, or the DAB fails to give a decision within the due period, it can give notice of dissatisfaction within 28 days.

- If no notice of dissatisfaction has been given within the due period, subject only to Sub-Clauses 20.7 [*Failure to Comply with Dispute Adjudication Board's Decision*] and 20.8 [*Expiry of Dispute Adjudication Board's Appointment*], the decision becomes final and binding on the Parties and they lose the right to refer that dispute to arbitration.

The requirement that both parties must comply immediately with the decision means that, if the DAB requires a sum of money to be paid, it must be paid immediately. The procedure for payment is not stated but some DABs may require that payment be made within, say, seven days. This may be impossible within the Employer's internal procedures and it is probably more realistic for payments during the construction period to be included in the next monthly payment certificate. In such case the Contractor should include the payment in his next Statement under Sub-Clause 14.3 [*Application for Interim Payment Certificates*]. If for any reason payment is not certified or is not paid then the Contractor has the remedies under Sub-Clauses 14.8 [*Delayed Payment*], 16.1 [*Contractor's Entitlement to Suspend Works*] and 16.2 [*Termination by Contractor*].

Since the introduction of DABs in the mid-90s many projects have successfully used the concept to avoid or mitigate conflicts and disputes. However, unlike arbitration, there was always the question of legal enforceability of any DAB decision rendered. Recent developments have shown that DAB decisions can also be enforced through the courts. The details of this go beyond the scope of this book but the authors would like to draw particular attention to a case in Singapore (i.e. CRW Joint Operation v. PT Perusahaan Gas Negera (Persero) TBK [2011] SGCA33).

The decision of the High Court of Singapore illustrates, a party in whose favour a DAB decision has been made should be advised to bring a second set of DAB proceedings should the other party issue a notice of dissatisfaction and then ignore the first DAB decision. That second referral will be in respect of what is essentially a separate dispute, namely the losing party's failure to comply with that decision, in breach of Sub-Clause 20.4 [*Obtaining Dispute Adjudication Board's Decision*]. A second DAB is of course likely to delay by 4–5 months (i.e. the time for the DAB to make a decision and the 56-day period of amicable settlement under Sub-Clause 20.5 [*Amicable Settlement*]) the commencement of arbitration, which may be critical in some jurisdictions where the limitation period is particularly short. It is however an essential step if the winning party wishes to enforce the first DAB decision.

The FIDIC Contracts Committee has issued a Guidance Note dealing with the powers of, effect of and the enforcement of the DAB decisions. The purpose of the Guidance Note is to clarify Clause 20 [*Claims, Disputes and Arbitration*] of the General Conditions in the 1999 Conditions of Contract. The contractual machinery at Clause 20 makes a distinction between DAB decisions that are binding and those that are final and binding. There is then a dual pathway to arbitration, depending upon whether a decision has become binding or final and binding. If one party issues a notice of dissatisfaction within 28 days then the decision is simply binding. In the absence of a notice of dissatisfaction the binding decision becomes final and binding.

The Guidance Note sets out a new Sub-Clause 20.4, and amends the wording to Sub-Clause 20.7 as well as providing further provisions at Sub-Clauses 14.6 and 14.7. The amendments can be used for the RED, YLW and Silver Books. The Gold Book adopts a different approach, and so the amendments could not be used in their current state.

FIDIC's recommendation is the introduction of a new penultimate paragraph of Sub-Clause 20.4: '*If the decision of the DAB requires a payment by one Party to the other Party, the DAB may require the payee to provide an appropriate security in respect of such payment.*' In effect, this is simply giving the DAB a further power. It is a contractual power to order one party to provide security. The DAB cannot force a party to comply, and so once again a party may seek to rely upon arbitration in order to obtain an appropriate sanction and then seek to enforce that award in an appropriate court.

In relation to the payment provisions in Clause 14 [*Contract Price and Payment*], a payment under Sub-Clause 14.6 [*Issue of Interim Payment Certificates*] 'shall' now include any amounts due to or from the Contractor in accordance with the DAB's decision. Sub-Clause 14.7 [*Payment*] requires amounts due under a DAB decision to be included within any Interim Payment Certificate that is to be issued. This new approach therefore requires any amount ordered by the DAB to be paid should be included within an assessment of payment made by the Engineer or the Employer's Representative, and then included within the Interim Payment. Failure to do so is simply a further breach. More importantly, Sub-Clause 20.7 [*Failure to Comply with Dispute Adjudication Board's Decision*] is deleted and replaced with a new Sub-Clause 20.7 as follows:

'*In the event that a Party fails to comply with any decision of the DAB, whether binding or final and binding, then the other Party may, without prejudice to any other rights it may have, refer the failure itself to arbitration under Sub-Clause 20.6 [Arbitration] for summary or other expedited relief, as may be appropriate. Sub-Clause 20.4 [Obtain Dispute Adjudication Board's Decision] and Sub-Clause 20.5 [Amicable Settlement] shall not apply to this reference.*'

Sub-Clause 20.7 [*Failure to Comply with Dispute Adjudication Board's Decision*] relates to decisions that are either binding or final and binding. So regardless of any notice of dissatisfaction, or more importantly any arguments or issues as to the adequacy or timing of any notice of dissatisfaction, a valid referral can be made to arbitration. The amendment also clarifies that the parties expect a summary or expedited relief to be used if and as appropriate (Gould, 2013).

The amendment to the FIDIC provisions should be included in future contracts, and should assist a party to obtain an arbitration award that could avoid the problems of the *Singapore CRW* v. *PT Perusahaan* case.

26.1.   Alternative to Sub-Clause 20.4 [*Obtaining Dispute Adjudication Board's decision*]

The FIDIC *Guidance for the Preparation of Particular Conditions* includes an alternative paragraph for Sub-Clause 20.4 for use if any disputes are to be referred to the Engineer for an initial decision. This requires the following changes.

- Delete Sub-Clauses *20.2 and 20.3.*
- Delete the second paragraph of Sub-Clause 20.4 and substitute: *The Engineer shall act as the DAB in accordance with this Sub-Clause 20.4, acting fairly, impartially and at the cost of the Employer. In the event that the Employer intends to replace the Engineer, the Employer's notice under Sub-Clause 3.4 shall include detailed proposals for the appointment of a replacement DAB.*

In this case the Engineer is required to follow the DAB procedures, acting fairly and impartially, but the Employer pays his fees. The DAB procedures impose duties and obligations on the Engineer that are different to his role under this Contract. It is difficult to envisage that an Engineer can realistically perform DAB duties having already given a determination under Sub-Clause 3.5 [*Determinations*] and a decision on the principles of the claim under Sub-Clause 20.1 [*Contractor's Claims*]. Also, the Employer pays the Engineer DAB, whereas the fact that DAB members are appointed and paid by both Parties adds considerably to their status and the creditability of their decision. An Engineer DAB who has had several decisions changed by an arbitration tribunal may find himself exposed to accusations of bad faith, impartiality or even negligence.

20.5 Amicable Settlement
Where notice of dissatisfaction has been given under Sub-Clause 20.4 above, both Parties shall attempt to settle the dispute amicably before the commencement of arbitration. However, unless both Parties agree otherwise, arbitration may be commenced on or after the fifty-sixth day after the day on which notice of dissatisfaction was given, even if no attempt at amicable settlement has been made.

Following issue of a notice of dissatisfaction there is a 56-day period before either Party can commence arbitration. Arbitration does not have to be started immediately. When the notice has been issued it may be preferable to leave arbitration proceedings until the construction has been completed. This may enable all disputes to be dealt with in a single arbitration. Alternatively, there may be reasons why a claimant wants a final decision on the dispute as soon as possible and the arbitration can be started immediately after the amicable settlement period.

The amicable settlement period enables both Parties to consider their position, consult with their advisers and senior management, and put forward proposals to settle the dispute in order to avoid the additional cost and loss of management time, together with the frustration and deterioration in relationships that can result from arbitration.

The FIDIC Conditions require that the Parties shall attempt to settle the dispute amicably, but do not give guidance as to the procedure that might be used. When an Employer has a preference for a particular procedure, as a part of a sequence of dispute procedures, then details should be included in the Particular Conditions. Possible procedures include the following.

- *Direct negotiation.* The Contract procedures require the Engineer to consider claims and consult with both Parties. The Engineer has been closely involved in the analysis of the claim and any discussions. If the Contract procedures have failed to achieve agreement then direct negotiations between the Parties, without the involvement of the Engineer, may help to break the deadlock. The negotiations may be conducted between senior personnel from head office, rather than people who have been involved with the project on Site.

- *The Engineer.* The DAB's reasons for its decision may cause the Engineer to reconsider or revise his previous analysis and recommendations to his client. This may enable the Engineer to assist the Parties to reach a settlement.

- *Mediation.* Mediation is a procedure under which an independent person is appointed to help the Parties overcome their dispute and find an agreed settlement. The mediator meets the Parties, together and separately in confidence, discusses their views and the reasons why each Party is not prepared to accept the DAB decision. The mediator is looking for a solution that will be acceptable to both Parties, as distinct from the correct decision in accordance with the Contract.

- *Conciliation.* Conciliation can be similar to mediation, but is generally conducted under a set of rules or a published procedure. The Institution of Civil Engineers' Conciliation Procedure starts with a procedure similar to mediation, but if the Parties are not able to reach an agreed settlement the conciliator gives them his recommendation as to how the dispute should be settled. The recommendation is based on practical commercial considerations rather than the legal contractual situation. Some regional arbitration centres and Chambers of Commerce have published conciliation procedures. These may provide for either a tribunal or single conciliator who proposes and discusses alternative ways of settling the dispute.

While the DAB has already performed a function as an independent tribunal, the mediator or conciliator is not just another independent person repeating a similar procedure. The DAB has collated and analysed the evidence in order to establish the rights and obligations of the Parties in accordance with the Contract. The mediator or conciliator is looking for a commercial agreement and has the additional benefit of confidential discussions with each Party separately and so can consider their aims and problems as well as their rights. Also, of course, one or both Parties may think that the DAB reached the wrong decision. A mediator or conciliator may have suggestions as to how any such situation could be resolved.

If provision for mediation or conciliation is included in the Particular Conditions it is necessary to name an appointing authority in case the Parties are not able to agree on a suitable person.

It must always be considered that any DAB decision has been made with due care and consideration taking account of all relevant facts and circumstances surrounding the dispute. Over the years experience has shown that only very few DAB decisions have ever been overturned in a subsequent arbitration procedure. As such it is unlikely that the 'winner' in a DAB decision is even willing to compromise on the decision. Similarly for the 'loser' in the DAB decision, it is important to realise these statistics before embarking on even more expenditure in his pursuit of what he rightly or wrongly perceives to be justice.

20.6 Arbitration

Unless settled amicably, any dispute in respect of which the DAB's decision (if any) has not become final and binding shall be finally settled by international arbitration. Unless otherwise agreed by both Parties:

(a) the dispute shall be finally settled under the Rules of Arbitration of the International Chamber of Commerce;

(b) the dispute shall be settled by three arbitrators appointed in accordance with these Rules; and

(c) the arbitration shall be conducted in the language for communications defined in Sub-Clause 1.4 [Law and Language].

The arbitrator(s) shall have full power to open up, review and revise any certificate, determination, instruction, opinion or valuation of the Engineer, and any decision of the DAB, relevant to the dispute. Nothing shall disqualify the Engineer from being called as a witness and giving evidence before the arbitrator(s) on any matter whatsoever relevant to the dispute.

Neither Party shall be limited in the proceedings before the arbitrator(s) to the evidence or arguments previously put before the DAB to obtain its decision, or to the reasons for dissatisfaction given in its notice of dissatisfaction. Any decision of the DAB shall be admissible in evidence in the arbitration.

Arbitration may be commenced prior to or after completion of the Works. The obligations of the Parties, the Engineer and the DAB shall not be altered by reason of any arbitration being conducted during the progress of the Works.

If either Party is not prepared to accept the DAB decision and has issued a notice of dissatisfaction within the proper time period then the dispute will finally be settled by arbitration. It is the dispute itself that is referred to arbitration and not the decision of the DAB.

A detailed review of the procedures and legal aspects of arbitration is outside the scope of this book. Arbitration is a legal process, which leads to an enforceable award, and is subject to the applicable law. However, most legal systems recognise an agreement to refer disputes to arbitration and, when the Contract includes such an agreement, the Courts will not consider a dispute until it has been decided by arbitration.

Most international construction Contracts incorporate an arbitration agreement and most international Contractors prefer an independent arbitration to having to refer disputes to the local Courts of the Country of the project. While most Court systems are reliable there is always likely to be some feeling that the Employer is from the same Country as the Courts. Also, arbitration is generally conducted in the same language as the Contract administration, whereas referral to the local Courts may require translation of a substantial number of documents and the use of interpreters during a hearing. All this can add to the costs and many international Contractors will increase their Tender price if the Contract does not include provision for arbitration. Furthermore, and as opposed to a normal court procedure, there is a distinct advantage that the Parties have a say in the appointment of the arbitral tribunal, and in doing so can appoint persons that have a distinct set of skills and experience relevant to the dispute. Nevertheless and subsequent to recent court decisions on disputes brought before an arbitral tribunal involving the FIDIC 1999 dispute resolution procedures (DAB – Amicable Settlement – Arbitration) it must be considered that there are two pathways to arbitration under the standard FIDIC form. The first (under Sub-Clause 20.6) is in order to resolve disputed DAB decisions or where no DAB decision has been issued, and the second (under Sub-Clause 20.7) is to deal with the situation where there has been a failure to comply with a DAB decision. In terms of enforcing a DAB decision this means that the party referring the matter to arbitration has to select the applicable arbitration clause, and draft a referral that reflects the requirements of that provision.

In order to commence arbitration under Sub-Clause 20.6 a notice of dissatisfaction must be issued within time. If the notice is not issued, then the DAB's decision becomes '*final and binding*', and any failure of either party to '*promptly give effect*' to the DAB's decision can be referred to arbitration under Sub-Clause 20.7. It is '*the failure itself*' to comply that is referred to arbitration. Nonetheless, the reference under Sub-Clause 20.7 is made without the need to return to the DAB or to engage in amicable dispute resolution.

A referral to arbitration under Sub-Clause 20.7 is merely one of checking to see if the DAB itself had jurisdiction to decide the dispute, and then simply confirm the DAB decision in arbitration award without considering the merits. However, the '*open up review and revise*' provisions in Sub-Clause 20.6 could still apply, even to a reference under Sub-Clause 20.7 (Gould, 2011).

In another landmark arbitral procedure (ICC Case No. 15956/GZ) the enforcement of a binding but not final 'ex parte' decision of a DAB under a 1999 FIDIC RED Book was sought. This case was broadly based on ICC Case No. 10619 concerning the 1987 fourth edition predecessor of the current 1999 RED Book. In this case the arbitral tribunal agreed with the claimant's primary argument that a DAB decision is enforceable directly and finally under a partial award. In reaching that view the tribunal made it clear that the subject matter of a DAB decision is, of course, able to be opened up, reviewed and revised by the arbitral tribunal later in the arbitration in accordance with express power to do so granted by Sub-Clause 20.6 [*Arbitration*] of the RED Book. Accordingly the result is to be seen as interim but, nonetheless, immediately enforceable (di Folco and Tiggeman, 2010).

In summary it must be said that when a dispute involving a prior DAB decision is considered to be settled under arbitration care must be taken, and professional advice must be sought.

The FIDIC Contract includes provision for arbitration under the Arbitration Rules of the International Chamber of Commerce in Paris, but the FIDIC *Guidance for the Preparation of Particular Conditions* allows that other rules or administering authorities can be named.

The procedures to be followed in arbitration are controlled by the arbitration rules designated in the Contract and arbitration law of the Country that is chosen as the place of arbitration. Arbitration rules vary considerably in the detailed procedures and must be studied carefully before being designated in the Contract. Most rules require that certain details are either stated in the Contract, agreed by the Parties after the dispute has arisen or, if the Parties cannot agree, decided by the administering authority. The details to be stated include the following.

- *Arbitration rules and administering authority.* Some organisations provide their rules and also act as administering authority. However, if the Contract designates rules such as the UNCITRAL Arbitration Rules, it is necessary to designate a separate administering authority.
- *The number of arbitrators in the tribunal.* The tribunal may be a single person, but is normally three people for a major construction dispute. Each Party nominates one arbitrator and the Chairman is chosen by agreement or by the appointing authority.
- *The place of arbitration.* This should not be the Country of the Employer or the Contractor. Careful consideration must be given to the choice of the place of arbitration because the arbitration law of that country will control the administration of the arbitration and the enforcement of the award. A country should be chosen which has adopted a modern arbitration law and which permits the tribunal to proceed completely independently of the national Courts. Many countries have adopted the UNCITRAL Model Arbitration Law, or have modified the Model Law to suit their requirements. Any such modifications must be studied with care.

A country should be chosen which has ratified the 1958 New York Convention on the Recognition and Enforcement of Foreign Arbitral Awards, which will assist with the eventual enforcement of the tribunal's award.

20.7 Failure to Comply with Dispute Adjudication Board's Decision
In the event that:

(a) *neither Party has given notice of dissatisfaction within the period stated in Sub-Clause 20.4 [Obtaining Dispute Adjudication Board's Decision];*
(b) *the DAB's related decision (if any) has become final and binding; and*
(c) *a Party fails to comply with this decision;*

then the other Party may, without prejudice to any other rights it may have, refer the failure itself to arbitration Under Sub-Clause 20.6 [Arbitration]. Sub-Clause 20.4 [Obtaining Dispute Adjudication Board's Decision] and Sub-Clause 20.5 [Amicable Settlement] shall not apply to this reference.

Sub-Clause 20.7 covers the situation when the DAB decision has become final and binding but has not been implemented. This failure can be referred to arbitration without having to follow the procedures of Sub-Clause 20.4 [*Obtaining Dispute Adjudication Board's Decision*] and Sub-Clause 20.5 [*Amicable Settlement*].

The fact that FIDIC has singled out a specific event that does not need to go through the procedures of Sub-Clauses 20.4 and 20.5 before being referred to arbitration, indicates that other events that may be considered for arbitration must necessarily first be subjected to a DAB decision followed by an amicable settlement attempt.

20.8 Expiry of Dispute Adjudication Board's Appointment

If a dispute arises between the Parties in connection with, or arising out of, the Contract or the execution of the Works and there is no DAB in place, whether by reason of the expiry of the DAB's appointment or otherwise:

(a) Sub-Clause 20.4 [Obtaining Dispute Adjudication Board's Decision] and Sub-Clause 20.5 [Amicable Settlement] shall not apply; and

(b) the dispute may be referred directly to arbitration under Sub-Clause 20.6 [Arbitration].

If there is no DAB in place, for whatever reason, then any dispute can be referred directly to arbitration. This is a necessary provision in order to maintain a route for the resolution of disputes if there is no DAB.

REFERENCES

Di Folco G and Tiggeman M (2010) Enforcement of a DAB decision through an ICC Final Partial Award 1-8.

Gould N (2011) Enforcing a Dispute Board's Decision: Issues and Considerations. DRBF European Regional Conference, Brussels, 17–18 November 2011.

Gould N (2013) FIDIC issues Guidance on DAB Decisions [Online]. Available at: http://www. fenwickelliott.com/files/nick_gould_guidance_on_dab_decisions.indd_.pdf [accessed 28/05/2013].

FIDIC Users' Guide
ISBN 978-0-7277-5856-9

Chapter 27
Appendix: General Conditions of Dispute Adjudication Agreement

Appendix A1
General Conditions of Dispute Adjudication Agreement
1 Definitions
Each 'Dispute Adjudication Agreement' is a tripartite agreement by and between:

(a) the 'Employer';
(b) the 'Contractor'; and
(c) the 'Member' who is defined in the Dispute Adjudication Agreement as being:
    (i) the sole member of the 'DAB' (or 'adjudicator') and, where this is the case, all references to the 'Other Members' do not apply, or
    (ii) one of the three persons who are jointly called the 'DAB' (or 'dispute adjudication board') and, where this is the case, the other two persons are called the 'Other Members'.

The Employer and the Contractor have entered (or intend to enter) into a contract, which is called the 'Contract' and is defined in the Dispute Adjudication Agreement, which incorporates this Appendix. In the Dispute Adjudication Agreement, words and expressions which are not otherwise defined shall have the meanings assigned to them in the Contract.

2 General Provisions
Unless otherwise stated in the Dispute Adjudication Agreement, it shall take effect on the latest of the following dates:

(a) the Commencement Date defined in the Contract;
(b) when the Employer, the Contractor and the Member have each signed the Dispute Adjudication Agreement; or
(c) when the Employer, the Contractor and each of the Other Members (if any) have respectively each signed a dispute adjudication agreement.

When the Dispute Adjudication Agreement has taken effect, the Employer and the Contractor shall each give notice to the Member accordingly. If the Member does not receive either notice within six months after entering into the Dispute Adjudication Agreement, it shall be void and ineffective.

This employment of the Member is a personal appointment. At any time, the Member may give not less than 70 days' notice of resignation to the Employer and to the Contractor, and the Dispute Adjudication Agreement shall terminate upon the expiry of this period.

No assignment or subcontracting of the Dispute Adjudication Agreement is permitted without the prior written agreement of all the parties to it and of the Other Members (if any).

3 Warranties

The Member warrants and agrees that he/she is and shall be impartial and independent of the Employer, the Contractor and the Engineer. The Member shall promptly disclose, to each of them and to the Other Members (if any), any fact or circumstance which might appear inconsistent with his/her warranty and agreement of impartiality and independence.

When appointing the Member, the Employer and the Contractor relied upon the Member's representations that he/she is:

(a) experienced in the work which the Contractor is to carry out under the Contract;

(b) experienced in the interpretation of contract documentation; and

(c) fluent in the language for communications defined in the Contract.

4 General Obligations of The Member

The Member shall:

(a) have no interest financial or otherwise in the Employer, the Contractor or the Engineer, nor any financial interest in the Contract except for payment under the Dispute Adjudication Agreement;

(b) not previously have been employed as a consultant or otherwise by the Employer, the Contractor or the Engineer, except in such circumstances as were disclosed in writing to the Employer and the Contractor before they signed the Dispute Adjudication Agreement;

(c) have disclosed in writing to the Employer, the Contractor and the Other Members (if any), before entering into the Dispute Adjudication Agreement and to his/her best knowledge and recollection, any professional or personal relationships with any director, officer or employee of the Employer, the Contractor or the Engineer, and any previous involvement in the overall project of which the Contract forms part;

(d) not, for the duration of the Dispute Adjudication Agreement, be employed as a consultant or otherwise by the Employer, the Contractor or the Engineer, except as may be agreed in writing by the Employer, the Contractor and the Other Members (if any);

(e) comply with the annexed procedural rules and with Sub-Clause 20.4 of the Conditions of Contract;

(f) not give advice to the Employer, the Contractor, the Employer's Personnel or the Contractor's Personnel concerning the conduct of the Contract, other than in accordance with the annexed procedural rules;

(g) not while a Member enter into discussions or make any agreement with the Employer, the Contractor or the Engineer regarding employment by any of them, whether as a

consultant or otherwise, after ceasing to act under the Dispute Adjudication Agreement;

(h) ensure his/her availability for all site visits and hearings as are necessary;

(i) become conversant with the Contract and with the progress of the Works (and of any other parts of the project of which the Contract forms part) by studying all documents received which shall be maintained in a current working file;

(j) treat the details of the Contract and all the DAB's activities and hearings as private and confidential, and not publish or disclose them without the prior written consent of the Employer, the Contractor and the Other Members (if any); and

(k) be available to give advice and opinions, on any matter relevant to the Contract when requested by both the Employer and the Contractor, subject to the agreement of the Other Members (if any).

5 General Obligations of the Employer and the Contractor

The Employer, the Contractor, the Employer's Personnel and the Contractor's Personnel shall not request advice from or consultation with the Member regarding the Contract, otherwise than in the normal course of the DAB's activities under the Contract and the Dispute Adjudication Agreement, and except to the extent that prior agreement is given by the Employer, the Contractor and the Other Members (if any). The Employer and the Contractor shall be responsible for compliance with this provision, by the Employer's Personnel and the Contractor's Personnel respectively.

The Employer and the Contractor undertake to each other and to the Member that the Member shall not, except as otherwise agreed in writing by the Employer, the Contractor, the Member and the Other Members (if any):

(a) be appointed as an arbitrator in any arbitration under the Contract;

(b) be called as a witness to give evidence concerning any dispute before arbitrator(s) appointed for any arbitration under the Contract; or

(c) be liable for any claims for anything done or omitted in the discharge or purported discharge of the Member's functions, unless the act or omission is shown to have been in bad faith.

The Employer and the Contractor hereby jointly and severally indemnify and hold the Member harmless against and from claims from which he/she is relieved from liability under the preceding paragraph.

Whenever the Employer or the Contractor refers a dispute to the DAB under Sub-Clause 20.4 of the Conditions of Contract, which will require the Member to make a site visit and attend a hearing, the Employer or the Contractor shall provide appropriate security for a sum equivalent to the reasonable expenses to be incurred by the Member. No account shall be taken of any other payments due or paid to the Member.

6 Payment

The Member shall be paid as follows, in the currency named in the Dispute Adjudication Agreement:

(a) a retainer fee per calendar month, which shall be considered as payment in full for:

    (i)   being available on 28 days' notice for all site visits and hearings;

    (ii)   becoming and remaining conversant with all project developments and maintaining relevant files;

    (iii) all office and overhead expenses including secretarial services, photocopying and office supplies incurred in connection with his duties; and

    (iv) all services performed hereunder except those referred to in subparagraphs (b) and (c) of this Clause.

The retainer fee shall be paid with effect from the last day of the calendar month in which the Dispute Adjudication Agreement becomes effective; until the last day of the calendar month in which the Taking-Over Certificate is issued for the whole of the Works.

With effect from the first day of the calendar month following the month in which Taking-Over Certificate is issued for the whole of the Works, the retainer fee shall be reduced by 50%. This reduced fee shall be paid until the first day of the calendar month in which the Member resigns or the Dispute Adjudication Agreement is otherwise terminated.

(b) a daily fee which shall be considered as payment in full for:

    (i)   each day or part of a day up to a maximum of two days' travel time in each direction for the journey between the Member's home and the site, or another location of a meeting with the Other Members (if any);

    (ii)   each working day on site visits, hearings or preparing decisions; and

    (iii) each day spent reading submissions in preparation for a hearing.

(c) all reasonable expenses incurred in connection with the Member's duties, including the cost of telephone calls, courier charges, faxes and telexes, travel expenses, hotel and subsistence costs: a receipt shall be required for each item in excess of five percent of the daily fee referred to in sub-paragraph (b) of this Clause;

(d) any taxes properly levied in the Country on payments made to the Member (unless a national or permanent resident of the Country) under this Clause 6.

The retainer and daily fees shall be as specified in the Dispute Adjudication Agreement. Unless it specifies otherwise, these fees shall remain fixed for the first 24 calendar months, and shall thereafter be adjusted by agreement between the Employer, the Contractor and the Member, at each anniversary of the date on which the Dispute Adjudication Agreement became effective.

The Member shall submit invoices for payment of the monthly retainer and air fares quarterly in advance. Invoices for other expenses and for daily fees shall be submitted following the conclusion of a site visit or hearing. All invoices shall be accompanied by a brief description of activities performed during the relevant period and shall be addressed to the Contractor.

The Contractor shall pay each of the Member's invoices in full within 56 calendar days after receiving each invoice and shall apply to the Employer (in the Statements under the Contract) for reimbursement of one-half of the amounts of these invoices. The Employer shall then pay the Contractor in accordance with the Contract.

If the Contractor fails to pay to the Member the amount to which he/she is entitled under the Dispute Adjudication Agreement, the Employer shall pay the amount due to the Member and *any other amount which may be required to maintain the operation of the DAB; and without prejudice to the Employer's rights or remedies. In addition to all other rights arising from this default, the Employer shall be entitled to reimbursement of all sums paid in excess of one-half of these payments, plus all costs of recovering these sums and financing charges calculated at the rate specified in Sub-Clause 14.8 of the Conditions of Contract.*

If the Member does not receive payment of the amount due within 70 days after submitting a valid invoice, the Member may (i) suspend his/her services (without notice) until the payment is received, and/or (ii) resign his/her appointment by giving notice under Clause 7.

7 Termination

At any time: (i) the Employer and the Contractor may jointly terminate the Dispute Adjudication Agreement by giving 42 days' notice to the Member; or (ii) the Member may resign as provided for in Clause 2.

If the Member fails to comply with the Dispute Adjudication Agreement, the Employer and the Contractor may, without prejudice to their other rights, terminate it by notice to the Member. The notice shall take effect when received by the Member.

If the Employer or the Contractor fails to comply with the Dispute Adjudication Agreement, the Member may, without prejudice to his/her other rights, terminate it by notice to the Employer and the Contractor. The notice shall take effect when received by them both.

Any such notice, resignation and termination shall be final and binding on the Employer, the Contractor and the Member. However, a notice by the Employer or the Contractor, but not by both, shall be of no effect.

8 Default of the Member

If the Member fails to comply with any obligation under Clause 4, he/she shall not be entitled to any fees or expenses hereunder and shall, without prejudice to their other rights, reimburse each of the Employer and the Contractor for any fees and expenses received by the Member and the Other Members (if any), for proceedings or decisions (if any) of the DAB which are rendered void or ineffective.

9 Disputes

Any dispute or claim arising out of or in connection with this Dispute Adjudication Agreement, or the breach, termination or invalidity thereof, shall be finally settled under the Rules of Arbitration of the International Chamber of Commerce by one arbitrator appointed in accordance with these Rules of Arbitration.

The Dispute Adjudication Agreement is an agreement between the Employer, the Contractor and each member of the DAB (known as a tripartite agreement). In addition to the General Conditions, FIDIC has published standard Dispute Adjudication Agreements for one-person and three-person DABs. These are the actual documents to be signed by the Employer, the Contractor and each member of the DAB and are printed at the end of the FIDIC document.

The General Conditions for the Agreement lay out extremely onerous requirements for the behaviour and duties of the DAB member, including the following.

- The Member must be impartial and independent of the Employer, the Contractor and the Engineer and must disclose anything that might appear inconsistent with this impartiality. This requirement is developed in a list of detailed stipulations.
- The Member must be fluent in the language defined in the Contract and experienced in the type of work and the interpretation of Contract documentation.
- The Member should study documents, become conversant with the progress of the Works and be available to give advice and opinions, but only when requested by both the Employer and the Contractor, with the agreement of the other Members.
- The Member is not permitted to be appointed arbitrator or to give evidence in any arbitration under the Contract except with the agreement of the Employer, Contractor and other Members.
- The Agreement includes figures for a monthly retainer and daily rate. The retainer is paid for keeping conversant with developments, being available for visits and for office expenses. The daily rate and expenses are paid for Site visits, hearings and preparing decisions, plus payment for travel time.
- The financial penalties for non-compliance with the obligations at Clause 4 are extremely severe. This emphasises the importance of the Member disclosing any previous involvement with either Party. Any problems or procedural changes must be confirmed in writing.

Further recommended reading for Dispute Boards is as follows.

- G Owen and B Totterdill (2007) *Dispute Boards: Procedure and Practice.* Thomas Telford, London, UK
- C Chern (2008) *Chern on Dispute Boards*. Blackwell Publishing Ltd, UK.

Annex: Procedural Rules

1   *Unless otherwise agreed by the Employer and the Contractor, the DAB shall visit the site at intervals of not more than 140 days, including times of critical construction events, at the request of either the Employer or the Contractor. Unless otherwise agreed by the Employer, the Contractor and the DAB, the period between consecutive visits shall not be less than 70 days, except as required to convene a hearing as described below.*

2   *The timing of and agenda for each site visit shall be as agreed jointly by the DAB, the Employer and the Contractor, or in the absence of agreement, shall be decided by the DAB. The purpose of site visits is to enable the DAB to become and remain acquainted with the progress of the Works and of any actual or potential problems or claims.*

3   *Site visits shall be attended by the Employer, the Contractor and the Engineer and shall be co-ordinated by the Employer in co-operation with the Contractor. The Employer shall ensure the provision of appropriate conference facilities and secretarial and copying services. At the conclusion of each site visit and before leaving the site, the DAB shall prepare a report on its activities during the visit and shall send copies to the Employer and the Contractor.*

4   *The Employer and the Contractor shall furnish to the DAB one copy of all documents which the DAB may request, including Contract documents, progress reports, variation instructions, certificates and other documents pertinent to the performance of the Contract. All communications between the DAB and the Employer or the Contractor shall be copied to the other Party. If the DAB comprises three persons, the Employer and the Contractor shall send copies of these requested documents and these communications to each of these persons.*

5   *If any dispute is referred to the DAB in accordance with Sub-Clause 20.4 of the Conditions of Contract, the DAB shall proceed in accordance with Sub-Clause 20.4 and these Rules. Subject to the time allowed to give notice of a decision and other relevant factors, the DAB shall:*

   (a) act fairly and impartially as between the Employer and the Contractor, giving each of them a reasonable opportunity of putting his case and responding to the other's case; and

   (b) adopt procedures suitable to the dispute, avoiding unnecessary delay or expense.

6   *The DAB may conduct a hearing on the dispute, in which event it will decide on the date and place for the hearing and may request that written documentation and arguments from the Employer and the Contractor be presented to it prior to or at the hearing.*

7   *Except as otherwise agreed in writing by the Employer and the Contractor, the DAB shall have power to adopt an inquisitorial procedure, to refuse admission to hearings or audience at hearings to any persons other than representatives of the Employer, the Contractor and the Engineer, and to proceed in the absence of any party who the DAB is satisfied received notice of the hearing; but shall have discretion to decide whether and to what extent this power may be exercised.*

8   *The Employer and the Contractor empower the DAB, among other things, to:*

   (a) establish the procedure to be applied in deciding a dispute;

   (b) decide upon the DAB's own jurisdiction, and as to the scope of any dispute referred to it;

    (c) *conduct any hearing as it thinks fit, not being bound by any rules or procedures other than those contained in the Contract and these Rules;*

    (d) *take the initiative in ascertaining the facts and matters required for a decision;*

    (e) *make use of its own specialist knowledge, if any;*

    (f) *decide upon the payment of financing charges in accordance with the Contract;*

    (g) *decide upon any provisional relief such as interim or conservatory measures; and*

    (h) *open up, review and revise any certificate, decision, determination, instruction, opinion or valuation of the Engineer, relevant to the dispute.*

9   *The DAB shall not express any opinions during any hearing concerning the merits of any arguments advanced by the Parties. Thereafter, the DAB shall make and give notice to its decision in accordance with Sub-Clause 20.4, or as otherwise agreed by the Employer and the Contractor in writing. If the DAB comprises three persons:*

    (a) *it shall convene in private after a hearing, in order to have discussions and prepare its decision;*

    (b) *it shall endeavour to reach a unanimous decision: if this proves impossible the applicable decision shall be made by a majority of the Members, who may require the minority Member to prepare a written report for submission to the Employer and the Contractor; and*

    (c) *if a Member fails to attend a meeting or hearing, or to fulfil any required function, the other two Members may nevertheless proceed to make a decision, unless:*

       (i) *either the Employer or the Contractor does not agree that they do so, or*

      (ii) *the absent Member is the chairman and he/she instructs the other Members to not make a decision.*

The DAB has considerable powers to decide its own procedures and may decide to hold a hearing on a dispute, with written and oral arguments. The DAB may adopt an inquisitorial procedure, unless otherwise agreed in writing by the Employer and the Contractor. The alternative is presumably for the Parties to agree on an adversarial procedure with submissions from both sides. The DAB would then judge between the submissions, rather than reach a decision based on its own enquiries. This would create difficulties for the DAB and increase the length and cost of the hearing.

FIDIC Users' Guide
ISBN 978-0-7277-5856-9

ICE Publishing: All rights reserved
http://dx.doi.org/10.1680/fug.58569.289

Chapter 28
Annexes and Forms

The Annexes and Forms which are included in the FIDIC Document have been reviewed under the Sub-Clause to which they relate and are reproduced in full for the RED Book with a description of differences for the YLW Book in Part VI Appendix 8.

The Pink Book

FIDIC Users' Guide
ISBN 978-0-7277-5856-9

Chapter 29
Introduction to the Pink Book

The Multilateral Development Banks (MDB) Harmonised Edition of the FIDIC *Conditions of Contract for Construction*, third edition 2010 (referred to here as 'the Pink Book' or PNK book) incorporates standard changes that have been agreed by a group of Multilateral Development Banks (or MDBs). It was first published in May 2005, followed by the second edition in March 2006 and the current (third) edition in June 2010. Terms and Condition of Use are included at the beginning of the document stating that this must only be used for contracts that are financed by one of the Participating Banks. The Participating Banks are listed below

- African Development Bank
- Asian Development Bank
- Black Sea Trade and Development Bank
- Caribbean Development Bank
- Council of Europe Development Bank
- European Bank for Reconstruction and Development
- Inter-American Development Bank
- International Bank for Reconstruction and Development (The World Bank).

The PNK Book follows the same layout as the RED Book and many of the Clauses are identical. Changes have been introduced as required by the MDBs and agreed by FIDIC. Some of these changes are obviously based on experience in the use of the 1999 FIDIC Contracts as well as the specific requirements of the MDBs. Some of the changes are improvements that, where appropriate, could also be incorporated into the Particular Conditions for other FIDIC Contracts.

There are a number of Sub-Clauses that are written so that the country in which the project is executed benefits commercially: (e.g. Sub-Clause 4.4 [*Subcontractors*], Sub-Clause 6.1 [*Engagement of Staff and Labour*]). These provisions encourage the Contractor to channel back funds into the Country thereby improving the local economy, reducing the financial risk for the bank, and allowing the bank to fulfil its social obligations.

The following review of the changes must be read in conjunction with the published PNK Book (available on the FIDIC website) and the review of the RED Book in Chapters 7 to 28. Sub-Clause numbers and references to the wording refer to the 2010 PNK Book or the 1999 RED Book as appropriate.

Only Sub-Clauses having significant changes to the RED Book are reproduced and commented here. Such comments are made only if they are necessary for correctly understanding the fundamental difference with the RED Book. Semantic and/or grammatical changes that do not significantly add to or change the intended purpose under the RED Book are shown without further comments. If not listed, the text is either identical to the RED Book or only have minor editing changes (e.g. Contract Data versus Appendix to Tender and DB versus DAB).

FIDIC Users' Guide
ISBN 978-0-7277-5856-9

Chapter 30
Clause 1: General Provisions

1.1 Definitions

The Particular Conditions now consist of Parts A and B where Part A is the '*Contract Data*' and Part B is the '*Special Provisions*'. The Contract Data is similar to the 'Appendix to Tender' as incorporated into the RED Book.

- 1.1.1.4 Now states that the 'Letter of Tender' may be referred to as the '*Letter of Bid*'.
- 1.1.1.9 (RED Book 1.1.1.10) It refers to '*Schedule of Payment Currencies*' as required by Sub-Clause 14.15 [*Currencies of Payment*].
- 1.1.1.10 (RED Book 1.1.1.9) The Appendix to Tender is now called '*Contract Data*'. It is Part A of the Particular Conditions and consists of information provided by the Employer. This is a logical change because the Employer has determined the information in the document.
- 1.1.2.9 The word 'Adjudication' is deleted from 'Dispute Adjudication Board', so the DAB becomes '*Dispute Board*' (DB) throughout. This is regretted as the DB still acts as Adjudicators, making a decision which must be implemented. Further, the words '*so named in the Contract, or other persons*' has been deleted from the first sentence. The appointment of a DB under the PNK Book is discussed under Sub-Clause 20.2 [*Appointment of Dispute Board*].

Two new defined terms are added:

'*1.1.2.11 'Bank' means the financing institution (if any) named in the Contract Data.*'

and

'*1.1.2.12 'Borrower' means the person (if any) named as the borrower in the Contract Data.*'

These two additional Sub-Clauses give the financing Bank a named status in the Contract, although it is not a Party to the Contract. The Borrower is also named, as the organisation who has the financial agreement with the Bank and who may, or may not, be the Employer.

1.1.3.6 The definition for '*Tests after Completion*' has amended by replacing the phrase '*in accordance with the provisions of the Particular Conditions*' by '*in accordance with the Specification*'. This is logical because tests after completion are a technical requirement and the correct location for such information is the Specification.

1.1.3.7 The phrase '*which extends over 365 days except if otherwise stated in the Contract Data*' is added defining the period of the '*Defects Notification Period*'. The Contract Data already includes 365 days that can be amended if required. It is unclear why this information is duplicated.

1.1.5.5 The definition of '*Plant*' now includes Employers vehicles. The phrase '*including vehicles purchased for the Employer and relating to the construction or operation of the Works*' is added to the end of the sentence. All obligations with respect to Plant are now applicable to Employer's vehicles.

1.1.6.7 The definition of '*Site*' is amended by adding the phrase '*including storage and working areas*' after the word '*executed,*'. All obligations with respect to the Site are now applicable to storage and working areas.

1.1.6.8 The definition for '*Unforeseeable*' is amended by replacing the phrase '*date for submission of the Tender*' by '*Base Date*' which is a more logical time for applying any test of reasonableness.

A new definition is added:

'1.1.6.10 '*Notice of Dissatisfaction*' means the notice given by either Party to the other under Sub-Clause 20.4 [*Obtaining Dispute Board's Decision*] indicating its dissatisfaction and intention to commence arbitration. This statement is in contradiction to the requirement under Sub-Clause 20.5 [*Amicable Settlement*] for the Parties to seek amicable settlement before commencing arbitration proceedings.

1.2 Interpretation
A new subparagraph (e) is added:

'*(e) the word 'tender' is synonymous with 'bid', and 'tenderer' with 'bidder' and the words 'tender documents' with 'bidding documents'.*'

A new paragraph is added (paragraph 3):

'*In these Conditions, provisions including the expression 'Cost plus profit' require this profit to be one-twentieth (5%) of this Cost unless otherwise indicated in the Contract Data.*'

The RED Book includes the phrase '*Cost plus reasonable profit*', in relation to certain claims Clauses. The PNK Book requires profit to be 5% of Cost unless otherwise indicated in the Contract Data. This means that wherever the Contract refers to '*Cost plus reasonable profit*', the word '*reasonable*' is deleted. The choice of 5% seems to be a reasonable compromise and the fact of stating a figure will avoid potential disagreement. The word '*Cost*' is defined under Sub-Clause 1.1.4.3 [*Cost*], and means all expenditure reasonably incurred including overhead and similar charges.

1.3 Communications
The words '*and discharges*' are added after '*requests*' in the first paragraph.

The term 'discharges' refers to other obligations to act or to perform duties required by the Contract and is, in our view, considered to be an improvement to the overall sense of the Sub-Clause. Unfortunately, the concept is not followed through for the rest of the Sub-Clause. The requirement for the various communications not to be unreasonably withheld or delayed is limited to approvals, certificates, consents and determinations. The terms 'notice', 'requests' and 'discharges' are not included in the list, for reasons that are not entirely apparent. The law should be checked to see if and to what extent the timing of such communications is regulated.

1.4 Law and Language

The brackets are removed from the phrase 'or other jurisdiction' in the first paragraph for reasons that can only be assumed to be of a grammatical nature rather than any contractual or even legal connotation. A new second paragraph is added as 'The ruling language of the Contract shall be that stated in the Contract Data.' Even though the Sub-Clause header may suggest otherwise, in the RED Book there is no clear unambiguous statement on the Language of the Contract. This omission has been rectified in the PNK Book. It also implies that there can be a Language for the Contract that is different from the language for communications.

The wording of the Sub-Clause has been simplified such that the ruling language and language for communication shall both be stated in the Contract Data.

1.5 Priority of Documents

The change from Appendix to Tender to Particular Conditions – Part A (i.e. Contract Data) means a change in the order of priority. Under the RED Book the Appendix to Tender forms part of the Letter of Tender (ref. RED Book Sub-Clause 1.1.1.9 [Appendix to Tender]), and is thus in the order for priority of documents in third position (c), whereas the Contract Data under the PNK Book forms part of the Particular Conditions – Part A in fourth position (d). This means that the Contractor in his bid or tender can deviate from the Employer's preferences as indicated in the Contract Data (assuming such deviation does not constitute a non-qualifying bid).

1.6 Contract Agreement

The phrase 'they agree otherwise' is deleted and substituted by 'unless the Particular Conditions establish otherwise.'

1.8 Care and Supply of Documents

The phrase 'of a technical nature' is deleted from the fourth paragraph.

As a result, all errors or defects, of whatever type, must be reported. This does not avoid the problem that the Contractor may notice an omission or conflict between documents, but will not be aware of an error in a statement that appears to be reasonable but does not reflect the Employer's intentions.

1.9 Delayed Drawings or Instructions

In the first paragraph, the words 'details of ' are removed before 'the nature and amount of delay'. This seems to be a matter of wording only, but the implication could be that under the PNK Book the notice is not required to give any details as to how the amount of delay was established. However, this would certainly be required under Sub-Clause 20.1 [*Contractor's Claims*] if the Contractor claims for the incurred delay.

1.12 Confidential Details
The Sub-Clause is amended and includes a new second paragraph as follows:

'*The Contractor's and the Employer's Personnel shall disclose all such confidential and other information as may be reasonably required in order to verify compliance with the Contract and allow its proper implementation.*

Each of them shall treat the details of the Contract as private and confidential, except to the extent necessary to carry out their respective obligations under the Contract or to comply with applicable Laws. Each of them shall not publish or disclose any particulars of the Works prepared by the other Party without the previous agreement of the other Party. However, the Contractor shall be permitted to disclose any publicly available information, or information otherwise required to establish his qualifications to compete for other projects.'

This is a far stricter requirement to keep all details of the Contract and the Works strictly confidential. This requirement is obviously added by the MDBs so as to protect and safeguard as a minimum their financial arrangements relative to the project from becoming public knowledge. This requirement is also imposed on any Subcontractor through Sub-Clause 4.4 [*Subcontractors*].

1.15 Inspections and Audit by the Bank
A new Sub-Clause is included:

'*The Contractor shall permit the Bank and/or persons appointed by the Bank to inspect the Site and/or the Contractor's accounts and records relating to the performance of the Contract and to have such accounts and records audited by auditors appointed by the Bank if required by the Bank.*'

Financing institutions have always taken an interest in financial control and have often appointed their own consultant(s) to work as a part of the Employer's team. This Sub-Clause increases the Bank's power to investigate and check the Contractor's use of money that has been provided by the Bank. The Sub-Clause reflects the Bank's agreement with the Borrower.

FIDIC Users' Guide
ISBN 978-0-7277-5856-9

ICE Publishing: All rights reserved
http://dx.doi.org/10.1680/fug.58569.299

Chapter 31
Clause 2: The Employer

2.1 Right of Access to the Site
In the second paragraph the words '*may be*' have been deleted. The words '*without disruption*' have been added before '*in accordance with*'.

Provided there is no time of access stated in the Appendix to Tender or Contract Data as appropriate, then under the RED Book the words '*may be*' would indicate that the Employer shall give access to the Site at times where the Contractor indicates that he would want to commence work without actually having to do so. By implication the PNK Book wording would make it difficult for the Contractor to claim delay, if he did not start his work immediately after the Employer actually gave possession of the Site.

Also refer to Sub-Clause 4.13 [*Rights of Way and Facilities*].

2.2 Permits, Licenses or Approvals
The Sub-Clause is amended as follows:

'*The Employer shall provide, at the request of the Contractor, such reasonable assistance as to allow the Contractor to obtain properly:*

(a) copies of the Laws of the Country which are relevant to the Contract but are not readily available, and
(b) any permits, licences or approvals required by the Laws of the Country:
    (i) which the Contractor is required to obtain under Sub-Clause 1.13 [Compliance with Laws],
    (ii) for the delivery of Goods, including clearance through customs, and
    (iii) for the export of Contractor's Equipment when it is removed from the Site.'

The overall wording of the Sub-Clause is improved with stronger emphasis placed on the Employer to provide the necessary requested assistance.

2.4 Employer's Financial Arrangements
The Sub-Clause is amended with a new second paragraph as follows:

'*The Employer shall submit, before the Commencement Date and thereafter within 28 days after receiving any request from the Contractor, reasonable evidence that financial arrangements have been made and are being maintained which will enable the Employer to pay the*

Contract Price punctually (as estimated at that time) in accordance with Clause 14 [Contract Price and Payment]. Before the Employer makes any material change to his financial arrangements, the Employer shall give notice to the Contractor with detailed particulars.

In addition, if the Bank has notified to the Borrower that the Bank has suspended disbursements under its loan, which finances in whole or in part the execution of the Works, the Employer shall give notice of such suspension to the Contractor with detailed particulars, including the date of such notification, with a copy to the Engineer, within 7 days of the Borrower having received the suspension notification from the Bank. If alternative funds will be available in appropriate currencies to the Employer to continue making payments to the Contractor beyond a date 60 days after the date of Bank notification of the suspension, the Employer shall provide reasonable evidence in his notice of the extent to which such funds will be available.'

The second paragraph imposes an obligation on the Employer to notify the Contractor if the Bank suspends disbursements under the loan. Sub-Clauses 16.1 [*Contractor's Entitlement to Suspend Work*] and 16.2 [*Termination by Contractor*] have also been amended to enable the Contractor to suspend work or terminate the Contract in such circumstances.

2.5 Employer's Claims

In the second paragraph the first sentence now states '*The notice shall be given as soon as practicable and no longer than 28 days after the Employer became aware, or should have become aware, of the event or circumstances giving rise to the claim.*'

The phrases '*and no longer than 28 days*' and '*or should have become aware*' have been added bringing the Employer's claim procedure one step closer to the Contractor's claims procedure. However, the 28-day time bar does not impose sanctions on the Employer.

FIDIC Users' Guide
ISBN 978-0-7277-5856-9

ICE Publishing: All rights reserved
http://dx.doi.org/10.1680/fug.58569.301

Chapter 32
Clause 3: The Engineer

3.1 Engineer's Duties and Authority

The final sentence of the third paragraph has been changed to '*The Employer shall promptly inform the Contractor of any change to the authority attributed to the Engineer.*'

For the Employer to change, and possibly reduce, the authority of the Engineer during the course of the project could have a serious impact on the administration of the Contract and lead to complaints and claims from the Contractor. However, if the Employer intends to change the Engineer under Sub-Clause 3.4 [*Replacement of the Engineer*], he now has the opportunity to change the authority during the notice period. Any such change could be construed by the Contractor as a Variation or a unilateral change to the Contract terms.

An additional subparagraph is added to the fifth paragraph as '*(d) any act by the Engineer in response to a Contractor's request except as otherwise expressly specified shall be notified in writing to the Contractor within 28 days of receipt.*'

This places a time restraint on the Engineer to act within 28 days of receiving any request from the Contractor.

The sixth and seventh paragraphs are new as follows:

'*The following provisions shall apply:*

The Engineer shall obtain the specific approval of the Employer before taking action under the following Sub-Clauses of these Conditions:

(A) Sub-Clause 4.12: agreeing or determining an extension of time and/or additional cost.
(B) Sub-Clause 13.1: instructing a Variation, except;
  (i) in an emergency situation as determined by the Engineer, or
  (ii) if such a Variation would increase the Accepted Contract Amount by less than the percentage specified in the Contract Data.
(C) Sub-Clause 13.3: approving a proposal for Variation submitted by the Contractor in accordance with Sub-Clause 13.1 or 13.2.
(D) Sub-Clause 13.4: specifying the amount payable in each of the applicable currencies.

Notwithstanding the obligation, as set out above, to obtain approval, if, in the opinion of the Engineer, an emergency occurs affecting the safety of life or of the Works or of adjoining property, he may, without relieving the Contractor of any of his duties and responsibility under the Contract, instruct the Contractor to execute all such work or to do all such things as may, in the opinion of the Engineer, be necessary to abate or reduce the risk. The Contractor shall forthwith comply, despite the absence of approval of the Employer, with any such instruction of the Engineer. The Engineer shall determine an addition to the Contract Price, in respect of such instruction, in accordance with Clause 13 and shall notify the Contractor accordingly, with a copy to the Employer.'

These provisions restrict the power of the Engineer to commit the Employer to additional payments, or extension of time, without prior approval, except in an emergency. It is often necessary for the Engineer to convince the Employer to give his approval when the Contractor has a contractual entitlement to additional time or money. The Employer must ensure he has the capacity to make timely decisions on these issues otherwise claims may ensue.

3.4 Replacement of the Engineer
The Sub-Clause is amended as follows:

'If the Employer intends to replace the Engineer, the Employer shall, not less than 21 days before the intended date of replacement, give notice to the Contractor of the name, address and relevant experience of the intended replacement Engineer.

If the Contractor considers the intended replacement Engineer to be unsuitable, he has the right to raise objection against him by notice to the Employer, with supporting particulars, and the Employer shall give full and fair consideration to this objection.'

The PNK Book significantly shortens the notice period from 42 days to 21 days and places increased emphasis on the Contractor's right to object. The Employer's legal requirements, particularly for public authorities, will have lengthy procurement procedures frequently making 21 days impracticable. The process of changing an Engineer therefore requires very careful planning.

3.5 Determinations
The second paragraph adds the phrase *'within 28 days from the receipt of the corresponding claim or request except when otherwise specified'* at the end of the first sentence. The 28-day time-bar would only apply where it is not overruled by another Sub-Clause.

When the Engineer makes a determination under Sub-Clauses 4.12 [*Unforeseeable Physical Conditions*], 13.1 [*Right to Vary*], 13.3 [*Variation Procedure*] and 13.4 [*Payment in Applicable Currencies*], he must, according to Sub-Clause 3.1 [*Engineer's Duties and Authority*], obtain the specific approval of the Employer. It is essential therefore, that Employers have established appropriate procedures to enable timely decisions.

FIDIC Users' Guide
ISBN 978-0-7277-5856-9

ICE Publishing: All rights reserved
http://dx.doi.org/10.1680/fug.58569.303

Chapter 33
Clause 4: The Contractor

4.1 Contractor's General Obligations
A new third paragraph is added:

'All equipment, material, and services to be incorporated in or required for the Works shall have their origin in any eligible source country as defined by the Bank.'

Individual MDBs have their own preferences as to allowable source countries that must be included in the Contract.

The words '*if applicable*' are added before '*operation and maintenance manuals*' in the final paragraph. This is merely to cover the fact that projects, when constructed, do not necessarily have any operations and/or maintenance manuals.

4.2 Performance Security
The first paragraph is amended as follows:

'The Contractor shall obtain (at his cost) a Performance Security for proper performance, in the amount stated in the Contract Data and denominated in the currency(ies) of the Contract or in a freely convertible currency acceptable to the Employer. If an amount is not stated in the Contract Data, this Sub-Clause shall not apply.'

The second paragraph is amended as follows:

'The Contractor shall deliver the Performance Security to the Employer within 28 days after receiving the Letter of Acceptance, and shall send a copy to the Engineer. The Performance Security shall be issued by a reputable bank or financial institution selected by the Contractor, and shall be in the form annexed to the Particular Conditions, as stipulated by the Employer in the Contract Data, or in another form approved by the Employer.'

Provided the Performance Security is submitted in a form contained within the Particular Conditions, the choice of bank or financial institution is for the Contractor. It must however, be reputable, thereby introducing a potential disagreement.

In the fifth paragraph the list of situations in which the Employer can claim on the Performance Security has been deleted [i.e. RED Book subparagraphs (a) to (d)]. This deletion widens the scope for the Employer to make a claim. However, it still retains the indemnity by the Employer in the event that the Employer makes a claim to which he was not entitled.

A new paragraph has been added (new paragraph 7):

'Without limitation to the provisions of the rest of this Sub-Clause, whenever the Engineer determines an addition or a reduction to the Contract Price as a result of a change in cost and/or legislation, or as a result of a Variation amounting to more than 25 percent of the portion of the Contract Price payable in a specific currency, the Contractor shall at the Engineer's request promptly increase, or may decrease, as the case may be, the value of the Performance Security in that currency by an equal percentage.'

The new final paragraph requires the Contractor to increase or decrease the value of the Performance Security if the Contract Price changes by more than 25% in certain circumstances.

4.3 Contractor's Representative
The second paragraph has been clarified by the addition of the words '*in terms of Sub-Clause 6.9 [Contractor's Personnel]*' that are added after '*If consent is withheld or subsequently revoked*'.

The seventh paragraph has been amended:

'The Contractor's Representative shall be fluent in the language for communications defined in Sub-Clause 1.4 [Law and Language]. If the Contractor's Representative's delegates are not fluent in the said language, the Contractor shall make competent interpreters available during all working hours in a number deemed sufficient by the Engineer.'

This amendment deals with the fact that under the RED Book, there is no real sanction on the Contractor for failing to ensure that his Representative's delegates have the necessary language capabilities. The Employer would be faced with an onerous route through Sub-Clause 2.5 [*Employer's Claims*] to seek recourse. Under the PNK Book, the Contractor is simply obliged to provide interpreters at his cost.

4.4 Subcontractors
The word '*solely*' is added after '*suppliers*' in paragraph (a).

New paragraphs 3 and 4 are added:

'The Contractor shall ensure that the requirements imposed on the Contractor by Sub-Clause 1.12 [Confidential Details] apply equally to each Subcontractor.

Where practicable, the Contractor shall give fair and reasonable opportunity for contractors from the Country to be appointed as Subcontractors.'

The new third paragraph obliges all Sub-contractors to keep details private and confidential as equally imposed on the Contractor through Sub-Clause 1.12 [*Confidential Details*].

The new fourth paragraph attempts to encourage the use of local companies as Subcontractors. However, the Contractor is still responsible for his Subcontractors and will still want to submit an economic Tender in order to obtain the Contract, so the practical effect of this Sub-Clause is uncertain.

4.6 Co-operation

The phrase '*suffer delays and/or to*' is added before '*incur Unforeseeable Cost*' in the second paragraph. Under the RED Book the Contractor is seemingly limited to claim for additional costs under this Sub-Clause. However, for delays his recourse would in any event have been Sub-Clause 8.4(e) [*Extension of Time for Completion*]. The added words in the PNK Book are thus merely stating the obvious.

4.12 Unforeseeable Physical Conditions

The fourth paragraph wording is improved by replacing '*After receiving such notice...*' with '*Upon receiving such notice...*'

The final (seventh) paragraph is reworded as follows:

'*The Engineer shall take account of any evidence of the physical conditions foreseen by the Contractor when submitting the Tender, which shall be made available by the Contractor, but shall not be bound by the Contractor's interpretation of any such evidence.*'

The substitution of the second word '*may*' with the word '*shall*' places an obligation on the Contractor to provide evidence to the Engineer of physical conditions foreseen at time of tender when accounting for any changes (refer to comments made on the RED Book in Chapter 10). Similarly, the Engineer must take this evidence into account before making any determination.

4.13 Rights of Way and Facilities

The Sub-Clause is amended as follows:

'*Unless otherwise specified in the Contract the Employer shall provide effective access to and possession of the Site including special and/or temporary rights-of-way which are necessary for the Works. The Contractor shall obtain, at his risk and cost, any additional rights of way or facilities outside the Site which he may require for the purposes of the Works.*'

This Sub-Clause must be interpreted together with Sub-Clause 2.1 [*Rights of Access to the Site*].

The Employer is required to provide effective access and possession to Site plus any special and/or temporary rights of way necessary for the Works. It is essential that, prior to Tender, the Employer identifies all land which it considers to be '*necessary for the Works*' otherwise the potential for disputes is high. The wording '*effective access*' is of course open to interpretation and could lead to disputes if not discussed and agreed at an early stage.

A Contractor using facilities at a location remote from the Site would need the agreement of the Engineer under the requirements of Sub-Clause 4.23 [*Contractor's Operations on Site*].

4.15 Access Route

The phrase '*at Base Date*' is added after '*routes to the Site*' in the first sentence. For rapidly changing environments, the time at which a test is carried out for suitability and availability is important.

4.19 Electricity, Water and Gas

The phrase '*for his construction activities and to the extent defined in the Specifications, for the tests*' is added after '*all power, water and other services he may require*' in the first paragraph.

4.23 Contractor's Operations on Site

The word '*additional*' is added before '*working areas*' at the end of the first sentence of the first paragraph.

Under Sub-Clause 4.13 [*Rights of Way and Facilities*], the Site is extended by the inclusion of special and/or temporary rights-of-way that are necessary for the Works. It is essential therefore, that additional working areas are agreed with the Engineer. No mention of this is made in Sub-Clause 4.13 [*Rights of Way and Facilities*] which refers to the Contractor obtaining any facilities '*at his risk and cost*' outside the Site.

FIDIC Users' Guide
ISBN 978-0-7277-5856-9

ICE Publishing: All rights reserved
http://dx.doi.org/10.1680/fug.58569.307

Chapter 34
Clause 5: Nominated Subcontractors

5.1 Definition of 'nominated Subcontractor'
The words '*subject to Sub-Clause 5.2 [Objection to Notification]*' are added to the end of the paragraph. They strengthen the Contractors right to object to any Subcontractor nominee.

5.2 Objection to Nomination
The words '*in writing*' are added before '*to indemnify*' in the second sentence.

The phrase '*the nominated Subcontractor does not accept to indemnify*' is added at the beginning of subparagraph (b).

A new subparagraph (c) (iii) is added:

'*(iii) be paid only if and when the Contractor has received from the Employer payments for sums due under the Subcontract referred to under Sub-Clause 5.3 [Payment to nominated Subcontractors].*'

There is an overall improvement in the way this Sub-Clause is written with an addition to the list of objections whereby the Subcontractor does not accept to be paid if and when the Contractor is paid. This so-called 'paid-if-paid' scenario should be checked against the relevant laws as this may be void under certain jurisdictions.

5.3 Payments to nominated Subcontractors
The phrase '*shown on the nominated Subcontractor's invoices approved by the Contractor*' is added before '*which the Engineer certifies*' in the first sentence.

This additional provision ensures that the Contractor only pays the Subcontractor the amounts that he has approved. He is no longer purely dependent upon the Engineer's certified amount.

FIDIC Users' Guide
ISBN 978-0-7277-5856-9

ICE Publishing: All rights reserved
http://dx.doi.org/10.1680/fug.58569.309

Chapter 35
Clause 6: Staff and Labour

6.1 Engagement of Staff and Labour

The phrase '*payment, feeding, transport and, when appropriate, housing*' replaces '*payment, housing, feeding and transport*'.

A new second paragraph is added:

'*The Contractor is encouraged, to the extent practicable and reasonable, to employ staff and labour with appropriate qualifications and experience from sources within the Country.*'

Housing is now only provided where appropriate thereby encouraging Contractors to use local labour. This theme is continued in a new paragraph (second) that encourages, rather than requires, the use of local staff and labour.

6.2 Rates of Wages and Conditions of Labour

A new paragraph (second) has been added:

'*The Contractor shall inform the Contractor's Personnel about their liability to pay personal income taxes in the Country in respect of such of their salaries, wages, allowances and any benefits as are subject to tax under the Laws of the Country for the time being in force, and the Contractor shall perform such duties in regard to such deductions thereof as may be imposed on him by such Laws.*'

This requires the Contractor to inform his personnel and to carry out any obligations with respect to local taxation.

6.7 Health and Safety

Three new paragraphs are added:

'*HIV-AIDS Prevention. The Contractor shall conduct an HIV-AIDS awareness programme via an approved service provider, and shall undertake such other measures as are specified in this Contract to reduce the risk of the transfer of the HIV virus between and among the Contractor's Personnel and the local community, to promote early diagnosis and to assist affected individuals.*

The Contractor shall throughout the contract (including the Defects Notification Period): (i) conduct Information, Education and Communication (IEC) campaigns, at least every other month, addressed to all the Site staff and labour (including all the Contractor's employees, all Subcontractors and any other Contractor's or Employer's personnel, and all truck drivers and crew making deliveries to Site for construction activities) and to the immediate local communities, concerning the risks, dangers and impact, and appropriate avoidance behaviour with respect to, of Sexually Transmitted Diseases (STD) – or Sexually Transmitted Infections (STI) in general and HIV/AIDS in particular; (ii) provide male or female condoms for all Site staff and labour as appropriate; and (iii) provide for STI and HIV/AIDS screening, diagnosis, counselling and referral to a dedicated national STI and HIV/AIDS programme, (unless otherwise agreed) of all Site staff and labour.

The Contractor shall include in the programme to be submitted for the execution of the Works under Sub-Clause 8.3 an alleviation programme for Site staff and labour and their families in respect of Sexually Transmitted Infections (STI) and Sexually Transmitted Diseases (STD) including HIV/AIDS. The STI, STD and HIV/AIDS alleviation programme shall indicate when, how and at what cost the Contractor plans to satisfy the requirements of this Sub-Clause and the related specification. For each component, the programme shall detail the resources to be provided or utilised and any related sub-contracting proposed. The programme shall also include provision of a detailed cost estimate with supporting documentation. Payment to the Contractor for preparation and implementation of this programme shall not exceed the Provisional Sum dedicated for this purpose.'

The MDBs are particularly concerned with combating the spread of HIV/AIDS. Labour camps are widely acknowledged as being a major source of the spread of the disease. Under these new requirements the Contractor is obliged to implement awareness programmes, reduce the risk of transfer, promote early diagnosis and assist affected individuals. This must be carried out by an approved service provider achieved, in the majority of cases, through a specialist Subcontractor. Interestingly this shall also be done during the Defects Notification Period.

35.1. Sub-Clauses 6.12–6.22 for additional staff and labour

The PNK Book includes additional Sub-Clauses that cover a variety of aspects for the employment of local staff and labour. The Sub-Clauses must be considered in relation to the governing law and regulations, which may cover similar requirements. Several of these Sub-Clauses have been in use in other MDB contracts and are included in the *Guidance for the Preparation of Particular Conditions* in the RED Book.

6.12 Foreign Personnel

The Contractor may bring in to the Country any foreign personnel who are necessary for the execution of the Works to the extent allowed by the applicable Laws. The Contractor shall ensure that these personnel are provided with the required residence visas and work permits. The Employer will, if requested by the Contractor, use his best endeavours in a timely and expeditious manner to assist the Contractor in obtaining any local, state, national, or government permission required for bringing in the Contractor's personnel.

The Contractor shall be responsible for the return of these personnel to the place where they were recruited or to their domicile. In the event of the death in the Country of any of these personnel or members of their families, the Contractor shall similarly be responsible for making the appropriate arrangements for their return or burial.

This Sub-Clause seems to be a logical extension to the RED Book for international projects. Although the meaning of the Sub-Clause is easy to understand, the actual wording may be ambiguous. As per the second paragraph, through the use of the word '*or*', and in the event of a death in the Country, the Contractor has the option of either making arrangements for the burial or the return (of the body). It is reasonable to assume that the drafters would have had both in mind, and a relevant change from '*or*' into '*and*' would be advisable through the Particular Conditions – Part B.

6.13 Supply of Foodstuffs
The Contractor shall arrange for the provision of a sufficient supply of suitable food as may be stated in the Specification at reasonable prices for the Contractor's Personnel for the purposes of or in connection with the Contract.

6.14 Supply of Water
The Contractor shall, having regard to local conditions, provide on the Site an adequate supply of drinking and other water for the use of the Contractor's Personnel.

6.15 Measures against Insect and Pest Nuisance
The Contractor shall at all times take the necessary precautions to protect the Contractor's Personnel employed on the Site from insect and pest nuisance, and to reduce the danger to their health. The Contractor shall comply with all the regulations of the local health authorities, including use of appropriate insecticide.

6.16 Alcoholic Liquor or Drugs
The Contractor shall not, otherwise than in accordance with the Laws of the Country, import, sell, give, barter or otherwise dispose of any alcoholic liquor or drugs, or permit or allow importation, sale, gift, barter or disposal thereto by Contractor's Personnel.

6.17 Arms and Ammunition
The Contractor shall not give, barter, or otherwise dispose of, to any person, any arms or ammunition of any kind, or allow Contractor's Personnel to do so.

6.18 Festivals and Religious Customs
The Contractor shall respect the Country's recognized festivals, days of rest and religious or other customs.

Nevertheless, if and to the extent required for the proper and timely execution of the Works, the parties are free to agree on reasonable deviations.

6.19 Funeral Arrangements
The Contractor shall be responsible, to the extent required by local regulations, for making any funeral arrangements for any of his local employees who may die while engaged upon the Works.

6.20 Forced Labour
The Contractor shall not employ forced labour, which consists of any work or service, not voluntarily performed, that is exacted from an individual under threat of force or penalty,

and includes any kind of involuntary or compulsory labour, such as indentured labour, bonded labour or similar labour-contracting arrangements.

6.21 Child Labour
The Contractor shall not employ children in a manner that is economically exploitative, or is likely to be hazardous, or to interfere with, the child's education, or to be harmful to the child's health or physical, mental, spiritual, moral, or social development. Where the relevant labour laws of the Country have provisions for employment of minors, the Contractor shall follow those laws applicable to the Contractor. Children below the age of 18 years shall not be employed in dangerous work.

6.22 Employment Records of Workers
The Contractor shall keep complete and accurate records of the employment of labour at the Site. The records shall include the names, ages, genders, hours worked and wages paid to all workers. These records shall be summarised on a monthly basis and submitted to the Engineer. These records shall be included in the details to be submitted by the Contractor under Sub-Clause 6.10 [Records of Contractor's Personnel and Equipment].

6.23 Workers' Organisations
In countries where the relevant labour laws recognise workers' rights to form and to join workers' organisations of their choosing without interference and to bargain collectively, the Contractor shall comply with such laws. Where the relevant labour laws substantially restrict workers' organisations, the Contractor shall enable alternative means for the Contractor's Personnel to express their grievances and protect their rights regarding working conditions and terms of employment. In either case described above, and where the relevant labour laws are silent, the Contractor shall not discourage the Contractor's Personnel from forming or joining workers' organisations of their choosing or from bargaining collectively, and shall not discriminate or retaliate against the Contractor's Personnel who participate, or seek to participate, in such organisations and bargain collectively. The Contractor shall engage with such workers' representatives. Workers' organisations are expected to fairly represent the workers in the workforce.

6.24 Non-Discrimination and Equal Opportunity
The Contractor shall not make employment decisions on the basis of personal characteristics unrelated to inherent job requirements. The Contractor shall base the employment relationship on the principle of equal opportunity and fair treatment, and shall not discriminate with respect to aspects of the employment relationship, including recruitment and hiring, compensation (including wages and benefits), working conditions and terms of employment, access to training, promotion, termination of employment or retirement, and discipline. In countries where the relevant labour laws provide for non-discrimination in employment, the Contractor shall comply with such laws. When the relevant labour laws are silent on nondiscrimination in employment, the Contractor shall meet this Sub-Clause's requirements. Special measures of protection or assistance to remedy past discrimination or selection for a particular job based on the inherent requirements of the job shall not be deemed discrimination.

FIDIC Users' Guide
ISBN 978-0-7277-5856-9

ICE Publishing: All rights reserved
http://dx.doi.org/10.1680/fug.58569.313

Chapter 36
Clause 7: Plant, Materials and Workmanship

7.4 Testing
The phrase 'Except as specified in the Contract' is added at the beginning of the second paragraph.

7.7 Ownership of Plant and Materials
The phrase 'Except as specified in the Contract' is added at the beginning of the first paragraph.

The word 'incorporated' replaces 'delivered' in subparagraph (a).

The phrase 'paid the corresponding' replaces 'entitled to payment of the' in subparagraph (b).

In the case where a Party is in financial difficulty, the ownership of items becomes extremely important. Under the RED Book, the Employer's ownership of Plant or Material commences when it is delivered to Site. Under the PNK Book, the Contractor retains ownership until it is incorporated within the Works.

FIDIC Users' Guide
ISBN 978-0-7277-5856-9

ICE Publishing: All rights reserved
http://dx.doi.org/10.1680/fug.58569.315

Institution of Civil Engineers

publishing

Chapter 37
Clause 8: Commencement, Delays and Suspension

8.1 Commencement of Works

The first paragraph is extensively amended as follows:

'Except as otherwise specified in the Particular Conditions of Contract, the Commencement Date shall be the date at which the following precedent conditions have all been fulfilled and the Engineer's notification recording the agreement of both Parties on such fulfilment and instructing to commence the Work is received by the Contractor:

(a) signature of the Contract Agreement by both Parties, and if required, approval of the Contract by relevant authorities of the Country;

(b) delivery to the Contractor of reasonable evidence of the Employer's Financial arrangements (under Sub-Clause 2.4 [Employer's Financial Arrangements]);

(c) except if otherwise specified in the Contract Data, effective access to and possession of the Site given to the Contractor together with such permission(s) under (a) of Sub-Clause 1.13 [Compliance with Laws] as required for the commencement of the Works;

(d) receipt by the Contractor of the Advance Payment under Sub-Clause 14.2 [Advance Payment] provided that the corresponding bank guarantee has been delivered by the Contractor.'

A new paragraph is added (new paragraph 2):

'If the said Engineer's instruction is not received by the Contractor within 180 days from his receipt of the Letter of Acceptance, the Contractor shall be entitled to terminate the Contract under Sub-Clause 16.2 [Termination by Contractor].'

The Engineer may only instruct the Commencement of Works after the fulfilment of four precedent conditions. These are

- signed Contract Agreement by both Parties
- Contractor receives evidence of Employer's financial arrangements
- access and possession of Site provided in accordance with the Contract Data
- Contractor receives Advance payment on condition that he has delivered a Guarantee.

The Engineer instructs Commencement of Works and records the agreement between the Parties that all conditions precedent has been fulfilled. This must occur within 180 days from receipt of the Letter of Acceptance otherwise the Contractor has a right to terminate under Sub-Clause 16.2 [*Termination by Contractor*].

8.4 Extension of Time for Completion

The phrase '*on the Site*' have been deleted from subparagraph (e). The Employer will take responsibility for delays caused by his other contractors, whether or not they are working on the same Site.

8.6 Rate of Progress

A new third paragraph has been added:

'*Additional costs of revised methods including acceleration measures, instructed by the Engineer to reduce delays resulting from causes listed under Sub-Clause 8.4 [Extension of Time for Completion] shall be paid by the Employer, without generating, however, any other additional payment benefit to the Contractor.*'

In the event of an Employer's delay, the Contractor may recover additional Costs through accelerated working methods.

8.7 Delay Damages

The first sentence of the first paragraph is improved by the addition of the words '*notice under*' before '*Sub-Clause 2.5*'.

8.12 Resumption of Work

The phrase added at the end of the paragraph '*after receiving from the Engineer an instruction to this effect*' under Clause 13 [*Variations and Adjustments*] is an improvement to the wording of the Sub-Clause.

FIDIC Users' Guide
ISBN 978-0-7277-5856-9

ICE Publishing: All rights reserved
http://dx.doi.org/10.1680/fug.58569.317

Chapter 38
Clause 9: Tests on Completion

This Clause is unchanged from the RED Book. Please refer to Chapter 15.

FIDIC Users' Guide
ISBN 978-0-7277-5856-9

ICE Publishing: All rights reserved
http://dx.doi.org/10.1680/fug.58569.319

Chapter 39
Clause 10: Employer's Taking Over

This Clause is unchanged from the RED Book. Please refer to Chapter 16.

FIDIC Users' Guide
ISBN 978-0-7277-5856-9

Chapter 40
Clause 11: Defects Liability

11.3 Extension of Defects Notification Period
The phrase '*by reason of damage attributable to the Contractor*' replaces the word '*damage*' at the end of the first sentence in the first paragraph. Culpability for any damage, lacking in the RED Book, is now clearly defined.

11.11 Clearance of Site
The phrase '*receipt by the Contractor*' replaces the phrase '*the Employer receives a copy*' in the first sentence in paragraph 2. It is more logical that the 28-day time frame begins when the Contractor receives the Performance Certificate.

FIDIC Users' Guide
ISBN 978-0-7277-5856-9

ICE Publishing: All rights reserved
http://dx.doi.org/10.1680/fug.58569.323

Chapter 41
Clause 12: Measurement and Evaluation

12.1 Works to be measured
The first paragraph is expanded by the addition of a second sentence as follows:

'The Contractor shall show in each application under Sub-Clauses 14.3 [Application for Interim Payment Certificates], 14.10 [Statement on Completion] and 14.11 [Application for Final Payment Certificate] the quantities and other particulars detailing the amounts which he considers to be entitled under the Contract.'

The phrase *'and certify the payment of the undisputed part'* is added at the end of the second sentence in the last paragraph (fifth paragraph).

This addition is necessary to ensure payment for undisputed amounts are not delayed.

12.3 Evaluation
A new paragraph has been added (paragraph 3), which confirms the position already taken by many Engineers:

'Any item of work included in the Bill of Quantities for which no rate or price was specified shall be considered as included in other rates and prices in the Bill of Quantities and will not be paid for separately.'

In subparagraphs (a) (i) and (ii) the figures 10% and 0.10% are replaced by 25% and 0.25% respectively. These are recognition that the RED Book figures are not realistic for the requirement to consider a new rate or price.

The phrase *'as soon as the concerned work commences'* is added to the end of the final paragraph (paragraph 6).

This new wording provides a deadline for the Engineer to act, enforceable through the requirements of Sub-Clause 1.3 [*Communications*] (i.e. '*Approvals, certificates, consents and determinations shall not be unreasonably withheld or delayed*').

FIDIC Users' Guide
ISBN 978-0-7277-5856-9

ICE Publishing: All rights reserved
http://dx.doi.org/10.1680/fug.58569.325

Chapter 42
Clause 13: Variations and Adjustments

13.1 Right to Vary
The second paragraph is amended as follows:

The phrase '*or (ii) such Variation triggers a substantial change in the sequence or progress of the Works*' is added to the end of the second paragraph with the original reason for objection being numbered '*(i)*'.

This additional reason for objection to a Variation provides the Contractor with a valuable tool where progress will be disrupted. The change must be substantial but applies to both the sequence and/or the progress.

13.7 Adjustments for Changes on Legislation
A new final paragraph has been added (paragraph 4) to avoid duplication between this and other Sub-Clauses.

'*Notwithstanding the foregoing, the Contractor shall not be entitled to an extension of time if the relevant delay has already been taken into account in the determination of a previous extension of time and such Cost shall not be separately paid if the same shall already have been taken into account in the indexing of any inputs to the table of adjustment data in accordance with the provisions of Sub-Clause 13.8 [Adjustments for Changes in Cost].*'

13.8 Adjustments for Changes in Cost
The phrase '*for local and foreign currencies included in the Schedules*' replaces the phrase '*included in the Appendix to Tender*' in the first paragraph.

The fourth paragraph is changed by the deletion of the phrase '*quoted in the fourth and fifth columns respectively of the table*' after '*stated dates*'.

The fifth paragraph is changed by the deletion of the phrase '*(stated in the table)*' after '*currency of index*'.

These amendments reflect the change from the adjustment data being stated in a Table within the Appendix to Tender (RED Book) to being quoted within the Contract Data.

FIDIC Users' Guide
ISBN 978-0-7277-5856-9

ICE Publishing: All rights reserved
http://dx.doi.org/10.1680/fug.58569.327

Chapter 43
Clause 14: Contractor's Price and Payment

14.1 The Contract Price
A new final paragraph has been added (paragraph 2):

'Notwithstanding the provisions of subparagraph (b), Contractor's Equipment, including essential spare parts therefore, imported by the Contractor for the sole purpose of executing the Contract shall be exempt from the payment of import duties and taxes upon importation.'

The MDB and the Government of the Country will have entered into an agreement for the project to be excluded from Import duties and taxes. This additional paragraph contractually supports this agreement.

14.2 Advance Payment
The phrase *'and cash flow support'* is added after *'mobilisation'* in the first paragraph.

The phrase *'shall deliver to the Employer and to the Contractor an Interim Payment Certificate for the advance payment or its first instalment'* replaces *'issue an Interim Payment Certificate for the first instalment'* in the third paragraph.

The phrase *'a reputable bank or financial institution selected by the Contractor'* replaces the phrase *'an entity within a country (or other jurisdiction) approved by the Employer'* in the third paragraph.

The fifth paragraph is reworded as follows:

'Unless stated otherwise in the Contract Data, the advance payment shall be repaid through percentage deductions from the interim payments determined by the Engineer in accordance with Sub-Clause 14.6 [Issue of Interim Payment Certificates], as follows:

(a) deductions shall commence in the next interim Payment Certificate following that in which the total of all certified interim payments (excluding the advance payment and deductions and repayments of retention) exceeds 30 percent (30%) of the Accepted Contract Amount less Provisional Sums; and
(b) deductions shall be made at the amortisation rate stated in the Contract Data of the amount of each Interim Payment Certificate (excluding the advance payment and

327

deductions for its repayments as well as deductions for retention money) in the currencies and proportions of the advance payment until such time as the advance payment has been repaid; provided that the advance payment shall be completely repaid prior to the time when 90 percent (90%) of the Accepted Contract Amount less Provisional Sums has been certified for payment.'

The phrase '*and in case of termination under Clause 15 [Termination by Employer], except for Sub-Clause 15.5 [Employer's Entitlement to Termination for Convenience]*' is added before '*payable by the Contractor to the Employer*' at the end of paragraph 6.

The wording of the repayment procedures has been improved with, no doubt, the intention of improving cash flow to the Contractor provided it is in a form approved by the Employer. Other issues of note include the choice of financial institution for the advance payment guarantee now being decided by the Contractor. The commencement is changed from 10% in the RED Book to 30% in the PNK Book with the amortisation rate stated in the Contract Data. This provides more favourable terms of repayment for the Contractor.

There is an error in the final paragraph where the words '*Clause 19 [Force Majeure] …*' have been changed to '*Clause 19.6 [Force Majeure]…*'. Although the reference to '19.6' is more precise, it should be worded 'Sub-Clause 19.6 [*Optional Termination, Payment and Release*]'.

14.3 Application for Interim Payment Certificates
Subparagraph (d) of the second paragraph has been amended as follows:

The phrase '*(if more than one instalment) and to be deducted for its repayments*' replaces the words '*and replacements*' in subparagraph (d).

14.4 Schedule of Payments
The words '*or more*' are added twice in subparagraph (c). Instalments may increase as well as decrease (under the RED Book only decrease) when actual progress differs from that stated in the schedules.

14.5 Plant and Materials intended for the Works
Three references to the '*Appendix to Tender*' at the second paragraph have been changed to refer to '*Schedules*'. The term '*(80%)*' is added after '*eighty percent*' in the third paragraph.

14.6 Issue of Interim Payment Certificates
The phrase '*deliver to the Employer and to the Contractor*' replaces the phrase '*issue to the Employer*' in the first paragraph.

The phrase '*all supporting particulars for any reduction or withholding made by the Engineer on the Statement if any*' replaces the phrase '*with supporting particulars*' at the end of the first paragraph.

14.7 Payment
The following phrase is added to the end of subparagraph (b) '*or, at a time when the Bank's loan or credit (from which part of the payments to the Contractor is being made) is*

suspended, the amount shown on any statement submitted by the Contractor within 14 days after such statement is submitted, any discrepancy being rectified in the next payment to the Contractor; and'.

The following phrase is added to the end of subparagraph (c) *'or, at a time when the Bank's loan or credit (from which part of the payments to the Contractor is being made) is suspended, the undisputed amount shown in the Final Statement within 56 days after the date of notification of the suspension in accordance with Sub-Clause 16.2 [Termination by Contractor].'*

This change allows for when the Bank's loan or credit has been suspended. In this situation the Employer must pay the full amount of the Contractor's Statement, within 14 days of the Statement.

14.8 Delayed Payment

The phrase *'or if not available, the interbank offered rate'* is added before the phrase *'and shall be paid in such currency'* in the second paragraph.

Unfortunately there may be more than one rate quoted in some countries and, in such cases, should be clarified prior to tender.

14.9 Payment of Retention Money

The proportion of retention money to be repaid to the Contractor in respect of Sections has been increased from 40% to 50% upon Taking Over and expiry of the DNP in the first and second paragraphs.

Two paragraphs have been added (paragraphs 5 and 6).

'Unless otherwise stated in the Particular Conditions, when the Taking-Over Certificate has been issued for the Works and the first half of the Retention Money has been certified for payment by the Engineer, the Contractor shall be entitled to substitute a guarantee, in the form annexed to the Particular Conditions or in another form approved by the Employer and issued by a reputable bank or financial institution selected by the Contractor, for the second half of the Retention Money. The Contractor shall ensure that the guarantee is in the amounts and currencies of the second half of the Retention Money and is valid and enforceable until the Contractor has executed and completed the Works and remedied any defects, as specified for the Performance Security in Sub-Clause 4.2. On receipt by the Employer of the required guarantee, the Engineer shall certify and the Employer shall pay the second half of the Retention Money. The release of the second half of the Retention Money against a guarantee shall then be in lieu of the release under the second paragraph of this Sub-Clause. The Employer shall return the guarantee to the Contractor within 21 days after receiving a copy of the Performance Certificate.

If the Performance Security required under Sub-Clause 4.2 is in the form of a demand guarantee, and the amount guaranteed under it when the Taking-Over Certificate is issued is more than half of the Retention Money, then the Retention Money guarantee will not be required. If the amount guaranteed under the Performance Security when the Taking-Over Certificate is issued is less than half of the Retention Money, the Retention Money guarantee will only be required for the difference between half of the Retention Money and the amount guaranteed under the Performance Security.'

These provisions entitle the Contractor to substitute guarantees for the release of the Retention Money.

14.11 Application for Final Payment Certificate

The words *'within 28 days from receipt of said draft'* are added to the first sentence of the second paragraph. The Engineer now has a 28-day time limit in which to respond to a draft final statement submitted by the Contractor. Any agreement of the changes should also include a time for submission of the final statement.

14.13 Issue of Final Payment Certificate

In the first paragraph, the Engineer must now *'deliver to the Employer and the Contractor the Final Payment Certificate'*. Subparagraph (a) now states *'the amount which he fairly determines is finally due,'*.

14.15 Currencies of Payment

The words *'Unless otherwise stated in the Particular Conditions'* are deleted from the first paragraph and references to the *'Appendix to Tender'* are changed to refer to the *'Schedule of Payment Currencies'*.

FIDIC Users' Guide
ISBN 978-0-7277-5856-9

http://dx.doi.org/10.1680/fug.58569.331

Chapter 44
Clause 15: Termination by Employer

15.5 Employer's Entitlement to Termination for Convenience

The Sub-Clause heading has been changed to reflect its content by the addition of the words '*for Convenience*'.

The words '*or to avoid a termination of the Contract by the Contractor under Sub-Clause 16.2 [Termination by Contractor]*' are added to the end of the first paragraph. The Employer is therefore, no longer able to terminate the Contract to avoid a Contractor's termination under Sub-Clause 16.2.

The second paragraph is amended as follows:

The phrase '*Sub-Clause 16.4 [Payment on Termination]*' replaces the phrase '*Sub-Clause 19.6 [Optional Termination, Payment and Release]*' at the end of the second paragraph.

The change in payment in the event of a Sub-Clause 15.5 [*Employer's Entitlement to Termination for Convenience*] for termination results in the Contractor receiving more favourable terms.

15.6 Corrupt or Fraudulent Practices

Sub-Clause 15.6 [*Corrupt or Fraudulent Practices*] is new and is produced in six different versions. Pages 55 and 56 are repeated six times for each of the following Banks

- African Development Bank
- Asian Development Bank
- Black Sea Trade and Development Bank or European Bank for Reconstruction and Development
- Caribbean Development Bank
- Inter-American Development Bank
- The World Bank.

The Council of Europe Development Bank, stated within the Terms and Conditions of Use as a Participating Bank, is not included and would need to state which version applies.

Only the version for the African Development Bank is reproduced below.

'If the Employer determines, based on reasonable evidence, that the Contractor has engaged in corrupt, fraudulent, collusive or coercive practices, in competing for or in executing the Contract, then the Employer may, after giving 14 days notice to the Contractor, terminate the Contract and expel him from the Site, and the provisions of Clause 15 shall apply as if such termination had been made under Sub-Clause 15.2 [Termination by Employer].

Should any employee of the Contractor be determined, based on reasonable evidence, to have engaged in corrupt, fraudulent or coercive practice during the execution of the work then that employee shall be removed in accordance with Sub-Clause 6.9 [Contractor's Personnel].

[For contracts financed by the African Development Bank:]

For the purposes of this Sub-Clause:

(a) *'corrupt practice' means the offering, giving, receiving or soliciting of any thing of value to influence the action of a public official in the procurement process or in the contract execution; and*

(b) *'fraudulent practice' means a misrepresentation of facts in order to influence a procurement process or the execution of the Contract to the detriment of the borrower, and includes collusive practice among bidders (prior to or after bid submission) designed to establish bid prices at artificial non-competitive levels and to deprive the borrower of the benefits of free and open competition.'*

FIDIC Users' Guide
ISBN 978-0-7277-5856-9

Chapter 45
Clause 16: Suspension and Termination by Contractor

16.1 Contractor's Entitlement to Suspend Work
A new second paragraph has been added:

'Notwithstanding the above, if the Bank has suspended disbursements under the loan or credit from which payments to the Contractor are being made, in whole or in part, for the execution of the Works, and no alternative funds are available as provided for in Sub-Clause 2.4 [Employer's Financial Arrangements], the Contractor may by notice suspend work or reduce the rate of work at any time, but not less than 7 days after the Borrower having received the suspension notification from the Bank.'

This provides the Contractor with the right to suspend work if the Bank has suspended disbursements under its loan.

16.2 Termination by Contractor
The phrase 'in such manner as to materially and adversely affect the economic balance of the Contract and/or the ability of the Contractor to perform the Contract' is added to the end of subparagraph (d).

The economic balance of the Contract would require a subjective test hence the potential for dispute is high.

A new subparagraph (h) has been added.

'(h) the Contractor does not receive the Engineer's instruction recording the agreement of both Parties on the fulfilment of the conditions for the Commencement of Works under Sub-Clause 8.1 [Commencement of Works].'

It will be a brave Employer/Engineer who commences work without recording a written agreement. In such a case, the Contractor is not subject to a time bar leaving an option to terminate at a time of his choice.

16.3 Cessation of Work and Removal of Contractor's Equipment
The reference to Sub-Clause 15.5 [Employer's Entitlement to Termination for Convenience] includes the words 'for Convenience' reflecting the revised title.

16.4 Payment on Termination

The phrase '*of profit or other loss*' is deleted from subparagraph (c).

The removal of the provision for loss of profit is unfair to the Contractor because the termination was caused by a default by the Employer. This should, however, be checked against the provisions of the respective legal system under which the project is being executed (Law of the Country), as in many jurisdictions this issue is addressed in a mandatory form.

FIDIC Users' Guide
ISBN 978-0-7277-5856-9

Chapter 46
Clause 17: Risks and Responsibilities

17.1 Indemnities
Paragraph (b) has been changed to read:

'(b) damage to or loss of any property, real or personal (other than the Works), to the extent that such damage or loss arises out of or in the course of or by reason of the Contractor's design (if any), the execution and completion of the Works and the remedying of any defects, unless and to the extent that any such damage or loss is attributable to any negligence, wilful act or breach of the Contract by the Employer, the Employer's Personnel, their respective agents, or anyone directly or indirectly employed by any of them.'

The revised paragraph is more logical than the RED Book in that the indemnity by the Contractor does not require negligence, wilful act or breach of Contract by the Contractor, but is excluded by negligence, wilful act or breach of Contract by the Employer.

17.3 Employer's Risks
The phrase *'insofar as they directly affect the execution of the Works in the Country'* is added to the first paragraph.

Subparagraph (c) includes the phrase *'sabotage by persons other than the Contractor's Personnel'*.

Subparagraph (d) deletes the reference to *'and other employees of the Contractor and Subcontractor'*.

The risks only apply if they directly affect the execution of the Works in the Country. This seems a logical clarification.

Subparagraphs (b) and (c) are changed to exclude sabotage by the Contractor's Personnel. The change to subparagraph (c) only deletes a repetition because the definition of Contractor's Personnel, at Sub-Clause 1.1.2.7 [*Contractor's Personnel*], includes employees of the Contractor and Subcontractors.

17.6 Limitation of Liability
The first and second paragraphs are revised as follows.

'Neither Party shall be liable to the other Party for loss of use of any Works, loss of profit, loss of any contract or for any indirect or consequential loss or damage which may be suffered by the

other Party in connection with the Contract, other than as specifically provided in Sub-Clause 8.7 [Delay Damages]; Sub-Clause 11.2 [Cost of Remedying Defects]; Sub-Clause 15.4 [Payment after Termination]; Sub-Clause 16.4 [Payment on Termination]; Sub-Clause 17.1 [Indemnities]; Sub-Clause 17.4(b) [Consequences of Employer's Risks] and Sub-Clause 17.5 [Intellectual and Industrial Property Rights].

The total liability of the Contractor to the Employer, under or in connection with the Contract other than under Sub-Clause 4.19 [Electricity, Water and Gas], Sub-Clause 4.20 [Employer's Equipment and Free-Issue Materials], Sub-Clause 17.1 [Indemnities] and Sub-Clause 17.5 [Intellectual and Industrial Property Rights], shall not exceed the sum resulting from the application of a multiplier (less or greater than one) to the Accepted Contract Amount, as stated in the Contract Data, or (if such multiplier or other sum is not so stated), the Accepted Contract Amount.'

This Sub-Clause provides a limit to the amount of liability a party may suffer. Exceptions to this limit contained within the RED Book include Sub-Clause 16.4 [*Payment on Termination*] and Sub-Clause 17.1 [*Indemnities*]. These exceptions are extended under the PNK Book by the addition of Sub-Clause 8.7 [*Delay Damages*], Sub-Clause 11.2 [*Cost of Remedying Defects*], Sub-Clause 15.4 [*Payment after Termination*], Sub-Clause 17.4b [*Consequences of Employer's Risks*] (Occupation of Works or Design by the Employer)] and Sub-Clause 17.5 [*Intellectual and Industrial Property Rights*] thereby significantly increasing the Contractor's risk. The Sub-Clause remains controversial because such limitation may not be acceptable under some jurisdictions.

17.7 Use of Employer's Accommodation/Facilities
A new Sub-Clause that is self-explanatory is added as follows:

'The Contractor shall take full responsibility for the care of the Employer-provided accommodation and facilities, if any, as detailed in the Specification, from the respective dates of hand-over to the Contractor until cessation of occupation (where hand-over or cessation of occupation may take place after the date stated in the Taking-Over Certificate for the Works).

If any loss or damage happens to any of the above items while the Contractor is responsible for their care arising from any cause whatsoever other than those for which the Employer is liable, the Contractor shall, at his own cost, rectify the loss or damage to the satisfaction of the Engineer.'

FIDIC Users' Guide
ISBN 978-0-7277-5856-9

Chapter 47
Clause 18: Insurance

In the Introduction to the PNK Book, FIDIC suggest that the Employer should consider a dedicated single policy for the project which would cover all the insurance requirements. Details of any such policy should be included in the invitation to tender.

18.1 General Requirements for Insurances
Paragraph 3 is amended as follows:

'Wherever the Employer is the insuring Party, each insurance shall be effected with insurers and in terms acceptable to the Contractor. These terms shall be consistent with any terms agreed by both Parties before the date of the Letter of Acceptance. This agreement of terms shall take precedence over the provisions of this Clause.'

This rewording provides the Contractor with some control over whom the Employer chooses as insurer.

A new paragraph has been added at the end.

'The Contractor shall be entitled to place all insurances relating to the Contract (including, but not limited to the insurance referred to Clause 18) with insurers from any eligible source country.'

It is not clear whether the statement that insurance may be placed with insurers from any eligible source country would ban insurers from all other countries. The list of eligible source countries will depend on the particular MDB and should be stated.

18.2 Insurance for Works and Contractor's Equipment
The phrase *'to the Party actually bearing the costs'* replaces the phrase *'between the Parties for the sole purpose'* in the fourth paragraph, subparagraph (b).

The phrase *'to the extent specifically required in the bidding documents of the Contract'* is included before *'loss or damage'* in the fourth paragraph, subparagraph (d).

Under new wording in subparagraph (b) the Party responsible for repairing the damage or replacing the loss now receives the payment from the insurer.

Under subparagraph (d), the extent of loss or damage attributable to the Employers use or occupation should be included within the bidding documents.

18.4 Insurance for Contractor's Personnel

The second paragraph is reworded as follows:

'The insurance shall cover the Employer and the Engineer against liability for claims, damages, losses and expenses (including legal fees and expenses) arising from injury, sickness, disease or death of any person employed by the Contractor or any other of the Contractor's Personnel, except that this insurance may exclude losses and claims to the extent that they arise from any act or neglect of the Employer or of the Employer's Personnel.'

This amendment indemnifies the employees of the Employer and the Engineer from any insurance claim involving persons employed by the Contractor.

FIDIC Users' Guide
ISBN 978-0-7277-5856-9

ICE Publishing: All rights reserved
http://dx.doi.org/10.1680/fug.58569.339

Chapter 48
Clause 19: Force Majeure

19.1 Definition of Force Majeure

The phrase '*sabotage by persons other than the Contractor's Personnel*' is added after '*terrorism*' in the second paragraph, subparagraph (ii).

The phrase '*and other employees of the Contractor and Subcontractor*' is deleted from the second paragraph, subparagraph (iii).

Subparagraph (ii) now includes sabotage by persons other than the Contractor's Personnel. The change to subparagraph (iii) only deletes the reference to other employees of the Contractor or Subcontractors, who are already included in Contractor's Personnel.

19.2 Notice of Force Majeure

In the first paragraph Force Majeure is restricted to events that prevent a Party from performing '*its substantial obligations*' rather than '*any of its obligations*'. In the second paragraph the phrase '*such obligations*' is changed to '*its obligations*'. These changes restrict the application of Force Majeure to really serious events.

The meaning of the word '*substantial*' indicates that the Party is prevented from performing an obligation, and therefore has no choice but to stop performance. The precise meaning and extent of the performance that must stop will then be the subject of debate, and possibly a dispute.

19.4 Consequences of Force Majeure

The first paragraph of the Sub-Clause has been changed to include the word '*substantial*' before the word '*obligations*' as in Sub-Clause 19.2 [*Notice of Force Majeure*].

Subparagraph (b) is amended as follows:

The phrase '*including the costs of rectifying or replacing the Works and/or Goods damaged or destroyed by Force Majeure, to the extent they are not indemnified through the insurance policy referred to in Sub-Clause 18.2 [Insurance for Works and Contractor's Equipment]*' is added to the end of subparagraph (b).'

19.6 Optional Termination, Payment and Release

The word '*any*' is deleted from the beginning of subparagraph (c). Also the Cost or liability must be '*necessarily*', as well as '*reasonably*', incurred. This is a stricter requirement than the RED Book and may exclude some of the Costs that would otherwise be claimed by the Contractor.

FIDIC Users' Guide
ISBN 978-0-7277-5856-9

ICE Publishing: All rights reserved
http://dx.doi.org/10.1680/fug.58569.341

ice
Institution of Civil Engineers

publishing

Chapter 49
Clause 20: Claims, Disputes and Arbitration

20.1 Contractor's Claims

The last sentence of the sixth paragraph refers to '*the above defined time period*' rather than '*such time*'.

A new paragraph seven has been included as follows:

'*Within the above defined period of 42 days, the Engineer shall proceed in accordance with Sub-Clause 3.5 [Determinations] to agree or determine (i) the extension (if any) of the Time for Completion (before or after its expiry) in accordance with Sub-Clause 8.4 [Extension of Time for Completion], and/or (ii) the additional payment (if any) to which the Contractor is entitled under the Contract.*'

The words '*additional payment*' replaces the word '*amounts*' in the first sentence of the eighth paragraph (RED Book, paragraph 7).

Paragraph 9 (RED Book, paragraph 8) is reworded as follows:

'*If the Engineer does not respond within the timeframe defined in this Clause, either Party may consider that the claim is rejected by the Engineer and any of the Parties may refer to the Dispute Board in accordance with Sub-Clause 20.4 [Obtaining Dispute Board's Decision].*'

These changes improve the claims procedure by requiring the Engineer to respond within a specified time ensuring differences of opinion are settled quickly.

20.2 Appointment of the Dispute Board

This clause is extensively amended and therefore reproduced in full:

'*Disputes shall be referred to a DB for decision in accordance with Sub-Clause 20.4 [Obtaining Dispute Board's Decision]. The Parties shall appoint a DB by the date stated in the Contract Data.*

The DB shall comprise, as stated in the Contract Data, either one or three suitably qualified persons ('the members'), each of whom shall be fluent in the language for communication defined in the Contract and shall be a professional experienced in the type of construction involved in the Works and with the interpretation of contractual documents. If the number is not so stated and the Parties do not agree otherwise, the DB shall comprise three persons.

If the Parties have not jointly appointed the DB 21 days before the date stated in the Contract Data and the DB is to comprise three persons, each Party shall nominate one member for the approval of the other Party. The first two members shall recommend and the Parties shall agree upon the third member, who shall act as chairman.

However, if a list of potential members has been agreed by the Parties and is included in the Contract, the members shall be selected from those on the list, other than anyone who is unable or unwilling to accept appointment to the DB.

The agreement between the Parties and either the sole member or each of the three members shall incorporate by reference the General Conditions of Dispute Board Agreement contained in the Appendix to these General Conditions, with such amendments as are agreed between them.

The terms of the remuneration of either the sole member or each of the three members, including the remuneration of any expert whom the DB consults, shall be mutually agreed upon by the Parties when agreeing the terms of appointment. Each Party shall be responsible for paying one-half of this remuneration.

If at any time the Parties so agree, they may jointly refer a matter to the DB for it to give its opinion. Neither Party shall consult the DB on any matter without the agreement of the other Party.

If a member declines to act or is unable to act as a result of death, disability, resignation or termination of appointment, a replacement shall be appointed in the same manner as the replaced person was required to have been nominated or agreed upon, as described in this Sub-Clause.

The appointment of any member may be terminated by mutual agreement of both Parties, but not by the Employer or the Contractor acting alone. Unless otherwise agreed by both Parties, the appointment of the DB (including each member) shall expire when the discharge referred to in Sub-Clause 14.12 [Discharge] shall have become effective.'

Sub-Clause 20.2 includes several significant changes that are presumably based on the MDBs' experience of the use of DBs. Most of the changes, such as the changes to the selection procedures, are improvements. The 2005 version of the PNK Book removed the right of the Parties to refer a matter to the DB for an opinion. However, this was reinstated in the later versions in its original RED Book form.

The main changes can be summarised as follows.

■ Some of the qualities for the DB member, which are given in the Appendix (i.e. paragraph 3), have been repeated in the Sub-Clause.

- The Parties must first consider together who shall serve as the DB, and it is only if they have not jointly appointed the DB 21 days before the date stated in the Contract Data that each Party proposes a member for a three-person DB. This enables the Parties to agree on a balance of skills in the DB.
- The first two members have an obligation to recommend someone to be appointed as the third member, who will act as chairman for the Parties' agreement.
- The provision for a list of potential members for a three-person board to be included in the Contract was deleted in the 2005 version of the PNK Book but was reinstated in later versions. However, it no longer relies upon the list being inserted by the Employer as there is an added provision for an agreement to be reached.
- There is a new requirement at Rule 2 in the Annex that, during the Site visits, the DB shall endeavour to prevent potential problems or claims from becoming disputes.
- The wording of the procedure for the appointment of a replacement member has been improved.

20.3 Failure to Agree on the Composition of the Dispute Board
The heading of the Sub-Clause has been changed to reflect its content. Subparagraph (b) has been changed to provide for reference to the appointing entity in the case where one Party fails to approve a member nominated by the other Party. This closes a gap in the RED Book provisions.

20.4 Obtaining Dispute Board's Decision
The fifth paragraph has been reworded as follows:

'If either Party is dissatisfied with the DB's decision, then either Party may, within 28 days after receiving the decision, give a Notice of Dissatisfaction to the other Party indicating its dissatisfaction and intention to commence arbitration. If the DB fails to give its decision within the period of 84 days (or as otherwise approved) after receiving such reference, then either Party may, within 28 days after this period has expired, give a Notice of Dissatisfaction to the other Party.'

The Notice of Dissatisfaction is defined in 1.1.6.10. It is now a requirement for the Party to give its intention to commence arbitration. As stated before, it is not clear as to how this requirement shall be interpreted in the light of the Sub-Clause 20.5 [*Amicable Settlement*] requirement to attempt an amicable settlement after the DB procedure and before commencement of Arbitration.

20.5 Amicable Settlement
The phrase *'the Party giving a Notice of Dissatisfaction in accordance with Sub-Clause 20.4 [Obtaining Dispute Board's Decision] above should move to commence arbitration'* is added before *'after the fifty-sixth day'* in the second sentence.

It is for the Party who issues the Notice of Dissatisfaction to commence arbitration, a requirement not stated in the RED Book.

20.6 Arbitration
The whole Sub-Clause is reworded as follows:

'*Any dispute between the Parties arising out of or in connection with the Contract not settled amicably in accordance with Sub-Clause 20.5 above and in respect of which the DB's decision (if any) has not become final and binding shall be finally settled by arbitration. Arbitration shall be conducted as follows:*

(a) if the Contract is with foreign contractors,
  (i) for contracts financed by all participating Banks except under subparagraph (a)(ii) below:
  international arbitration (1) with proceedings administered by the arbitration institution designated in the Contract Data, and conducted under the rules of arbitration of such institution; or, if so specified in the Contract Data, (2) international arbitration in accordance with the arbitration rules of the United Nations Commission on International Trade Law (UNCITRAL); or (3) if neither an arbitration institution nor UNCITRAL arbitration rules are specified in the Contract Data, with proceedings administered by the International Chamber of Commerce (ICC) and conducted under the ICC Rules of Arbitration; by one or more arbitrators appointed in accordance with said arbitration rules.
  (ii) for contracts financed by the Asian Development Bank: international arbitration (1) with proceedings administered by the arbitration institution specified in the Contract Data and conducted under the rules of arbitration of such institution unless it is specified in the Contract Data that the arbitration shall be conducted under the rules of the United Nations Commission on International Trade Law (UNCITRAL) and if UNCITRAL Rules are so specified then the named arbitration institution shall be the appointing authority and shall administer the arbitration); or (2) if an arbitration institution is not specified in the Contract Data, with proceedings administered by the Singapore International Arbitration Centre (SIAC) and conducted under the SIAC Rules, by one or more arbitrators appointed in accordance with the said arbitration rules.
(b) if the Contract is with domestic contractors, arbitration with proceedings conducted in accordance with the laws of the Employer's country.

The place of arbitration shall be the neutral location specified in the Contract Data; and the arbitration shall be conducted in the language for communications defined in Sub-Clause 1.4 [Law and Language].

The arbitrators shall have full power to open up, review and revise any certificate, determination, instruction, opinion or valuation of the Engineer, and any decision of the DB, relevant to the dispute. Nothing shall disqualify representatives of the Parties and the Engineer from being called as a witness and giving evidence before the arbitrators on any matter whatsoever relevant to the dispute.

Neither Party shall be limited in the proceedings before the arbitrators to the evidence or arguments previously put before the DB to obtain its decision, or to the reasons for dissatisfaction given in its Notice of Dissatisfaction. Any decision of the DB shall be admissible in evidence in the arbitration.

Arbitration may be commenced prior to or after completion of the Works. The obligations of the Parties, the Engineer and the DB shall not be altered by reason of any arbitration being conducted during the progress of the Works.'

Different rules apply for foreign contractors (i.e. based in a country other than the Country) and domestic contractors (i.e. based in the Country). The Asian Development Bank (ADB) and other Banks (i.e. the remainder) each have further modifications.

If it concerns a foreign contractor and a Bank other than the ADB then Arbitration rules shall be

- an institution designated in the Contract Data
- UNICITRAL rules if specified
- ICC rules if neither of the above are specified.

If it concerns a foreign contractor and the Contract is financed by the ADB then Arbitration rules shall be

- an institution designated in the Contract Data or, if specified, UNICITRAL rules in which case the named arbitration institution shall appoint and administer the arbitration; or
- SIAC rules if an arbitration institution is not specified.

If it concerns a domestic contractor, the arbitration and proceedings may be conducted in accordance with the laws of the Employer's country.

If it concerns a foreign contractor in dispute with a local Subcontractor the PNK Book is silent on the issue. The parties should therefore agree on a suitable procedure (e.g. Sub-Clause 20.7 [*Subcontract Arbitration*] of the SUB Book).

20.7 Failure to Comply with Dispute Board's Decision
This Sub-Clause is amended as follows:

'In the event that a Party fails to comply with a final and binding DB decision, then the other Party may, without prejudice to any other rights it may have, refer the failure itself to arbitration under Sub-Clause 20.6 [Arbitration]. Sub-Clause 20.4 [Obtaining Dispute Board's Decision] and Sub-Clause 20.5 [Amicable Settlement] shall not apply to this reference.'

This Sub-Clause has been revised such that a failure to comply with a DB decision, which has become final and binding as the final paragraph of Sub-Clause 20.4 [*Obtaining Dispute Adjudication Board's Decision*], can be referred to arbitration, whether or not a notice of dissatisfaction has been issued.

FIDIC Users' Guide
ISBN 978-0-7277-5856-9

ICE Publishing: All rights reserved
http://dx.doi.org/10.1680/fug.58569.347

Chapter 50
Appendix: General Conditions of Dispute Board Agreement

The heading of the Appendix has been changed to '*General Conditions of Dispute Board Agreement*', consistent with the general omission of the word '*Adjudication*'. These changes are made throughout the Appendix and not mentioned further in this narrative.

General Provisions
The PNK Book has omitted the paragraph which states that the Agreement shall be void and ineffective if the Member has not been notified by both Parties, within six months of signing. This notice should be a formality, if copies of the signed Agreement have been circulated to the Member and the Parties. However, if the Member has not received the signed copies he could be uncertain, perhaps for some considerable time, as to whether the Agreement has come into effect.

The ban on assignment or subcontracting has also been deleted (paragraph 4 in the RED Book). However, the previous paragraph is clear that this employment is a personal appointment so assignment or subcontracting would not be possible.

5 General Obligations of the Employer and the Contractor
The phrase '*and except to the extent that prior agreement is given by the Employer, the Contractor and the other Members (if any)*' is deleted from the end of the first sentence in the first paragraph.

This phrase is superfluous, as the Member is already required to give advice under Rule 4(k).

6 Payment
Subparagraph (a) (iv) has been revised so that the retainer is only reduced by one third, rather than 50%, after the Taking-Over Certificate has been issued.

Subparagraph (c) has been reworded as follows:

'*(c) all reasonable expenses including necessary travel expenses (air fare in less than first class, hotel and subsistence and other direct travel expenses) incurred in connection with the Member's duties, as well as the cost of telephone calls, courier charges, faxes and telexes: a receipt shall be required for each item in excess of five percent of the daily fee referred to in sub-paragraph (b) of this Clause;*'

This imposes restrictions on the expenses.

A new paragraph has been added (new paragraph 3).

'If the parties fail to agree on the retainer fee or the daily fee, the appointing entity or official named in the Contract Data shall determine the amount of the fees to be used.'

This addition aims to resolve a situation if the Parties and the Member cannot agree to the retainer or daily fee. The appointing entity or official named in the Contract Data will determine it.

8 Default of the Member
This section has been amended as follows:

'If the Member fails to comply with any of his obligations under Clause 4 (a)–(d) above, he shall not be entitled to any fees or expenses hereunder and shall, without prejudice to their other rights, reimburse each of the Employer and the Contractor for any fees and expenses received by the Member and the Other Members (if any), for proceedings or decisions (if any) of the DB which are rendered void or ineffective by the said failure to comply.

If the Member fails to comply with any of his obligations under Clause 4 (e)–(k) above, he shall not be entitled to any fees or expenses hereunder from the date and to the extent of the non-compliance and shall, without prejudice to their other rights, reimburse each of the Employer and the Contractor for any fees and expenses already received by the Member, for proceedings or decisions (if any) of the DB which are rendered void or ineffective by the said failure to comply.'

This section has been revised to separate the procedural parts of Clause 4 General Obligations, items (e) to (k), from the parts that refer to the Member's relations with the Parties, items (a) to (d). The penalty in respect of failure to comply with items (e) to (k) is restricted to fees and expenses from the date and to the extent of the non-compliance. Also, the penalties only apply to proceedings or decisions that were rendered ineffective by the failure to comply.

9 Disputes
This section has been amended as follows:

'Any dispute or claim arising out of or in connection with this Dispute Board Agreement, or the breach, termination or invalidity thereof, shall be finally settled by institutional arbitration. If no other arbitration institute is agreed, the arbitration shall be conducted under the Rules of Arbitration of the International Chamber of Commerce by one arbitrator appointed in accordance with these Rules of Arbitration.'

The Parties to the Dispute Board Agreement can agree on an arbitration institution or, if no institution is agreed, that the Rules of the ICC will apply.

50.1. Changes to Annex: Procedural Rules

Rule 2 has been altered as follows:

'The timing of and agenda for each Site visit shall be as agreed jointly by the DB, the Employer and the Contractor, or in the absence of agreement, shall be decided by the DB. The purpose of Site visits is to enable the DB to become and remain acquainted with the progress of the Works and of any actual or potential problems or claims, and, as far as reasonable, to endeavour to prevent potential problems or claims from becoming disputes.'

The rule has been revised to include provision for the Site visit to, as far as possible, endeavour to prevent potential problems or claims from becoming disputes. This requires much wider, more proactive actions to be undertaken by the DB than under the RED Book that relies on the provision for an opinion to be given when requested under Sub-Clause 20.2 *[Appointment of the Dispute Adjudication Board]*. A DB may exercise this role by asking questions. Such a technique makes it clear what the DB considers to be important without giving any unrequested opinions.

FIDIC Users' Guide
ISBN 978-0-7277-5856-9

ICE Publishing: All rights reserved
http://dx.doi.org/10.1680/fug.58569.351

Chapter 51
Particular Conditions and Standard Forms

51.1. Changes to standard forms

The PNK Book includes standard forms for the Particular Conditions, the Letters of Bid and Acceptance, the Agreements, and the Forms of Security and Guarantee. The following notes should be read together with the full text in the PNK Book.

- *'The Particular Conditions – Part A: Contract Data'*
  The Contract Data replaces the Appendix to Tender in other FIDIC Contracts. It includes the same items, modified to suit the wording of the PNK Book. However, all the information must be provided by the Employer unlike the RED Book where some information must be entered by the tenderer.
- *'The Particular Conditions – Part B: Specific Provisions'*
  Additional specific provisions may be added to suit the particular requirements of the project.
- *'Sample Forms'*
  The PNK Book includes 11 sample forms. These are similar to the FIDIC sample forms, modified to suit the MDB requirements. The differences should be studied to check whether they could be of value for the particular requirements of other FIDIC Contracts. The PNK Book standard forms are provided in three sections as follows.
  1. Forms of letter of Bid, Letter of Acceptance, Contract Agreement and Dispute Board Agreement.
     Annex A Letter of Bid
     Annex B Letter of Acceptance
     Annex C Contract Agreement
     Annex D Dispute Board Agreement (sole Member DB)
     Annex E Dispute Board Agreement (each Member of a three-member DB).
  2. Forms of Security annexed to the Particular Conditions.
     Annex F Performance Security: Demand Guarantee
     Annex G Performance Security: Performance Bond
     Annex H Advance Payment Security: Demand Guarantee
     Annex I Retention Money Security: Demand Guarantee.
  3. Forms of Guarantee annexed to the Instructions to Tenderers.
     Annex J Parent Company Guarantee
     Annex K Bid Security: Bank Guarantee.

Part 3

The FIDIC Conditions of Contract for Plant and Design Build for Electrical and Mechanical Plant, and for Building and Engineering Works, designed by the Contractor (Yellow Book, YLW Book)

FIDIC Users' Guide
ISBN 978-0-7277-5856-9

ICE Publishing: All rights reserved
http://dx.doi.org/10.1680/fug.58569.355

Chapter 52
Introduction to Part 3

The purpose for the Yellow Book (YLW) is stated on the title page:

The *Conditions of Contract for Plant and Design-Build* is for electrical and mechanical plant, and for building and engineering works, designed by the Contractor.

Part 3 of this book includes a detailed review of the Sub-Clauses in the YLW Book. As per the Foreword to the YLW Book interpretation must necessarily take account of the following:

In the preparation of these Conditions of Contract for Plant and Design-Build, it was recognised that, while there are many sub-clauses which will be generally applicable, there are some sub-clauses which must necessarily vary to take account of the circumstances relevant to the particular contract. The sub-clauses which were considered to be applicable to many (but not all) contracts have been included in the General Conditions, in order to facilitate their incorporation into each contract.

The General Conditions and the Particular Conditions will together comprise the Conditions of Contract governing the rights and obligations of the parties. It will be necessary to prepare the Particular Conditions for each individual contract, and to take account of those in the General Conditions which mention the Particular Conditions.

For this publication, the General Conditions were prepared on the following basis:

(i) *interim payments, in respect of the lump sum Contract Price, will be made as work proceeds, and will typically be based on instalments specified in a schedule;*
(ii) *if the wording in the General Conditions necessitates further data, then (unless it is so descriptive that it would have to be detailed in the Employer's Requirements) the sub-clause makes reference to this data being contained in the Appendix to Tender, the data either being prescribed by the Employer or being inserted by the Tenderer;*
(iii) *where a sub-clause in the General Conditions deals with a matter on which different contract terms are likely to be applicable for different contracts, the principles applied in writing the sub-clause were:*
    (a) *users would find it more convenient if any provisions which they did not wish to apply could simply be deleted or not invoked, than if additional text had to be written (in the Particular Conditions) because the General Conditions did not cover their requirements; or*

(b) in other cases, where the application of (a) was thought to be inappropriate, the sub-clauses contains the provisions which were considered applicable to most contracts.

For example, Sub-Clause 14.2 [Advance Payment] is included for convenience, not because of any FIDIC policy in respect of advance payments. This Sub-Clause becomes inapplicable (even if it is not deleted) if it is disregarded by not specifying the amount of the advance. It should therefore be noted that some of the provisions contained in the General Conditions may not be appropriate for an apparently-typical contract.

This commentary only considers those Sub-Clauses that are different to the RED Book. For those Sub-Clauses not mentioned below, please refer to the commentary on the RED Book. The commentary includes suggestions for possible changes that could be made in the Particular Conditions, together with cross-references to other Sub-Clauses that may be relevant.

This review does not attempt to give a detailed legal analysis of every Sub-Clause, but to draw readers' attention to problems that may arise in the interpretation and use of the Sub-Clause. These comments and any references to the content of the Sub-Clauses must be read in relation to the complete Sub-Clause, which is reproduced in either Part 3 or Part 2 of this book.

Part 1 of this book includes sixteen flow charts (Figures 5.1 to 5.16 in Chapter 5). With the exception of Figure 5.5 these flow charts together with the following three flow charts, show the procedures described within the YLW Book.

- Figure 52.1 Errors in Employer's Requirements: Sub-Clauses 1.9, 5.1, 8.1, 13.1, 20.1
- Figure 52.2 Design procedures: Sub-Clauses 1.1.6.1, 1.13, 5.1
- Figure 52.3 Procedures during Defects Notification Period: Sub-Clauses 2.5, 10.1, 11.1, 11.3, 11.6, 11.8, 11.9, 12.

Figure 52.1 Errors in Employer's Requirements

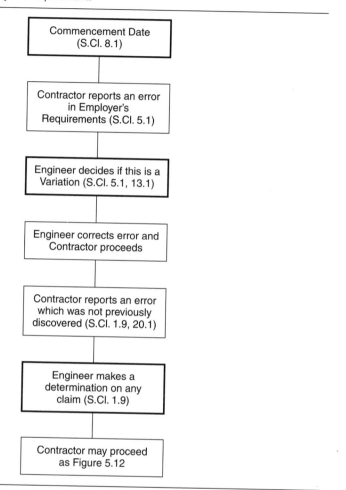

Figure 52.2 Design procedures

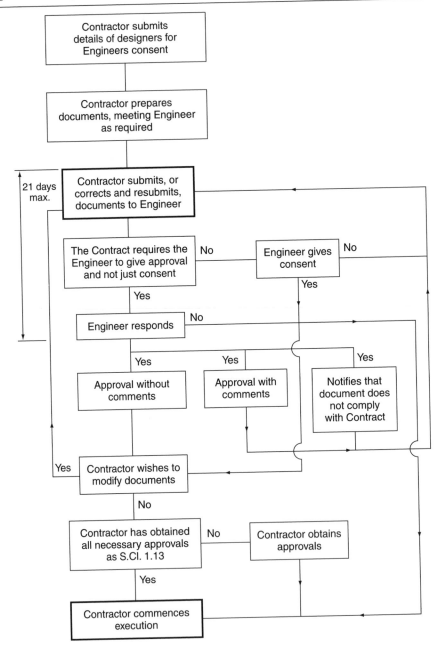

Notes:
1. All boxes refer to SCl. 5.1 (YLW) unless noted otherwise.
2. 'Document' refers to Contractor's Documents as S.Cl. 1.1.6.1 (YLW)

Figure 52.3 Procedures during Defects Notification Period

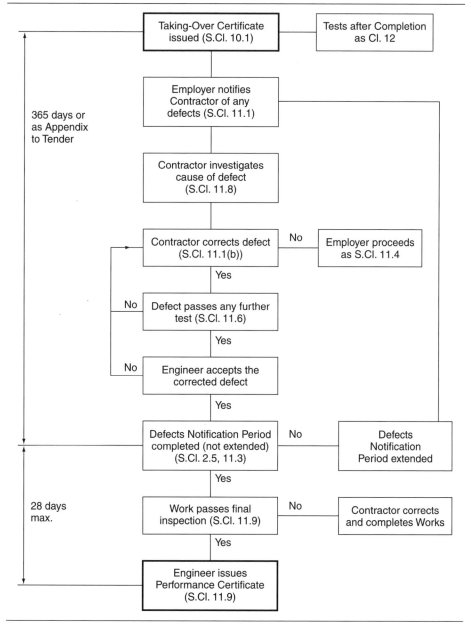

FIDIC Users' Guide
ISBN 978-0-7277-5856-9

ICE Publishing: All rights reserved
http://dx.doi.org/10.1680/fug.58569.361

Chapter 53
Clause 1: General Provisions

Clause 1 covers general matters such as definitions, the law and language of the Contract and various matters concerning the documents.

The definitions 1.1.1.5, 1.1.1.6, 1.1.1.7, 1.1.1.10, 1.1.6.3 and 1.1.6.9 have been changed and now read:

1.1.1.5 *'Employer's Requirements'* means the document entitled employer's requirements, as included in the Contract, and any additions and modifications to such document in accordance with the Contract. Such document specifies the purpose, scope, and/or design and/or other technical criteria, for the Works.

1.1.1.6 *'Schedules'* means the document(s) entitled schedules, completed by the Contractor and submitted with the Letter of Tender, as included in the Contract. Such document may include data, lists and schedules of payments and/or prices.

1.1.1.7 *'Contractor's Proposal'* means the document entitled proposal, which the Contractor submitted with the Letter of Tender, as included in the Contract. Such document may include the Contractor's preliminary design.

1.1.1.10 *'Schedule of Guarantees'* and *'Schedule of Payments'* mean the documents so named (if any) which are comprised in the Schedules.

1.1.6.3 *'Employer's Equipment'* means the apparatus, machinery and vehicles (if any) made available by the Employer for the use of the Contractor in the execution of the Works, as stated in the Employer's Requirements; but does not include Plant which has not been taken over by the Employer.'

The word *'Specification'* as stated in the RED Book is changed to *'Employer's Requirements'* above as directed by the errata contained inside the back cover of the YLW Book.

1.1.6.9 *'Variation'* means any change to the Employer's Requirements or the Works, which is instructed or approved as a variation under Clause 13 [*Variations and Adjustments*].

1.5 Priority of Documents
The documents (f), (g) and (h) under RED or (g), (h) and (i) under PNK are replaced by:

'(f) the Employer's requirements

(g) the Schedules

(h) the Contractor's Proposal and any other documents forming part of the Contract.'

It is significant that, although the Contractor's Proposal was studied and approved before the Contract was awarded, it remains a lower priority than the Employer's Requirements. If the Contractor's Proposals include any details that are different to the Employer's Requirements, but the Employer decided he preferred the Contractor's details, then it must be incorporated in the Contract as an amendment to the Employer's Requirements.

Similarly, if there are any errors in the Employer's Requirements, or the Engineer approves a detail in the Contractor's design that is different to the Employer's Requirements, then the change must be recorded as a Variation. This situation will be discussed at Sub-Clauses 1.9 [*Errors in the Employer's Requirements*], 13.1 [*Right to Vary*] and 13.2 [*Value Engineering*].

1.8 Care and Supply of Documents

A Specification and Drawings provided by the Employer are not relevant to construction with Contractor's design and so this first paragraph has been deleted. Under the YLW Book the design is performed by the Contractor and there is a greater need for a detailed definition of the Contractor's Documents. This extended definition can be found at Sub-Clause 1.1.6.1 [*Contractor's Documents*], and in greater detail under Sub-Clause 5.2 [*Contractor's Documents*].

When the Employer, or the Engineer acting as Employer's Personnel, become aware of any error or defect in a document '*which was prepared for use in executing the Works*' they are obligated to promptly inform the Contractor accordingly.

1.9 Errors in the Employer's Requirements

This Sub-Clause is specifically drafted for contracts where the Contractor executes the design. There is no equivalent Sub-Clause under either RED or PNK Books.

If the Contractor suffers delay and/or incurs Cost as a result of an error in the Employer's Requirements, and an experienced contractor exercising due care would not have discovered the error when scrutinising the Employer's Requirements under Sub-Clause 5.1 [General Design Obligations], the Contractor shall give notice to the Engineer and shall be entitled subject to Sub-Clause 20.1 [Contractor's Claims] to

(a) an extension of time for any such delay, if completion is or will be delayed, under Sub-Clause 8.4 [Extension of Time for Completion]; and

(b) Payment of any such Cost plus reasonable profit, which shall be included in the Contract Price.

After receiving this notice, the Engineer shall proceed in accordance with Sub-Clause 3.5 [Determinations] to agree or determine (i) whether and (if so) to what extent the error could not reasonably have been so discovered, and (ii) the matters described in subparagraphs (a) and (b) above related to this extent.

Under a Contractor's design Contract the Employer's Requirements are fixed when the Contract is agreed, subject only to Clause 13 [*Variations*]. The Engineer can issue instructions, '*which may be necessary for the execution of the Works*', under Sub-Clause 3.3 [*Instructions of the Engineer*], but he cannot issue additional Drawings to develop the basic information which was contained in the Tender Drawings.

Sub-Clause 1.9 [*Errors in the Employer's Requirements*] enables the Contractor to claim for any delay or additional Cost as a result of an error in the Employer's Requirements. The Employer must give notice to the Engineer and then follow the same procedures as contained within the RED Book. The error must be such that it could not have been discovered by an experienced Contractor during the initial check of the Employer's Requirements under the second paragraph of Sub-Clause 5.1 [*General Design Obligations*].

1.11 Contractor's Use of Employer's Documents
The phrase '*Specification, the Drawings*' is replaced by '*Employer's Requirements*'. The Contractor may copy and use the Employer's documents, but only for the purposes of the Contract.

1.13 Compliance with Laws
'*Specification*' is replaced by '*Employer's Requirements*' in paragraph (a) and '*design*' is added before '*execution*' in paragraph (b).

The addition of the word '*design*' at paragraph (b) confirms that the Contractor must obtain the necessary design approvals. This is in addition to the consent or approval by the Engineer under Sub-Clause 5.1 [*General Design Obligations*]. The Contractor must include a suitable allowance in the Sub-Clause 8.3 [*Programme*].

FIDIC Users' Guide
ISBN 978-0-7277-5856-9

Chapter 54
Clause 2: The Employer

This Clause is unchanged from the RED Book. Please refer to Chapter 8.

FIDIC Users' Guide
ISBN 978-0-7277-5856-9

ICE Publishing: All rights reserved
http://dx.doi.org/10.1680/fug.58569.367

Chapter 55
Clause 3: The Engineer

3.3 Instructions of the Engineer

The provisions for the Engineer to issue additional or modified Drawings and to give oral instructions have been deleted. Any instructions or additional Drawings from the Engineer will be Variations, in accordance with Clause 13 [*Variations and Adjustments*] because the Contractor is responsible for the design.

FIDIC Users' Guide
ISBN 978-0-7277-5856-9

ICE Publishing: All rights reserved
http://dx.doi.org/10.1680/fug.58569.369

Chapter 56
Clause 4: The Contractor

4.1 Contractor's General Obligations

References within the RED Book to limited design and the requirements in the final paragraph have been deleted and are replaced by requirements contained within Clause 5 [*Design*]. The remaining provisions under this Sub-Clause are identical to the RED Book, and users are directed to the relevant sections in this book.

4.4 Subcontractors

Paragraph (d) has been deleted.

4.5 Nominated Subcontractors

The equivalent section in the RED or PNK Book is under Clause 5 [*Nominated Subcontractors*]. This Sub-Clause is replaced as follows:

'In this Sub-Clause, 'nominated Subcontractor' means a Subcontractor whom the Engineer, under Clause 13 [Variations and Adjustments], instructs the Contractor to employ as a Subcontractor. The Contractor shall not be under any obligation to employ a nominated Subcontractor against whom the Contractor raises reasonable objection by notice to the Engineer as soon as practicable, with supporting particulars.'

The YLW Book does not include provision for a nominated Subcontractor to be named in the Contract but the Engineer may issue instructions nominating a Subcontractor as a Variation. However, the Contractor is not obliged to use a Subcontractor against which he raises a reasonable objection. The RED Book Clause 5 [*Nominated Subcontractors*] gives detailed provisions (e.g. payment conditions) when using nominated subcontractors. The YLW Book is silent on these and other detailed matters. The Parties are advised to carefully consider this issue, and when prudent to do so, incorporate the RED Book Clause 5 provisions into the YLW Book through the Particular Conditions.

The provisions concerning the Contractor's liability for the actions of Subcontractors apply to Subcontractors that have been designated by the Employer, as well as to Subcontractors that have been selected by the Contractor.

4.6 Co-operation

The final word '*Specification*' is changed to '*Employer's Requirements*' and a paragraph is added, as the penultimate paragraph of the Sub-Clause.

The Contractor shall be responsible for his construction activities on the Site, and shall co-ordinate his own activities with those of other contractors to the extent (if any) specified in the Employer's Requirements.

This additional paragraph depends on the inclusion in the Employer's Requirements of any co-ordination that is required from the Contractor. He should take into account the potential difficulties that may occur when coordinating his activities with others with whom he has no legal or contractual relationship. Any such requirements would need to be shown in the Sub-Clause 8.3 [*Programme*].

4.10 Site Data

For a YLW Book Contract each tenderer must make his own preliminary design in order to prepare his Tender. Different tenderers will locate foundations, and even structures, in different locations on the Site. It is important that the data on sub-surface and hydrological conditions provided by the Employer covers all possible foundation locations. In order that the Employer can compare Tenders on an equal basis they must be prepared based on the same information. If one Tender was prepared based on boreholes and another was based only on trial pits then the borehole Tender might be based on the expectation of softer material at a greater depth than was checked by the trial pits, and so result in a more expensive Tender. The cheaper tenderer might then be awarded the Contract only to encounter problems later and submit claims for unforeseeable physical conditions under Sub-Clause 4.12 [*Unforeseeable Conditions*].

4.11 Sufficiency of the Accepted Contract Amount

Paragraph (b) is extended to include '*any further data relevant to the contractor's design*'.

4.12 Unforeseeable Physical Conditions

Physical conditions situations frequently require a change to the design or method of working. Under the YLW Book it may be necessary for the Contractor to change his design and this will be taken into account in any claim.

Sub-Clause 4.12 [*Unforeseeable Conditions*] also includes a provision to enable the Engineer to take into account physical conditions at other parts of the Site that were more favourable than could reasonably have been foreseen. This issue is discussed in Part 2 under the RED Book. However, the situation may become even more complex under the YLW Book. The Contractor may claim either that his design took these conditions into account, or that he intends to change his design to take the unforeseeable better conditions into account and so claim an extension of time. To avoid conflict and dispute on this controversial provision, it is important that data provided in accordance with Sub-Clause 4.10 [*Site Data*] is as accurate as possible, and that the parties work in transparent cooperation.

4.18 Protection of the Environment

References to '*Specification*' are replaced by '*Employer's Requirements*'.

4.19 Electricity, Water and Gas

References to '*Specification*' are replaced by '*Employer's Requirements*'.

4.20 Employer's Equipment and Free-Issue Material

References to '*Specification*' are replaced by '*Employer's Requirements*'.

4.21 Progress Reports

Information to be included in the progress report has been changed to suit the requirements for a Contract with a Contractor's design. Hence, in paragraph (a) '*(if any)*' after design and the reference to nominated Subcontractors have been omitted, a reference to commissioning and trial operation has been added and paragraph (f) requires a list of Variations in addition to the lists of claim notices.

FIDIC Users' Guide
ISBN 978-0-7277-5856-9

ICE Publishing: All rights reserved
http://dx.doi.org/10.1680/fug.58569.373

Chapter 57
Clause 5: Design

Clause 5 covers the procedures for design carried out by the Contractor and is completely different to Clause 5 in the RED Book. The design must comply with the Employer's Requirements and must comply with any applicable regulations. The decision to make the Contractor responsible for the design is not taken lightly. It is usually done because the project involves a high degree of specialist or even propriety engineering knowledge. The advantage is that such specialist/propriety knowledge is made available to the Employer directly from those who have developed/own it. The disadvantage is that to a certain extent the Employer will lose his involvement in the design process. The trick is to get the balance right within the Employer's Requirements: too detailed and the design process is too prescriptive to enable the Contractor to use his specialist/propriety knowledge, too general and the Employer is exposed to increased claim potential during the design/execution process. In practice, it is often the main source of disputes and contributes to the success or failure of the project.

The preparation of the Employer's Requirements, therefore, needs particular care and attention. It must precisely define

- function
- operation and maintenance
- operating costs
- quality
- tests
- suppliers including nominated Subcontractors (if any)
- designers qualifications
- training requirements and manuals
- spare parts requirements
- amplify the Sub-Clauses shown in Table 3.4 in Chapter 3
- and so on.

At the same time they must allow flexibility to the tenderer to enable his specialist knowledge to be utilised without providing too much detail that will result in uneconomic tenders or will compromise the purpose for which the works are intended.

Whatever happens, the design must satisfy the Engineer that it complies with the Contract. Whether the Engineer's satisfaction can be achieved by a review, or requires detailed study and approval, must be stated in the Employer's Requirements. The Engineer must also recognise that the design is the Contractor's responsibility. The Engineer's task is to check that it complies with the Contract and not to impose his personal preferences for design methods and details. If the Engineer requires any changes to the design or Contractor's Documents then the Contractor will check whether the change is a Variation.

This chapter reviews the procedures and requirements for the Contractor to prepare the design and Contractor's Documents which relate to the design, construction, installation and operation of the Works that have been designed. If part of the Works is required to be constructed in accordance with design prepared by the Employer then this must be clearly stated in the Employer's Requirements. The extent of the Contractor's obligations and liability must be defined and it may be necessary to import some sub-clauses from the RED Book.

The Contract procedures for design must be linked to the procedures for obtaining the necessary permissions and approvals from the relevant authorities. Under some legal systems it is necessary to obtain design approval before the Employer can obtain approval to proceed with the project. The compliance with Laws is also referred to in Sub-Clause 1.13 [*Compliance with Laws*], which is reviewed in Chapters 7 and 53. The sequence of the events described in this chapter is given in Figure 52.2. Although this Clause is headed '*Design*', many of the requirements, such as the undertaking at Sub-Clause 5.3 [*Contractor's Undertaking*], also refer to the execution and the completed Works.

5.1 General Design Obligations

The Contractor shall carry out, and be responsible for, the design of the Works. Design shall be prepared by qualified designers who are engineers or other professionals who comply with the criteria (if any) stated in the Employer's Requirements. Unless otherwise stated in the Contract, the Contractor shall submit to the Engineer for consent the name and particulars of each proposed designer and design Subcontractor.

The Contractor warrants that he, his designers and design Subcontractors have the experience and capability necessary for the design. The Contractor undertakes that the designers shall be available to attend discussions with the Engineer at all reasonable times, until the expiry date of the relevant Defects Notification Period.

Upon receiving notice under Sub-Clause 8.1 [Commencement of Works], the Contractor shall scrutinise the Employer's Requirements (including design criteria and calculations, if any) and the items of reference mentioned in Sub-Clause 4.7 [Setting Out]. Within the period stated in the Appendix to Tender, calculated from the Commencement Date, the Contractor shall give notice to the Engineer of any error, fault or other defect found in the Employer's Requirements or these items of reference.

After receiving this notice, the Engineer shall determine whether Clause 13 [Variations and Adjustments] shall be applied, and shall give notice to the Contractor accordingly. If and to the extent that (taking account of cost and time) an experienced contractor exercising due care would have discovered the error, fault or other defect when examining the Site and the Employer's Requirements before submitting the Tender, the Time for Completion shall not be extended and the Contract Price shall not be adjusted.

Sub-Clause 5.1 [*General Design Obligations*] requires the Contractor to submit the names and details of each proposed designer and design Subcontractor for the Engineer's consent. All the designers must be qualified professionals who meet the criteria contained within the Employer's Requirements. The Laws may also require them to be qualified in the Country. However, if they were named in the Tender and so were accepted by the Employer, then consent has already been given. The designers must be available to attend discussions with the Engineer, through to the end of the Defects Notification Period.

The third paragraph of Sub-Clause 5.1 [*General Design Obligations*] requires the Contractor to carry out a detailed study of the Employer's Requirements and the setting out information that has been provided under Sub-Clause 4.7 [*Setting Out*]. Any errors, faults or other defects must be reported to the Engineer within the period stated in the Appendix to Tender. The Appendix to Tender refers to these as being '*unforeseeable*', which presumably means that they were unforeseeable when the Contractor prepared his Tender. The problem for the Contractor is that he does not know the details of exactly what the Employer requires. He only knows what is stated in the Employer's Requirements, which has formed the basis of his calculations in preparing his Tender. The Contractor will probably raise questions and ask for instructions. The Engineer will then decide whether the Contractor should have noticed the problem when preparing his Tender. The Contractor will, of course, know whether any other tenderer raised the matter, all questions raised having been answered to *all* tenderers. The Engineer will inform the Contractor whether he intends to issue a Variation instruction in accordance with Clause 13 [*Variations and Adjustments*]. If the Contractor is not satisfied with the Engineer's reply then he must give notice of a claim, under Sub-Clause 20.1 [*Contractor's Claims*], and the relevant claim procedures will apply.

The requirements for this review are wider than just a simple design check. The review must cover everything in the Employer's Requirements, including the requirements for obtaining permissions from authorities, as described at Sub-Clause 1.13 [*Compliance with Laws*].

If the Contractor suffers delay and/or incurs cost due to an error in the Employer's Requirements that was not noticed under Sub-Clause 5.1 [*General Design Obligations*] then he must give notice under Sub-Clause 1.9 [*Errors in the Employer's Requirements*], reviewed in Chapters 7 and 53. If there is a problem with the setting out information, which was not noticed under Sub-Clause 5.1, then the Contractor may give notice under Sub-Clause 4.7 [*Setting Out*], reviewed in Chapter 10. In both cases, he must also comply with the requirements of Sub-Clause 20.1 [*Contractor's Claims*].

When a query is raised under any of these Sub-Clauses it is essential that the Engineer gives clear instructions as to what is required for the Works. This is necessary so that the work can proceed in the design office or on Site with the least disruption or delay (if any) while any liability is decided.

5.2 Contractor's Documents
The Contractor's Documents shall comprise the technical documents specified in the Employer's Requirements, documents required to satisfy all regulatory approvals, and the documents described in Sub-Clause 5.6 [As-Built Documents] and Sub-Clause 5.7

[Operation and Maintenance Manuals]. Unless otherwise stated in the Employer's Require-ments, the Contractor's Documents shall be written in the language for communications defined in Sub-Clause 1.4 [Law and Language].

The Contractor shall prepare all Contractor's Documents, and shall also prepare any other documents necessary to instruct the Contractor's Personnel. The Employer's Personnel shall have the right to inspect the preparation of all these documents, wherever they are being prepared.

If the Employer's Requirements describe the Contractor's Documents which are to be submitted to the Engineer for review and/or for approval, they shall be submitted accord-ingly, together with a notice as described below. In the following provisions of this Sub-Clause, (i) 'review period' means the period required by the Engineer for review and (if so specified) for approval, and (ii) 'Contractor's Documents' exclude any documents which are not specified as being required to be submitted for review and/or for approval.

Unless otherwise stated in the Employer's Requirements, each review period shall not exceed 21 days, calculated from the date on which the Engineer receives a Contractor's Document and the Contractor's notice. This notice shall state that the Contractor's Document is considered ready, both for review (and approval, if so specified) in accordance with this Sub-Clause and for use. The notice shall also state that the Contractor's Document complies with the Contract, or the extent to which it does not comply.

The Engineer may, within the review period, give notice to the Contractor that a Contractor's Document fails (to the extent stated) to comply with the Contract. If a Contractor's Document so fails to comply, it shall be rectified, resubmitted and reviewed (and, if specified, approved) in accordance with this Sub-Clause, at the Contractor's cost.

For each part of the Works, and except to the extent that the prior approval or consent of the Engineer shall have been obtained:

(a) in the case of a Contractor's Document which has (as specified) been submitted for the Engineer's approval:

    (i) the Engineer shall give notice to the Contractor that the Contractor's Document is approved, with or without comments, or that it fails (to the extent stated) to comply with the Contract;

    (ii) execution of such part of the Works shall not commence until the Engineer has approved the Contractor's Document; and

    (iii) the Engineer shall be deemed to have approved the Contractor's Document upon the expiry of the review periods for all the Contractor's Documents which are relevant to the design and execution of such part, unless the Engineer has previously notified otherwise in accordance with sub-paragraph (i);

(b) execution of such part of the Works shall not commence prior to the expiry of the review periods for all the Contractor's Documents which are relevant to its design and execution;

(c) *execution of such part of the Works shall be in accordance with these reviewed (and, if specified, approved) Contractor's Documents; and*

(d) *if the Contractor wishes to modify any design or document which has previously been submitted for review (and, if specified, approval), the Contractor shall immediately give notice to the Engineer. Thereafter, the Contractor shall submit revised documents to the Engineer in accordance with the above procedure.*

If the Engineer instructs that further Contractor's Documents are required, the Contractor shall prepare them promptly.

Any such approval or consent, or any review (under this Sub-Clause or otherwise), shall not relieve the Contractor from any obligation or responsibility.

Sub-Clause 5.2 [*Contractor's Documents*] gives the procedures and requirements for the production of the Contractor's Documents. This refers to documents which the Contract requires the Contractor to prepare, either in the Employer's Requirements, elsewhere in the Contract, or to satisfy a statutory requirement. Documents that the Contractor prepares as instructions to his personnel, or for other internal purposes, do not have to follow these procedures. The Employer's Requirements must describe the documents that the Contractor is required to submit to the Engineer, whether they are to be submitted for review and/or for approval and whether the review period will exceed 21 days.

The procedures and notices are described in detail in the Sub-Clause and the sequence of events is summarised in Figure 52.2. The periods for reviews and other submissions, approvals and consents must be included in the Contractor's programme as Sub-Clause 8.3 [*Programme*]. It is the Contractor's responsibility to ensure that his documents are accurate and in accordance with the Employer's Requirements. Sub-Clause 1.5 [*Priority of Documents*] is clear that the Employer's Requirements have priority over any document prepared by the Contractor. If the Engineer has agreed to, or instructed, anything that differs from the Employer's Requirements then the Contractor must ensure that this is properly recorded, in accordance with Sub-Clause 3.3 [*Instructions of the Engineer*]. The YLW Book, unlike the RED Book, does *not* permit oral instructions. If the Contractor receives an oral instruction he must immediately confirm the instruction in writing and ask for a written instruction. Failure to obtain a written instruction could lead to the instruction being denied, with serious possible problems and disputes at a later stage of the project.

The potential problem from the review procedures is that the Engineer has the power to ask for the resubmission of documents and to delay the construction until he is satisfied. While the Engineer can only object that the document fails to comply with the Contract, to the extent stated, most design is a subjective process, subject to the preferences of the particular designer. The Contractor is ultimately responsible for his design and the Engineer must not impose his own design preferences on to the Contractor's design.

5.3 Contractor's Undertaking
The Contractor undertakes that the design, the Contractor's Documents, the execution and the completed Works will be in accordance with:

(a) the Laws in the Country; and

(b) the documents forming the Contract, as altered or modified by Variations.

This is a clear statement to confirm that the Contractor must meet his obligations.

5.4 Technical Standards and Regulations

The design, the Contractor's Documents, the execution and the completed Works shall comply with the Country's technical standards, building, construction and environmental Laws, Laws applicable to the product being produced from the Works, and other standards specified in the Employer's Requirements, applicable to the Works, or defined by the applicable Laws.

All these Laws shall, in respect of the Works and each Section, be those prevailing when the Works or Section are taken over by the Employer under Clause 10 [Employer's Taking Over]. References in the Contract to published standards shall be understood to be references to the edition applicable on the Base Date, unless stated otherwise.

If changed or new applicable standards come into force in the Country after the Base Date, the Contractor shall give notice to the Engineer and (if appropriate) submit proposals for compliance. In the event that:

(a) the Engineer determines that compliance is required, and

(b) the proposals for compliance constitute a variation,

then the Engineer shall initiate a Variation in accordance with Clause 13 [Variations and Adjustments].

This Sub-Clause highlights a difference in procedure between changes in the Laws and changes in technical standards, during the period from the Base Date to the date of the Taking-Over Certificate. If there is a change in the Laws, which include regulations, then the Contractor must be aware of the change and can claim under Sub-Clause 13.7 [*Adjustments for Changes in Legislation*]. However, if there is a change to an applicable standard the Contractor must give notice and submit proposals to the Engineer. The Engineer then decides whether he requires compliance with the new standard and whether the proposals require a Variation order. If the Contractor is not satisfied with the actions by the Engineer then he may initiate claims procedures.

5.5 Training

The Contractor shall carry out the training of Employer's Personnel in the operation and maintenance of the Works to the extent specified in the Employer's Requirements. If the Contract specifies training which is to be carried out before taking-over, the Works shall not be considered to be completed for the purposes of taking-over under Sub-Clause 10.1 [Taking Over of the Works and Sections] until this training has been completed.

Training of Employer's Personnel is an important part of any plant or design-build Contract. The Employer's Requirements must specify the extent and timing of the required training. The Contractor's Proposal will add further details. It is important that the parties agree on the level of training as well as the level of experience and/or prior knowledge of the people to be trained. This is even more important if and when the outcome of the training is a specified target, for example, the skills to operate a specific unit of a complex power plant. Often Contractors have priced their training activities based on the assumption that reasonably experienced people from the Employer will attend, only to find that the level of experience by the attendees is forcing longer than budgeted training activities. If the training is to be carried out before taking-over then it must be included in the Contractor's programme as per Sub-Clause 8.3 [*Programme*] and in the lists of work to be completed at Sub-Clauses 8.2 [*Time for Completion*] and 10.1 [*Taking Over of the Works and Sections*]. The period for trial operation may be coordinated with the training requirement, although the Contractor will not want compliance with the performance requirements to be compromised by the use of inexperienced personnel undergoing training.

5.6 As-Built Documents

The Contractor shall prepare, and keep up-to-date, a complete set of 'as-built' records of the execution of the Works, showing the exact as-built built locations, sizes and details of the work as executed. These records shall be kept on the Site and shall be used exclusively for the purposes of this Sub-Clause. Two copies shall be supplied to the Engineer prior to the commencement of the Tests on Completion.

In addition, the Contractor shall supply to the Engineer as-built drawings of the Works, showing all Works as executed, and submit them to the Engineer for review under Sub-Clause 5.2 [Contractor's Documents]. The Contractor shall obtain the consent of the Engineer as to their size, the referencing system, and other relevant details.

Prior to the issue of any Taking-Over Certificate, the Contractor shall supply to the Engineer the specified numbers and types of copies of the relevant as-built drawings, in accordance with the Employer's Requirements. The Works shall not be considered to be completed for the purposes of taking-over under Sub-Clause 10.1 [Taking Over of the Works and Sections] until the Engineer has received these documents.

The as-built records must be submitted to the Engineer for review before the issue of the Taking-Over Certificate. They are normally required before commissioning and trial operation. If this is not the case, than the Parties may decide to amend this Sub-Clause, and allow the Contractor to produce them as part of the outstanding work to be performed during the Defects Notification Period in accordance with Sub-Clause 10.1(a) [*Taking Over of the Works and Sections*] and Sub-Clause 11.1(a) [*Completion of Outstanding Work and Defects*].

5.7 Operation and Maintenance Manuals

Prior to commencement of the Tests on Completion, the Contractor shall supply to the Engineer provisional operation and maintenance manuals in sufficient detail for the Employer to operate, maintain, dismantle, reassemble, adjust and repair the Plant.

The Works shall not be considered to be completed for the purposes of taking-over under Sub-Clause 10.1 [Taking Over of the Works and Sections] until the Engineer has received final operation and maintenance manuals in such detail, and any other manuals specified in the Employer's Requirements for these purposes.

The details and timing of the operation and maintenance manuals should be given in the Employer's Requirements and included in the Contractor's Sub-Clause 8.3 [*Programme*].

5.8 Design Error

If errors, omissions, ambiguities, inconsistencies, inadequacies or other defects are found in the Contractor's Documents, they and the Works shall be corrected at the Contractor's cost, notwithstanding any consent or approval under this Clause.

This Sub-Clause confirms and repeats other statements (e.g. Sub-Clause 1.8 [*Care and Supply of Documents*], Sub-Clause 1.9 [*Errors in the Employer's Requirements*] and Sub-Clause 5.1 [*General Design Obligations*] that the Contractor's work is his own responsibility, regardless of any consents or approvals. Sub-Clause 11.9 [*Performance Certificate*] is also relevant and includes the statement that '*Only the Performance Certificate shall be deemed to constitute acceptance of the Works*'.

FIDIC Users' Guide
ISBN 978-0-7277-5856-9

ICE Publishing: All rights reserved
http://dx.doi.org/10.1680/fug.58569.381

Chapter 58
Clause 6: Staff and Labour

6.1 Engagement of Staff and Labour
Reference to '*Specification*' is replaced by '*Employer's Requirements*'.

6.6 Facilities for Staff and Labour
Reference to '*Specification*' is replaced by '*Employer's Requirements*'.

6.8 Contractor's Superintendence
The words '*design and*' are added before '*execution*'.

FIDIC Users' Guide
ISBN 978-0-7277-5856-9

ICE Publishing: All rights reserved
http://dx.doi.org/10.1680/fug.58569.383

Chapter 59
Clause 7: Plant, Materials and Workmanship

For a YLW Book Contract the Contractor's Proposal may include a recommendation relating to the quality of Plant, Materials and workmanship which are different to the provisions in the Employer's Requirements. The Employer's Requirements will prevail, in accordance with the priority at Sub-Clause 1.5 [*Priority of Documents*].

7.2 Samples

The phrase '*for the Engineer's consent prior to using the Materials in or for the Works*' is replaced by '*to the Engineer for review in accordance with the procedures for Contractor's Documents described in Sub-Clause 5.2*' [*Contractor's Documents*]. Thus, unlike the RED Book, there is no provision for the Engineer to give consent to any Materials prior to their use in or for the Works. The reason is that a design made by the Contractor should not be compromised by arbitrary requirements from either the Engineer or the Employer. In a situation where the Contractor is forced to use (or denied to use) a specific Material in works designed by him, his ability to assume the corresponding risk for such design may be affected. If any substandard material were to be used by the Contractor, he is still liable for the overall performance of the Works in accordance with the Contract as per Sub-Clause 5.3 [*Contractor's Undertaking*].

7.4 Testing

Sub-Clause 7.4 [*Testing*] gives the procedures for tests specified in the Contract and additional tests instructed under Clause 13 [*Variations and Adjustments*]. Tests on Completion are covered at Clause 9 [*Tests on Completion*], which refers back to this Sub-Clause 7.4. Clause 12 [*Tests after Completion*] requires different procedures because, after completion, the project is occupied and controlled by and at the risk of the Employer.

7.5 Rejection

The word '*workmanship*' is replaced by the phrase '*design or workmanship*' in the first sentence of the first and second paragraph. This is a necessary amendment to reflect the design responsibility of the Contractor.

7.8 Royalties

'*Specification*' is replaced by '*Employer's Requirements*'.

FIDIC Users' Guide
ISBN 978-0-7277-5856-9

ICE Publishing: All rights reserved
http://dx.doi.org/10.1680/fug.58569.385

Chapter 60
Clause 8: Commencement, Delays and Suspension

8.1 Commencement of Work

The words '*design and*' have been added before '*execution*'. Again this is a necessary amendment reflecting the design responsibilities of the Contractor.

8.2 Time for Completion

Under Sub-Clause 8.2 [*Time for Completion*] the Contractor is obliged to complete all the work which is required for taking over under Sub-Clause 10.1 [*Taking Over of the Works and Sections*], including passing the Tests on Completion as per Clause 9 [*Tests on Completion*], before expiry of the Time for Completion. Sub-Clause 10.1 [*Taking Over of the Works and Sections*] just refers back to Sub-Clause 8.2 [*Time for Completion*], which is not very helpful. The Particular Conditions should be expanded to include a more detailed list of the work that must be completed, in addition to the actual construction and tests. Under the YLW Book this includes training, commissioning and trial operation.

8.3 Programme

To reflect the design responsibilities of the Contractor paragraphs (a) and (b) from the RED and PNK books at Sub-Clause 8.3 [*Programme*] have been changed as follows:

'*(a) the order in which the Contractor intends to carry out the Works, including the anticipated timing of each stage of design, Contractor's Documents, procurement, manufacture, inspection, delivery to Site, construction, erection, testing, commissioning and trial operation;*

(b) *the periods for reviews under Sub-Clause 5.2 [Contractor's Documents] and for any other submissions, approvals and consents specified in the Employer's Requirements.*'

The timing and details for commissioning and trial operation need to be clarified within the Employer's Requirements and the Contractor's Proposal. Commissioning is not a defined term so will presumably have its usual meaning of 'bringing into operation'. Nevertheless, when the commissioning involves complex stages, for example, power plants' hot and cold commissioning, it may be advisable for the Parties to define this term in more detail under the Particular Conditions. Commissioning is usually executed and completed before trial operation. The sequence of post-installation events including testing, training, submission of documents, commissioning and trial operation should be included within the Sub-Clause 8.3 [*Programme*].

8.4 Extension of Time for Completion

The reference to a '*change in the quantity*' in paragraph (a) is not relevant and has been omitted.

FIDIC Users' Guide
ISBN 978-0-7277-5856-9

ICE Publishing: All rights reserved
http://dx.doi.org/10.1680/fug.58569.387

Chapter 61
Clause 9: Tests on Completion

The YLW Book includes additional requirements for commissioning and trial operation. The tests before taking-over are inevitably more extensive than the RED or PNK books, and the criteria for acceptance are more complex.

9.1 Contractor's Obligations

The reference to Sub-Clause 4.1 [*Contractor's General Obligations*] is changed to refer to Sub-Clause 5.6 [*As-Built Documents*] and Sub-Clause 5.7 [*Operation and Maintenance Manuals*]. Furthermore, the YLW Book includes additional paragraphs concerning the sequence of the various functional tests as prescribed in the Employer's Requirements.

'*Unless otherwise stated in the Particular Conditions, the Tests on Completion shall be carried out in the following sequence:*

(a) *pre-commissioning tests, which shall include the appropriate inspections and ('dry' or 'cold') functional tests to demonstrate that each item of Plant can safely undertake the next stage, (b);*

(b) *commissioning tests, which shall include the specified operational tests to demonstrate that the Works or Section can be operated safely and as specified, under all available operating conditions; and*

(c) *trial operation, which shall demonstrate that the Works or Section perform reliably and in accordance with the Contract.*

During trial operation, when the Works are operating under stable conditions, the Contractor shall give notice to the Engineer that the Works are ready for any other Tests on Completion, including performance tests to demonstrate whether the Works conform with criteria specified in the Employer's Requirements and with the Schedule of Guarantees.

Trial operation shall not constitute a taking-over under Clause 10 [Employer's Taking Over]. Unless otherwise stated in the Particular Conditions, any product produced by the Works during trial operation shall be the property of the Employer.

In considering the results of the Tests on Completion, the Engineer shall make allowances for the effect of any use of the Works by the Employer on the performance or other characteristics of the Works. As soon as the Works, or a Section, have passed each of the Tests on Completion described in sub-paragraph (a), (b) or (c), the Contractor shall submit a certified report of the results of these Tests to the Engineer.'

These additional requirements are essential when the Works include Plant, as distinct from Materials.

FIDIC Users' Guide
ISBN 978-0-7277-5856-9

ICE Publishing: All rights reserved
http://dx.doi.org/10.1680/fug.58569.389

Chapter 62
Clause 10: Employer's Taking Over

Clause 10 [*Employer's Taking Over*] is the same as in the RED Book however; the detailed requirements for work that must be completed before the Works can be taken over as listed at Sub-Clause 8.2 [*Time for Completion*], include the Tests on Completion, as per Clause 9 [*Tests on Completion*]. The test requirements are more extensive, including commissioning and trial operation. It is also necessary to check that the Contractor's design complies with the Employer's Requirements.

FIDIC Users' Guide
ISBN 978-0-7277-5856-9

ICE Publishing: All rights reserved
http://dx.doi.org/10.1680/fug.58569.391

Chapter 63
Clause 11: Defects Liability

While a period of one year will generally be suitable for civil engineering projects, a longer period may be required for electrical, mechanical or building services work. For example, performance tests on air conditioning plant must be carried out during hot weather, may be specified as Tests after Completion and are often followed by balancing and adjustment of the plant. For the Defects Notification Period to include a full hot-weather season after the completion of the balancing may require a two-year or 730-day period.

Failure to meet the performance specifications during the Defects Notification Period would be considered a defect.

11.2 Cost of Remedying Defects
The list of items of work for which the Contractor is liable for the cost of remedying defects is extended to include the additional responsibilities for the design and performance of the Works. The RED and PNK books paragraphs (a) to (c) are replaced by the following:

'(a) the design of the Works, other than a part of the design for which the Employer is responsible (if any);
(b) Plant, Materials or workmanship not being in accordance with the Contract;
(c) improper operation or maintenance which was attributable to matters for which the Contractor is responsible (under Sub-Clauses 5.5 to 5.7 or otherwise); or
(d) failure by the Contractor to comply with any other obligation.'

Paragraphs (a) to (d) list the types of defects that must be repaired at the Contractor's Cost. The Employer must pay for any other work. While category (c) may seem to cover a wide range of problems, it is still necessary for the Employer to state and prove under which obligation the Contractor has failed to comply.

11.6 Further Tests
The phrase 'including Tests on Completion and/or Tests after Completion.' is added at the end of the first sentence in the first paragraph.

This provision is particularly important in a YLW Book Contract. Unlike a RED Book Contract that is defined through design and specification, a YLW Book Contract may have uncertainties in the final product. Satisfactory completion of Tests on Completion and/or Tests after Completion provides assurance to the Employer that he is getting the product he requested.

11.7 *Right of Access*

Until the Performance Certificate has been issued, the Contractor shall have the right of access to all parts of the Works and to records of the operation and performance of the Works, except as may be inconsistent with the Employer's reasonable security restrictions.

Obviously the Contractor needs access to the Site in order to carry out his obligations. Times and arrangements must be agreed to suit both the Employer and the Contractor. The Contractor may also need office and storage facilities on Site in order to meet his obligations during the Defects Notification Period. The details must be agreed between the Contractor and the Engineer. The removal of these facilities and any reinstatement would be covered by Sub-Clauses 10.4 [*Surfaces Requiring Reinstatement*] and 11.11 [*Clearance of Site*] concerning surfaces requiring reinstatement and clearance of the Site.

The Contractor is entitled to additional access to the Works during the Defects Notification period. This enables him to observe and check the operation of the Works and the performance records that are being kept by the Employer. The Contractor's design responsibilities are based on the requirement that the performance shall comply with the Employer's Requirements. If the Works fail to meet these requirements then there would be a defect requiring correction. It is therefore important that the Contractor has the opportunity to check that the Works are being operated and maintained in accordance with his design and training. Any problems or failures in operation or maintenance should be reported by the Contractor in order to avoid allegations of design failure or failure to pass the Tests after Completion.

FIDIC Users' Guide
ISBN 978-0-7277-5856-9

ICE Publishing: All rights reserved
http://dx.doi.org/10.1680/fug.58569.393

Chapter 64
Clause 12: Tests after Completion

The procedure for payment by measurement of work executed is not appropriate for a Contract with design by the Contractor and so the RED Book, Clause 12 [*Measurement and Evaluation*] is omitted. The YLW Book, Clause 12 [*Tests After Completion*] refers to additional tests that are to be carried out after the Works have been occupied by the Employer, following the issue of the Taking-Over Certificate. The term '*Tests after Completion*' is used in several other Sub-Clauses' including 7.4 [*Testing*], 11.6 [*Further Tests*], 12.1 [*Procedure for Tests after Completion*], 12.2 [*Delayed Tests*], 12.3 [*Retesting*], 12.4 [*Failure to Pass Tests after Completion*], 14.9 [*Payment of Retention Money*] and 18.2 [*Insurance for Works and Contractor's Equipment*]. The procedures described in this Clause do not just relate to individual items of Plant but to tests of the Works as a whole, or to a Section that the Contract requires to be taken over as a separate entity. The basic requirement is that the completed project must meet the Employer's Requirements, so the tests must be carried out when the Employer is using the project in order to confirm that performance requirements are being met. The procedures for these tests differ from the Tests on Completion because the Site is now occupied and controlled by the Employer and not by the Contractor.

12.1 Procedure for Tests after Completion
If Tests after Completion are specified in the Contract, this Clause shall apply. Unless other-wise stated in the Particular Conditions, the Employer shall:

(a) provide all electricity, equipment, fuel, instruments, labour, materials, and suitably qualified and experienced staff, as are necessary to carry out the Tests after Completion efficiently; and

(b) carry out the Tests after Completion in accordance with the manuals supplied by the Contractor under Sub-Clause 5.7 [Operation and Maintenance Manuals] and such guidance as the Contractor may be required to give during the course of these Tests; and in the presence of such Contractor's Personnel as either Party may reasonably request.

The Tests after Completion shall be carried out as soon as is reasonably practicable after the Works or Section have been taken over by the Employer. The Employer shall give to the Contractor 21 days' notice of the date after which the Tests after Completion will be carried out. Unless otherwise agreed, these Tests shall be carried out within 14 days after this date, on the day or days determined by the Employer.

If the Contractor does not attend at the time and place agreed, the Employer may proceed with the Tests after Completion, which shall be deemed to have been made in the Contractor's presence, and the Contractor shall accept the readings as accurate.

The results of the Tests after Completion shall be compiled and evaluated by both Parties. Appropriate account shall be taken of the effect of the Employer's prior use of the Works.

By definition, the Tests after Completion are carried out after the project has been handed over and occupied by the Employer in accordance with Sub-Clause 10.1 [*Taking Over of the Works and Sections*]. Hence, the Employer is required to carry out these tests and to provide all the Materials, services and staff required. The tests are to be carried out in accordance with the operation and maintenance manuals that have been provided by the Contractor under Sub-Clause 5.7 [*Operation and Maintenance Manuals*]. However, the requirements for the tests should have been given in the Employer's Requirements, with more details in the Contractor's Proposal.

This Sub-Clause includes the procedures for the Employer to inform the Contractor when he is ready to carry out the tests and for the results to be evaluated by both Parties. The Contractor is entitled to be present during these tests. It is important that these tests are carried out as soon as possible after handover. The tests may reveal problems that require further action and it is essential that all problems are resolved before the end of the Defects Notification Period.

12.2 Delayed Tests

If the Contractor incurs Cost as a result of any unreasonable delay by the Employer to the Tests after Completion, the Contractor shall (i) give notice to the Engineer and (ii) be entitled subject to Sub-Clause 20.1 [Contractor's Claims] to payment of any such Cost plus reasonable profit, which shall be included in the Contract Price.

After receiving this notice, the Engineer shall proceed in accordance with Sub-Clause 3.5 [Determinations] to agree or determine this Cost and profit.

If, for reasons not attributable to the Contractor, a Test after Completion on the Works or any Section cannot be completed during the Defects Notification Period (or any other period agreed upon by both Parties), then the Works or Section shall be deemed to have passed this Test after Completion.

This Sub-Clause follows the usual claims procedures if the Contractor has incurred Cost due to the Employer having delayed these tests.

Sub-Clause 14.9 [*Payment of Retention Money*] requires that the Tests after Completion must have been completed before the first half of the Retention Money is certified by the Engineer for payment to the Contractor. Hence, if the Employer delays the tests then the Contractor's claim will include the Cost of the delay in the payment of the Retention Money.

12.3 Retesting

If the Works, or a Section, fail to pass the Tests after Completion:

(a) sub-paragraph (b) of Sub-Clause 11.1 [Completion of Outstanding Work and Remedying Defects] shall apply; and

(b) either Party may then require the failed Tests, and the Tests after Completion on any related work, to be repeated under the same terms and conditions.

If and to the extent that this failure and retesting are attributable to any of the matters listed in sub-paragraphs (a) to (d) of Sub-Clause 11.2 [Cost of Remedying Defects] and cause the Employer to incur additional costs, the Contractor shall subject to Sub-Clause 2.5 [Employer's Claims] pay these costs to the Employer.

In the event of a failed Test after Completion in accordance with Sub-Clause 12.1 [*Procedure for Tests after Completion*] the Contractor must execute any work that is necessary to correct the problem that caused the Works or Section to fail the test. Either Party may require that the failed tests be repeated.

If the failure is attributable to the Contractor then the Employer may claim additional Costs that have been incurred. In order to prove that the failure was due to some cause attributable to the Contractor, it is necessary to establish exactly why the tests failed and which Party is liable for the failure. This may be easily established or may need detailed investigation. The Contract does not give procedures for establishing the cause and liability for the failure but the Sub-Clause includes a reference to Sub-Clause 11.1 [*Completion of Outstanding Work and Remedying Defects*].

It is therefore logical to assume that the procedures at Sub-Clause 11.8 [*Contractor to Search*] should also apply. This requires the Contractor to search for the cause of a defect, under the direction of the Engineer, and to be paid for the Cost of the search unless the defect is to be remedied at the cost of the Contractor.

12.4 Failure to Pass Tests after Completion
If the following conditions apply, namely:

(a) the Works, or a Section, fail to pass any or all of the Tests after Completion;
(b) the relevant sum payable as non-performance damages for this failure is stated (or its method of calculation is defined) in the Contract; and
(c) the Contractor pays this relevant sum to the Employer during the Defects Notification Period;

then the Works or Section shall be deemed to have passed these Tests after Completion.

If the Works, or a Section, fail to pass a Test after Completion and the Contractor proposes to make adjustments or modifications to the Works or such Section, the Contractor may be instructed by (or on behalf of) the Employer that right of access to the Works or Section cannot be given until a time that is convenient to the Employer. The Contractor shall then remain liable to carry out the adjustments or modifications and to satisfy this Test, within a reasonable period of receiving notice by (or on behalf of) the Employer of the time that is convenient to the Employer. However, if the Contractor does not receive this notice during the relevant Defects Notification Period, the Contractor shall be relieved of this obligation and the Works or Section (as the case may be) shall be deemed to have passed this Test after Completion.

If the Contractor incurs additional Cost as a result of any unreasonable delay by the Employer in permitting access to the Works or Plant by the Contractor, either to investigate the causes of a failure to pass a Test after Completion or to carry out any adjustments or modifications, the Contractor shall (i) give notice to the Engineer and (ii) be entitled subject to Sub-Clause 20.1 [Contractor's Claims] to payment of any such Cost plus reasonable profit, which shall be included in the Contract Price.

After receiving this notice, the Engineer shall proceed in accordance with Sub-Clause 3.5 [Determinations] to agree or determine this Cost and profit.

This Sub-Clause applies when the Works have failed to pass the Tests after Completion, and the Contractor was unable to remedy the cause of the failure. It includes a reference to non-performance damages for failure to pass any of the Tests after Completion. The sum payable, or its method of calculation, must be stated within the Contract, but unfortunately there is no provision for this in the Appendix to Tender. This provision must also be considered with the provisions at Sub-Clause 11.4 [*Failure to Remedy Defects*] for the situation if the Contractor fails to remedy any defects. Failure to pass a test could be considered to be a defect and be subject to further discussion on this basis.

The wording of this Sub-Clause is clear that it is the Works or Section, as a whole, which passes or fails the Tests after Completion. If any one test fails then the whole of the Works or Section has failed.

FIDIC Users' Guide
ISBN 978-0-7277-5856-9

ICE Publishing: All rights reserved
http://dx.doi.org/10.1680/fug.58569.397

Chapter 65
Clause 13: Variations and Adjustments

Under the YLW Book the scope of the changes that may be instructed by the Engineer is restricted, which reflects the Contractor's responsibility for the design and the performance of the Works at completion.

13.1 Right to vary

'Variations may be initiated by the Engineer at any time prior to issuing the Taking-Over Certificate for the Works, either by an instruction or by a request for the Contractor to submit a proposal. A Variation shall not comprise the omission of any work which is to be carried out by others.

The Contractor shall execute and be bound by each Variation, unless the Contractor promptly gives notice to the Engineer stating (with supporting particulars) that (i) the Contractor cannot readily obtain the Goods required for the Variation, (ii) it will reduce the safety or suitability of the Works, or (iii) it will have an adverse impact on the achievement of the Schedule of Guarantees. Upon receiving this notice, the Engineer shall cancel, confirm or vary the instruction.'

The YLW Book definition of a Variation covers an instruction to change either the Employer's Requirements or the Works, which has been confirmed by the Engineer to be a Variation. That is, the instruction has time or payment implications that will be valued by the Engineer under Sub-Clause 13.3 [*Variation Procedure*]. As with the RED or PNK books, the Variation may be initiated by the Engineer either as an instruction or a request for a proposal but shall not include the omission of any work that is to be carried out by others.

Because the Contractor has extra responsibilities, there are additional reasons by which the Contractor can object to the Variation.

Point (i) refers to difficulties in obtaining the necessary Goods, which is the same as the RED or PNK books.

Point (ii) refers to a reduction in the safety of the Works, which clearly refers to the Contractor's responsibilities as the designer of the Works. The reference to suitability is not so clear but presumably refers to the designer's responsibility to meet the performance specification.

Point (iii) refers to the Schedule of Guarantees, which must be provided by the Contractor.

The Engineer may then either cancel, confirm or vary the instruction. If the Engineer confirms an instruction following objections by the Contractor then the Employer may ultimately have to take responsibility for the potential problems that have been notified by the Contractor depending upon the applicable law. If the Variation has design or safety implications then it could also bring additional responsibilities to both the Employer and the Engineer under the applicable law.

It is essential that the Contractor assesses any impact created by a proposed Variation and, if necessary, proposes modifications taking account of time and additional Cost. Detailed discussions are required to ensure that difficulties/impossibilities are highlighted, discussed and clarified or agreed.

13.2 Value Engineering
The Contractor may, at any time, submit to the Engineer a written proposal which (in the Contractor's opinion) will, if adopted, (i) accelerate completion, (ii) reduce the cost to the Employer of executing, maintaining or operating the Works, (iii) improve the efficiency or value to the Employer of the completed Works, or (iv) otherwise be of benefit to the Employer.

The proposal shall be prepared at the cost of the Contractor and shall include the items listed in Sub-Clause 13.3 [Variation Procedure].

The Contractor makes a proposal that will benefit the Employer without compromising quality. Under the RED or PNK books, any savings are split 50/50 between the Contractor and the Employer. This is not the case in the YLW Book where the split on savings would need to be agreed during negotiation. If the saving is for time, then both Parties may benefit by reduced construction establishment costs for the Contractor and quicker production start-up for the Employer.

Any value-engineering proposal becomes a Variation and is valued as such by the Engineer.

13.3 Variation Procedure
The YLW Book includes provision for a design proposal under subparagraph (a). The different payment terminology at subparagraph (c) is also shown by the replacement of the final paragraph of Sub-Clause 13.3 [*Variation Procedure*] by the following:

'*Upon instructing or approving a Variation, the Engineer shall proceed in accordance with Sub-Clause 3.5 [Determinations] to agree or determine adjustments to the Contract Price and the Schedule of Payments. These adjustments shall include reasonable profit, and shall take account of the Contractor's submissions under Sub-Clause 13.2 [Value Engineering] if applicable.*'

The Engineer will value Variations and carry out a determination in accordance with Sub-Clause 3.5 [*Determinations*].

The procedure for the valuation of omissions, which is given at Sub-Clause 12.4 [*Omissions*] of the RED and PNK books, has been omitted. However, the requirement for fairness, at Sub-Clause 3.5 [*Determinations*], should result in a similar procedure.

13.5 Provisional Sums

Reference to a nominated Subcontractor at subparagraph (b) has been omitted. The use of a nominated Subcontractor for purchasing Plant, Materials or services would not be appropriate under the YLW Book. Nevertheless, if it is considered to prescribe a nominated supplier/ subcontractor for certain Plant items then this should not be done as a Provisional Sum. It should be stated in the Employer's Requirements enabling the tendering contractor to take this into account for his Tender price.

FIDIC Users' Guide
ISBN 978-0-7277-5856-9

ICE Publishing: All rights reserved
http://dx.doi.org/10.1680/fug.58569.401

Chapter 66
Clause 14: Contract Price and Payment

References to re-measurement are omitted and replaced by references to the Accepted Contract Amount as a lump sum. Payments are made in accordance with a Schedule and are subject to adjustment in accordance with the Contract. The provisions of Sub-Clause 14.4 [*Schedule of Payments*] will apply where provision for some payments are to be made as instalments of a lump sum.

14.1 The Contract Price
Unless otherwise stated in the Particular Conditions:

(a) *the Contract Price shall be the lump sum Accepted Contract Amount and be subject to adjustments in accordance with the Contract;*

(b) *the Contractor shall pay all taxes, duties and fees required to be paid by him under the Contract, and the Contract Price shall not be adjusted for any of these costs, except as stated in Sub-Clause 13.7 [Adjustments for Changes in Legislation];*

(c) *any quantities which may be set out in a Schedule are estimated quantities and are not to be taken as the actual and correct quantities of the Works which the Contractor is required to execute; and*

(d) *any quantities or price data which may be set out in a Schedule shall be used for the purposes stated in the Schedule and may be inapplicable for other purposes.*

However, if any part of the Works is to be paid according to quantity supplied or work done, the provisions for measurement and evaluation shall be as stated in the Particular Conditions. The Contract Price shall be determined accordingly, subject to adjustments in accordance with the Contract.

For a lump sum contract it is essential that full details of the Employer's Requirements are given in the Tender documents to enable the Contractor to submit a realistic Tender. Any uncertainties in the Tender documents, or additional information issued by the Engineer, are likely to result in claims. Sub-Clause 14.1(d) [*The Contract Price*] requires that the Schedules will include the purposes for which the quantities or price data should be used.

The final paragraph of Sub-Clause 14.1 [*The Contract Price*] requires the Particular Conditions to include the provisions if any measurement is required. The *Guidance for the Preparation of Particular Conditions* includes a suggested Sub-Clause, which states that if the value is not agreed then the Engineer will make a determination under Sub-Clause 3.5 [*Determinations*]. This is consistent with the procedure for the valuation of Variations, as per Sub-Clause 13.3 [*Variation Procedure*]. The suggested Sub-Clause then repeats Sub-Clauses 12.1 [*Works to be Measured*] and 12.2(a) [*Method of Measurement*] from the RED Book, but Sub-Clause 12.2(b) is not repeated. This is presumably because the YLW Book does not normally include a Bill of Quantities. However, if a YLW Book Contract includes for the measurement of work, which may be controversial, then some method of measurement should be included.

14.3 Application for Interim Payment Certificates
The phrase '*after the end of each month,*' is replaced by '*after the end of the period of payment stated in the Contract (if not stated, at the end of each month)*'. Interim payments are made as required by the Schedules, which may not be monthly. Hence, Sub-Clause 14.3 [*Application for Interim Payment Certificates*] allows for the periods for payment to be stated in the Contract.

14.4 Schedule of Payments
The Schedule of Payments is particularly important for a YLW Book Contract. Whereas the RED and PNK books include provisions, at Clause 12 [*Measurement and Evaluation*], for the Works to be measured and valued for payment, the YLW Book relies only on the Schedule of Payments in order to calculate the estimated contract value for interim payments as Sub-Clause 14.3(a) [*Application for Interim Payment Certificates*].

14.9 Payment of Retention Money
The whole Sub-Clause is replaced by the following:

'*When the Taking-Over Certificate has been issued for the Works, and the Works have passed all specified tests (including the Tests after Completion, if any), the first half of the Retention Money shall be certified by the Engineer for payment to the Contractor. If a Taking-Over Certificate is issued for a Section, the relevant percentage of the first half of the Retention Money shall be certified and paid when the Section passes all tests.*

Promptly after the latest of the expiry dates of the Defects Notification Periods, the outstanding balance of the Retention Money shall be certified by the Engineer for payment to the Contractor. If a Taking-Over Certificate was issued for a Section, the relevant percentage of the second half of the Retention Money shall be certified and paid promptly after the expiry date of the Defects Notification Period for the Section.

However, if any work remains to be executed under Clause 11 [Defects Liability] or Clause 12 [Tests after Completion], the Engineer shall be entitled to withhold certification of the estimated cost of this work until it has been executed.

The relevant percentage for each Section shall be the percentage value of the Section as stated in the Appendix to Tender. If the percentage value of a Section is not stated in the Appendix to Tender, no percentage of either half of the Retention Money shall be released under this Sub-Clause in respect of such Section.'

The payment of the first half of the Retention Money is not due until the Works have passed the Tests after Completion in accordance with Clause 12 [*Tests after Completion*]. The timing of the Tests after Completion depends on the Employer being ready for the tests, which may not be until after the Employer has completed preparations and the Works have been occupied or taken into commercial operation. The Contractor's cash flow will clearly suffer from any delay and the inclusion of a Retention Money Guarantee, as the example form at Annex F, must be recommended. This early release of Retention Money requires an additional Sub-Clause, as discussed for Sub-Clause 14.9 (RED and PNK books) [*Payment of Retention Money*].

The arrangement for the calculation of Retention Money in relation to a Section of the Works is different from the RED and PNK books, due to the different method of valuing parts of the Works.

FIDIC Users' Guide
ISBN 978-0-7277-5856-9

ICE Publishing: All rights reserved
http://dx.doi.org/10.1680/fug.58569.405

Chapter 67
Clause 15: Termination by Employer

15.4 Payment after Termination

There is no change to the wording of this Sub-Clause, however problems may occur if a dispute arises due to the termination procedure that must be referred to the DAB. Under Sub-Clause 20.2 (YLW) [*Appointment of a Dispute Adjudication Board*] for an ad hoc DAB the appointment of the DAB expires when it has given its decision on the dispute for which it was appointed.

FIDIC Users' Guide
ISBN 978-0-7277-5856-9

ICE Publishing: All rights reserved
http://dx.doi.org/10.1680/fug.58569.407

Chapter 68
Clause 16: Suspension and Termination by Contractor

This Clause is unchanged from the RED Book. Please refer to Chapter 22.

FIDIC Users' Guide
ISBN 978-0-7277-5856-9

Chapter 69
Clause 17: Risk and Responsibility

17.3 Employer's Risks

The words '*if any*' are added after '*responsible*' in subparagraph (g). This is a logical amendment to the RED and PNK books, because the YLW Book is drafted in principle for projects designed by the Contractor only.

17.5 Intellectual and Industrial Property Rights

The word '*Contract*' in subparagraph (a) is replaced by '*Employer's Requirements*' and the fourth paragraph is replaced by:

'*The Contractor shall indemnify and hold the Employer harmless against and from any other claim which arises out of or in relation to (i) the Contractor's design, manufacture, construction or execution of the Works, (ii) the use of Contractor's Equipment, or (iii) the proper use of the Works.*'

FIDIC Users' Guide
ISBN 978-0-7277-5856-9

ICE Publishing: All rights reserved
http://dx.doi.org/10.1680/fug.58569.411

Chapter 70
Clause 18: Insurance

When the Contractor is carrying out all or part of the design, then Professional Indemnity Insurance to cover the design liability may also be required. This may cause a problem because of the fitness for purpose requirement at Clause 5 [*Design*] in the YLW Book.

18.2 Insurance for Works and Contractor's Equipment
The phrase '*and Clause 12 [Tests after Completion]*' have been added at the end of the second paragraph.

The exclusions at subparagraph (e) are the same in the RED and the YLW books, including the reference to design defects, although design may be the responsibility of either the Employer or the Contractor. This requires further analysis when preparing the Particular Conditions in order to establish the Employer's requirements for this insurance.

FIDIC Users' Guide
ISBN 978-0-7277-5856-9

ICE Publishing: All rights reserved
http://dx.doi.org/10.1680/fug.58569.413

Chapter 71
Clause 19: Force Majeure

This Clause is unchanged from the RED Book. Please refer to Chapter 25.

FIDIC Users' Guide
ISBN 978-0-7277-5856-9

ICE Publishing: All rights reserved
http://dx.doi.org/10.1680/fug.58569.415

Chapter 72
Clause 20: Claims, Disputes and Arbitration

20.1 Contractor's Claims

Claims under the YLW Book may include additional Costs incurred in the design office. This will be away from the Site so the immediate submission of contemporary records becomes even more important.

20.2 Appointment of the Dispute Adjudication Board

The entire Sub-Clause is replaced by the following:

'*Disputes shall be adjudicated by a DAB in accordance with Sub-Clause 20.4 [Obtaining Dispute Adjudication Board's Decision]. The Parties shall jointly appoint a DAB by the date 28 days after a Party gives notice to the other Party of its intention to refer a dispute to a DAB in accordance with Sub-Clause 20.4.*

The DAB shall comprise, as stated in the Appendix to Tender, either one or three suitably qualified persons ('the members'). If the number is not so stated and the Parties do not agree otherwise, the DAB shall comprise three persons.

If the DAB is to comprise three persons, each Party shall nominate one member for the approval of the other Party. The Parties shall consult both these members and shall agree upon the third member, who shall be appointed to act as chairman.

However, if a list of potential members is included in the Contract, the members shall be selected from those on the list, other than anyone who is unable or unwilling to accept appointment to the DAB.

The agreement between the Parties and either the sole member ('adjudicator') or each of the three members shall incorporate by reference the General Conditions of Dispute Adjudication Agreement contained in the Appendix to these General Conditions, with such amendments as are agreed between them.

The terms of the remuneration of either the sole member or each of the three members shall be mutually agreed upon by the Parties when agreeing the terms of appointment. Each Party shall be responsible for paying one-half of this remuneration.

If at any time the Parties so agree, they may appoint a suitably qualified person or persons to replace any one or more members of the DAB. Unless the Parties agree otherwise, the appointment will come into effect if a member declines to act or is unable to act as a result of death, disability, resignation or termination of appointment. The replacement shall be appointed in the same manner as the replaced person was required to have been nominated or agreed upon, as described in this Sub-Clause.

The appointment of any member may be terminated by mutual agreement of both Parties, but not by the Employer or the Contractor acting alone. Unless otherwise agreed by both Parties, the appointment of the DAB (including each member) shall expire when the DAB has given its decision on the dispute referred to it under Sub-Clause 20.4 [Obtaining Dispute Adjudication Board's Decision], unless other disputes have been referred to the DAB by that time under Sub-Clause 20.4, in which event the relevant date shall be when the DAB has also given decisions on those disputes.'

A comma after *'each of the members'* in the sixth paragraph was deleted as required by an ERRATA to the first edition to the YLW Book.

Under the YLW Book the DAB is appointed after a dispute has arisen and is referred to as an *'ad hoc'* DAB. The ad hoc DAB is only appointed for that particular dispute, so its jurisdiction is limited to that particular dispute. After rendering its decision the DAB appointment has expired. The FIDIC Contracts Guide advises that the two types of DAB are not intended to be particularly suitable for the RED and YLW books, respectively. If a YLW Book Employer wishes to have a full term DAB, then the RED Book DAB should be specified in the Particular Conditions. Similarly, a YLW Book DAB could be specified in RED Book Particular Conditions. This is not just a matter of introducing the wording of each Sub-Clause 20.2 [*Appointment of the Dispute Board*] respectively; all the DAB provisions must be reviewed and changed where necessary.

It is recommended to amend Clause 2 of the *General Conditions of Dispute Adjudication Agreement* (or DAA) by removing the requirement for the Employer and the Contractor to give notice after the Agreement has taken effect. Refer to comments made on this issue in Chapter 26 for the RED Book.

An ad hoc DAB may appear to be cheaper, because it saves the cost of the monthly retainer and Site visits. However the following problems could arise.

- It is always more difficult to agree on the members of a DAB *after* a dispute has arisen.
- There will be a further delay before the DAB can actually consider the dispute.
- The DAB will charge for the time necessary to mobilise and familiarise itself with the background to the dispute.
- Perhaps most importantly, the Parties have lost the potential for the DAB to help to prevent the dispute from arising.

Sub-Clause 20.2 (YLW) [*Appointment of the Dispute Adjudication Board*] requires the DAB to be appointed within 28 days of one Party giving notice to the other Party of its intention to refer a dispute to a DAB in accordance with Sub-Clause 20.4 [*Obtaining Dispute Adjudication Board's Decision*].

In accordance with the final paragraph of Sub-Clause 20.2 (RED) [*Appointment of the Dispute Adjudication Board*], a full term DAB's appointment will expire when the Contractor's discharge under Sub-Clause 14.12 [*Discharge*] becomes effective. This will normally be when the Contractor has received back his Performance Security, together with the amount due under the Final Payment Certificate. However, if the Engineer and the Contractor cannot agree on the Final Payment Certificate then the resulting dispute can be referred to the DAB, amicable settlement or arbitration. The discharge does not then become effective and the DAB would remain in place. The Parties might then agree to terminate the appointment of the DAB.

An ad hoc DAB's appointment will expire when it has given a decision on any disputes that have been referred to it in accordance with Sub-Clause 20.2 (YLW) [*Appointment of the Dispute Adjudication Board*].

The provision for the DAB to give an opinion is not relevant to an ad hoc DAB so it has been omitted from Sub-Clause 20.2 (YLW) [*Appointment of the Dispute Adjudication Board*].

Sub-Clause 20.2 (YLW) also omits the provision that the terms of remuneration of the DAB shall include the remuneration of any expert whom the DAB consults. The reason for this omission is not apparent and an ad hoc DAB would be well advised to ensure that the provision is included in their Agreements. An ad hoc DAB may be less likely to require expert advice but it is still a possibility.

20.4 Obtaining Dispute Adjudication Board's Decision

For an ad hoc DAB the payment procedures require an advance payment to the DAB, so the fourth paragraph of Sub-Clause 20.4 [*Obtaining Dispute Adjudication Board's Decision*] has been changed to the following:

'*Within 84 days after receiving such reference, or the advance payment referred to in Clause 6 of the Appendix – General Conditions of Dispute Adjudication Agreement, whichever date is later, or within such other period as may be proposed by the DAB and approved by both Parties, the DAB shall give its decision, which shall be reasoned and shall state that it is given under this Sub-Clause. However, if neither of the Parties has paid in full the invoices submitted by each member pursuant to Clause 6 of the Appendix, the DAB shall not be obliged to give its decision until such invoices have been paid in full. The decision shall be binding on both Parties, who shall promptly give effect to it unless and until it shall be revised in an amicable settlement or an arbitral award as described below. Unless the Contract has already been abandoned, repudiated or terminated, the Contractor shall continue to proceed with the Works in accordance with the Contract.*'

Two minor changes were made in an erratum to the first edition to the YLW Book and are incorporated above. The word '*advanced*' was changed to '*advance*' and '*General*

Conditions of the Dispute Adjudication Agreement' was changed to '*General Conditions of Dispute Adjudication Agreement*' in the first sentence.

The YLW Book utilises the '*ad hoc*' DAB that is described in Chapter 4. The Procedural Rules for this type of DAB repeat much of the information that is given in Clause 20 and the *General Conditions of Dispute Adjudication Agreement*, with further details of procedures, including the following.

■ Site visits by the full term DAB will normally be at intervals of not more than 140 days, with not less than 70 days between consecutive visits.

■ Site visits enable the DAB to be informed about progress and any problems or claims. Before leaving the Site the DAB must prepare a report and send copies to both the Employer and the Contractor

Part 4

The FIDIC Conditions of Subcontract for Construction for Building and Engineering Works Designed by the Employer (SUBCONTRACT Book)

FIDIC Users' Guide
ISBN 978-0-7277-5856-9

Chapter 73
Introduction to Part 4

Since the 1999 introduction of the new Rainbow Suite of Contracts by FIDIC, there has been an increased demand for a suitable Form of Contract that could be used in a back-to-back situation with subcontractors. It was considered that the 1994 first edition of the FIDIC *Conditions of Subcontract for Works of Civil Engineering Construction* was no longer suitable. In answer to these calls FIDIC decided to investigate the needs of the modern day international construction industry, with specific emphasis on a market tendency to increasingly utilise specialist companies for key project elements in a contractor-subcontractor relationship. This culminated in the 2011 publishing of the first edition of FIDIC *Conditions of Subcontract for Construction for Building and Engineering Works Designed by the Employer.* This Form of Contract is specifically drafted for use in conjunction with the FIDIC RED Book (1999, first edition). It may also be used with the PNK Book or MDB Harmonised version(s), but in that case it is recommended to carefully consider and introduce the necessary amendments to reflect the significant differences between the RED and the PNK books (refer to Part 2 of this book which highlights the differences between the two contract forms).

The mandate of the FIDIC Task-Group involved in the drafting of this new Form of Subcontract was to produce a document that needed to comply with, amongst others, two basic principles: (i) the document should follow the FIDIC RED Book as closely as possible (i.e. back-to-back), and (ii) wherever possible, it should leave the rights and obligations of the various parties involved in the overall project (i.e. accrued under the Main Contract) intact. The latter principle refers mainly to the Employer, who is not a legal Party to the Subcontract but who will be affected by any activity between the parties to such Subcontract. The Employer necessarily has more rights in relation to 'owning' the project than both the Contractor and the Subcontractor. There is thus a significant difference in the relationship between the Employer and the Contractor as opposed to the Contractor and the Subcontractor. These principles are especially important in situations of potential Termination and/or Assignment. Another situation of importance is the Employer's recourse for defects under the Subcontract after completion of the subcontract works. These principles are further detailed in the relevant section of this book.

In general the interpretation of the Subcontract in relation to the Main Contract (RED or PNK books) is stated in Sub-Clause 1.3 [*Subcontract Interpretation*]. As with the other FIDIC Forms of Contract, the 'Subcontract' or SUB Book includes helpful flow charts. They show in visual form the critical sequences of activities specific to the Subcontract Form of Contract. They facilitate a proper understanding of the procedures and also terminology used in the Subcontract. These flow charts are

- typical sequence of Principal Events during Subcontracts for Construction
- typical sequence of Payment Events envisaged in Clause 14
- sequence of Subcontractor's Claims and Disputes envisaged in Clause 20
  - □ sequence of Subcontractor Claims and Disputes envisaged in First Alternative Clause 20 given in the *Guidance for the Preparation of Particular Conditions of Subcontract*
  - □ sequence of Subcontractor Claims envisaged in the Second Alternative Clause 20 given in the *Guidance for the Preparation of Particular Conditions of Subcontract*
  - □ sequence of Subcontractor Disputes envisaged in the Second Alternative Clause 20 given in the *Guidance for the Preparation of Particular Conditions of Subcontract.*

While there are numerous clauses that are generally applicable, there are some clauses that must necessarily vary through appropriate amendments in the Particular Conditions of Subcontract to take account of the circumstances and locality of the Subcontract Works. These particular conditions are linked with the General Conditions of Subcontract by the corresponding numbering of the clauses, so that the General and the Particular Conditions together make up the Conditions of Subcontract outlining the rights and obligations of the Parties. The Particular Conditions of Subcontract must be drafted to suit each individual project. It is for this reason that the book contains an extensive part on Guidance for the Preparation of Particular Conditions of Subcontract. But caution is required. It is recommended to seek professional assistance and guidance when considering the particular conditions as any uncontrolled amendment may lead to unwanted problems during the execution of the project.

The SUB Book introduces a series of Annexes without which the Form of Contract cannot be used effectively. These are

- Annex A: Particular of the Main Contract
- Annex B: Scope of Subcontract Work and Schedule of Subcontract Documents
- Annex C: Incentive(s) for Early Completion, Taking-Over by the Contractor and Subcontract Bill of Quantities
- Annex D: Equipment, Temporary Works, Facilities, and Free-Issue Materials to be provided by the Contractor
- Annex E: Insurances
- Annex F: Subcontract Programme
- Annex G: Other Items.

Users are advised to carefully consider these annexes, as they necessarily contain potentially contentious information. In particular, Annex A has proven to be a regular source for disputes.

FIDIC Users' Guide
ISBN 978-0-7277-5856-9

ICE Publishing: All rights reserved
http://dx.doi.org/10.1680/fug.58569.423

Chapter 74
Clause 1: Definitions and Interpretation

This Clause under the SUB Book is drafted in a similar way to the corresponding RED and PNK Books. It contains most of the standard definitions of capitalised words in the Sub-Clause text blocks. The significant difference is that the definitions are listed alphabetically rather than in sub-categories, which is easier to use. All words and expressions shall have the same meanings as are respectively assigned to them in the Main Contract, except where the context requires otherwise. It is thus important to note that any capitalised word used in the SUB Book must at all times be checked against the provisions of the Main Contract. For example, as the word '*Cost*' is not stated in this Sub-Clause, this term as used in the SUB Book has the meaning as stated under the Main Contract Clause 1.1.4.3. In addition some words and expressions shall have amended meanings especially assigned to them for use in the SUB Book. Words with different meanings are as follows.

1.1 "Subcontract Definitions"
In the Conditions of Subcontract, which include Particular Conditions of Subcontract, Annexes to the Particular Conditions of Subcontract and these General Conditions of Subcontract, all words and expressions shall have the same meanings as are respectively assigned to them in the Main Contract, except where the context otherwise requires and except that the following words and expressions shall have the meanings hereby assigned to them:

1.1.1 "Accepted Subcontract Amount"
means the amount accepted in the Contractor's Letter of Acceptance for the execution and completion of the Subcontract Works and the remedying of any defects.

This is logical and self-explanatory. The reference is to the Contractor's Letter of Acceptance, which is further defined under Sub-Clause 1.1.6 the equivalent Sub-Clause under the RED and PNK books is Sub-Clause 1.1.4.1

1.1.2 "Annex"
means the document entitled annex attached to the Particular Conditions of Subcontract, completed by the Contractor and/or the Subcontractor, as included in the Subcontract. "Annex A" means the document entitled Annex A attached to the Particular Conditions of Subcontract, and similarly for "Annex B", "Annex C", and so on. All the annexes attached to the Particular Conditions of Subcontract are referred to jointly as the "Annexes".

This definition allows the Parties to include specialised information pertinent to the sub-contract works, for which there is no adequate provision under the standard text, and for which it would be cumbersome or even impossible to amend these through the Particular Conditions.

1.1.3 *"Appendix to the Subcontractor's Offer"*
means the completed pages entitled appendix to the subcontractor's offer which are appended to and form part of the Subcontractor's Offer.

These are the equivalent pages to the 'Appendix to Tender' under the RED Book, and the 'Contract Data' under the PNK Book.

1.1.4 *"Contractor"*
means the person named as contractor in the Appendix to Subcontractor's Offer and the legal successors in title to such person, but not (except with the consent of the Subcontractor) any assignee of such person.

The contractor should by implication be the same as the one under the Main Contract although there is no explicit requirement for this to be the case.

1.1.5 *"Contractor's Instruction"*
means an instruction given by the Contractor's Subcontract Representative in accordance with Sub-Clause 3.1 [Contractor's Instructions].

For this definition there is no equivalent provision under the RED or PNK books.

1.1.6 *"Contractor's Letter of Acceptance"*
means the letter of formal acceptance signed by the Contractor of the Subcontractor's Offer, including any appended memoranda comprising agreements between and signed by both Parties.

The equivalent Sub-Clause under the RED and PNK books is Sub-Clause 1.1.1.3. There is no default provision in the event that there is no Contractor's Letter of Acceptance like under the RED and PNK books. The Letter of Acceptance is for various reasons an important contractual document, and it was felt by the Task Group that such a document should at all times exist. As such the wording of the related Sub-Clauses is amended to encourage the Parties to actually create it, without prescribing it (refer Sub-Clause 1.5 (2) [*Priority of Documents*]).

1.1.7 *"Contractor's Subcontract Representative"*
means the person named by the Contractor in the Appendix to the Subcontractor's Offer or appointed from time to time by the Contractor under Sub-Clause 6.3 [Contractor's Subcontract Representative], who acts on behalf of the Contractor.

The equivalent Sub-Clause under the RED and PNK books is Sub-Clause 1.1.2.5.

1.1.8 *"Employer"*
means the person named as employer in Part A of Annex A and the legal successors in title to, or assignees of such person.

The equivalent Sub-Clause under the RED and PNK books is Sub-Clause 1.1.2.2. Under the SUB Book the definition includes any assignee to the Employer.

1.1.9 *"Engineer"*

means the person appointed by the Employer to act as the Engineer for the purposes of the Main Contract and named in Part A of Annex A, or other person as appointed from time to time by the Employer and notified to the Contractor under the Main Contract, and notified thereafter to the Subcontractor by the Contractor.

The equivalent Sub-Clause under the RED and PNK books is Sub-Clause 1.1.2.4. This definition refers to the Engineer under the Main Contract as further defined under Sub-Clause 1.1.11.

1.1.10 *"Letter of Subcontractor's Offer"*

means the document entitled letter of subcontractor's offer, which was completed by the Subcontractor and includes the signed offer to the Contractor for the Subcontract Works.

The equivalent Sub-Clause under the RED and PNK books is Sub-Clause 1.1.1.4 *Letter of Tender.*

1.1.11 *"Main Contract"*

means the contract entered into between the Employer and the Contractor in respect of the Main Works, brief particulars of which are given in Part A of Annex A.

This refers to the Contract as entered into between the Contractor and the Employer for the execution of the Works of which the SUB Book is a part. The relevant related Sub-Clause under the RED and PNK books are Sub-Clause 1.1.1.1.

1.1.12 *"Main Contract DAB"*

means the DAB as defined under the Main Contract.

This refers to the DAB or DB as appointed under the Main Contract and is related to Sub-Clauses 1.1.2.9 under the RED and PNK books respectively.

1.1.13 *"Main Contract Tests on Completion"*

means the tests which the Subcontractor is to carry out on completion of the Subcontract Works in accordance with Sub-Clause 9.1 [Subcontract Tests on Completion] which constitute Tests on Completion under the Main Contract.

The equivalent Sub-Clause under the RED and PNK books is Sub-Clause 1.1.3.4. Also refer to the relevant detailed section on this in Chapter 15.

1.1.14 *'Main Works'*

means the Works as defined in the Main Contract.

This definition refers to the works to be constructed under the Main Contract Sub-Clause 1.1.5.8.

1.1.15 "Party"
means the Contractor or the Subcontractor, as the context requires. The Contractor and the Subcontractor are referred to jointly as "the Parties".

The equivalent Sub-Clause under the RED and PNK books is Sub-Clause 1.1.2.1. The SUB Book within the formal definition includes a sub-definition for the Parties, referring jointly to the Contractor and the Subcontractor.

1.1.16 "Subcontract"
means the agreement between the Parties comprising the documents listed at Sub-Clause 1.5 [Priority of Subcontract Documents].

1.1.17 "Subcontract Agreement"
means the subcontract agreement referred to in Sub-Clause 1.9 [Subcontract Agreement].

The equivalent Sub-Clause under the RED and PNK books is Sub-Clause 1.1.1.2.

1.1.18 "Subcontract Bill of Quantities"
means the document entitled bill of quantities, completed by the Subcontractor and submitted with the Subcontractor's Offer, as included in Annex C.

The equivalent Sub-Clause under the RED and PNK books is Sub-Clause 1.1.1.10. However, the reference to a day-work schedule has been removed from the definition, and is also not explicitly referred to in Annex C.

1.1.19 'Subcontract Commencement Date'
means the date notified under Sub-Clause 8.1 [Commencement of Subcontract Works].

The equivalent Sub-Clause under the RED and PNK Books is Sub-Clause 1.1.3.2.

1.1.20 "Subcontract DAB"
means either one or three persons as stated in the Appendix to the Subcontractor's Offer, or other person(s) appointed under Sub-Clause 20.5 [Appointment of the Subcontract DAB].

The equivalent Sub-Clause under the RED and PNK books is 1.1.2.9, whereas in deviation the default is for a one-member DAB or DB (PNK Book).

1.1.21 "Subcontract Defects Notification Period"
means the period for notifying defects in accordance with Sub-Clause 11.2 [Subcontract Defects Notification Period].

The equivalent Sub-Clause under the RED and PNK books is Sub-Clause 1.1.3.7. Significantly, this period starts with the taking over of the subcontract works and finishes with the end of the relevant defects notification period under the Main Contract.

1.1.22 *"Subcontract Drawings"*

means the drawings of the Subcontract Works as included in the Subcontract, and any additional and/or modified drawings issued by the Contractor in accordance with the Subcontract.

The equivalent Sub-Clause under the RED and PNK books is Sub-Clause 1.1.1.6.

1.1.23 *"Subcontract Goods"*

means the Subcontractor's Equipment, Subcontract Plant, the materials intended to form or forming part of the Subcontract Works (including supply only materials (if any) to be supplied by the Subcontractor under the Subcontract) and the Subcontractor's temporary works, or any of them as appropriate.

The equivalent Sub-Clauses under the RED and PNK books are Sub-Clause 1.1.5.2 [*Goods*] and Sub-Clause 1.1.5.3 [*Materials*]. Under the SUB Book the definitions have been combined.

1.1.24 *"Subcontract Performance Security"*

means the security (or securities, if any) under Sub-Clause 4.2 [Subcontract Performance Security].

The equivalent Sub-Clause under the RED and PNK books is Sub-Clause 1.1.6.6.

1.1.25 *"Subcontract Plant"*

means the apparatus, machinery and vehicles intended to form or forming part of the permanent works to be executed by the Subcontractor under the Subcontract.

The equivalent Sub-Clause under the RED and PNK books is Sub-Clause 1.1.5.5.

1.1.26 *"Subcontract Price"*

means the price defined in Sub-Clause 14.1 [The Subcontract Price}, and includes adjustments in accordance with the Subcontract.

The equivalent Sub-Clause under the RED and PNK books is Sub-Clause 1.1.4.2.

1.1.27 *"Subcontract Programme"*

means the programme defined in Sub-Clause 8.3 [Subcontract Programme] and Annex F.

There is no equivalent definition under the RED and PNK books.

1.1.28 *"Subcontract Section"*

means a part of the Subcontract Works as defined in the Appendix to the Subcontractor's Offer.

The equivalent Sub-Clause under the RED and the PNK books is Sub-Clause 1.1.5.6.

1.1.29 "Subcontract Specification"

means the document entitled specification, as included in the Subcontract, and any additions and/or modifications made in accordance with the Subcontract. This document specifies the Subcontract Works and may include calculations and technical information of a like nature.

The equivalent Sub-Clause under the RED and PNK books is Sub-Clause 1.1.1.5. The definition is more detailed to enhance interpretation.

1.1.30 "Subcontract Tests on Completion"

means the tests which are specified in the Subcontract or agreed by the Parties which the Subcontractor is to carry out on completion of the Subcontract Works in accordance with Sub-Clause 9.1 [Subcontract Tests on Completion] which do not constitute any Main Contract Tests on Completion.

This definition should be interpreted in conjunction with Sub-Clauses 1.1.13, 1.1.30 and 9.1.

1.1.31 "Subcontract Time for Completion"

means the time for completion of the Subcontract Works under Sub-Clause 8.2 [Subcontract Time for Completion], as stated in the Appendix to the Subcontractor's Offer (with any extension under Sub-Clause 8.4 [Extension of Subcontract Time for Completion]), calculated from the Subcontract Commencement Date.

The equivalent Sub-Clause under the RED and PNK books is Sub-Clause 1.1.3.3.

1.1.32 "Subcontract Variation"

means any change to the Subcontract Works which is instructed or approved as a variation under Clause 13 [Subcontract Variations and Adjustments].

The equivalent Sub-Clause under the RED and PNK books is Sub-Clause 1.1.6.9.

1.1.33 "Subcontract Works"

means the permanent works to be executed and completed by the Subcontractor under the Subcontract, and the Subcontractor's temporary works required for the execution and completion of these permanent works and the remedying of any defects, or either these permanent works or temporary works as appropriate.

The equivalent Sub-Clauses under the RED and PNK books are Sub-Clause 1.1.5.7 [*Temporary Works*] and Sub-Clause 1.1.5.8 [*Works*]. The wording has been amended to enhance the interpretation especially in terms of the obligation to remedy any defects.

1.1.34 "Subcontractor"

means the person named as subcontractor in the Appendix to Subcontractor's Offer accepted by the Contractor and the legal successors in title to that person, but not (except with the consent of the Contractor) any assignee of that person.

The equivalent Sub-Clause under the RED and PNK books is Sub-Clause 1.1.2.3.

1.1.35 *"Subcontractor's Documents"*

means the calculations, computer programmes and other software, drawings, manuals, models and other documents of a technical nature (if any) supplied by the Subcontractor under the Subcontract.

The equivalent Sub-Clause under the RED and PNK books is Sub-Clause 1.1.6.1.

1.1.36 *"Subcontractor's Equipment"*

means all apparatus, machinery, vehicles and other things required for the execution and completion of the Subcontract Works and the remedying of any defects. Subcontractor's Equipment excludes the Subcontractor's temporary works, the Contractor's Equipment (if any), the Employer's Equipment (if any), the Subcontract Plant, materials and any other things intended to form or forming part of the Permanent Works.

The equivalent Sub-Clause under the RED and PNK books is Sub-Clause 1.1.5.1.

1.1.37 *"Subcontractor's Offer"*

means the Letter of Subcontractor's Offer and all other documents which the Subcontractor submitted with this offer including the Appendix to the Subcontractor's Offer, as accepted by the Contractor's Letter of Acceptance.

The equivalent Sub-Clauses under the RED and PNK books are Sub-Clause 1.1.1.4 [*Letter of Tender*], Sub-Clause 1.1.1.8 [*Tender*] and Sub-Clause 1.1.1.9 [*Appendix to Tender*] (Sub-Clause 1.1.1.10 [*Contract Data*] for PNK).

1.1.38 *"Subcontractor's Personnel"*

means the Subcontractor's Representative and all personnel whom the Subcontractor utilises on Site, who may include the staff, labour, and other employees of the Subcontractor and of the Subcontractor's subcontractors (if any), and any other personnel assisting the Subcontractor in the execution of the Subcontract Works.

The equivalent Sub-Clause under the RED and PNK books is Sub-Clause 1.1.2.7.

1.1.39 *"Subcontractor's Representative"*

means the person named by the Subcontractor in the Appendix to Subcontractor' Offer or appointed from time to time by the Subcontractor under Sub-Clause 6.4 [Subcontractor's Representative], who acts on behalf of the Subcontractor.

The equivalent Sub-Clause under the RED and PNK books is Sub-Clause 1.1.2.5.

1.2 Headings and Marginal Notes

The headings and marginal notes in the Conditions of Subcontract shall not be taken into consideration in the interpretation or construction of the Subcontract.

Under the RED and PNK books this Sub-Clause is embedded in Sub-Clause 1.2 [*Interpretation*]. There is no clear reason as to why the drafters decided to isolate this provision.

1.3 Subcontract Interpretation

In the Subcontract, except where the context requires otherwise:

(a) *if it is stated that a Sub-Clause or provision of the Main Contract shall apply then that Sub-Clause or provision of the Main Contract shall be read with necessary changes, including that any reference to the Employer and/or Engineer shall be read as a reference to the Contractor, any reference to the Contractor shall be read as a reference to the Subcontractor, any reference to the Works shall be read as a reference to the Subcontract Works, and any reference to another Main Contract Clause or defined term therein shall be read as a reference to the equivalent clause or Sub-Clause or defined term of these Conditions;*

(b) *words indicating one gender include all genders;*

(c) *words indicating the singular also include the plural and vice versa where the context requires;*

(d) *words indicating persons or parties shall include corporations and other legal entities;*

(e) *provisions including the word 'agree', 'agreed' or 'agreement' require the agreement to be recorded in writing;*

(f) *'written' or 'in writing' means hand-written, type-written, printed or electronically made, and resulting in or being capable of resulting in a permanent record;*

(g) *any reference to 'Main Contract Clause' shall be read as a reference to the clause or Sub-Clause of the Conditions of the Main Contract. Unless expressly stated to be a Main Contract Clause, any reference to a clause or Sub-Clause shall be read as a reference to the clause or Sub-Clause of these Conditions.*

Subparagraph (a) and (g) provide for the necessary directions as to how to correctly interpret the various relevant portions of the Main Contract when seen in relation to the Subcontract.

1.4 Subcontract Communications

All communications under the Subcontract shall be delivered, sent or transmitted to the address for the recipient's communications as stated in the Appendix to the Subcontractor's Offer. However, if the recipient gives written notice of another address for the purpose of this Sub-Clause, communications shall thereafter be delivered accordingly.

Under the RED and PNK books the equivalent Sub-Clause (Sub-Clause 1.3 [*Communications*]) contains additional provisions, which the drafters decided to omit from this Sub-Clause. What remains is a logical provision for the user when having to formally communicate with the other Party. The remainder of the provisions under the RED and PNK books are included in Sub-Clause 1.6 [*Notices, Consents, Approvals, Certificates, Confirmations, Decisions and Determinations*].

1.5 Priority of Subcontract Documents

The several documents forming the Subcontract are to be taken as mutually explanatory of one another. For the purposes of Interpretation the priority of the documents forming the Subcontract shall be as follows:

(1) The Subcontract Agreement (if any);

(2) The Contractor's Letter of Acceptance (if any);

(3) The Letter of Subcontractor's Offer;

(4) The Particular Conditions of Subcontract and Annexes, except any part of any Annex that is referred to elsewhere in this listed priority of Subcontract documents;

(5) The General Conditions of Subcontract;

(6) The Subcontract Specification;

(7) The Subcontract Drawings;

(8) The Subcontract Bill of Quantities and other schedules of rates and prices in the Subcontract, including the daywork schedule (if any) and schedule of payments (if any); and

(9) Any other document forming part of the Subcontractor's Offer as accepted by the Contractor, and listed in the Contractor's Letter of Acceptance.

If an ambiguity or discrepancy is found in the Subcontract documents, the Contractor shall issue any necessary clarification or Contractor's Instruction.

If a Party becomes aware of an error or defect in a document which was prepared for use in executing the Subcontract Works, the Party shall promptly give notice to the other Party of this error or defect.

This Sub-Clause is very similar to the equivalent provisions under the RED and PNK books (Sub-Clause 1.5 [*Priority of Documents*]). Daywork schedule is no longer a defined term, but may be included by the Subcontractor by inclusion in Annex C. By signing the Contractor's Letter of Acceptance such daywork schedule would be incorporated into and form part of the Subcontract, except when explicitly excluded. Any other documents that the Subcontractor included in his Offer must be individually listed as accepted in the Contractor's Letter of Acceptance for them to be included into and form part of the Subcontract.

1.6 Notices, Consents, Approvals, Certificates, Confirmations, Decisions, and Determinations

Wherever in the Subcontract provision is made for the giving or issuing of any notice, consent, approval, request, certificate, confirmation, decision or determination by any person, unless otherwise specified such notice, consent, approval, request, certificate, confirmation, decision, or determination shall be in writing and the words 'notify', 'consent', 'approve', 'request', 'certify', 'confirm', 'decide', or 'determine' shall be construed accordingly.

Any notice, consent, approval, request, certificate, confirmation, decision, or determination under the Subcontract shall not be unreasonably withheld or delayed.

In the RED Book, this provision is contained in Sub-Clause 1.3 [*Communications*] which is not the obvious place when looking for it. Due to its importance, the Task-Group for the SUB Book has included it in its own dedicated Sub-Clause. Furthermore, the list of possible communications has been increased by two more options: confirmations and decisions.

Unlike the equivalent provisions under RED and PNK books, notices and request should now not be unreasonably withheld. This difference could prove problematic when applied under Sub-Clause 3.4 [*Employer's Claims*] in connection with the Main Contract].

1.7 Joint and Several Liability under the Subcontract

If one Party constitutes (under applicable Laws) a joint venture, consortium or other unincorporated grouping of two or more persons:

(a) *these persons shall be deemed to be jointly and severally liable to the other Party for the performance of the Subcontract;*
(b) *these persons shall notify the other Party of their leader who shall have authority to bind that other Party and each of these persons; and*
(c) *that Party shall not alter its composition or legal status without the prior consent of the other Party.*

The equivalent Sub-Clause under RED and PNK books is Sub-Clause 1.14 [*Joint and Several Liability*].

To allow for the fact that both the Contractor as well as the Subcontractor may be a consortium or joint venture (or other unincorporated grouping), the wording has been amended so that the joint and several liabilities can be applied in both directions. Unfortunately this has potentially created an even bigger problem: Sub-Clause 1.7(b) states that the notified leader of a Party shall have the authority to bind that other Party (capitalised!). In accordance with the definition under Sub-Clause 1.1.15 this would indicate that, for example, the leader of a subcontractor consortium should have the authority to bind the Contractor (as the other Party)! Realistically this would never occur, and it is thus recommended that under the Special Conditions the wording of Sub-Clause 1.7(b) is amended to read '...*who shall have authority to bind such consortium, joint venture or other unincorporated grouping, and each of these persons; and*'.

1.8 Subcontract Law and Language

The law of the country (or other jurisdiction) which governs the Main Contract shall govern the Subcontract.

The ruling language and the language for communications under the Subcontract shall be the ruling language and the language for communications of the Main Contract.

If part of a Subcontract document is written in a language other than the ruling language under this Sub-Clause:

(a) *that part shall be interpreted according to the ruling language; and*
(b) *if there is any ambiguity or discrepancy in interpretation between the language of that part and the ruling language, the Contractor shall issue any necessary clarification or Contractor's Instruction.*

If there are versions of any part of the Subcontract which are written in more than one language, the version which is in the ruling language shall prevail.

The equivalent Sub-Clause under RED and PNK books is Sub-Clause 1.4 [*Law and Language*].

It is stated that the law shall be the law that governs the Main Contract, and that in turn is defined in the respective section of the Appendix to Tender. It may not always be preferable to have the law of the Main Contract governing the Subcontract. This can easily be amended by a relevant amendment under the Particular Conditions of Subcontract.

Furthermore, if the governing law of the Subcontract is to be that of a member state of the European Union and/or the Subcontract Works are to be carried out in a member state of the European Union, then, before inviting tenders, the Contractor should verify that the provisions of this Sub-Clause are consistent with Regulation (EC) No. 593/2008 of the European Parliament and of the Council, or any re-enactment or amendment of this Regulation which is in force at the time of tender.

Although the wording of paragraph 2, dealing with the language for communications, is problematic (it may be interpreted that the language to be used under the Subcontract shall be the same as the Main Contract), the only realistic interpretation is for the Main Contract language to apply to the Subcontract. Again, when required, this can easily be amended through the Particular Conditions of Subcontract.

1.9 Subcontract Agreement
The Parties shall enter into and execute a Subcontract Agreement within 28 days after the Subcontractor receives the Contractor's Letter of Acceptance, unless they agree otherwise. The Subcontract Agreement shall be based on the form annexed to the Particular Conditions of Subcontract. The costs of stamp duties and similar charges (if any) imposed by law in connection with entry into the Subcontract Agreement shall be borne by the Contractor.

The equivalent Sub-Clause under RED and PNK books is Sub-Clause 1.6 [*Contract Agreement*].

If under applicable law, or under the Main Contract, the validity of the Subcontract is subject to the Employer's consent than such requirement must be taken into account through suitable Particular Conditions of Subcontract.

1.10 No Privity of Contract with Employer
Nothing stated in the Subcontract shall be construed as creating any privity of contract between the Subcontractor and the Employer.

Before inviting tenders, the Contractor should verify that the wording of this Sub-Clause is consistent with the law governing the Subcontract. If under the terms of the Main Contract the Subcontractor is to give a collateral warranty to the Employer the wording of this Sub-Clause should be amended to read: '*Save in relation to any collateral warranty given by the Subcontractor to the Employer, nothing stated in the Subcontract shall be construed as creating any privity of contract between the Subcontractor and the Employer.*'

1.11 Subcontract Sections

Where Subcontract Section(s) are defined in the Appendix to the Subcontractor's Offer, any reference to 'the Subcontract Works' and any Clause of these Conditions shall be read as referring also to a Subcontract Section.

Whereas under the RED and PNK books reference to Sections is generally made within the appropriate Sub-Clauses, this has been simplified in that any reference to the Subcontract Works will imply a reference to sections provided that sections have been defined in the Subcontractor's Offer.

If the Parties require confidentiality, an additional Sub-Clause may be added. The following suggestion is taken from the *Guidance for the Preparation of Particular Conditions of Subcontract*:

1.12 Details to be Confidential

The Subcontractor shall treat the details of the Subcontract as private and confidential, except to the extent necessary to carry out obligations under it or to comply with applicable Laws. The Subcontractor shall not publish, permit to be published or disclose any particulars of the Subcontract Works or the Main Works in any trade or technical paper or elsewhere without the previous agreement of the Contractor.

FIDIC Users' Guide
ISBN 978-0-7277-5856-9

ICE Publishing: All rights reserved
http://dx.doi.org/10.1680/fug.58569.435

Chapter 75
Clause 2: The Main Contract

This section deals with the Main Contract. It is a document (or group of documents) that, as per Sub-Clause 1.5 [*Priority of Subcontract Documents*], does not expressly form part of the Subcontract Documents, but which nevertheless is important in the effective administration of any project where the subcontract is used.

2.1 Subcontractor's Knowledge of Main Contract
The Contractor shall make all the documents of the Main Contract available to the Sub-contractor for inspection, save that the Subcontractor shall have no right to inspect the Contractor's prices as stated in the Main Contract and the confidential parts of the Main Contract listed in Part A of Annex A. The Contractor shall provide the Subcontractor with a copy of the Appendix to Tender of the Main Contract together with the Particular Conditions of the Main Contract and details of any other contractual provisions which apply to the Main Contract and differ from the General Conditions of the Main Contract (except the Contractor's prices and the confidential parts of the Main Contract listed in Part A of Annex A). If so requested by the Subcontractor, the Contractor shall provide the Subcontractor with a true copy of the Main Contract (except the Contractor's prices and the confidential parts of the Main Contract listed in Part A of Annex A), at the cost of the Subcontractor who may make or request further copies at his own cost. The Subcontractor shall be deemed to have full knowledge of the relevant provisions of the Main Contract (except the Contractor's prices and the confidential parts of the Main Contract listed in Part A of Annex A).

The Subcontractor shall promptly give notice to the Contractor of any ambiguity or discrepancy which he discovers when reviewing the Subcontract and the Main Contract or executing the Subcontract Works. If any ambiguity or discrepancy is found, the Contractor shall issue any necessary clarification or Contractor's Instruction.

Apart from the prices and any listed confidential parts of the Main Contract all documents forming part thereof shall be made available to the Subcontractor for inspection. As per the definition under RED and PNK books (Sub-Clause 1.1.1.1) these documents include the Letter of Acceptance signed by the Employer, The Letter of Tender as completed by the Contractor, the specifications, drawings, schedules and any other listed documents in the Contract Agreement or Letter of Acceptance. This would give the Subcontractor an enormous insight in the particular details as agreed by the Contractor and the Employer. It is not stated, but implied, that the Contractor shall make these documents available to the subcontractor prior to submission of Subcontractor's Offer. In reality any contractor would want to limit his exposure in this way, and would aim to maximise exclusion of confidential information. Care should be taken in this approach as any information not given, but submitted later as a clarifying Instruction under Sub-Clause 3.1 [*Contractor's Instructions*] may constitute a Variation under Sub-Clause 13.

2.2 Compliance with Main Contract

The Subcontractor shall, in relation to the Subcontract Works, perform and assume all the obligations and liabilities of the Contractor under the Main Contract other than where the provisions of the Subcontract otherwise require, and save that the Subcontractor shall have no obligations in respect of:

(i)   *Main Contract Clause 2.2 [Permits, Licences or Approvals];*
(ii) *Main Contract Clause 4.7 [Setting Out];*
(iii) *sub-paragraphs (d) and (e) of Main Contract Clause 4.8 [Safety Procedures];*
(iv) *Main Contract Clause 4.9 [Quality Assurance];*
(v)   *Main Contract Clause 4.13 [Rights of Way and Facilities];*
(vi) *sub-paragraphs (a) and (b) of Main Contract Clause 4.15 [Access Route];*
(vii) *Main Contract Clause 4.19 [Electricity, Water and Gas];*
(viii) *sub-paragraph (a) of Main Contract Clause 4.22 [Security of the Site];*
(ix) *Main Contract Clause 7.8 [Royalties];*
(x)   *overall co-ordination and project management of the Main Works; and*
(xi) *Those exclusions (if any) expressly set out in Part B of Annex A.*

Save where the provisions of the Subcontract otherwise require, the Subcontractor shall design (to the extent provided for by the Subcontract), execute and complete the Subcontract Works and remedy any defects in such good time and in such a manner that no act or omission of his shall constitute, cause or contribute to any breach by the Contractor of any of his obligations under the Main Contract.

Subject to Sub-Clause 8.7 [Subcontract Damages for Delay] and Sub-Clause 17.3 [Subcontract Limitation of Liability], if the Subcontractor commits any breach(es) of the Subcontract he shall indemnify and hold the Contractor harmless against and from all damages for which the Contractor becomes liable under the Main Contract as a result of such breach(es). Without prejudice to any other remedy of the Contractor for such breach(es) but subject to Sub-Clause 3.3 [Contractor's Claims in connection with the Subcontract], the Contractor may recover these damages from monies otherwise due to the Subcontractor under the Subcontract.

The Subcontractor assumes the same obligations as the main Contractor under the Main Contract in relation to the Subcontract Works, except for stated exclusions and for the obligations under the listed Sub-Clauses (i–xi).

This is an onerous clause for the Subcontractor and it pre-assumes that the Main Contract has been studied and understood as per Sub-Clause 2.1 [*Right of Access to the Site*] in total before committing to the Subcontract.

2.3 Instructions and Determinations under Main Contract

The Subcontractor shall, in relation to the Subcontract Works, comply with all instructions and determinations of the Engineer which are notified to him as a Contractor's Instruction, irrespective of whether the instructions and determinations were validly given under the Main Contract. This obligation shall be subject to Sub-Clause 3.1 [Contractor's Instructions]. If the Subcontractor shall receive any direct instruction from the Employer or the Engineer:

(i)   he shall immediately inform the Contractor's Subcontract Representative and shall supply him with a copy of the direct instruction if given in writing; and

(ii) he shall have no obligation to comply with any such direct instruction unless and until it has been confirmed in writing as a Contractor's Instruction.

If any instruction or determination of the Engineer notified by the Contractor constitutes a Subcontract Variation, Clause 13 [Subcontract Variations and Adjustments] shall apply.

Only the Contractor's Subcontract Representative, and subject to Sub-Clause 3.1 [*Contractor's Instructions*], shall have the authority to issue instructions to the Subcontractor. Neither the Engineer nor the Employer has a right to do this. The Subcontractor shall comply with any instruction received from the Contractor's Subcontract Representative. This applies even if there are questions over the validity of the instruction under the Main Contract. This follows the equivalent obligation placed on the Contractor under the Main Contract in accordance with Sub-Clause 3.1 [*Engineer's Duties and Authority*].

2.4 Rights, Entitlements and Remedies under Main Contract

The Subcontractor shall have like rights, entitlements and remedies that the Contractor has under the Main Contract with respect to the Subcontract Works and the Contractor shall take all reasonable steps to secure from the Employer (including the Engineer) such rights, entitlements and remedies. This obligation shall be subject to the Subcontractor's obligations under Sub-Clause 20.1 [Notices] and Sub-Clause 20.2 [Subcontractor's Claims].

If the Contractor accrues a right or entitlement under the Main Contract relating to the Subcontract Works, than the Subcontractor shall also accrue that same right. In practice this means that the Contractor should pass on such rights and entitlements to the Subcontractor, subject to Sub-Clause 20.1 [*Notices*] and Sub-Clause 20.2 [*Subcontractor's Claims*]. The latter proviso seems to indicate that if the Subcontractor does not comply with these stated Sub-Clauses (i.e. 20.1 and 20.2) he would lose his rights to the accrued rights and entitlements. This would remain the case even if the Contractor still managed to realise these rights and entitlements, but could avoid passing them on due to the Subcontractor's failure under Sub-Clauses 20.1 and 20.2. This appears to create an imbalance in the risk allocation established within the original wording. Consideration should be given to amending this through a suitable change in the Particular Conditions to Subcontract.

2.5 Main Contract Documents

If any document of the Subcontract was made by (or on behalf of) the Employer, Main Contract Clause 1.11 [Contractor's Use of Employer's Documents] shall apply.

This Sub-Clause protects the Employer's intellectual rights to documents made by him for use in the Subcontract or by the Subcontractor. It is a logical extension of Sub-Clause 1.11 [*Contractor's Use of Employer's Documents*] from the Main Contract.

FIDIC Users' Guide
ISBN 978-0-7277-5856-9

ICE Publishing: All rights reserved
http://dx.doi.org/10.1680/fug.58569.439

Chapter 76
Clause 3: The Contractor

This Clause deals with the Contractor in relation to the Subcontract Works. It would be the equivalent to Clause 2 under the RED and PNK books [*The Employer*].

3.1 Contractor's Instructions
The Subcontractor shall take instructions only from the Contractor's Subcontract Representative who shall have the like authority in relation to the Subcontract Works to give instructions as the Engineer has under Main Contract Clause 3.3 [Instructions of the Engineer]. The Subcontractor shall comply with all instructions, given or confirmed in writing, of the Contractor's Subcontract Representative on any matter related to the Subcontract.

Provided that any instruction of the Contractor's Subcontract Representative shall be effective for the purposes of the Subcontract only if it is given or confirmed in writing and it expressly states that it is an instruction given in accordance with this Sub-Clause.

As stated before, the Subcontractor shall only take instructions from the Contractor's Subcontract Representative who is appointed under Sub-Clause 6.3 [*Contractor's Subcontract Representative*]. This appointment shall give the representative all authority necessary to act on the Contractor's behalf under the Subcontract. Under the Main Contract there is a provision for the Engineer to delegate his authority (Sub-Clause 3.2 [*Delegation of the Engineer*] Main Contract). No such provision is included in the SUB Book (i.e. Sub-Clause 6.3). This may be impracticable on large projects and the Contractor may wish to consider using the wording of Sub-Clause 3.2 of the Main Contract as a basis for introducing delegated powers of the Contractor's Subcontract Representative.

3.2 Access to the Site
The Contractor shall give the Subcontractor right of access to and possession of so much of the Site within the times that shall be required to enable the Subcontractor to proceed with execution of the Subcontract Works in accordance with the Subcontract Programme pursuant to Sub-Clause 8.3 [Subcontract Programme].

The Contractor shall not be bound to make any part of the Sire available exclusively to the Subcontractor.

This Sub-Clause provides for the Subcontractor to be given right of access to so much of the Site as may be ascertained from the Subcontract Programme. This highlights the necessity for the Subcontractor to comply at all times with Sub-Clause 8.3 [*Subcontract Programme*], as

439

the Contractor is entitled to rely upon such Programme to co-ordinate the Main Works. Any potential delay claim by the Subcontractor due to lack of access, but for which no suitable programme was available to deduce such required access from, would fail in principle.

3.3 Contractor's Claims in connection with the Subcontract

If the Contractor considers himself to be entitled to any payment under any Clause of these Conditions or otherwise in connection with the Subcontract, the Contractor shall give notice to the Subcontractor describing the event or circumstance giving rise to the claim. The notice shall be given as soon as practicable and not later than 28 days after the Contractor became aware of the event or circumstance giving rise to the claim and shall specify the basis of the claim.

As soon as practicable and not later than 28 days after giving notice, the Contractor shall send to the Subcontractor detailed particulars of the claim which includes substantiation of the amount to which the Contractor considers himself to be entitled. The Contractor shall consult with the Subcontractor in an endeavour to reach agreement on this amount. If agreement is not reached, the Contractor shall make a fair decision as to the appropriate and applicable amount, taking due account of the Subcontractor's views, the extent to which the claim has been reasonably substantiated, and all other relevant circumstances. The Contractor shall give notice, with reasons and supporting particulars, to the Subcontractor of this decision.

Provided that the Contractor has given notice of his decision no later than 7 days before the date the Subcontractor is due to be paid, this amount may be deducted from sums otherwise due to the Subcontractor. The Contractor shall only be entitled to make a deduction or otherwise to claim against the Subcontractor or under the Subcontract Performance Security, in accordance with this Sub-Clause.

The Contractor shall pay financing charges (at the rate referred to in Sub-Clause 14.9 [Delayed Payment under the Subcontract]) and shall indemnify and hold harmless the Subcontractor against and from all claims, damages, losses and expenses (including legal fees and expenses) in respect of any deduction of an amount to which the Contractor was not entitled.

This Sub-Clause provides procedures for the Contractor if he is of the opinion that he is entitled to any payment under any clause of the Subcontract in connection with that Subcontract. It requires a notice to be given by no later than 28 days after the Contractor became aware of the event or circumstance giving rise to the claim. There is an implied time-bar imposed on the Contractor, but there is no sanction for non-compliance, in other words there is no explicit loss of entitlement for failing to notify within the said 28 days. However, the Contractor would nevertheless be in breach of the conditions in the SUB Book.

After notifying the claim under this Sub-Clause the Contractor has a further 28 days (after having given the notice) to submit a detailed claim with particulars. He shall than negotiate the claim with the Subcontractor and failing to reach an amicable agreement, the Contractor shall make a fair decision as to the appropriate and applicable amount he should receive from the Subcontractor. This theoretical approach will inevitably lead to disputes in practice, as it is unlikely that the Parties will easily agree on these issues. If the Subcontractor does not agree with such arbitrary decision on the appropriate amounts, he can seek recourse through the relevant provisions of the DAB through Sub-Clause 20.5 [*Appointment of the Subcontract DAB*].

Provided the Contractor has given notice of his decision on the appropriate amounts no later than 7 days before the Subcontractor is due a payment, the decided amounts may be deducted from such due payment.

3.4 Employer's Claims in connection with the Main Contract

If the Contractor receives from the Employer or the Engineer any notice and particulars of an Employer's claim which concerns the Subcontractor, the Contractor shall immediately send a copy to the Subcontractor. The Subcontractor shall then provide all reasonable assistance to the Contractor in relation to the Employer's claim. To the extent that the Contractor considers himself entitled to pass on the claim to the Subcontractor, Sub-Clause 3.3 [Contractor's Claims in connection with the Subcontract] shall apply.

The Contractor shall pass on to the Subcontractor any claim he receives from the Employer that concerns the Subcontractor. This is a general statement and could lead to problems in situations where the Employer's claim refers only in part to the Subcontractor. In such a situation the Contractor would be unwilling to pass on the entire claim received from the Employer. Instead he would want to limit this to those portions for which he could seek recourse from the Subcontractor. To avoid misinterpretation clear and unambiguous wording may be needed through the Particular Conditions to the Subcontract.

In addition to passing on the claim particulars received from the Employer, a suitable notification in terms of Sub-Clause 3.3 [*Contractor's Claims in connection with the Subcontract*] is required when passing on the claim to the Subcontractor,

3.5 Co-ordination of Main Works

The Contractor shall be responsible for overall co-ordination and project management of the Main Works. Subject to Sub-Clause 6.1 [Co-operation under the Subcontract], the Contractor shall be responsible for co-ordination of the Subcontract Works with the works of the Contractor and the works of any other subcontractors employed by the Contractor.

The Subcontractor shall, whenever required by a Contractor's Instruction, submit details of the arrangements and methods which the Subcontractor proposes to adopt for the execution of the Subcontract Works, and no significant alteration to these arrangements and methods shall be made without the Contractor's prior consent.

The Contractor is responsible for the overall coordination of the Main Works. To enable him to comply with this obligation the Subcontractor shall, when requested by an instruction, submit full details and particulars of the arrangements to be adopted for the execution of the works, for example, method statements, and so on. Once submitted the Subcontractor cannot significantly deviate from these without the Contractor's prior consent, which shall, in accordance with Sub-Clause 1.6 [*Notices, Consents, Approvals, Certificates, Confirmations, Decisions, and Determinations*], be in writing. In this context also refer to Sub-Clause 4.1 [*Contractor's General Obligations*] paragraph 4 under the RED or PNK book conditions (Main Contract).

FIDIC Users' Guide
ISBN 978-0-7277-5856-9

ICE Publishing: All rights reserved
http://dx.doi.org/10.1680/fug.58569.443

Chapter 77
Clause 4: The Subcontractor

This Clause deals with specific provisions concerning the Subcontractor in terms of executing the Subcontract. Although he is obligated to comply with the provisions of the Main Contract as if he were the Main Contractor (refer Sub-Clause 2.2 [*Compliance with Main Contract*]), there are a number of issues that require special provisions for them to be effective in the project's execution. Therefore this Sub-Clause shall always be interpreted in conjunction with the relevant Sub-Clauses under the Main Contract.

4.1 Subcontractor's General Obligations

The Subcontractor shall design (to the extent provided for by the Subcontract), execute and complete the Subcontract Works and remedy any defects in accordance with the Subcontract and with the Contractor's Instructions. If design by the Subcontractor of any part of the Subcontract Works is necessary for the Subcontractor's execution, completion and/or remedying of any defects under the Subcontract, then the Subcontractor shall be responsible for this part and this part shall, when the Subcontract Works are completed, be fit for the purposes for which the part is intended as are indicated by or reasonably to be inferred from the Subcontract.

Save as specified in the Annex D, the Subcontractor shall provide all personnel, superintendence, labour, Subcontract Plant, Subcontractor's Equipment, Subcontractor's Documents, and all other things, whether of a temporary or permanent nature, required in and for the design, execution, completion, and remedying of any defects.

The Subcontractor shall be responsible for the adequacy, stability and safety of all his Site operations and methods of construction.

The equivalent Sub-Clause under the Main Contract (RED or PNK books) is Sub-Clause 4.1 [*Contractor's General Obligations*]. As stated before, any Sub-Clause under the SUB Book shall be construed and interpreted in addition to the various relevant Sub-Clauses under the Main Contract. Thus, any Contractor's obligation under the Main Contract concerning the Subcontract, but not specifically mentioned in the Subcontract, is still an obligation on the Subcontractor, except when the Subcontract provides otherwise. Such a situation occurs in the situation where, under the RED or PNK books, the Contractor receives an Engineer's instruction as envisaged under Sub-Clause 4.1, paragraph 1, concerning the Subcontract, and which the Subcontractor is to comply with provided that such instruction was duly passed on to the Subcontractor in terms of Sub-Clause 2.3 [*Instructions and Determinations under the Main Contract*].

443

Any design that is to be carried out by the Subcontractor as part of the Subcontract Works shall be expressly referred to in Annex B, and shall be fit for such purposes for which the part is intended as specified in the Contract (Main Contract Sub-Clause 4.1 (c) [*Contractor's General Obligations*]), and specified or reasonably inferred as specified under the Subcontract (Subcontract Sub-Clause 4.1, paragraph 1). It is thus extremely important for the Parties, before agreeing on any design part to be executed by the Subcontractor, what the ultimate purpose of this part shall be, not only in terms of the Subcontract, but also in terms of the Main Contract.

4.2 Subcontract Performance Security

Where an amount in respect of a Subcontract Performance Security is stated in the Appendix to the Subcontractor's Offer, the Subcontractor shall obtain (at his cost) a Subcontract Performance Security for proper performance in the amount and currencies stated in the Appendix to the Subcontractor's Offer. The Subcontractor shall deliver the Subcontract Performance Security to the Contractor within 28 days of receiving the Contractor's Letter of Acceptance. The Subcontract Performance Security shall be in a similar form to that of the Performance Security under the Main Contract or in the form that may be agreed between the Contractor and the Subcontractor, and shall be issued by an entity and from within a country (or other jurisdiction) approved by the Contractor which approval shall not be unreasonably withheld or delayed. In all other respects, Main Contract Clause 4.2 [Performance Security] shall apply to this performance security.

Only if an amount in respect of the Subcontract Performance Security has been stated in the Appendix to the Subcontractor's Offer shall this Sub-Clause apply. However, it should be noted that the last sentence of this Sub-Clause makes reference to the provisions of the Main Contract Sub-Clause 4.2 [*Performance Security*]. The effect of this reference when read together with Sub-Clause 1.3 [*Subcontract Interpretation*], is that the Subcontractor is required to maintain the Subcontract Performance Security as valid and enforceable until he receives the Subcontract Performance Certificate. Under Sub-Clause 11.3 [*Subcontract Performance Certificate*] the date of such certificate shall be the date stated in the Performance Certificate issued by the Engineer under the Main Contract.

If the date corresponding to the Subcontract Time for Completion is considerably earlier than the probable date of the Performance Certificate of the Main Contract, then it may be contemplated to allow the expiry date for the Subcontract Performance Certificate to be earlier than the corresponding certificate issued under the Main Contract. This will have undoubtedly a positive effect on the willingness and price that any potential subcontractor at time of tender may display to get involved in the project. Obviously any such contemplation should be adequately addressed through a suitable amendment in the Particular Conditions of Subcontract.

If the form of the Subcontract Performance Certificate is not to be similar to that of the equivalent certificate under the Main Contract, then the acceptable form should be included in the tender documents annexed to the Particular Conditions of Subcontract.

4.3 Access to the Subcontract Works

The Subcontractor shall permit the Employer's Personnel and the Contractor's Personnel to have full access at all reasonable times to examine, inspect, measure, and test the materials and workmanship, and to check the progress of the Subcontract Works whether within the Site or elsewhere, and to the places where any of the materials or Subcontract Plant for the Subcontract Works are manufactured, produced, used, or stored.

The requirements under this Sub-Clause are related to Sub-Clause 7.3 [*Inspection*] under the RED and PNK books. They are to be construed and interpreted as additional. The obligation to provide access is related to the Employer's Personnel and the Contractor's Personnel. Both are capitalised in the wording, but are not expressly defined in the Subcontract. As a consequence the definition must be obtained from the relevant definitions under the Main Contract (RED and PNK books) in Sub-Clauses 1.1.2.6 and 1.1.2.7 respectively. As the Employer is not limited in any way to either increase or decrease the number of people listed under the definition, it is of the utmost importance that the Contractor continuously informs the Subcontractor of any such changes whenever they occur. Failing this the Subcontractor may refuse access to persons that have not been properly notified to him but who are entitled to be there under the Main Contract. The problem gets worse in relation to the Contractor's Personnel; firstly there is no express requirement for the Contractor to notify the Subcontractor who should be regarded as falling under the definition, and secondly as per the definition this may also include each and every employee of the Contractor or of his subcontractors on the project. It is doubtful whether the Subcontractor would want to allow other subcontractors on the project to come and inspect his workmanship, his facilities, and so on. It is recommended to regulate this issue through suitable amendments in the Particular Conditions to Subcontract.

4.4 Subcontractor's Documents

The Subcontractor's Documents shall become part of the Contractor's Documents, save that the Subcontractor shall retain the copyright and other intellectual property rights in the Subcontractor's Documents. In all other respects, Main Contract Clause 1.10 [Employer's Use of Contractor's Documents] shall apply to the Subcontractor's Documents.

This Sub-Clause is self-explanatory and in line with the equivalent Sub-Clauses under the Main Contract.

If the Subcontractor is also to be responsible for obtaining any permit, license or approval that is required by the law that governs the Subcontract, an additional Sub-Clause may be added:

4.5 Subcontract Permits Licenses or Approvals

The provisions of Main Contract Sub-Clause 2.2 [Permits, Licenses or Approvals] shall apply to the Subcontract.

FIDIC Users' Guide
ISBN 978-0-7277-5856-9

ICE Publishing: All rights reserved
http://dx.doi.org/10.1680/fug.58569.447

Chapter 78
Clause 5: Assignment of the Subcontract and Subcontracting

In modern construction projects it is no longer unusual that companies make abundant use of specialist subcontractors. It is for this reason that the FIDIC *Conditions of Subcontract for Construction* were developed in the first place. This is particularly true for larger projects where many subcontractors likewise subcontract portions of their work. What is created is a sub-subcontractor, for which there is also a need of properly prepared and suitable contractual conditions.

In addition the increased international character of the construction industry has seen a proliferation of financial mechanisms used by companies to secure early liquidity or other benefits accruing out of their construction activities. These mechanisms are generally related to assigning future benefits to others, in return for factored monetary benefits. As construction contracts are expressly between an employer and a contractor, any assignment by any of the parties may upset this balance of expecting a quality product in return for an agreed fee. For this reason the Subcontract generally does not allow assignments, except for specific circumstances. This Clause 5 deals with the issues of assignment and sub-subcontracting.

5.1 Assignment of Subcontract
Neither Party shall, without the prior consent of the other Party (which consent, notwith-standing Sub-Clause 1.6 [Notices, Consents, Approvals, Certificates, Confirmations, Decisions, and Determinations], shall be at the sole discretion of the other Party), assign the whole or any part of the Subcontract, or any benefit or interest in or under the Subcontract.

However, no prior consent shall be required in the following instances:

(a) either Party may, as security in favour of a bank or financial Institution, assign its right to any monies due or to become due under the Subcontract; and/or

(b) if the Engineer instructs the assignment of the Subcontract under Main Contract Clause 4.5 [Assignment of Benefit of Subcontract], or if the Employer terminates the Main Contract under Main Contract Clause 15.2 [Termination by Employer] and if required by the Employer to do so, then the Contractor shall be entitled to assign the Subcontract to the Employer.

Assigning the benefits of the Subcontract is only allowed with the express agreement of the other Party with the following exceptions. An assignment may be made by either Party with a bank or other financial institution, to facilitate a positive cash flow for the duration of the project. An assignment may also be made by the Employer who seeks to obtain the benefits of the Subcontract in terms of Sub-Clause 4.5 to the Main Contract. Assignments are also acceptable when terminating the Main Contract due to the fault of the Contractor but the Employer wishes to maintain the Subcontractor and his continued performance on the project.

5.2 Subcontracting

The Subcontractor shall not subcontract the whole or any part of the Subcontract without the prior consent of the Contractor. Any consent shall not relieve the Subcontractor from any liability or obligation under the Subcontract and the Subcontractor shall be responsible for the acts, defaults and neglects of any of his subcontractors, his agents or employees as fully as if they were the acts, defaults or neglects of the Subcontractor.

Provided that the Subcontractor shall not be required to obtain consent for:

(a) the provision of labour;
(b) suppliers of materials named in the Subcontract; or
(c) a subcontract for which the subcontractor is named in the Subcontract.

The Subcontractor shall give the Contractor not less than 14 days' notice of the intended date of the commencement of each of his subcontractors' work, and of the commencement of that work on the Site.

Each of the Subcontractor's subcontracts shall include provisions which would entitle the Contractor to require that the benefits of the subcontractor's obligations under the subcontract be assigned to the Contractor:

(i) if these obligations extend beyond the expiry date of the Defects Notification Period and the Contractor requests the Subcontractor prior to this date to so do; or
(ii) if the Subcontract is terminated under Sub-Clause 15.6 [Termination of Subcontract by the Contractor].

This Sub-Clause is related to Sub-Clause 4.4 [*Subcontractors*] under the RED and PNK books (Main Contract).

There is a potential problem in that the Main Contract is silent on the issue of sub-subcontracting. It could thus be construed that the Contractor has no obligation under the Main Contract to comply with Sub-Clause 4.4 (b) (RED and PNK books) when his approved subcontractor wishes to engage a subcontractor himself. This in turn would create a backdoor for the Contractor if he, for example, wishes to use a certain subcontractor who is actually disapproved under Main Contract Sub-Clause 4.4, but is engaged by another subcontractor as a sub-subcontractor to do the same work. As this undoubtedly cannot be to the mutual intention of the parties to the Main Contract, it stands to reason that the Engineer should also approve a sub-subcontractor. The consequence is that for every sub-subcontractor the Contractor must go through Sub-Clause 4.4 and obtain prior consent by the Engineer. Any sub-subcontractor to be included in the Subcontract under Sub-Clause 5.2 (c) [*Objection to Nomination*], should first be checked for acceptance by the Engineer under the Main Contract. One dare not contemplate the situation where a sub-sub-subcontractor is envisaged.

This still leaves the question whether a notification of start of works by a subcontractor under Sub-Clause 4.4 (c) (Main Contract) would also by implication require an equivalent notification for a sub-subcontractor. Assuming this to be the case the next potential problem arises concerning the timing of such notifications. Under Sub-Clause 5.2 [*Subcontracting*] the Subcontractor shall give the Contractor not less than 14 days notice for any works to start. Subsequently under the Main Contract Sub-Clause 4.4 (c) the Contractor shall notify the Engineer with not less than 28 days notice of the sub-subcontractor's start of works. If the Subcontractor issues his notice in contractual compliance between 15 and 27 days before the works are to start than by implication the Contractor is in breach of his obligation to notify the Engineer not less than 28 days.

When subcontractors and sub-subcontractors are contemplated for use on a project it is recommended to amend the relevant Sub-Clauses under the Main Contract as well as the Subcontract so as to avoid serious problems and related disputes.

The RED Book is silent on the issue of sub-subcontractors. Discussion with the Task Force revealed that it could be implied that the Main Contract does not contain an obligation on the Contractor in the sub-subcontract situation. This however, would not alleviate the problem described above (i.e. the Engineer's disapproval of Subcontractor) but would avoid the situation whereby the Contractor would be in breach due to notification timing. Having created the Subcontract, the authors would welcome a clarifying opinion from FIDIC as to how to deal with this situation.

FIDIC Users' Guide
ISBN 978-0-7277-5856-9

ICE Publishing: All rights reserved
http://dx.doi.org/10.1680/fug.58569.451

Chapter 79
Clause 6: Co-operation, Staff and Labour

In addition to the various Sub-Clauses in the Main Contract (RED and PNK books) there will be special requirements for an effective implementation of the Subcontract in relation to matters dealing with staff, labour and co-operation with others. These are covered under Clause 6, which should be construed and interpreted with the various relevant Sub-Clauses under the Main Contract.

6.1 Co-operation under the Subcontract

The Subcontractor shall, as specified in the Subcontract or as required by a Contractor's Instruction, co-operate with and allow appropriate opportunities for carrying out work to:

(a) the Employer's Personnel;
(b) any other contractors employed by the Employer;
(c) the Contractor and the Contractor's Personnel;
(d) any other subcontractors employed by the Contractor; and
(e) the personnel of any legally constituted public authorities,

who may be employed in the execution on or near the Site of any work not included in the Subcontract. Provided that nothing in this Sub-Clause shall prejudice the Contractor's responsibilities under Sub-Clause 3.5 [Co-ordination of Main Works].

The Contractor shall ensure that the Contractor's Personnel and any other subcontractors employed by the Contractor co-operate with and allow appropriate opportunities for carrying out work to the Subcontractor.

If any of the above persons do not co-operate with the Subcontractor and this non-cooperation affects execution of the Subcontract Works, then the Subcontractor shall immediately notify the Contractor of this non-cooperation. If the Subcontractor is delayed, impeded or prevented from performing any of his obligations under the Subcontract by the non-cooperation of any of the above persons and suffers delay and/or incurs Cost, the Subcontractor shall give notice to the Contractor. Provided that the Subcontractor has used all reasonable endeavours to facilitate co-operation by that person, the Subcontractor shall be entitled, subject to Sub-Clause 20.2 [Subcontractor's Claims], to an extension of time under Sub-Clause 8.4 [Extension of Subcontract Time for Completion] and payment of such Cost which shall be included in the Subcontract Price.

451

The equivalent Sub-Clause under the Main Contract (RED and PNK books) is Sub-Clause 4.6 [*Co-operation*].

If the Subcontractor is delayed or hindered in the execution of his Works due to non-cooperation of any of the listed persons (a–e) than he may submit a claim under Sub-Clause 20.2 [*Subcontractor's Claims*], provided he has used all reasonable endeavours to facilitate co-operation. The phrase '*reasonable endeavours*' is misleading and may create an impression that something specific and meaningful has been stated. It will be onerous and difficult to prove such endeavours as they are by implication undefined. Only good record keeping and contract administration by the Parties may avoid potential disputes. Co-operation by the Subcontractor may also be instructed through an appropriate Contractor's Instruction under Sub-Clause 3.1 [Contractor's Instructions].

6.2 Persons in the Service of Others
The Subcontractor shall not recruit, or attempt to recruit, staff and labour from amongst the Contractor's Personnel or the Employer's Personnel.

The Contractor shall not recruit, or attempt to recruit, staff and labour from amongst the Subcontractor's Personnel.

The equivalent Sub-Clause under the Main Contract is Sub-Clause 6.3 [*Persons in the Service of Employer*], and is logically amended to include Contractor's Personnel, as well an obligation on the Contractor not to recruit or attempt to recruit from the Subcontractor's staff.

As there is no contractual provision for the Contractor to notify the Subcontractor of whom he considers part of his Contractor's Personnel, the implementation of this Sub-Clause may be open to debate or disputes. To alleviate this problem, a provision should be included in the Particular Conditions to Subcontract which requires the Contractor to notify the Subcon-tractor whom he considers to be Contractor's Personnel similar to the definition contained in the Main Contract for Employer's Personnel (i.e. Sub-Clause 1.1.2.6). Furthermore these provisions should also be included in the Contractor's contracts with other subcontractors on the site.

6.3 Contractor's Subcontract Representative
The Contractor shall appoint the Contractor's Subcontract Representative and shall give him all authority necessary to act on the Contractor's behalf under the Subcontract. The Contractor shall, prior to the Subcontract Commencement Date, notify the Subcontractor of the name and particulars of the person the Contractor has appointed as Contractor's Subcontract Representative.

If the Contractor's Subcontract Representative is to be temporarily absent from the Site during the execution of the Subcontract Works, a suitable replacement person shall be appointed and the Subcontractor shall be notified accordingly.

The Contractor's Subcontract Representative shall, on behalf of the Contractor, issue any Contractor's Instructions.

The Contractor's Subcontract Representative shall be fluent in the language for communications defined in Sub-Clause 1.8 [Subcontract Law and Language].

Similar to the Engineer being an instrument to act on behalf of the Employer (refer to Main Contract Sub-Clause 3.1(a) [*Engineer's Duties and Authority*]), the Subcontractor also needs a Contractor's Subcontract Representative with whom he can communicate in his day-to-day execution of the project. Whereas under the Main Contract there is an implied provision for the Engineer to have precise duties and authorities assigned to him, there is no express similar provision under the Subcontract. It merely states that the Contractor's Subcontract Representative shall be given all authority necessary to act on the Contractor's behalf. It is left open to interpretation what this would mean in reality. It may be advisable to include more precise details in the Particular Conditions of Subcontract.

6.4 Subcontractor's Representative
The Subcontractor shall appoint the Subcontractor's Representative and shall give him all authority necessary to act on the Subcontractor's behalf under the Subcontract.

The whole of the time of the Subcontractor's Representative shall be given to directing the Subcontractor's performance of the Subcontract. If the Subcontractor's Representative is to be temporarily absent from the Site during the execution of the Subcontract Works, the Subcontractor shall, subject to the Contractor's prior consent, appoint a suitable replacement person.

The Subcontractor's Representative shall, on behalf of the Subcontractor, receive the Contractor's Instructions.

The Subcontractor's Representative shall be fluent in the language for communications defined in Sub-Clause 1.8 [Subcontract Law and Language].

The equivalent Sub-Clause under the Main Contract (RED and PNK books) is Sub-Clause 4.3 [*Contractor's Representative*].

Unlike the equivalent provision under the Main Contract there is no express requirement for the Contractor to approve the Subcontractor's Representative, and it is thus up to the Subcontractor to appoint a person he deems suitable for his own purposes. The Appendix to the Subcontractor's Offer makes provision for the Subcontractor to include the details of the person he wants to appoint in that position. There is only a requirement for this person to be fluent in the language for communications defined in Sub-Clause 1.8 [*Subcontract Law and Language*]. Interestingly, any replacement (temporary or permanent) shall be subject to prior approval by the Contractor. The Subcontractor's Representative is the only person authorised to receive a valid Contractor's Instruction as envisaged under Sub-Clause 3.1 [Contractor's Instructions].

FIDIC Users' Guide
ISBN 978-0-7277-5856-9

ICE Publishing: All rights reserved
http://dx.doi.org/10.1680/fug.58569.455

Chapter 80
Clause 7: Equipment, Temporary Works, Other Facilities, Plant and Materials

It is normal practice in the international construction industry that the various parties to a project share certain resources such as equipment, temporary works and the like. The equivalent provision under the Main Contract (RED and PNK books) would be Sub-Clause 4.20 [*Employer's Equipment and Free Issue Material*]. Note that in the Subcontract the '*Free-Issue Materials*' provisions have been moved into a special Sub-Clause 7.2 [*Free Issue Materials*]. However, in a subcontract scenario there are additional resources that also can be shared to optimise the economical effectiveness of the project's execution. They are generally referred to as the Contractor's Equipment. Sharing all these resources cannot be done without some form of formal procedure, as explained further below.

7.1 Subcontractor's Use of Equipment, Temporary Works, and/or Other Facilities
The Contractor shall make the Employer's Equipment, the Contractor's Equipment, the Temporary Works and/or other facilities (if any) specified in Annex D available to the Subcontractor within the times that shall be required to enable the Subcontractor to proceed with execution and completion of the Subcontract Works in accordance with the Subcontract Programme.

The availability of such equipment, temporary works and/or other facilities shall be accordance with the details and arrangements and upon the terms and conditions (if any), specified in Annex D and, unless expressly stated therein, they shall not be provided for the exclusive use of the Subcontractor.

When made available by the Contractor the Subcontractor shall visually inspect such equipment, temporary works and/or other facilities and shall promptly give notice to the Contractor of any shortage, defect or default in them. Unless the Parties agree or the Contractor instructs otherwise, the Contractor shall immediately rectify the notified shortage, defect or default.

The Contractor shall be responsible for the Employer's Equipment (while in his possession and/or under his control), the Contractor's Equipment, the Temporary Works, and/or other facilities (if any), except that the Subcontractor shall be responsible for each item while any of the Subcontractor's Personnel is operating it, driving it, directing it, using it, or in control of it.

The Subcontractor shall not remove from the Site any items of such Employer's Equipment or Contractor's Equipment without the consent of the Contractor. However, consent shall not be required for vehicles transporting Goods or Subcontractor's Personnel off site.

The principle of sharing resources governed by the stipulations of the Subcontract only apply if and to the extent that such resources have been included in Annex 'D'. The purpose, quality and terms and conditions for the provision of these shared resources should ideally be included in this Annex 'D' so as to avoid any future disputes. There is a strict requirement for the Contractor to make these agreed resources available in accordance with the Sub-Clause 8.3 [*Subcontract Programme*]. There is thus no explicit requirement for the timing of such resources to be included in Annex 'D'. In actual fact Sub-Clause 8.3, paragraph 6, states that the Contractor shall be entitled to rely upon the current Subcontract Programme when planning his activities, thus including the timely issue of shared resources. The Subcontractor must carefully check the issued resources upon receipt, failing this any later recourse would be waived.

7.2 Free-Issue Materials
The Contractor shall supply to the Subcontractor, free of charge, the free-issue materials (if any) specified in Annex D, at the place(s) specified in Annex D and within the times that shall be required to enable the Subcontractor to proceed with execution and completion of the Subcontract Works in accordance with the Subcontract Programme. The supply of such free-issue materials shall be in accordance with the details and arrangements and upon the terms and conditions (if any), stated in Annex D.

When made available by the Contractor, the Subcontractor shall visually inspect the free-issue materials and shall promptly give notice to the Contractor of any shortage, defect or default in them. Unless the Parties agree or the Contractor instructs otherwise, the Contractor shall immediately rectify the notified shortage, defect or default. After this visual inspection, the free-issue materials shall come under the care, custody and control of the Subcontractor. The Subcontractor's obligations of inspection, care, custody, and control shall not relieve the Contractor of liability for any shortage, defect or default not apparent from a visual inspection.

The provisions for the issue of Free Materials are to a large extent equal to those for sharing resources under Sub-Clause 7.1 [*Subcontractor's Use of Equipment, Temporary Works, and/ or Other Facilities*]. There is a more stringent provision for checking the delivered material by the Subcontractor as to quality and quantity. Unless instructing the Subcontractor otherwise, the Contractor shall rectify any defect, default or shortage immediately. There is no strict requirement for the visual inspection to be carried out mutually by the Parties, however it stands to reason that this would make absolute sense.

7.3 Indemnity for Misuse
The Subcontractor shall indemnify and hold the Contractor harmless against and from all damages to or loss of any property, real or personal, arising from the misuse by the Subcontractor of the Employer's Equipment, the Contractor's Equipment, the Temporary Works, the free-issue materials (if any), and/or other facilities made available by the Contractor.

Whereas there is no equivalent provision under the RED and PNK books, the SUB Book includes an express provision stating that the Subcontractor shall indemnify and hold harmless the Contractor against any damage or loss of property (real or personal) through the misuse of the resources, materials or other items made available to the Subcontractor as envisaged under Sub-Clause 7.1 [*Subcontractor's Use of Equipment, Temporary Works, and/or Other Facilities*] and Sub-Clause 7.2 [*Free Issue Materials*].

7.4 Ownership of Subcontract Plant and Materials

Each item of Subcontract Plant and of the materials intended to form or forming part of the Permanent Works shall, to the extent consistent with the Laws of the Country, become the property of the Contractor at whichever is the earlier of the following times, free from liens and other encumbrances:

(a) *when it is delivered to the Site;*
(b) *when the Contractor is entitled to payment of value of the Plant and Materials under Main Contract Clause 8.10 [Payment for Plant and Materials in Event of Suspension].*

The equivalent provision under the Main Contract (RED and PNK books) would be Sub-Clause 7.7 [*Ownership of Plant and Materials*].

Again also here the principle of a true back-to-back scenario is incorporated into the Subcontract. The only question that arises is whether the true meaning can be achieved, seeing that the requirements under (a) and (b) are identical in both the Subcontract and the Main Contract. The effect will always be that the ownership of Plant and Materials intended to form part of the Permanent Works will automatically go to the Employer if any of the two requirements is met. Overall however, this is consistent with the aim of the task-group to create a document that would recognise the status of the Employer as project owner whose legitimate interests should not be compromised by the Subcontract. Interestingly under the PNK Book requirement (b) is not triggered by an entitlement to payment but rather by such payment itself.

7.5 Subcontractor's Equipment and Subcontract Plant

The Subcontractor shall be responsible for all Subcontractors' Equipment. When brought on to the Site, the Subcontractor's Equipment shall be deemed to be exclusively intended for the execution of the Subcontract Works. The Subcontractor shall not remove from the Site any major items of Subcontractor's Equipment without the consent of the Contractor. However, consent shall not be required for vehicles transporting Subcontract Goods or Subcontractor's Personnel off Site.

The Subcontract Plant shall be taken to be included in the definition of Plant under the Main Contract and the provisions of the Main Contract concerning the Plant shall apply to the Subcontract Plant.

The Subcontract Plant is deemed to be included in the definition of Plant under the Main Contract as defined in Sub-Clause 1.1.5.5. Therefore, such Plant shall be subject to the provisions of the Main Contract concerning Plant. In the first instance this would refer to

Main Contract Sub-Clause 7.1 [*Manner of Execution*] specifying that the manufacture of Plant shall be in the manner specified in the Contract. In accordance with Subcontract Sub-Clause 1.3(a) [*Subcontract Interpretation*] stating that if a Main Contract Clause shall apply it shall be read with the necessary changes, that is, any reference to a defined term in the Main Contract shall be read as a reference to the equivalent defined term of the Subcontract. Thus the reference to the '*Contract*' under Main Contract Sub-Clause 7.1(a) should be read as being the Subcontract.

This could be a source of misinterpretation and subsequent dispute in that it is seemingly not clear as to what specification is to be used for the manufacture of Plant. The issue is resolved through Sub-Clause 2.2 [*Compliance with Main Contract*] where it is stated in paragraph 2 that the Subcontractor shall do nothing that constitutes, causes or contributes to any breach by the Contractor of any of his obligations under the Main Contract. It is thus imperative for the Subcontractor to make absolutely certain that he has knowledge of and understands the provisions of the Main Contract, all in accordance with Sub-Clause 2.1 [*Subcontractor's Knowledge of Main Contract*].

FIDIC Users' Guide
ISBN 978-0-7277-5856-9

ICE Publishing: All rights reserved
http://dx.doi.org/10.1680/fug.58569.459

Chapter 81
Clause 8: Commencement and Completion

This is the equivalent to Clause 8 under the Main Contract (RED and PNK books). The title of the Clause is slightly different ('*Commencement and Completion*' versus '*Commencement, Delays and Suspension*'), however, it covers the same basic elements concerning the time for execution of the project. These basic elements are

■ Commencement Date
■ Time for Completion
■ Extension of Time
■ Programme
■ Delay Damages
■ Suspension.

8.1 Commencement of Subcontract Works
The Contractor shall give the Subcontractor not less than 14 days' notice of the Subcontract Commencement Date.

The Subcontractor shall commence the execution of the Subcontract Works as soon as is reasonably practicable after the Subcontract Commencement Date, and shall proceed with the Subcontract Works with due expedition and without delay.

The Commencement Date is the most basic, but also the most important issue concerning the start and subsequent execution of any project. It is the date when the clock starts ticking. It is thus a date that requires some prior preparation by both Parties to ensure a smooth and effective project execution. The Main Contract states under Sub-Clause 8.1 [*Commencement of Works*] that preparation time shall be not less than 7 days. Under the Subcontract, the Subcontractor is given not less than 14 days notice. Thereafter the Subcontractor shall commence the Works without unreasonable delay, and shall proceed with due expedition.

8.2 Subcontract Time for Completion
The Subcontractor shall complete the Subcontract Works in accordance with Sub-Clause 10.1 [Completion of Subcontract Works] within the Subcontract Time for Completion, or the extended time that may be allowed under Sub-Clause 8.4 [Extension of Subcontract Time for Completion].

The Time for Completion of the Subcontract Works is defined under Sub-Clause 1.1.31 where it is stated that this shall be the time as included in the Appendix to the Subcontractor's Offer. This is usually stated in calendar days.

8.3 Subcontract Programme

The provisions of Annex F shall apply to programming of the Subcontract Works and the Subcontract Programme shall be the programme defined therein.

The Subcontractor shall submit a detailed programme for the execution of the Subcontract Works to the Contractor within 14 days of receipt of the Contractor's Letter of Acceptance or the Contractor's programme submitted under Main Contract Clause 8.3 [Programme], whichever is the later. The form and detail of this initial programme shall fully comply with:

(a) the programming and reporting requirements of the Main Contract; and
(b) the requirements set out in Part A of Annex F.

If a Contractor's Instruction regarding the programming and/or sequencing of the Subcontract Works constitutes a Subcontract Variation, Clause 13 [Subcontract Variations and Adjustments] shall apply.

Provided that if, at any time:

(i) actual progress is too slow to complete within the Subcontract Time for Completion, and/or
(ii) progress has fallen (or will fall) behind the Subcontract Programme,

other than as a result of a cause listed in Sub-Clause 8.4 [Extension of Subcontract Time for Completion], then the Contractor may issue a Contractor's Instruction requiring the Subcontractor to submit an updated programme and supporting report describing the revised methods which the Subcontractor proposes to adopt in order to expedite progress and complete the Subcontract Works within the Subcontract Time for Completion. Unless the Contractor notifies otherwise, the Subcontractor shall adopt these revised methods, which may require increases in the working hours and/or in the numbers of Subcontractor's Personnel and/or Subcontract Goods, at the risk and cost of the Subcontractor. If these revised methods cause the Contractor to incur any additional Cost, the Contractor shall, subject to Sub-Clause 3.3 [Contractor's Claims under the Subcontract], be entitled to deduct this Cost from the Subcontract Price.

The Contractor shall be entitled to rely upon the current Subcontract Programme when coordinating the Main Works and/or planning his activities and those of other subcontractors employed by the Contractor in respect of the Main Works.

The Contractor shall give the Subcontractor all reasonable co-operation and assistance in order that he may progress the Subcontract Works as required by the Subcontract Programme.

As with any project an effective and professional execution is only possible through proper planning. FIDIC has for many years, through their Standard Forms of Contract, propagated the importance of this often underestimated project management tool through critical contractual requirements as to submission of suitable Programmes. The Subcontract is no different. Sub-Clause 8.3 [*Subcontract Programme*] is well defined and has similar requirements to the Main Contract Programme. This Sub-Clause must be interpreted in conjunction and compliance with Annex 'F'.

The most basic requirement for the Subcontract Programme is that it shall fully comply with the programming and reporting requirements of the Main Contract and that it shall be produced using the software specified (if any) in the Main Contract [Annex 'F' – A (b)]. Sub-Clause 8.3 [*Subcontract Programme*] paragraph 2, further states that the Subcontract Programme shall be submitted within 14 days after receipt of the Contractor's Letter of Acceptance. Other important requirements are

- identify when and what approvals, consents, and so on, are required from the Contractor, the Engineer and/or the Employer. [A (g)]
- identify holiday periods [A (i)]
- key testing dates [A (k)]
- be fully and logically linked [A (l)]
- identify the critical path [A (m)]
- be supported by an adequate resource statement required to comply with the programme (A (q)).

The Contractor is to respond with comments within 14 days of receiving such programme from the Subcontractor, failing which the submitted programme becomes the Subcontract Programme. This programme shall be regularly updated. It shall be updated in any event upon the occurrence of any of the listed occasions under Annex 'F' – B (a) to (f).

The provision in Annex 'F' to comply with the Main Contract also refers to the reporting requirements for the Contractor. This logically refers to Main Contract Sub-Clause 4.21 [*Progress Reports*] that by definition include detailed descriptions of progress of the works, which includes the Subcontract and 8.3 [*Programme*]. This brings two potential problems.

1   There is now an implied requirement for the Subcontractor to update his Subcontract Programme whenever the Contractor updates the Main Contract programme. If such a Contractor's update is unrelated to the work performed by the Subcontractor, then, subject to Sub-Clause 20.2 [*Subcontractor's Claims*], the Subcontractor may have a claim for additional costs in preparing such an update.
2   There is no explicit requirement under the Subcontract to issue updates on a regular interval consistent with the Contractor's reporting requirements under the Main Contract. The Parties should agree an acceptable procedure through an appropriate amendment in the Particular Conditions to Subcontract, with due regard for the provisions of Sub-Clause 8.5 [*Subcontract Progress Reports*].

8.4 Extension of Subcontract Time for Completion

The Subcontractor shall be entitled subject to Sub-Clause 20.2 [Subcontractor's Claims], to an extension of the Subcontract Time for Completion if and to the extent that completion of the Subcontract Works is delayed by any of the following causes:

(a) a Subcontract Variation or other substantial change in the quantity of any item of work included in the Subcontract;

(b) a cause of delay giving an entitlement to an extension of the Subcontract Time for Completion under a Sub-Clause of these conditions;

(c) any delay, impediment or prevention caused by or attributable to the Contractor, the Contractor's Personnel, or the Contractor's other subcontractors; or

(d) any one of the causes set out in the Main Contract Clause 8.4 [Extension of Time for Completion].

When deciding each extension of time under Sub-Clause 20.2 [Subcontractor's Claims], the Contractor may review previous decisions and may increase, but shall not decrease, the total extension of time.

The equivalent Sub-Clause under the Main Contract (RED and PNK books) is Sub-Clause 8.4 *[Extension of Time for Completion]*.

Apart from the logical reasons for granting an extension of the Subcontract Time for Completion (refer Sub-Clause 8.4(a) and (b) *[Extension of Subcontract Time for Completion]*), there are additional provisions that may apply. These include when the Subcontractor is delayed by the Contractor [refer to Sub-Clause 8.4(c)] or when an event occurs which qualifies the Contractor for an extension to his time for completion under the Main Contract. Subcontract Clause 8 does not make express provision for delays caused by authorities. However, the latter provision [i.e. Sub-Clause 8.4(d)] would allow relief for this relief (refer to Main Contract Sub-Clause 8.5).

8.5 Subcontract Progress Reports

If required by the Contractor, the Subcontractor shall prepare and submit monthly progress reports to the Contractor. Each report shall be submitted no later than 5 days before the due date for submission of the Contractor's progress report under the Main Contract, which due date shall be notified to the Subcontractor by the Contractor. Each Subcontractor's progress report shall include the details as set out in sub-paragraphs (a) to (h) inclusive of Main Contract Clause 4.21 [Progress Reports].

The Contractor's reporting obligations under the Main Contract are dependent on the Subcontractor submitting relevant information relating to the Subcontract Works. The Contractor shall notify the Subcontractor when such detailed information must be available on a monthly basis, failing which it shall be 5 days before the due date under the main Contract (7 days after month's end), that is, 2 days after month's end! Two days is normally not enough time to prepare a report sufficiently detailed as envisaged under Main Contract Sub-Clause 4.21(a) to (h) inclusive *[Progress Reports]*. It is recommended that the Parties agree on a mutually acceptable period for preparation of such a report, even considering the amendment of the period as prescribed under the Main Contract.

8.6 Suspension of Subcontract Works by the Contractor

The Contractor may at any time issue a Contractor's Instruction requiring the Subcontractor to suspend progress of part or all of the Subcontract Works. In his notice, the Contractor shall state the reason or reasons for the Suspension. Main Contract Clause 8.9 [Consequences of Suspension], Main Contract Clause 8.10 [Payment for Plant and Materials in Event of Suspension] and Main Contract Clause 8.11 [Prolonged Suspension] shall apply to the suspended Subcontract Works unless the cause of the suspension is the responsibility of the Subcontractor. Main Contract Clause 8.12 [Resumption of Work] shall apply to any suspended Subcontract Works unless the cause of the suspension is the responsibility of the Subcontractor.

Subject to Sub-Clause 16.1 [Subcontractor's Entitlement to Suspend Work], the Subcontractor shall not suspend progress of part or all of the Subcontract Works unless and until required to do so by a Contractor's Instruction.

The equivalent Sub-Clause under the Main Contract (RED and PNK books) is Sub-Clause 8.8 [*Suspension of Work*].

The Subcontract Sub-Clause states that in the case of a suspension, the cause of which was not the responsibility of the Subcontractor, the Main Contract Sub-Clauses 8.10 [*Payment for Plant and Materials in Event of Suspension*], 8.11 [*Prolonged Suspension*] and 8.12 [*Resumption of Work*] shall apply to the Subcontract. There is no express reference to Main Contract Sub-Clause 8.9 [*Consequences of Suspension*] but any potential time and/or money consequences resulting from a suspension under the Subcontract would be dealt with through the appropriate procedures under Sub-Clauses 2.3 [*Instructions and Determinations under Main Contract*], 2.4 [*Rights, Entitlements and Remedies under Main Contract*] and 20.2 [*Subcontractor's Claims*].

8.7 Subcontract Damages for Delay

If the Subcontractor fails to comply with Sub-Clause 8.2 [Subcontract Time for Completion] and this failure causes or contributes to a failure by the Contractor to comply with Main Contract Clause 8.2 [Time for Completion] the Contractor shall, subject to Sub-Clause 3.3 [Contractor's Claims in connection with the Subcontract], be entitled to deduct delay damages from the Subcontract Price for this default. The liability of the Subcontractor to the Contractor for delay to the Subcontract Works shall be limited to the amount stated in the Appendix to the Subcontractor's Offer.

If no amount is stated in the Appendix to the Subcontractor's Offer, the liability of the Subcontractor to the Contractor for delay to the Subcontract Works shall be limited to 10% of the Accepted Subcontract Amount.

In the event that the Subcontractor is delayed due to his own fault and responsibility he shall be liable to pay the Contractor damages. These Subcontract damages for delay are only applicable to the extent that they also cause the Contractor to suffer a delay to the Time for Completion under the Main Contract. There is thus no automatic entitlement to delay damages by the Contractor in the event of a Subcontractor delay. The Contractor has an implied obligation to prove the knock-on effect under the Main Contract. As in all cases of potential delay damages, the most important matter is the quality of contemporaneous records kept (if any) by the parties. These records will ultimately determine the matter. Furthermore, unless stated otherwise, the liability for delay damages by the Subcontractor is limited to 10% of the Accepted Subcontract Amount.

FIDIC Users' Guide
ISBN 978-0-7277-5856-9

ICE Publishing: All rights reserved
http://dx.doi.org/10.1680/fug.58569.465

ice
Institution of Civil Engineers

publishing

Chapter 82
Clause 9: Tests on Completion

This section deals with the various contractual requirements to demonstrate that the Subcontract Works have been completed in compliance with the Subcontract. As the work to be carried out under the Subcontract is generally a part of the works under the Main Contract, it may be necessary for some portions of the Subcontract Works to demonstrate compliance with the Main Contract. This is considered in Sub-Clause 9.2 [*Main Contract Tests on Completion*] below.

9.1 Subcontract Tests on Completion
Insofar as the Subcontract specifies Subcontract Tests on Completion, the Subcontractor shall give reasonable notice to the Contractor of the date after which he shall be ready to carry out each of these tests. If the Subcontract Works fail to pass the Subcontract Tests on Completion, the Subcontractor shall as soon as practicable make good the defect in the Subcontract Works and repeat the tests under the same terms and conditions as specified in the Subcontract until the tests are passed.

This Sub-Clause only applies if the Subcontract specifically requires Tests on Completion to be undertaken. In such a case the Subcontractor shall give the Contractor reasonable notice of the date on which he shall be ready to conduct the relevant tests. A test failure obligates the Subcontractor to make good any defects and to repeat the tests until and, as often as may be required, to demonstrate compliance with the Subcontract, all at his own costs.

9.2 Main Contract Tests on Completion
Insofar as the Subcontract specifies or makes express reference to Main Contract Tests on Completion, Main Contract Clause 9 [Tests on Completion] shall apply. Save that the Subcontractor shall comply with this Sub-Clause in good time to enable the Contractor to comply with his obligations in respect of the Tests on completion under the Main Contract.

If there are any Tests on Completion under the scope of the Subcontract that would also have to be conducted as part of the Contractor's obligations under the Main Contract, then this should be stated in the Subcontract. This is achieved in various ways: (i) detailed within the Subcontract Specification, describing the tests which the Subcontractor is required to carry out on completion of the Subcontract Works or Subcontract Section (if applicable) which constitute Tests on Completion under the Main Contract, (ii) detailed within the Particular Conditions to Subcontract making express reference to the Tests on Completion described in the Specification under the Main Contract, or (iii) attached as an agreed list of tests to be carried out as Tests on Completion under the Main Contract in, for example, Annex G [*Other Items*]. The Main Contract Clause 9 shall only apply if there is specific reference to such tests under this Sub-Clause 9.2.

FIDIC Users' Guide
ISBN 978-0-7277-5856-9

ICE Publishing: All rights reserved
http://dx.doi.org/10.1680/fug.58569.467

Chapter 83
Clause 10: Completion and Taking-Over the Subcontract Works

Every project must come to an end. To achieve this in a construction project several steps have to be completed. One of the first significant steps is the taking over of the Works after completion. Clause 10, similar to the Main Contract, deals with this procedure.

10.1 Completion of the Subcontract Works
Completion of the Subcontract Works shall be achieved when:

(a) *these works have been completed in accordance with the Subcontract except for any minor outstanding work and defects which will not substantially affect the use of the Subcontract Works for their intended purpose;*

(b) *these works have passed the tests on completion specified in the Subcontract (if any); and*

(c) *if required by the Subcontract Specification, the 'as-built' documents and operation and maintenance manuals in respect of the Subcontract Works have been submitted by the Subcontractor.*

Not earlier than 7 days before the Subcontract Works will, in the Subcontractor's opinion, be complete the Subcontractor shall notify the Contractor. The Contractor shall, within 21 days after receipt of this notice:

(i) *notify the Subcontractor that completion of the Subcontract Works has been achieved, stating the date of completion; or*

(ii) *notify the Subcontractor of his opinion that completion of the Subcontract Works has not been achieved, giving reasons and specifying the work required to be done by the Subcontractor to achieve completion. The Subcontractor shall then complete this work before issuing a further notice under this Sub-Clause.*

Sub-Clause 10.1(a) to (c) inclusive states the condition for when Completion of the Subcontract Works is achieved. It is important to note that any requirement by the Contractor for the production of '*as-built*' documents and/or '*operation and maintenance manuals (if any)*' in respect of the Subcontract Works shall be expressly stated in the Specification for them to be included as a condition precedent to achieving completion (refer Sub-Clause 10.1(c)). As per Sub-Clause 10.1(a) the possibility of a so-called '*snag-list*' or '*list-of-open-points*' (LOP) is included as an option. When the Subcontractor is of the opinion that completion

467

is imminent he shall accordingly notify the Contractor not less than 7 days before this shall be achieved. In turn the Contractor shall within 21 days of receipt of such notice either: (i) confirm completion stating the date on which this happened, or (ii) notify the Subcontractor with reasons that he is not in agreement stating what still needs to be done.

Unlike under the Main Contract Sub-Clause 10.1 [*Taking-Over of the Works and Sections*], there is no default '*deemed taken over*' provision for when the Contractor (Engineer under the Main Contract) fails to act within the said 21 days. The reason for this will become apparent when seen in conjunction with Sub-Clause 10.2 below.

10.2 Taking-Over Subcontract Works
Where:

(a) *completion of the Subcontract Works has been achieved in accordance with Sub-Clause 10.1 [Completion of the Subcontract Works]; and*
(b) *there is no provision in Annex C for the taking-over of the Subcontract Works by the Contractor before taking-over by the Employer,*

the Subcontract Works shall be deemed to have been taken-over when a Taking-Over Certificate in respect of the Main Works, or a Section or part of the Main Works of which the Subcontract Works are part has been issued or deemed to have been issued under Main Contract Clause 10.1 [Taking-Over of the Works and Sections]. The date of taking-over of the Subcontract Works shall be the date stated in this Taking-Over Certificate and the Subcontractor shall cease to be liable for the care of the Subcontract Works in accordance with Sub-Clause 17.1 [Subcontractor's Risks and Indemnities] as from this date.

If taking-over of the Subcontract Works is delayed by causes which entitle the Contractor to claim additional payment under the Main Contract, the Subcontractor shall subject to Sub-Clause 20.2 [Subcontractor's Claims] be entitled to recover any Cost incurred by the Subcontractor as a result of such delay within 14 days after the Contractor has received such additional payment under the Main Contract or within 84 days after the expiry of the Subcontract Defects Notification Period, whichever is earlier.

If the delay to taking-over of the Subcontract Works is the Contractor's responsibility, any Cost incurred by the Subcontractor as a result of such delay shall subject to Sub-Clause 20.2 [Subcontractor's Claims] be recoverable from the Contractor

This Sub-Clause is likely to be the subject of controversy and potential disputes. The Subcontractor, when having achieved completion under Sub-Clause 10.1 [*Completion of Subcontract Works*], and in the absence of any provisions for earlier taking over by the Contractor under Sub-Clause 10.3 [*Taking-Over by the Contractor*], will remain responsible for his Works, and thus carry the risk, until such time that the Work has been taken over in accordance with the Main Contract Sub-Clause 10.1. This principle is confirmed within Sub-Clause 17.1 [*Subcontractor's Risks and Indemnities*]. Additionally, and in accordance with Sub-Clause 18.1 [*Subcontractor's Obligation to Insure*] there is an obligation to keep the Works insured for that period. In many projects this could be a significant time during which the

Subcontractor is fully responsible for his Works. For example a piling contractor having completed the foundation works in a large power station project may face up to 2–3 years of project risks and responsibility until taking over of his Works. Apart from the difficulties this will bring in actually quantifying such risk and responsibility at the time of submitting his Subcontractor's Offer, he must also take into account that the Subcontract Defects Notification Period as per Sub-Clause 11.2 [*Subcontract Defects Notification Period*] shall only commence on the date of such taking over. It is for this calculated effect that there is no '*deemed-taken-over*' provision, similar to the equivalent Main Contract Sub-Clause 10.1.

10.3 Taking-Over by the Contractor

If there is a requirement for the Contractor to take over the Subcontract Works before taking-over of the Main Works by the Employer, this shall be as stated in and shall be in accordance with Annex C.

The wisdom of forcing a Subcontractor to carry the Subcontract risks for an extended period beyond achieving completion should be questioned. The reason behind the provision may be linked to the mandate of achieving back-to-back provisions with the Main Contract. It may also provide some form of perceived safety for the Employer. However, there is a distinct probability that particularly smaller specialist subcontractors cannot and will not accept this risk.

Even if they do, it will inevitably lead to an increased Offer Price. This increased Price needs to be balanced against the perceived advantage of a much longer period of risk taken by the Subcontractor (as against the Contractor). The Parties should openly discuss this with the aim of reaching a mutually acceptable agreement, and have such agreement reflected in the Subcontract through a relevant amendment incorporated in the Particular Conditions to Subcontract.

FIDIC Users' Guide
ISBN 978-0-7277-5856-9

ICE Publishing: All rights reserved
http://dx.doi.org/10.1680/fug.58569.471

Chapter 84
Clause 11: Defects Liability

After taking over has been achieved the defects notification period starts. That is the period in which the Subcontractor is to repair, replace or make good any defects due to faulty workmanship or other cause attributable to him in the taken over works that have been duly reported to him. Such a period is usually limited in time, for example, 1 or 2 years depending on the type of project. The equivalent clause under the Main Contract is clause 11 [*Defects Liability*].

11.1 Subcontractor's Obligations after Taking-Over

Following taking-over of the Subcontract Works in accordance with Clause 10 [Completion of and Taking-Over the Subcontract Works], the provisions of Main Contract Clause 11.1 [Completion of Outstanding Work and Remedying Defects], Main Contract Clause 11.4 [Failure to Remedy Defects], Main Contract Clause 11.5 [Removal of Defective Work], Main Contract Clause 11.6 [Further Tests], Main Contract Clause 11.7 [Right of Access], and Main Contract Clause 11.8 [Contractor to Search] shall apply, subject to:

(a) Sub-Clause 11.2 [Subcontract Defects Notification Period]; and
(b) Sub-Clause 11.3 [Performance Certificate].

All work referred to in sub-paragraph (b) of Main Contract Clause 11.1 [Completion of Outstanding Work and Remedying Defects] shall be executed at the risk and cost of the Subcontractor, if and to the extent that the work is attributable to:

(i) Subcontract Plant, the materials or workmanship of the Subcontract Works not being in accordance with the Subcontract; or
(ii) failure by the Subcontractor to comply with any other obligation under the Subcontract.

Insofar as a defect or damage in the Subcontract Works which is the responsibility of the Subcontractor causes the Contractor to incur any additional Cost in complying with Main Contract Clauses 11.5 [Removal of Defective Work], 11.6 [Further Tests] and 11.8 [Contractor to Search], the Contractor shall subject to Sub-Clause 3.3 [Contractor's Claims under the Subcontract], be entitled to deduct this Cost from the Subcontract Price.

If and to the extent that the work referred to in sub-paragraph (b) of Main Contract Clause 11.1 [Completion of Outstanding Work and Remedying Defects] is attributable to any other cause, the Subcontractor shall be notified promptly by the Contractor, and this work shall be valued as a Subcontract Variation in accordance with Sub-Clause 13.2 [Valuation of Subcontract Variations].

Similar to the provisions for taking over after completion in accordance with Clause 10 [*Completion and Taking-Over of the Subcontract Works*], this Clause may be the cause of much controversy and disputes. The overall aim of the Task-Group was to achieve a true back-to-back scenario even though it may override the basic principle of perceived balance and fairness. The obligations of the Subcontractor, after taking-over of his works in accordance with Sub-Clause 10.1 [*Completion of Subcontract Works*], are logical and not unduly onerous. They are in accordance with the equivalent provisions of the Main Contract, and the relevant Sub-Clauses shall apply to the Subcontract. However, it is the reference to Sub-Clause 11.2 [*Subcontracts Defects Notification Period*] and 11.3 [*Performance Certificate*] that is seen by many as a cause for concern.

If any defect or damage in the Subcontract Works results in additional costs to the Contractor in complying with the Main Contract, (i.e. Sub-Clauses 11.5 [*Removal of Defective Work*], 11.6 [*Further Tests*] and 11.8 [*Contractor to Search*]), such costs can be deducted from the Subcontract Price, subject to Sub-Clause 3.3 [*Contractor's Claims under the Subcontract*]. The final extent of these Costs may only become evident after a relatively lengthy claims/ determination process resulting in significant delays and uncertainty of payment for the Subcontractor.

11.2 Subcontract Defects Notification Period

The Subcontract Defects Notification Period shall be the period for notifying defects in the Subcontract Works under Sub-Clause 11.1 [Subcontractor's Obligations after Taking-Over], which period shall be from the date on which the whole of the Subcontract Works have been taken-over under Clause 10 [Completion of and Taking-Over the Subcontract Works] to the date of expiry of the Defects Notification Period applicable to the Main Works or Section or part of the Main Works of which the whole of the Subcontract Works are part.

Insofar as a defect or damage in the Subcontract Works attributable to a Subcontractor default gives rise to an extension of any Defects Notification Period under Main Contract Clause 11.3 [Extension of Defects Notification Period], and this extension causes the Contractor to incur any additional Cost, the Contractor shall, subject to Sub-Clause 3.3 [Contractor's Claims in connection with the Subcontract], be entitled to deduct this Cost from the Subcontract Price.

This Sub-Clause is controversial because it dictates that the defects notification periods for the Subcontract and Main Contract are concurrent. This may result in significant time during which the Subcontractor is subjected to risk for the Subcontract Works (refer Sub-Clauses 17.1 [*Subcontractor's Risks and Indemnities*] and 18.1 [*Subcontractor's Obligation to Insure*]. It is important for the Parties to discuss and agree these issues transparently before signing the Subcontract, as any misinterpretations and/or claims for damages may be problematic when viewed in relation to Sub-Clauses 1.3 [*Subcontract Interpretation*], 2.1 [*Subcontractor's Knowledge of Main Contract*] and 2.2 [*Compliance with Main Contract*]. Depending on the type of project and the effective scope of the Subcontract Works in relation to the Main Contract works, it may potentially be beneficial to the Parties to agree on a different, less onerous arrangement through the Particular Conditions to Subcontract.

11.3 Performance Certificate

The Performance Certificate applicable to the Subcontract Works shall be that which is issued by the Engineer under the Main Contract. Performance of the Subcontractor's obligations shall not be considered to have been completed until the Engineer has issued the Performance Certificate to the Contractor, and the date stated in the Performance Certificate on which the Contractor's obligations under the Main Contract were completed shall be the date upon which the Subcontractor's obligations under the Subcontract were completed. Immediately upon receipt of the Performance Certificate from the Engineer, the Contractor shall forward a copy to the Subcontractor.

After the Performance Certificate has been issued, the provisions of Main Contract Clauses 11.10 [Unfulfilled Obligations] and 11.11 [Clearance of Site] shall apply.

With the issue of the Performance Certificate the Subcontractor's obligations under the Subcontract have been completed, save for any unfulfilled obligations as per the Main Contract Sub-Clause 11.10 [*Unfulfilled Obligations*] and the responsibility to clear the site of any remaining Subcontractor's Equipment, surplus Subcontract Material, wreckage, rubbish and Temporary Works as per Main Contract Sub-Clause 11.11 [*Clearance of Site*].

This Performance Certificate shall be the same certificate issued by the Engineer to the Contractor under the Main Contract Sub-Clause 11.9 [*Performance Certificate*]. The time for issuing this certificate shall be the latest of either: (i) the expiry of the Main Contract Defects Notification Period, or (ii) as soon thereafter as the Contractor has fulfilled all his obligations in terms of remedying defects, supplying all Contractor's Documents, and completing and testing the Main Contract works. This means that the final date for the Performance Certificate is not fixed and is to a large extent dependent on the Contractor's performance.

In all modern day projects the Employer will require some form of security after Taking-Over. This confirms the Contractor's compliance in remedying any defects during the defects notification period, or failing this to have funds available to have these defects repaired by others. Such a security can be in various forms, but the usual options are either the withholding of a retention fund as a percentage of the contract price or a warranty bond or guarantee similar to the performance or advance payment securities. The Subcontract makes provision for retention monies as per Sub-Clause 14.7 [*Payment of Retention Money under the Subcontract*], if and to the extent that these have been agreed and included in the Appendix to the Subcontractor's Offer. Any warranty bonds or guarantees are usually exchanged for the return of the Sub-Clause 4.2 [*Subcontract Performance Security*]. In both instances there will be a cost to the Subcontractor if the return of the Performance Certificate is delayed (e.g. opportunity costs due to retention monies being tied up longer, the obligation to extend the validity of any warranty bond or guarantee etc.). If the Subcontractor did not cause the delay, than subject to Sub-Clause 20.2 [*Subcontractor's Claims*] there may be an entitlement to recover any subsequent damages due to such delayed Performance Certificate.

FIDIC Users' Guide
ISBN 978-0-7277-5856-9

ICE Publishing: All rights reserved
http://dx.doi.org/10.1680/fug.58569.475

Chapter 85
Clause 12: Measurement and Evaluation

This Clause deals with the way the Subcontract Works are to be evaluated and measured. The equivalent clause under the Main Contract (RED and PNK books) will be Clause 12 [*Measurement and Evaluation*]. As the measurement of the Subcontract is in principle also based on a Bill of Quantities the methods used under the Main Contract have been substantially retained.

12.1 Measurement of Subcontract Works
The Subcontract Works shall be measured in accordance with Main Contract Clauses 12.1 [Works to be Measured] and 12.2 [Method of Measurement].

The Contractor shall permit the Subcontractor to attend to assist the Engineer and the Contractor in making the measurement in relation to the Subcontract Works. The Subcontractor shall supply the Contractor with any particulars requested by the Contractor and/or the Engineer. If the Subcontractor has been given reasonable notice to attend and does not attend, the measurement made by (or on behalf of) the Engineer shall be deemed to be accurate and to be accepted by the Subcontractor.

Wherever the Subcontract Works are to be measured by records, the Contractor shall permit the Subcontractor to attend with the Contractor to examine and agree the records with the Engineer. If the Subcontractor has been given reasonable notice to attend by the Contractor and does not attend, the records as agreed between the Contractor and the Engineer, or accepted by non-attendance or failure by the Contractor in accordance with Main Contract Clause 12.1 [Works to be Measured] shall be deemed to be accurate and to be accepted by the Subcontractor. If the Subcontractor examines and disagrees with the records, then he shall give notice to the Contractor within 7 days of the date of the examination of the respects in which the records are asserted to be inaccurate, which the Contractor shall then notify to the Engineer. If the Subcontractor does not give notice within the 7 days, the records shall be deemed to be accurate and to be accepted by the Subcontractor. The Contractor shall immediately notify the Subcontractor of any determination made by the Engineer in respect of disagreed records.

Notwithstanding local practice, the measurement shall be made on the net actual quantity of each item of the Subcontract Works, and the method of measurement shall be that which applies under the Main Contract.

If the Contractor does not give notice to the Subcontractor in accordance with this Sub-Clause, and/or if notice is given but it is not reasonable having regard to the notice given by the Engineer under Main Contract Clause 12.1 [Works to be Measured] and all relevant circumstances, the Contractor shall consult with the Subcontractor in an endeavour to reach agreement on the measurement of the Subcontract Works. If agreement is not reached, the Contractor shall make a fair decision as to the appropriate and applicable measurement, having due regard to the Subcontractor's views and all relevant circumstances. The Contractor shall give notice, with supporting particulars, to the Subcontractor of this decision.

In the first paragraph it is stated that the Subcontract Works shall be measured in accordance with Main Contract clauses 12.1 *[Works to be Measured]* and 12.2 *[Method of Measurement]*. This implies that the method of measurement shall be in accordance with the Main Contract Bill of Quantities or other applicable Main Contract Schedules (refer to Main Contract Sub-Clause 12.2(b)). It is thus of the utmost importance that the Subcontractor makes himself familiar with the relevant measurement methods pertinent to the Main Contract before making a formal Offer. This Main Contract method also supersedes any local practice with which the Subcontractor may be more familiar (refer to Main Contract Sub-Clause 12.1 *[Works to be Measured]* paragraph 4). Any Bill of Quantities prepared for the Subcontract should be based on the same measurement principles as the Main Contract.

The Subcontractor has a right to attend any measurement conducted by the Contractor and the Engineer. If he fails to attend then he must accept the results. This also applies if the Works are to be measured by records however, if he attends and disagrees with the results he shall give notice to the Contractor within 7 days of the date of the examination. Failure to give notice means the results of the examination are considered accurate and are accepted.

The Subcontractor's attendance in relation to this Sub-Clause is obviously dependent on the Contractor being given reasonable notice as to the date and venue. If unreasonable, or no notice is given to the Subcontractor then the Parties shall consult with each other in order to find a mutually acceptable agreement on the measurement. Failing any agreement, the Contractor shall make a fair decision having due regard to all the circumstances including the Subcontractor's views on the matter. If the Subcontractor does not agree with such notified decision he has the option of declaring a dispute and may seek a DAB resolution under Sub-Clause 20.6 *[Obtaining Subcontract's DAB decision]*.

12.2 Quantity Estimated and Quantity Executed

Any quantities set out in the Subcontract Bill of Quantities, or other schedule of rates and prices in the Subcontract, are the estimated quantities for the Subcontract Works, and they are not to be taken as the actual and correct quantities of the Subcontract Works which the Subcontractor is required to execute.

No Contractor's Instruction shall be required for any increase or decrease in the quantity of any work where the increase or decrease is not the result of an instruction given under Sub-Clause 13.1 [Variation of Subcontract Works] but is the result of the quantities exceeding or being less than those stated in the Subcontract Bill of Quantities, or other schedule of rates and prices in the Subcontract.

This Sub-Clause confirms that any quantities appearing in the Bill of Quantities are to be regarded as estimates only and any change in those quantities not related to a Contractor's Instruction given under Sub-Clause 13.1 *[Variation of Subcontract Works]* will not require a formal Contractor's Instruction for them to be varied from the original Bill of Quantities.

12.3 Evaluation under the Subcontract

The Contractor shall consult with the Subcontractor in an endeavour to reach agreement on the Subcontract Price by evaluating each item of the Subcontract Works applying:

(a) *the measurement agreed or determined in accordance with Sub-Clause 12.1 [Measurement of the Subcontract Works]; and*
(b) *the provisions of the second and third paragraphs of Main Contract Clause 12.3 [Evaluation]*

to the Subcontract.

If agreement is not reached, the Contractor shall make a fair evaluation having due regard to the Subcontractor's views and all relevant circumstances and shall promptly notify the Subcontractor of this evaluation with supporting particulars.

Each Party shall give effect to each agreement reached or evaluation made under this Sub-Clause unless and until revised under Clause 20 [Claims and Disputes].

Whenever the omission of any part of the Subcontract Works forms part (or all) of a Subcontract Variation, the value of which has not been agreed, Main Contract Clause 12.4 [Omissions] shall apply.

In ascertaining the Subcontract Price the Subcontract shall be evaluated in accordance with this Sub-Clause 12.3 *[Evaluation under the Subcontract]*. That means that the Contractor and the Subcontractor shall consult with each other to reach agreement on such Subcontract Price by applying: (i) the method of measurement in accordance with Sub-Clause 12.1 *[Measurement of Subcontract Works]* (refer to Sub-Clause 12.3(a)) and (ii) the provisions of the second and third paragraphs of Main Contract 12.3 *[Evaluation]* (refer to Sub-Clause 12.3(b)). The latter dictates that for each item of work, the appropriate rate or price as specified in the Subcontract shall be used, or, if there is no such rate or price, an equivalent value from similar work.

However, a new rate shall be used if: (i) the measured quantity varies by more than 10% (under PNK as Main Contract this value is 25%) from the stated quantity in the Subcontract Bill of Quantities or other agreed Schedule, and such quantity multiplied by such specified rate exceeds 0.01% (under PNK as Main Contract this value is 0.25%) of the Accepted Subcontract Amount, and this change in quantity directly affects the Cost per unit by more than 1% and the related Bill of Quantity item is not stated as a fixed price item, or (ii) it involves an instructed item under Clause 13 *[Subcontract Variations and Adjustments]* for which no rate or price is neither specified nor appropriate in the Subcontract. Such new rates shall be derived from any relevant rates or prices in the Subcontract, with reasonable adjustments. If this is not possible then the new rates shall be derived from reasonable cost of executing the work for which the rate is to be applied together with reasonable profit. Until such time that the Parties have an agreement on a new rate or price, the Contractor shall determine a provisional rate or price for the purpose of Interim Payment Certificates.

If there is an omission of any part of the Subcontract Works and it forms part of a Subcontract Variation (i.e. Clause 13 [*Subcontract Variations and Adjustments*]), Sub-Clause 12.4 [*Omissions*] from the Main Contract shall apply except if the Parties have reached an agreement on the cost of such an omission.

FIDIC Users' Guide
ISBN 978-0-7277-5856-9

ICE Publishing: All rights reserved
http://dx.doi.org/10.1680/fug.58569.479

Chapter 86
Clause 13: Subcontract Variations and Adjustments

This Clause includes provisions to manage changes or adjustments to the Subcontract. The equivalent clause under the Main Contract (RED and PNK books) is clause 13 [*Variations and Adjustments*].

13.1 Variation of Subcontract Works

The Subcontract Works shall be varied only by way of a Contractor's Instruction. Subcontract Variations may be initiated by the Contractor at any time prior to the date the Taking-Over Certificate is issued by the Engineer for the whole of the Main Works, either by a Contractor's Instruction or by a request for the Subcontractor to submit a proposal.

The Subcontractor shall execute and be bound by each Subcontract Variation whether instructed by:

(a) the Engineer, provided the instruction is notified to the Subcontractor as a Contractor's Instruction; or

(b) the Contractor's Subcontract Representative, as a Contractor's Instruction.

If the Subcontractor cannot readily obtain the Subcontract Goods required for the Subcontract Variation, he shall promptly give notice (with supporting particulars) to the Contractor who shall cancel, confirm or vary the Contractor's Instruction.

A Subcontract Variation may include any of the matters described in sub-paragraphs (a) to (f) of Main Contract Clause 13.1 [Right to Vary].

Any variation to the Subcontract can only be introduced prior to the date of the issued Taking-Over Certificate (refer to Sub-Clause 10.2 [*Taking-Over Subcontract Works*]) by a relevant Contractor's Instruction issued in accordance with Sub-Clause 3.1 [*Contractor's Instructions*]. The Subcontractor shall only accept such instructions from the Contractor's Subcontract Representative appointed under Sub-Clause 6.3 [*Contractor's Subcontract Representative*]. Any instruction issued directly by the Engineer to the Subcontractor is not a valid instruction under the terms of the Subcontract.

Similar to the provisions under the Main Contract, and subject to the above, the Subcontractor is bound by and shall execute any given instruction, unless he cannot readily obtain the required Goods to execute the variation. In such case he shall notify the Contractor who shall cancel, confirm or vary the instruction accordingly.

13.2 Valuation of Subcontract Variations

Each Subcontract Variation shall be evaluated in accordance with Sub-Clause 12.3 [Evaluation under the Subcontract], unless agreed otherwise by the Parties. The value of each Subcontract Variation shall then be added to or deducted from the Subcontract Price, as appropriate.

Provided that, if the Contractor instructs a Subcontract Variation at any time after the date upon which the Subcontract Works have been completed in accordance with Clause 10.1 [Completion of the Subcontract Works], the value of this Subcontract Variation shall take account of any additional Cost or liability which in the circumstances was reasonably incurred by the Subcontractor in remobilising his resources to the Site and/or in maintaining a presence on the Site in order to execute the Subcontract Variation.

As seen under Sub-Clause 10.1 [*Completion of the Subcontract Works*] it is possible for the Subcontractor to have completed his scope long before he actually achieves contractual Taking-Over. This may lead to the situation where the Subcontractor is physically no longer on the Site, but the Contractor is required to implement a Variation.

Sub-Clause 13.2, second paragraph, provides for the appropriate evaluation of such a Variation issued after completion of the Subcontract Works. In such an event the Subcontractor is entitled to be reimbursed for any remobilisation and other reasonably incurred costs due to the fact that he is no longer on the Site, or to maintain a presence on the Site purely for implementing the Variation.

13.3 Request for Proposal for Subcontract Variation

If the Contractor requests a proposal, or if the Contractor notifies the Subcontractor that the Engineer has requested a proposal, prior to instructing a Subcontract Variation, the Subcontractor shall respond in writing as soon as practicable, either by giving reasons why he cannot comply (if this is the case) or by submitting the Subcontractor's proposal:

(a) giving a description of the proposed work to be performed and a programme for its execution;

(b) stating any necessary modifications to the Subcontract Programme pursuant to Sub-Clause 8.3 [Subcontract Programme] and to the Subcontract Time for Completion; and

(c) evaluation of the Subcontract Variation.

The Contractor shall, as soon as practicable after receiving this proposal respond with approval, disapproval or comments.

The Subcontractor shall not delay any work while awaiting a response in respect of any proposal submitted under this Sub-Clause.

The equivalent Sub-Clause under the Main Contract (RED and PNK books) is Sub-Clause 13.3 [*Variation Procedure*].

This procedure provides for the Contractor to request a relevant proposal by the Subcontractor prior to instructing the Variation.

13.4 Subcontract Adjustments for Changes in Legislation

The Subcontract Price shall be adjusted to take account of any increase or decrease in the Cost incurred by the Subcontractor under the Subcontract resulting from a change in the Laws of the Country, or in the judicial or official governmental interpretation of the Laws, made after the date of submission of the Subcontractor's Offer which affects the Subcontractor's performance of his obligations under the Subcontract. If the Subcontractor suffers delay and/or incurs Cost as a result of such a change in the Laws, the Subcontractor shall give notice to the Contractor and shall be entitled, subject to Sub-Clause 20.2 [Subcontractor's Claims] to an extension of time under Sub-Clause 8.4 [Extension of Subcontract Time for Completion] and payment of any Cost which shall be included in the Subcontract Price.

The equivalent Sub-Clause under the Main Contract (RED and PNK books) is Sub-Clause 13.7 [*Adjustment for Changes in Legislation*].

13.5 Subcontract Adjustments for Changes in Cost

Where Main Contract Clause 13.8 [Adjustments for Changes in Cost] applies to the Main Contract, the Subcontract Price shall be adjusted for rises or falls in the cost of labour, Subcontract Goods, materials, and any inputs to the Subcontract Works. The adjustment shall be calculated using the formula set out in the third paragraph of Main Contract Clause 13.8 [Adjustments for Changes in Cost] and the data contained in the complete table of adjustments data included in the Appendix to the Subcontractor's Offer.

If there is no table of adjustment data included in the Appendix to the Subcontractor's Offer, then the Subcontract Price shall be adjusted by the same adjustment multiplier as applies to adjustment of the Contract Price under Main Contract Clause 13.8 [Adjustments for Changes in Cost].

The equivalent Sub-Clause under the Main Contract (RED and PNK books) is Sub-Clause 13.8 [*Adjustment for Changes in Cost*] which is only applicable if and to the extent that the Appendix to Tender (or Contract Data under the PNK Book) is duly completed with the relevant adjustment data. If no such data is included the Sub-Clause shall not apply. Thus, the application under the Subcontract is dependent on whether the Main Contract Sub-Clause applies or not. If it applies than any changes to the Subcontract Price shall be ascertained using the formulas dictated by the Main Contract Sub-Clause 13.8 [*Adjustments for Changes in Cost*] third paragraph.

13.6 Subcontract Daywork

Where a daywork schedule is included in the Subcontract, the Contractor may instruct that a Subcontract Variation shall be executed on a daywork basis. The work shall then be evaluated in accordance with the daywork schedule included in the Subcontract, and the procedure set out in Main Contract Clause 13.6 [Daywork].

If the Parties included a '*daywork schedule*' in the Subcontract, the Contractor may decide to instruct any or all Variations to be executed and evaluated in accordance with this '*daywork schedule*'. In such case the procedure under the equivalent Main Contract Sub-Clause 13.6 [*Daywork*] shall apply. The Subcontractor must obtain quotations prior to ordering Goods, and submit daily statements. These statements are submitted in duplicate for signature by the Contractor and shall include: (i) the names, occupations and time of Personnel, (ii) identification, type and time of Equipment and Temporary Works and (iii) the quantities and types of Plant and Materials used.

FIDIC Users' Guide
ISBN 978-0-7277-5856-9

ICE Publishing: All rights reserved
http://dx.doi.org/10.1680/fug.58569.483

Chapter 87
Clause 14: Subcontract Price and Payment

Similar to the Main Contract Clause 14 [*Contract Price and Payment*], this Clause is argu-ably the most important for both Parties because it regulates the commercial procedures including payment.

14.1 The Subcontract Price
The Subcontract Price shall be agreed or decided in accordance with Sub-Clause 12.3 [Evaluation under the Subcontract] and be subject to adjustments in accordance with the Subcontract. The Subcontractor shall pay all taxes, duties and fees required to be paid by him under the Subcontract, and the Subcontract Price shall not be adjusted for any of these costs unless there is provision for the Contract Price to be so adjusted under the Main Contract, in which case the Subcontract Price shall be adjusted to the like extent that the Contract Price shall be adjusted under the Main Contract.

The Subcontractor shall submit to the Contractor, within 7 days of the Contractor's request to do so, a breakdown of each lump-sum price in the Subcontract. The Contractor may take account of such a breakdown when considering the Subcontractor's monthly statements submitted under Sub-Clause 14.3 [Subcontractor's Monthly Statements].

The Subcontract Price is deemed to include any taxes, duties and fees required for the execution of the Subcontract Works, unless these costs are reimbursable to the Contractor under the Main Contract. In the latter event the Subcontractor shall be reimbursed to the like extent that the Main Contract Price shall be adjusted. In theory this seems plausible, but the practical implementation could be difficult. As per Sub-Clause 2.1 [*Subcontractor's Knowledge of Main Contract*] the Contractor has the option to withhold certain information from the Subcontractor. Such information is listed as confidential in Part A of Annex A. This confidential information is likely to apply to any price contained within the Main Contract. Therefore, if there is to be any reimbursement on a like for like basis, the Subcon-tractor has no real opportunity to verify what this would be for lack of pertinent data listed as confidential. If it is likely that there shall be any reimbursement for these costs, the Parties are advised to agree on a mutually acceptable method of computing and verifying this. Such agreement can than be included as Particular Conditions to Subcontract.

14.2 Subcontract Advance Payment
Where there is no amount for advance payment stated in the Appendix to the Subcontractor's Offer, the Contractor shall have no obligation to make an advance payment to the Subcontractor.

483

The advance payment shall be made in the instalments and in the applicable currencies and proportions stated in the Appendix to the Subcontractor's Offer.

The advance payment guarantee shall be in the amounts and currencies corresponding to the advance payment, shall be in a similar form to that of the guarantee approved by the Engineer under Main Contract Clause 14.2 [Advance Payment], or in the form that may be agreed between the Contractor and the Subcontractor, and shall be issued by an entity and from within a country (or other jurisdiction) approved by the Contractor. The Subcontractor shall ensure that this guarantee is valid and enforceable until the advance payment has been repaid, but its amount may be progressively reduced by the amount repaid by the Subcontractor as deductions in payments otherwise due.

The advance payment shall be repaid through percentage deductions in interim payments made to the Subcontractor under Sub-Clause 14.6 [Interim Subcontract Payments] at the amortisation rate of one-fifth (20%) of the amount of each interim payment (excluding the advance payment and deductions and repayments of retention) in the currencies and proportions of the advance payment, until such time as the advance payment has been repaid. Deductions shall commence when the total of all interim payments made to the Subcontractor (excluding the advance payment and deductions and repayments of retention) exceeds twenty percent (20%) of the Accepted Subcontract Amount.

If the advance payment has not been repaid prior to the taking-over of the Subcontract Works or prior to termination of the Subcontract, the whole of the balance then outstanding shall immediately become due and payable by the Subcontractor to the Contractor.

Cash flow is important for any contractor, and the requirements for the Subcontractor are no different. For this reason the various FIDIC Standard Forms of Contract make provision for such a procedure. The equivalent Sub-Clause under the Main Contract (RED and PNK books) is 14.2 [*Advance Payment*].

For this Sub-Clause to apply, the amount of the advance payment, representing a percentage of the Accepted Subcontract Amount, should be stated in the Appendix to the Subcontractor's Offer. If the Accepted Subcontract Amount is in more than one currency, then the applicable currencies and proportions of the advance payment should also be stated in the Appendix to the Subcontractor's Offer. If the number and timing of instalments of the advance payment are to differ from those stated in the Appendix to Tender/Contract Data of the Main Contract, then these should also be stated in the Appendix to the Subcontractor's Offer.

If the form of the Subcontract advance payment guarantee is not to be similar to that of the Advance Payment Guarantee under the Main Contract, then the acceptable form of Subcontract advance payment guarantee should be included in the tender documents annexed to the Particular Conditions, and this Sub-Clause should be amended accordingly. The example form of advance payment guarantee annexed at Annex E to the *Guidance for the Preparation of Particular Conditions* published with the FIDIC *Conditions of Contract for Construction for Building and Engineering Works Designed by the Employer* (first edition, 1999) may be used as a basis for the acceptable form of Subcontract advance payment guarantee.

14.3 Subcontractor's Monthly Statements

The Subcontractor shall prepare and submit monthly statements to the Contractor showing in detail the amounts to which the Subcontractor considers himself to be entitled, together with supporting documents which shall include the report on the progress during this month in accordance with Sub-Clause 8.5 [Subcontract Progress Reports]. Each statement shall be submitted no later than 7 days before the due date for submission of the Contractor's Statement under Main Contract Clause 14.3 [Application for Interim Payment Certificates], which due date shall be notified to the Subcontractor by the Contractor.

Each Subcontractor's monthly statement shall include the items as set out in subparagraphs (a) to (d) inclusive and sub-paragraphs (f) and (g) of Main Contract Clause 14.3 [Application for Interim Payment Certificates].

Just as an advance payment is critical to a positive cash flow, so is the regular income from work performed. This Sub-Clause provides for monthly statements from the Subcontractor, with supporting documents (e.g. report on progress as per Sub-Clause 8.5 [*Subcontract Progress Reports*]) indicating the amount to which entitlement is considered. The statement shall be submitted to the Contractor no later than 7 days before the Contractor submits his Contractor's Statement under the Main Contract Sub-Clause 14.3 [*Application for Interim Payment Certificates*]. This is misleading, as the said Sub-Clause under the Main Contract does not stipulate timing for the Contractor to submit his statements. It merely states that this shall be done after each month end. However the Contractor is obligated to notify the Subcontractor of any due date he may have, and this could be a self-imposed date. In accordance with Sub-Clause 14.5 [*Contractor's Application for Interim Payment Certificate*], the Contractor is only obligated to include any Subcontractor Statement for Interim Payment if the Subcontractor has complied with the provisions of Sub-Clause 14.3, for example, the timely submission of the monthly statements.

Interestingly the option to receive payment for materials delivered to Site before they are incorporated into the Permanent Works is expressly omitted from the Subcontract. This is in comparison to the Main Contract where payment is an option under Sub-Clause 14.3(e) if included into the Appendix to Tender (or Contract Data under PNK). There is thus a possibility for an astute Contractor to arrange for an alternative positive cash flow by including these payments into his own monthly statements, but without the obligation to pay the Subcontractor accordingly.

14.4 Subcontractor's Statement at Completion

The Subcontractor shall submit a statement no later than 7 days before the due date for submission of the Contractor's Statement at Completion under Main Contract Clause 14.10 [Statement at Completion]. Provided that, if Sub-Clause 10.3 [Taking-Over by the Contractor] applies, the Subcontractor shall submit a statement no later than 28 days after the whole of the Subcontract Works have been taken-over by the Contractor. The Subcontractor's statement shall show the information relevant to the Subcontract as set out in sub-paragraphs (a) to (c) inclusive of Main Contract Clause 14.10 [Statement at Completion].

The Main Contract Sub-Clause 14.10 [*Statement at Completion*] obligates the Contractor to submit his corresponding Statement at Completion within 84 days after receiving the

Taking-Over Certificate for the Main Contract Works. This would include a corresponding Statement by the Subcontractor for the Subcontract portion of the Main Contract Works. By implication the Subcontractor shall thus submit his Statement to the Contractor within 77 days after the Contractor received the Taking-Over Certificate. According to the Main Contract Sub-Clause 10.1 [*Taking Over of the Works and Sections*] there is a strict procedure for the Taking-Over certificate to be issued, or deemed to be issued.

If the Taking-Over Certificate is '*deemed to be issued*' then the issue date is not specific. In order for the Subcontractor to comply with his obligations he must necessarily be informed about the date on which the Contractor received the Certificate. The Parties are therefore advised to make necessary amendments through the Particular Conditions to Subcontract to avoid any future misunderstanding or subsequent disputes. This contractual obligation is however not sanctioned, and it is unclear as to what the repercussions for the Subcontractor would be, other than a breach of this Sub-Clause.

14.5 Contractor's Application for Interim Payment Certificate

The Contractor shall make appropriate provision for the amounts set out in each Subcontractor's monthly statement in the Contractor's next Statement under Main Contract Clause 14.3 [Application for Interim Payment Certificates]. This obligation shall be subject to the Subcontractor having submitted his monthly statement to the Contractor in accordance with Sub-Clause 14.3 [Subcontractor's Monthly Statements].

In respect of any Subcontractor's monthly statement and if so requested by the Subcontractor, the Contractor shall advise the Subcontractor without delay and providing substantiating documentation of the date on which:

the Contractor's next Statement was actually submitted to the Engineer in accordance with Main Contract Clause 14.3 [Application for Interim Payment Certificates];

and

the Contractor received an Interim Payment Certificate issued by the Engineer under Main Contract Clause 14.3 [Issue of Interim Payment Certificates] for that next Statement.

If, after receiving such a request from the Subcontractor, the Contractor fails to provide information as to whether and when the Engineer issued such an Interim Payment Certificate, such Interim Certificate shall be deemed to have been issued 35 days after the due date for submission of the Contractor's Statement under Main Contract Clause 14.3 [Application for Interim Payment Certificates].

Provided the Subcontractor has submitted his monthly Statement in accordance with Sub-Clause 14.3 [*Subcontractor's Monthly Statement*], the Contractor shall include this in his next Statement under the Main Contract. Upon the request of the Subcontractor, the Contractor shall furnish detailed substantiating information on when he actually submitted his Statement under the Main Contract, and if and when the Engineer issued a corresponding Interim Payment Certificate. As per Sub-Clause 1.6 [*Notices, Consents, Approvals,*

Certificates, Confirmations, Decisions and Determinations] such request shall be in writing. If the Contractor does not reply adequately to such a request the corresponding Interim Payment Certificate to be issued under the Main Contract is, for the purposes of the Subcontract, deemed to be issued 35 days after the due date under the Main Contract for the Contractor's Statement. Again it becomes imperative for the Subcontractor to know what this date is, and he should insist on the Contractor's compliance in this regard under Sub-Clause 14.3 [*Subcontractor's Monthly Statements*].

14.6 Interim Subcontract Payments
Within 70 days of receipt by the Contractor of:

(a) *the Subcontractor's monthly statement submitted under Sub-Clause 14.3
    [Subcontractor's Monthly Statements], or*
(b) *the Subcontractor's Statement at Completion under Sub-Clause 14.4 [Subcontractor's
    Statement at Completion],*

the amounts included in that statement, and any other sums to which the Subcontractor is entitled in the opinion of the Contractor, shall be due and payable to the Subcontractor. If a percentage of retention is stated in the Appendix to the Subcontractor's Offer, the Contractor shall be entitled to deduct an amount for retention, calculated by applying this percentage of retention to the payment otherwise due to the Subcontractor, until the amount so retained by the Contractor reaches the limit of retention money (if any) stated in the Appendix to the Subcontractor's Offer.

Provided that, subject to the obligation to pay pursuant to the last paragraph of Sub-Clause 14.8 [Final Subcontract Payment], the Contractor shall be entitled to withhold or defer payment of any sums in a Subcontractor's monthly statement to the extent that he notifies the Subcontractor within 70 days of his receipt of that monthly statement with reasons that:

(a) *where a minimum amount is stated in the Appendix to the Subcontractor's Offer, the amounts included in the Subcontractor's monthly statement together with any sums to which the Subcontractor might otherwise be entitled in the opinion of the Contractor, but after all retentions (if any) and deductions, are less in the aggregate than this minimum amount. Save that the aforesaid shall not apply to the amounts in the Subcontractor's monthly statement which are included in the Contractor's final statement under Main Contract Clause 14.11 [Application for Final Payment Certificate];*
(b) *the amounts included in the Subcontractor's monthly statement are not certified by the Engineer and only to the extent that the sum is not certified, provided that the Contractor has complied with Sub-Clause 14.5 [Contractor's Application for Interim Payment Certificate] in respect of the Subcontractor's monthly statement and provided that such failure to certify is not due to any act or default of the Contractor under the Main Contract;*
(c) *the amounts included in the Subcontractor's monthly statement have been certified by the Engineer but the Employer has failed to make payment in full to the Contractor in respect of these amounts and only to the extent of such non-payment; or*

(d) the Contractor fairly considers that any sum in the Subcontractor's monthly statement is not due in accordance with the Subcontract and/or the Contractor is entitled to deduct any sum pursuant to Sub-Clause 3.3 [Contractor's Claims in connection with the Subcontract].

Provided that the Contractor shall have no entitlement to defer or withhold payment of any sum in a Subcontractor's monthly statement under sub-paragraph (c) of this Sub-Clause if the failure by the Employer to make payment is due to:

(i) any act or default of the Contractor under the Main Contract; and/or

(ii) the Employer's bankruptcy or insolvency, going into liquidation, having a receiving or administration order made against him, compounding with his creditors, or carrying a business under a receiver, trustee or manager for the benefit of his creditors, or by any act which (under applicable Laws) has a similar effect to any of these acts or events.

In his notice to the Subcontractor that payment of any sum in a Subcontractor's monthly statement is to be withheld or deferred under sub-paragraph (b) or (c) of this Sub-Clause, the Contractor shall give full particulars and provide substantiating documentation of the amounts that have not been certified by the Engineer or the amounts for which the Employer has failed to make payment.

The Contractor shall not be entitled to defer or withhold payment of any sum in a Subcontractor's monthly statement unless the Subcontractor has first been notified in accordance with this Sub-Clause.

Payment by the Contractor to the Subcontractor of any amount in a Subcontractor's monthly statement which has previously been withheld or deferred by the Contractor shall be due 7 days after receipt by the Contractor of any payment from the Employer which includes a sum in respect of this amount.

If the Subcontractor is under obligation to provide a Subcontract Performance Security under the Subcontract, notwithstanding the terms of this Sub-Clause or any other term of the Subcontract, no amount shall become due and payable to the Subcontractor until the security in accordance with Sub-Clause 4.2 [Subcontract Performance Security] has been delivered to the Contractor.

Together with Clauses 10 [*Completion and taking-Over of the Subcontract Works*] and 11 [*Defects Liability*] this Sub-Clause 14.6 [*Interim Subcontract Payments*] is undoubtedly the most controversial provision of the Subcontract. This is because the wording provides that the Subcontractor shall be paid under the Subcontract only when payment to the Contractor is certified/made under the Main Contract, commonly known as a '*pay-when-paid*' provision. This means that, under certain circumstances, interim payments to the Subcontractor can be delayed until such time as the Employer pays the Contractor.

FIDIC stresses that this is not a '*pay-if-paid*' provision, as the Contractor can only withhold or defer payment to the Subcontractor until the date that is 84 days after the expiry of the

Subcontract Defects Notification Period, regardless of whether or not the Contractor is paid by the Employer, as provided by the last paragraph of Sub-Clause 14.8 [*Final Subcontract Payment*] of the General Conditions. This is only a theoretical reprieve for the Subcontractor, as the Expiry of the Subcontract Defects Notification Period can be a substantial period after Completion of the Subcontract Works. Also the Sub-Clause 14.9 [*Delayed Payment under the Subcontract*] provision for financing charges to be applied to any such deferred or withheld payment is in reality not a solution for the Subcontractor, as such financing charges are again payable when the deferred payment is also made.

Many legal jurisdictions do not allow this '*pay-when-paid*' provision. In those legal jurisdictions where the '*pay-when-paid*' terms are unenforceable, this Sub-Clause, and Sub-Clause 14.7 [*Payment of Retention Money under the Subcontract*], Sub-Clause 14.8 [*Final Subcontract Payment*], Sub-Clause 14.9 [*Delayed Payment under the Subcontract*] and Sub-Clause 15.3 [*Payment after Termination of Main Contract*] should be changed.

The Parties are advised to carefully consider the terms and implications of this Sub-Clause before concluding the Agreement. Under the section *Guidance for the Preparation of Particular Conditions of Subcontract* are included recommended amendments to the various affected Sub-Clauses for when the '*pay-when-paid*' provision is unenforceable or considered to be unsuitable.

14.7 Payment of Retention Money under the Subcontract

Where a percentage of retention is stated in the Appendix to the Subcontractor's Offer, the Contractor shall pay to the Subcontractor the retention money under the Subcontract in the same proportions that apply to the Retention Money under Main Contract Clause 14.9 [Payment of Retention Money], no later than 14 days after the Contractor has received payment from the Employer.

Provided that if Sub-Clause 10.3 [Taking-Over by the Contractor] applies:

(a) *to the whole of the Subcontract Works, the Contractor shall pay the Subcontractor the first half of the retention money under the Subcontract no later than 28 days alter the whole of the Subcontract Works have been (or are deemed to have been) taken-over by the Contractor; or*

(b) *to a part of the Subcontract Works, the Contractor shall pay the Subcontractor a proportion of the retention money under the Subcontract no later than 28 days after that part has been (or is deemed to have been) taken-over by the Contractor. This proportion shall be 40% of the proportion calculated by dividing the estimated subcontract value of the part by the estimated final Subcontract Price; and*

the Contractor shall pay the Subcontractor the outstanding balance of the retention money under the Subcontract no later than 7 days after the expiry of the Subcontract Defects Notification Period. However, if any work remains to be executed under Clause 11 [Defects Liability], the Contractor shall be entitled to withhold payment of the estimated cost of this work until it has been executed.

Many Employers nowadays favour a form of money guarantee or other similar instrument over a cash retention fund. If a money guarantee is to be considered, an additional Sub-Clause may be added. The example form of retention money guarantee annexed at Annex F to the *Guidance for the Preparation of Particular Conditions* published with the RED Book, may be used as a basis for an acceptable form. Again *Guidance for the Preparation of Particular Conditions of Subcontract* provide sample wording for suitable amendments for the parties to consider.

14.8 Final Subcontract Payment

The Subcontractor shall prepare and submit a draft final statement to the Contractor stating the sum which in the Subcontractor's opinion is the Subcontract Price finally due, showing in detail the value of all work done in accordance with the Subcontract, and any further sums which the Subcontractor considers to be due to him, together with supporting documents. This draft final statement shall be submitted no later than 28 days after the expiry of the Subcontract Defects Notification Period.

If the Contractor cannot verify any part of the Subcontractor's final statement, the Subcontractor shall submit the further information that the Contractor may reasonably require.

Within 56 days after the expiry of the Subcontract Defects Notification Period the Contractor shall pay to the Subcontractor balance of the Subcontract Price finally due.

Similar to Sub-Clause 14.3 [*Subcontractor's Monthly Statements*] the Subcontractor shall submit to the Contractor a draft final statement outlining what he considers himself to be entitled to, including verifiable valuations of work done in accordance with the Subcontract.

14.9 Delayed Payment under the Subcontract

If the Contractor fails to make payment of any sum properly due and payable to the Subcontractor in accordance with this Clause 14 [Subcontract Price and Payment], then Main Contract Clause 14.8 [Delayed Payment] shall apply provided that the period of delay shall be deemed to commence on the date for payment specified in this Clause 14 [Subcontract Price and Payment].

If payment is withheld or deferred pursuant to sub-paragraph (c) of Sub-Clause 14.6 [Interim Subcontract Payment], then the Contractor shall pay the Subcontractor the amount of financing charges applicable to the overdue sum at the rate payable by the Employer to the Contractor under Main Contract Clause 14.8 [Delayed Payment], provided that the period of delay shall be deemed to commence on the date for payment specified in Sub-Clause 14.6 [Interim Subcontract Payments]. Payment by the Contractor to the Subcontractor of this amount of financing charges for withheld or deferred payment shall be due within 14 days of payment by the Contractor to the Subcontractor of the amount which was previously withheld or deferred, or within 7 days after receipt by the Contractor of any payment from the Employer in accordance with Main Contract Clause 14.8 [Delayed Payment] which includes a sum in respect of this amount, whichever is earlier.

Any delayed payment to the Subcontractor will attract financing charges in accordance with Main Contract Sub-Clause 14.8 [*Delayed Payment*]. Such entitlement shall be automatic and without formal notice under Sub-Clause 20.2 [*Subcontractor's Claims*].

14.10 Cessation of the Contractor's Liability

The Contractor shall not be liable to the Subcontractor for any matter or thing arising out of or in connection with the Subcontract or execution of the Subcontract Works as from the date stated in the Performance Certificate issued under the Main Contract, which date shall be promptly notified to the Subcontractor by the Contractor, except to the extent that the Subcontractor has given a notice of claim in accordance with Sub-Clause 20.2 [Subcontractor's Claims] prior to that date.

The Subcontractor shall remain entitled to payment for the fulfilment of any obligation which remains unperformed after the date stated in the Performance Certificate under the Main Contract.

The equivalent Sub-Clause under the Main Contract (RED or PNK books) is Sub-Clause 14.14 [*Cessation of Employer's Liability*].

Any claims the Subcontractor wishes to lodge in accordance with Sub-Clause 20.2 [*Subcontractor's Claims*] must be submitted before the issue of the Performance Certificate under the Main Contract.

14.11 Subcontract Currencies of Payment

The Subcontract Price and any amount due under this Clause 14 [Subcontract Price and Payment] shall be paid in the currency or currencies stated in the Appendix to the Subcontractor's Offer.

The equivalent Sub-Clause under the Main Contract (RED or PNK books) is 14.15 [*Currencies of Payment*].

If the Subcontractor has any requirements for his work to be paid in certain currencies (can be more than one) the details of this shall be included in the Appendix to the Subcontractor's Offer.

FIDIC Users' Guide
ISBN 978-0-7277-5856-9

ICE Publishing: All rights reserved
http://dx.doi.org/10.1680/fug.58569.493

Chapter 88
Clause 15: Termination of the Main Contract and Termination of the Subcontract by the Contractor

This Clause deals with the issue of terminating the Subcontract by the Contractor or if the Main Contract is terminated. The equivalent clause under the main Contract is Clause 15 [*Termination by Employer*].

15.1 Termination of Main Contract
If the Main Contract is terminated or the Contractor and/or the Employer is released from performance of the Main Contract under Main Contract Clause 19.7 [Release from Performance under the Law], then the Contractor may by notice to the Subcontractor terminate the Subcontract immediately, save where he is required to assign the Subcontract to the Employer in accordance with sub-paragraph (b) of Sub-Clause 5.1 [Assignment of Subcontract].

Within 7 days after the date the Employer returns the Performance Security under the Main Contract, which date shall be promptly notified to the Subcontractor by the Contractor, or within 28 days after a notice of termination under this Sub-Clause has taken effect, whichever is earlier, the Contractor shall return the Subcontract Performance Security to the Subcontractor, unless the Main Contract has been terminated as a consequence of any breach of the Subcontract by the Subcontractor.

This Sub-Clause gives the Contractor certain rights to terminate the Subcontract in the event that: (i) the Contractor or the Employer are released from performance under Main Contract Sub-Clause 19.7 [*Release from Performance under Law*], or (ii) the Main Contract is terminated. This right to terminate the Subcontract is not applicable if the Employer, in his termination of the Main Contract, instructed the Contractor to have the Subcontract assigned to him under Sub-Clause 5.1(b) [*Assignment of Subcontract*]. Upon termination under this Sub-Clause the Contractor shall return the Subcontract Performance Security to the Subcontractor. Any payments due after such termination shall be in accordance with Sub-Clause 15.3 [*Payment after Termination of Main Contract*].

15.2 Valuation at Date of Subcontract Termination
As soon as practicable after a notice of termination under Sub-Clause 15.1 [Termination of Main Contract] or Sub-Clause 15.6 [Termination of Subcontract by the Contractor] has

taken effect, the Contractor shall promptly evaluate the Subcontract Works, Subcontract Goods and Subcontractor's Documents, and any other sums due to the Subcontractor for work executed in accordance with the Subcontract whether on or off the Site.

If the Subcontract is terminated under Sub-Clause 15.1 [Termination of Main Contract], then the Contractor's evaluation shall have regard to the Engineer's valuation under Main Contract Clause 15.3 [Valuation at Date of Termination].

If the Subcontract is terminated under Sub-Clause 15.6 [Termination of Subcontract by the Contractor] then the Contractor's evaluation shall have regard to Sub-Clause 12.3 [Evaluation under the Subcontract].

The Contractor shall give notice, with supporting particulars, to the Subcontractor of this evaluation.

This Sub-Clause provides additional provisions to Sub-Clause 12.3 [*Evaluation under the Subcontract*] in the event of a termination of the Subcontract.

15.3 Payment after Termination of Main Contract

If the termination of the Subcontract arises from termination of the Main Contract or from the release of performance by the Contractor and/or Employer of the Main Contract under Main Contract Clause 19.7 [Release from Performance under the Law], the following shall be due and payable to the Subcontractor, insofar as these amounts or Costs have not been covered by payments already made to the Subcontractor:

(a) *the value of the Subcontract Works, Subcontract Goods and Subcontractor's Documents, and any other sums due to the Subcontractor for work executed in accordance with the Subcontract, as evaluated in accordance with Sub-Clause 15.2 [Valuation at Date of Subcontract Termination];*

(b) *the Cost of removal of the Subcontractor's Equipment and temporary works from the Site and, if required by the Subcontractor, return thereof to the Subcontractor in his country, or to any other destination at no greater cost;*

(c) *the reasonable Cost of repatriation of all the Subcontractor's staff and labour employed wholly in connection with the Subcontract Works at the date of termination;*

(d) *any other Cost or liability which in the circumstances was reasonably incurred by the Subcontractor in the expectation of completing the Subcontract Works; and*

(e) *any loss of profit or other loss or damage sustained by the Subcontractor as a result of this termination.*

If the Main Contract has been terminated under Main Contract Clause 15.2 [Termination by Employer], unless the Main Contract has been terminated as a consequence of any breach of the Subcontract by the Subcontractor, then payment by the Contractor of any of these amounts or Costs shall be due immediately after termination of the Subcontract.

If Main Contract Clause 19.7 [Release from Performance under the Law] applies to the Main Contract or the Main Contract has been terminated under Main Contract Clause 16.2

[Termination by Contractor] or Main Contract Clause 19.6 [Optional Termination, Payment and Release], then payment by the Contractor to the Subcontractor of any of these amounts or Costs shall be due 7 days after receipt by the Contractor of any payment from the Employer which includes a sum in respect of this amount or 112 days after termination of the Subcontract, whichever is earlier. Any Subcontract Plant or materials shall become the property of (and be at the risk of) the Employer when paid for by the Employer and the Subcontractor shall place the same at the Employer's disposal.

Provided that:

(i) *if the Main Contract has been terminated as a consequence of any breach of the Subcontract by the Subcontractor, no payment shall become due under this Sub-Clause and Sub-Clause 15.4 [Termination of the Main Contract in Consequence of Subcontractor Breach] shall apply;*

(ii) *if the Main Contract has been terminated under Main Contract Clause 15.5 [Employer's Entitlement to Termination] or under Main Contract Clause 19.6 [Optional Termination, Payment and Release], or under Main Contract Clause 19.7 [Release from Performance under the Law] then the Subcontractor shall have no entitlement to payment of any loss of profit; and*

(iii) *nothing in this Sub-Clause shall affect any right of either Party to receive payment in respect of any breach of the Subcontract committed by the other Party prior to the termination, or any other right to payment under Subcontract.*

This Sub-Clause regulates what payments shall be due in the event of the Subcontract being terminated as a consequence that is not the responsibility of the Subcontractor. Under item (e) any loss of profit or other loss or damage as a result of the termination shall be payable. This may be unenforceable under certain legal jurisdictions and the wording of this Sub-Clause should be checked for compliance in this regard before inviting tenders.

15.4 Termination of Main Contract in Consequence of Subcontractor Breach

If the Main Contract is terminated as a consequence of any breach of the Subcontract by the Subcontractor, the Subcontractor shall be entitled to payment of:

(a) *the value of the Subcontract Works, Subcontract Goods and Subcontractor's Documents, and any other sums due to the Subcontractor for work executed in accordance with the Subcontract, as evaluated in accordance with Sub-Clause 15.2 [Valuation at Date of Subcontract Termination], insofar as this value has not already been covered by payments made to the Subcontractor; less*

(b) *any amounts recovered by the Employer from the Contractor under Main Contract Clause 15.4 [Payment after Termination] in respect of any extra Cost in executing, completing and remedying of any defects, damages for delay, and all other costs incurred by the Employer in completion of the Subcontract Works; and less*

(c) *any losses and damages incurred by the Contractor as a result of such termination of the Main Contract and any liabilities incurred by the Contractor to other subcontractors which are attributable to such termination of the Main Contract.*

Provided that this payment by the Contractor shall not become due until after the amounts under sub-paragraph (b) above have been ascertained under the Main Contract, which amounts shall be promptly notified to the Subcontractor by the Contractor.

If the amount of sub-paragraph (a) above is less than the aggregate sum of the amounts of sub-paragraphs (b) and (c) above, the Contractor shall be entitled to recover the balance from the Subcontractor.

This Sub-Clause regulates what payments shall be due in the event of the Subcontract being terminated as a consequence that is the responsibility of the Subcontractor. Again these provisions may be unenforceable under certain legal jurisdictions and the wording of this Sub-Clause should be checked for compliance in this regard before inviting tenders.

15.5 Notice to Correct under the Subcontract

If the Subcontractor fails to carry out any obligation under the Subcontract, the Contractor may by notice require the Subcontractor to make good the failure and to remedy it within a reasonable time specified by the Contractor.

The equivalent Sub-Clause under the Main Contract (RED or PNK books) is Sub-Clause 15.1 [*Notice to Correct*].

15.6 Termination of Subcontract by the Contractor

The Contractor shall be entitled to terminate the Subcontract, without prejudice to any other rights or remedies under the Subcontract or otherwise, if any one or more of the events or circumstances set out in sub-paragraphs (a) to (f) inclusive of Main Contract Clause 15.2 [Termination by Employer] are applicable to the Subcontractor's performance under the Subcontract.

In any of these events or circumstances the Contractor may, upon giving 14 days' notice to the Subcontractor, terminate the Subcontract and expel the Subcontractor from the Site, except in the case of sub-paragraph (e) or (f) of Main Contract Clause 15.2 [Termination by Employer] when the Contractor may by notice terminate the Subcontract immediately. In lieu of terminating the Subcontract, the Contractor may take part only of the Subcontract Works out of the hands of the Subcontractor and may execute and complete, or have executed and completed by others, this part of the Subcontract Works. In this event the Contractor may recover his Cost of so doing from the Subcontractor.

If the Contractor terminates the Subcontract in accordance with this Sub-Clause, the provisions of the last three paragraphs of Main Contract Clause 15.2 [Termination by Employer] shall apply.

If the Contractor terminates the Subcontract in accordance with this Sub-Clause the Subcontractor shall be entitled to payment of:

(i) the value of the Subcontract Works, Subcontract Goods and Subcontractor's Documents, and any other sums due to the Subcontractor for work executed in

accordance with the Subcontract, as evaluated in accordance with Sub-Clause 15.2 [Valuation at Date of Subcontract Termination], insofar as this value has not already been covered by payments made to the Subcontractor; less

(ii)   the amount of any losses and damages incurred by the Contractor, to the extent that such losses and damages are attributable to such termination of the Subcontract, and any extra Cost in executing, completing and remedying of any defects, damages for delay, and all other costs incurred by the Contractor in completion of the Subcontract Works.

Provided that this payment by the Contractor shall not become due until after the amount under sub-paragraph (ii) above has been ascertained.

If the amount of sub-paragraph (i) above is less than the amount of sub-paragraph (ii) above, the Contractor shall be entitled to recover the balance from the Subcontractor.

Provided that nothing in this Sub-Clause shall affect the right of either Party to receive payment in respect of any breach of the Subcontract committed by the other Party prior to the termination, or any other right to payment specified under the Subcontract.

The equivalent Sub-Clause under the Main Contract (RED or PNK books) is Sub-Clause 15.2 [Termination by Employer].

FIDIC Users' Guide
ISBN 978-0-7277-5856-9

ICE Publishing: All rights reserved
http://dx.doi.org/10.1680/fug.58569.499

Chapter 89
Clause 16: Suspension and Termination by the Subcontractor

The equivalent clause under the Main Contract (RED and PNK books) is clause 16 [*Suspension and Termination by Contractor*]. It provides the framework within which circumstances or events may give the Subcontractor an entitlement to either reduce or suspend the rate of work, or allow him to terminate the Subcontract.

16.1 Subcontractor's Entitlement to Suspend Work
If the Contractor fails to pay any amount which is due to the Subcontractor pursuant to:

(i)   *Clause 14 [Subcontract Price and Payment] but subject to sub-paragraphs (a) to (d) of Sub-Clause 14.6 [Interim Subcontract Payments]; or*
(ii) *a Subcontract DAB's decision,*

the Subcontractor may, after giving not less than 21 days' notice to the Contractor, describing such failure of payment, suspend work (or reduce the rate of work) unless and until the Subcontractor receives payment.

If the Subcontractor subsequently receives the payment described in the above notice before giving a notice of termination, the Subcontractor shall resume normal working as soon as is reasonably practicable.

If the Subcontractor suffers delay and/or incurs Cost as a result of suspending work (or reducing the rate of work) in accordance with this Sub-Clause, the Subcontractor shall give notice to the Contractor and shall be entitled, subject to Sub-Clause 20.2 [Subcontractor's Claims], to an extension of time under Sub-Clause 8.4 [Extension of Subcontract Time for Completion] and payment of any Cost plus reasonable profit which shall be included in the Subcontract Price.

This Sub-Clause may be invoked by the Subcontractor, if the Contractor fails to pay an amount to the Subcontractor, which is due, subject to subparagraphs (a) to (d) of Sub-Clause 14.6 [*Interim Subcontract Payments*]. The requirements of Subparagraphs (c) and (d) may have a particularly severe effect in practice. Subparagraph (c) is the controversial '*pay-when-paid*' provision, and subparagraph (d) allows the Contractor to arbitrarily consider that the monthly payment is not properly due under the terms of the Subcontract. Both subparagraphs are contestable (under law or otherwise), but the initial effect is most likely to be a non-payment by the Contractor until such time that the issue has been resolved through the agreed dispute resolution mechanisms under the Subcontract. These mechanisms are, by design, time consuming, and it will be difficult for a financially challenged subcontractor to wait for the outcome.

There is a serious danger of Contractors abusing the provision to starve a subcontractor into accepting lesser terms. This issue was given serious consideration by the Task-Group but was eventually accepted in order not to compromise on the overall back-to-back principle.

Any damage or delay suffered by the Subcontractor as a result of a reduction in work rate or suspension may, subject to Sub-Clause 20.2 [*Subcontractor's Claims*], be claimed from the Contractor.

16.2 Termination by Subcontractor

If:

(a) *within 28 days after giving notice in accordance with Sub-Clause 16.1 [Subcontractor's Entitlement to Suspend Work] the Subcontractor does not receive the payment described in that notice;*

(b) *the Contractor or the Employer becomes bankrupt or insolvent, goes into liquidation, has a receiving or administration order made against him, compounds with his creditors, or carries an business under a receiver, trustee or manager for the benefit of his creditors, or if any act is done or event occurs which (under applicable Laws) has a similar effect to any of these acts or events; or*

(c) *the Contractor substantially fails to perform his obligations under the Subcontract,*

then the Subcontractor may upon giving notice to the Contractor, without prejudice to any other rights or remedies under the Subcontract or otherwise, terminate the Subcontract. In the case of sub-paragraphs (a) and (c) of this Sub-Clause, the Subcontractor shall give not less than 14 days notice, and in the case of sub-paragraph (b) of this Sub-Clause the Subcontractor may by notice terminate the Subcontract immediately.

This Sub-Clause sets out the situations under which the Subcontractor may terminate the Subcontract. In the event of non-payment or other substantial failure to perform his obligations under the Subcontract by the Contractor, and subject to the provisions of Sub-Clause 16.1 [*Subcontractor's Entitlement to Suspend Work*], such termination shall be preceded by a notice period of 14 days. Only if either the Contractor or the Employer become bankrupt, go into liquidation or are affected by similar circumstances may a termination be immediate.

16.3 Payment on Termination by Subcontractor

After a notice of termination under Sub-Clause 16.2 [Termination by Subcontractor] has taken effect, the Contractor shall promptly:

(a) *return the Subcontract Performance Security to the Subcontractor; and*

(b) *pay the Subcontractor the amounts and Costs set out in sub-paragraphs (a) to (e) of Sub-Clause 15.3 [Payment after Termination of Main Contract] insofar as these amounts or Costs have not been covered by payments already made to the Subcontractor.*

Provided that nothing in this Sub-Clause shall affect the right of either Party to receive payment in respect of any breach of the Subcontract committed by the other Party prior to the termination, or any other right to payment under the Subcontract.

This Sub-Clause sets out what happens once a notice to terminate under Sub-Clause 16.2 [*Termination by Subcontractor*] has taken effect.

It is stated under this Sub-Clause, that, if a notice of termination has taken effect, then the Contractor shall cease further works, and so on. It is thus intended that no further work, which would attract further payment, shall be executed. This makes perfect sense if the reason for termination is non-payment for executed works. Therefore, the notice has taken effect when it was sent and not the end of the related notice period as is sometimes argued. However, this raises the question as to why there is a notice period as compared to the immediate termination possible under Sub-Clause 16.2(b).

The reason for giving a notice is to provide time for a defaulting party to correct a mistake. In the case where no notice is required [i.e. Sub-Clause 16.2(b)] the defaulting party is not in a position to correct the situation therefore no notice period is required.

FIDIC Users' Guide
ISBN 978-0-7277-5856-9

ICE Publishing: All rights reserved
http://dx.doi.org/10.1680/fug.58569.503

Chapter 90
Clause 17: Risk and Indemnities

The equivalent Clause under the Main Contract (RED or PNK books) is Sub-Clause 17 [*Risk and Responsibility*].

17.1 Subcontractor's Risks and Indemnities
The Subcontractor shall have full responsibility for the care of the Subcontract Works and Subcontract Goods from the Subcontract Commencement Date until the Subcontract Works have been taken-over under Sub-Clause 10.2 [Taking-Over the Subcontract Works] or Sub-Clause 10.3 [Taking-Over by the Contractor] when responsibility for the Subcontract Works shall pass to the Employer or the Contractor, respectively. Provided that, after responsibility has so passed to the Employer or the Contractor, the Subcontractor shall:

(a) take responsibility for the care of any work which is outstanding on the date of taking-over until this outstanding work has been completed;

(b) be liable for any loss or damage caused by any actions performed by the Subcontractor; and

(c) be liable for any loss or damage which occurs and which arose from a previous event for which the Subcontractor was liable.

If any loss or damage happens to the Subcontract Works during the period when the Subcontractor is responsible for their care, the Subcontractor shall without delay rectify the loss or damage so that the Subcontract Works conform with the Subcontract. Provided that, and to the extent that, if any such loss or damage arises from:

(i) any cause for which the Subcontractor is responsible under the Subcontract, the rectification work shall be at the Subcontractor's risk and cost;

(ii) any risk listed in Main Contract Clause 17.3 [Employer's Risks], the Subcontractor shall promptly give notice to the Contractor, shall rectify this loss or damage to the extent required by the Engineer as confirmed by a Contractor's Instruction, and in all other respects Main Contract Clause 17.4 [Consequences of Employer's Risks] shall apply;

(iii) any act or default of the Contractor or the Contractor's Personnel, the Subcontractor shall subject to Sub-Clause 20.3 [Subcontractor's Claims] be entitled to recover from the Contractor any Costs so incurred;

(iv) any cause, other than those described in sub-paragraphs (i) to (iii) of this Sub-Clause, the Subcontractor shall be entitled to claim for the cost of such rectification work under the relevant insurance policy set out in Annex E.

The first paragraph of Main Contract Clause 17.1 [Indemnities] shall apply to the Subcontractor's indemnities.

In accordance with Clause 10 [*Completion and Taking-Over the Subcontract Works*] of the General Conditions of Subcontract this Sub-Clause has been drafted so that the Subcontractor remains responsible and liable for the Subcontract Works in the period between his completion of the Subcontract Works and taking-over of the Subcontract Works. Please refer to comments made on this issue in Chapter 83 under Sub-Clause 10.2 [*Taking Over Subcontract Works*].

17.2 Contractor's Indemnities

The Contractor shall indemnify and hold harmless the Subcontractor against and from all claims, damages, losses, and expenses (including legal fees and expenses) in respect of:

(a) the matters as described in sub-paragraphs (d)(i), (ii) and (iii) of Main Contract Clause 18.3 [Insurance against Injury to Persons and Damage to Property];
(b) a fault, error, defect, or omission in any element of the Contractor's design (if any) of the Permanent Works under the Main Contract other than design carried out by the Subcontractor pursuant to his obligations under the Subcontract:
(c) bodily injury, sickness, disease, or death, which is attributable to any negligence or wilful act or breach of the Main Contract by the Employer, the Employer's Personnel, the Contractor, the Contractor's Personnel, or any of their respective agents;
(d) bodily injury, sickness, disease, or death, which is attributable to any negligence, wilful act or breach of the Subcontract by the Contractor, the Contractor's Personnel, or any of their respective agents.

Provided that the Contractor's indemnities under this Sub-Clause do not extend to matters arising as a result of the Subcontractor's design (if any), the execution and completion of the Subcontract Works and the remedying of any defects.

Under the provisions of this Sub-Clause the Subcontractor is protected from and against certain exclusions to insurances effected under the Main Contract, provided these exclusions are not the consequence of a Subcontractor's responsibility under the Subcontract.

17.3 Subcontract Limitation of Liability

Neither Party shall be liable to the other Party for loss of use of any Subcontract Works, loss of profit, loss of any contract or for any indirect or consequential loss or damage which may be suffered by the other Party in connection with the Subcontract, other than under Clause 15 [Termination of the Main Contract and Termination of the Subcontract by the Contractor], Sub-Clause 16.3 [Payment an Termination by Subcontractor], Sub-Clause 17.1 [Subcontractor's Risks and Indemnities] and/or Sub-Clause 17.2 [Contractor's Indemnities].

The total liability of the Subcontractor to the Contractor under or in connection with the Subcontract, other than under Sub-Clause 7.1 [Subcontractor's Use of Equipment, Temporary Works, and/or Other Facilities], Sub-Clause 7.2 [Free-Issue Materials], Sub-Clause 8.7 [Subcontract Damages for Delay] and Sub-Clause 17.1 [Subcontractor's Risks and Indemnities], shall not exceed the sum stated in the Particular Conditions of Subcontract or (if a sum is not so stated) the Accepted Subcontract Amount.

This Sub-Clause shall not limit liability in any case of fraud, deliberate default or reckless misconduct by the defaulting Party.

The equivalent Sub-Clause under the Main Contract (RED or PNK books) is 17.6 [*Limitation of Liability*], and is to be interpreted and construed in a similar way. However, before inviting tenders the Contractor is advised to check the wording for compliance with the Law governing the Subcontract.

FIDIC Users' Guide
ISBN 978-0-7277-5856-9

ICE Publishing: All rights reserved
http://dx.doi.org/10.1680/fug.58569.507

Chapter 91
Clause 18: Subcontract Insurances

The equivalent Clause under the Main Contract (RED or PNK books) is Clause 18 [*Insurance*]. It sets out the various insurance requirements to be effected under the terms of the Subcontract, and should, where stated, be interpreted in conjunction with the equivalent clause under the main Contract.

18.1 Subcontractor's Obligation to Insure

The Subcontractor shall effect and maintain insurance against the risks and in the sums and in the names specified in Annex E. The Subcontractor shall maintain these insurances from the Subcontract Commencement Date until the Subcontract Works have been taken-over in accordance with Sub-Clause 10.2 [Taking-Over of the Subcontract Works] or Sub-Clause 10.3 [Taking-Over by the Contractor].

Provided always that the Subcontractor shall insure the Subcontractor's Personnel and Main Contract Clause 18.4 [Insurance for Contractor's Personnel] shall apply.

Before inviting tenders, the Contractor should set out the detailed insurance requirements in Annex E. It is also a requirement under Sub-Clause 18.2 [*Insurance arranged by the Contractor and/or the Employer*] to include the information on the insurances taken out under the Main Contract. This should include the conditions, limits, exceptions, and deductibles; preferably in the form of a copy of each policy where available. It is important that the Subcontract and Main Contract insurances do not overlap.

The following should be specified in Annex E

(a) those risks against which the Subcontractor is required to effect and maintain insurances
(b) the sums in which the insurances are to be effected and maintained
(c) those persons for which the insurances are to be effected and maintained; and
(d) an alternative period, if the insurances are to be maintained other than for the period from when so much of the Site and access is made available to the Subcontractor, as may be required to enable him to commence and proceed with the execution of the Subcontract Works in accordance with the Subcontract, until he has finally performed his obligations under the Subcontract.

The Subcontractor's obligations extend to the date on which the relevant Works are taken over under the Main Contract. The Subcontractor is advised to bear this in mind before submitting his tender.

18.2 Insurance arranged by the Contractor and/or the Employer

The Contractor shall effect and maintain the insurances for which he is responsible under the Main Contract, the details of which are set out in Annex E.

The details of insurances (if any) to be arranged by the Employer under the Main Contract are as set out in Annex E. If the Employer fails to effect and keep in force such insurances, the Subcontractor shall subject to Sub-Clause 20.2 [Subcontractor's Claims] be entitled to recover from the Contractor any monies which should have been recoverable by the Subcontractor under such insurances.

Where the Subcontractor discovers any inadequacy in these insurances or duplication with these insurances when reviewing the insurance coverage of the Subcontract Works or executing the Subcontract Works, he shall immediately notify the Contractor.

If:

(a) the Subcontract Works, Subcontract Goods, or other things belonging to the Subcontractor are destroyed or damaged during the period that insurance under Main Contract Clause 18.2 [Insurance for Works and Contractor's Equipment] is required to be effected and maintained by the Contractor and/or the Employer; and

(b) a claim is established in respect of that which was destroyed or damaged under the relevant policy of insurance,

then, provided that this claim is paid by the relevant insurer(s), the Subcontractor shall be paid the amount of this claim less any deductible(s), or the amount of his loss, whichever is less, and shall apply this sum in replacing or repairing that which was destroyed or damaged.

This Sub-Clause makes provision for insurances to be provided by the Contractor or the Employer. If the Employer fails to effect insurance for which he is responsible, then, in the event of damage, the Subcontractor can claim those amounts from the Contractor. Any claim would be made in accordance with Sub-Clause 20.2 *[Subcontractor's Claims]*.

18.3 Evidence of Insurance and Failure to Insure

Where by virtue of this Clause 18 [Subcontract insurances] either Party is required to effect and maintain insurance, he shall if so requested by the other Party promptly provide evidence of that insurance and the receipt for the payment of the current premium.

If either Party fails to effect and maintain any insurance it is required to effect and maintain under the Subcontract, or fails to provide satisfactory evidence of any insurance and the receipt for the payment of the current premium, without delay following a request for this evidence by the other Party, then the other Party may (at its option and without prejudice to any other right or remedy) effect and maintain insurance for the relevant coverage and pay the premiums due. The Party in default shall then pay the amount of these premiums plus any extra expense incurred in effecting this insurance to the other Party as an adjustment to the Subcontract Price.

This Sub-Clause provides the sanction in the event any Party fails to effect any insurance for which it is contractually responsible. The other Party may effect such insurance at the cost of the Party in default.

FIDIC Users' Guide
ISBN 978-0-7277-5856-9

ICE Publishing: All rights reserved
http://dx.doi.org/10.1680/fug.58569.511

Chapter 92
Clause 19: Subcontract Force Majeure

In keeping with the objective to maintain a true back-to-back agreement with the Main Contract, the Subcontract recognises the principle of Force Majeure.

19.1 Subcontract Force Majeure
The provisions of Main Contract Clause 19 [Force Majeure] shall apply to the Subcontract.

With reference and subject to Sub-Clauses 1.3(a) and (g) [*Communications*] the principle of Force Majeure is dealt with as per the provisions of the Main Contract Clause 19 [*Force Majeure*].

FIDIC Users' Guide
ISBN 978-0-7277-5856-9

Chapter 93
Clause 20: Notices, Subcontractor's Claims and Disputes

The equivalent Clause under the Main Contract is Clause 20 [*Claim, Disputes and Arbitration*].

This is arguably the most often referred to clause in any contract. It is also the most controversial clause in all of the FIDIC Standard Forms of Contract. The Subcontract is no exception. As the Task Group appointed to create this Standard Form of Subcontract was specifically mandated to adhere as far as possible to the back-to-back principle, the issue of claims was always going to be a problem. Claims under the Subcontract have a high probability to be valid under the Main Contract. In other words the Contractor may well be in a position to submit a received claim from his Subcontractor as his own entitlement to the Employer. Obviously for that to happen, the Contractor must comply with the provisions of the Main Contract, most notably the so-called time-bar under Sub-Clause 20.1 [*Contractor's Claims*].

For the back-to-back principle to work it is essential that any claim submitted by the Subcontractor that can be considered and evaluated as a valid claim under the Main Contract, is notified under the Main Contract within the time-bar, that is, no later than 28 days after the Contractor became aware, or should have been aware, of the event or circumstance giving rise to the claim. The Contractor must rely on his Subcontractor to give timely notices and sufficient relevant information to allow him to notify his claim under the Main Contract. This starts with notices, and especially the timely submission thereof. Notices are an important and integral aspect of any FIDIC contract. However, they are also highly controversial and are often the cause of complex disputes.

20.1 Notices
Without prejudice to the generality of Clause 4 [The Subcontractor], whenever the Contractor is required by the terms of the Main Contract to give any notice or other information to the Engineer or to the Employer, or to keep contemporary records (whether in relation to a claim or otherwise), to the extent that these terms apply to the Subcontract Works the Subcontractor shall give a similar notice or other information in writing to the Contractor and keep the contemporary records that will enable the Contractor to comply with these terms of the Main Contract. The Subcontractor shall do so in good time to enable the Contractor to comply with these terms. Provided always that the Subcontractor shall be excused from any non-compliance with this requirement for so long as he could

not have reasonably known of the Contractor's need of the notice or information from him or the contemporary records.

Notwithstanding Sub-Clause 20.2 [Subcontractor's Claims] and Sub-Clause 3.3 [Contractor's Claims in connection with the Subcontract], each Party shall immediately give notice to the other Party of any delay event which has occurred, or specific probable future event(s) or circumstance(s), which may adversely affect the other Party's activities, or delay the execution of the Subcontract Works and/or the Main Works. The Subcontractor shall immediately give notice to the Contractor of any event which has occurred, or specific probable future event(s) or circumstance(s), which may increase the Subcontract Price and/or the Contract Price.

For the notice requirement under the Main Contract to work, there must be a similar provision under the Subcontract. To emphasise the importance of the notice requirements, the Task Group dedicated a specific Sub-Clause to this issue (i.e. Sub-Clause 20.1 [*Notices*]). The first paragraph of this Sub-Clause states that the Subcontractor shall issue a '*similar notice or other information in writing*' whenever the Contractor is required to give any notice (whether in relation to a claim or otherwise) in relation to the Subcontract Works to the Engineer or to the Employer. The Subcontractor shall do so in a manner that allows the Contractor to comply with his requirements under the Main Contract, for example, time-bar or the keeping of contemporary records. This highlights the importance of Sub-Clause 2.1 [*Subcontractor's Knowledge of the Main Contract*]. The Subcontractor shall do so for two specific events: (i) those that require a notice under the Main Contract and (ii) those that require the Contractor to submit '*other information*' to the Engineer or Employer under the Main Contract.

(*i*) Requirement for a notice under the Main Contract.

For every instance in relation to the Subcontract Works where the Contractor is to issue a notice under the Main Contract, the Subcontractor shall submit a similar notice under the Subcontract. There are two distinct problems with this provision. Firstly a '*similar*' notice is ambiguous as it allows misinterpretation. It wrongly assumes that both the Contractor and the Subcontractor have the same interpretation of an event that took place. Secondly there are numerous instances under the Main Contract where the Contractor is under obligation to issue a notice, or even a double notice (refer to Appendix 1, 2 and 3). It is imperative for the Subcontractor to be aware of all these instances in detail, failing which he may become in breach of contract. There is little incentive either for the Contractor to inform the Subcontractor of such a requirement in advance, as any breach by the Subcontractor would enable the Contractor to seek recourse for damages through the Subcontract. In the event that the Subcontractor fails to give a timely notice, and as a consequence the Contractor fails to recover damages under the Main Contract, he simply recovers such damages from the Subcontractor on the strength of a breach of this notice requirement (refer Sub-Clause 2.3 [*Failure to Comply*]). This is an easier option for the Contractor as he can arbitrarily decide on his own damage under Sub-Clause 3.3 [*Contractor's Claims in Connection with the Subcontract*], and subsequently recover the monies by off-setting against future payments. It is also most certainly a politically far less damaging option for the

Contractor when compared to submitting a claim to his client (the Employer) from which he may hope to gain future contracts.

(*ii*) Requirement for other information under the Main Contract.

Whereas the requirements for notices could arguably be obtained with relative ease from the Main Contract provisions (refer to Sub-Clause 2.1 [*Subcontractor's Knowledge of Main Contract*], the requirements for other information to be given to the Engineer or the Employer are far less obvious. It is almost impossible for the Subcontractor to know exactly if and when the Engineer or the Employer has requested additional information from the Contractor, but if he fails to submit in time he will be in breach of Subcontract! In addition such required information should normally be produced in a specific format as determined by the Engineer, of which the Subcontractor may have no knowledge.

The second paragraph of this Sub-Clause is also ambiguous. The first sentence states clearly that what follows is notwithstanding Sub-Clause 20.2 [*Subcontractor's Claims*] and Sub-Clause 3.3 [*Contractor's Claims in connection with the Subcontract*]. It further states that both parties shall immediately give notice to the other party of any delay event which has occurred or specific probable future event or circumstance(s) which may adversely affect the other party's activities. Furthermore, the Subcontractor shall notify the Contractor immediately of any event or probable future event or circumstance that may increase the Subcontract Price and/or the Contract Price.

The words '*immediately*' are seemingly in contradiction with the time-bar provisions under Sub-Clause 20.2 [*Subcontractor's Claims*] (i.e. 21 days) and Sub-Clause 3.3 [*Contractor's Claims in connection with the Subcontract*] (i.e. 28 days). By implication there could be a construed double notification requirement: (i) immediately when the event occurs (or when a party knows about a future event or circumstance), and (ii) within the time-bar to initiate a claim. It is advisable to any party to be vigilant with regard to events or circumstances that would initiate a notice requirement under this Sub-Clause. Further, when issuing such a notice, it should clearly be marked as being given under Sub-Clauses 20.1 [*Notices*], 20.2 [*Subcontractor's Claims*] as well as any other Sub-Clause that may be applicable.

The latter part of the second paragraph obligates the Subcontractor to give a notice if and when he knows of (or should have known of) or even suspects an event or circumstance that may increase the Contract Price. This clearly refers to the Price of the Main Contract. As the Contractor through Annex 'A' part 'A' has the option not to divulge certain information pertaining to the Main Contract Price (refer Sub-Clause 2.1 [*Subcontractor's Knowledge of the Main Contract*]) it seems difficult for the Subcontractor to comply with this provision. How will he know if and when the Contract Price is to change if he is not privy to the information that would indicate this to him?

It is not too difficult to understand the enormous potential for conflict and disputes this Sub-Clause will bring to any project. In all instances where the Subcontractor breaches this Sub-Clause, the Contractor has a relatively easy route of recovery for any damages sustained. This route would also steer him clear of any confrontation or subsequent animosity with his Employer. There is thus an incentive for the Contractor not to communicate with the

Subcontractor when dealing with notices or requests for information under the Main Contract.

To avoid future conflicts and disputes it is recommended that the parties make appropriate amendments through the Particular Conditions that will provide a more balanced and fair agreement.

20.2 Subcontractor's Claims

If the Subcontractor considers himself to be entitled to any extension of the Subcontract Time for Completion and/or any additional payment, under any Clause of these Conditions or otherwise in connection with the Subcontract, Main Contract Clause 20.1 [Contractor's Claims] shall apply to Subcontractor's claims. Save that:

(a) the period of notice applicable to Subcontractor's claims shall be not later than 21 days after the Subcontractor became aware (or should have become aware) of the event or circumstance giving rise to the claim. If the Subcontractor fails to give notice of a Claim within such period of 21 days, the Subcontract Time for Completion shall not be extended, the Subcontractor shall not be entitled to additional payment, and the Contractor shall be discharged from all liability in connection with the claim;

(b) the period for submission by the Subcontractor of a full detailed Claim shall be not later than 35 days after the Subcontractor became aware (or should have become aware) of the event or circumstance giving rise to the claim, or within such other period as may be proposed by the Subcontractor and approved by the Contractor;

(c) the reference to 'Payment Certificate' shall be replaced by 'interim payment made to the Subcontractor under Sub-Clause 14.6 [Interim Subcontract Payments]'; and

(d) in lieu of the penultimate paragraph of Main Contract Clause 20.1 [Contractor's Claims] the following shall apply:

The Contractor shall consult with the Subcontractor in an endeavour to reach agreement on the extension of the Subcontract Time for Completion and/or additional payment to which the Subcontractor may be entitled for his claim. If agreement is not reached, the Contractor shall within 49 days after receiving from the Subcontractor a fully detailed claim or any further particulars requested by the Contractor, or within such other time period agreed between the Parties:

(a) make a fair decision, having due regard to the Subcontractor's submissions, the extent to which his Claim for additional payment and/or extension of time has been substantiated, and all other relevant circumstances;

(b) notify the Subcontractor, with reasons and making reference to this sub-paragraph, of the appropriate and applicable additional payment (if any) and/or extension (if any) to the Subcontractor, and grant the extension (if any) of the Subcontract Time for Completion.

Each Party shall give effect to each agreement reached or decision made under this Sub-Clause unless and until revised under this Clause 20 [Notices, Subcontractor's Claims and Disputes].

516

Taking into account the comments under Sub-Clause 20.1 [*Notices*] this Sub-Clause merely brings into line the back-to-back principles between the Subcontract and the Main Contract. It does so by making Main Contract Sub-Clause 20.1 [*Contractor's Claims*] apply under the Subcontract with certain amendments. It reduces the time-bar for a notice to claim under the Subcontract from 28 to 21 days, thus leaving the Contractor potentially 7 days to submit his notice to claim under the Main Contract. Similarly, the periods for submitting claim particulars have been amended from 42 to 35 days. Analogue to Main Contract Sub-Clause 3.5 [*Determinations*] any claim submitted by the Subcontractor shall be discussed between the Parties in an endeavour to reach agreement, failing which the Contractor shall make a fair, albeit possibly arbitrary, decision on the justification and value of the claim.

Interestingly it is stated that both Parties shall give effect to each agreement or decision made under this Sub-Clause until revised under Clause 20 [*Notices, Subcontractor's Claims and Disputes*]. By implication this agreement can be changed at a later date through a DAB, an amicable settlement or arbitration. It puts doubt on the effectiveness of the obligatory attempt at consultation between the parties and questions the legality of such a provision under some jurisdictions.

It is regretted that the strong encouragement for the parties to seek final and binding agreement in other FIDIC contracts is lost by the inclusion of this provision in the SUB Book. The authors believe this concept to be wrong. How can a '*provisional agreement*' (as this is what it is) be reopened at a later date? This later date is at a time when a party has accepted a situation (e.g. ceased to keep contemporary records) and is unexpectedly compelled to refight his case through a DAB/amicable settlement/arbitration procedure. Or alternatively, is the concept of fairness and balance really served if a party is at first contractually allowed to '*accept*' an agreement, with the possibility to gain the benefit of hindsight, at which point he can back out of the earlier agreement to seek a better deal?

It is suggested to amend this part of the Sub-Clause such that only Contractor's decisions can be revised through the appropriate procedures incorporated into Clause 20. This would also provide some form of protection for the Subcontractor so that payment(s) for an agreement on a claim could not be delayed further through a simple retraction of such an agreement.

20.3 Failure to Comply

If by reason of any failure by the Subcontractor to comply with the first paragraph of Sub-Clause 20.1 [Notices] and/or the provisions of Sub-Clause 20.2 [Subcontractor's Claims] the Contractor is prevented from recovering any sum other than in respect of Subcontractor's claims from the Employer under the Main Contract in respect of the Subcontract Works, then, without prejudice to any other remedy of the Contractor for this failure the Contractor shall, subject to Sub-Clause 3.3 [Contractor's Claims], be entitled to deduct this sum from the Subcontract Price.

This Sub-Clause confirms the Contractor's entitlement to recover any damages he may have as a result of the Subcontractor's non-compliance with Clause 20, through a simple reduction of the Subcontract Price.

20.4 Subcontract Disputes

If a dispute (of any kind whatsoever) arises between the Contractor and the Subcontractor in connection with, or arising out of, the Subcontract or the execution of the Subcontract Works, then either Party may give a notice of this dispute to the other Party (the 'Notice of Dispute').

In any Notice of Dispute given by the Contractor or within 14 days of receiving a Notice of Dispute from the Subcontractor, the Contractor may notify the Subcontractor, with reasons, of his opinion that the dispute involves an issue or issues which is/are also involved in a dispute between the Contractor and the Employer under the Main Contract. If the Contractor so notifies the Subcontractor:

(a) the Parties shall defer any referral of the dispute to the Subcontract DAB until a date that is no earlier than 112 days, or other period as may be agreed between the Parties, after the date of the Notice of Dispute;

(b) if the subject of the Subcontract dispute has not previously been referred to the Main Contract DAB, the Contractor shall refer the subject of Subcontract dispute to the Main Contract DAB, with a copy to the Subcontractor, in accordance with Main Contract Clause 20.4 [Obtaining Dispute Adjudication Board's Decision] within 28 days, or other period as may be agreed between the Parties, of the Notice of Dispute. If Main Contract Clause 20.8 [Expiry of Dispute Adjudication Board's Appointment] applies under the Main Contract, the Contractor shall immediately give notice to the Subcontractor; and

(c) the Subcontractor shall, in good time, afford the Contractor all information and assistance that may reasonably be required to enable the Contractor to diligently pursue his dispute which includes the subject of the Subcontract dispute under the Main Contract

Provided that, if the Subcontract dispute is not referred to the Main Contract DAB within the time stated in sub-paragraph (b) of this Sub-Clause, or if Main Contract Clause 20.8 [Expiry of Dispute Adjudication Board's Appointment] applies under the Main Contract, either Party shall immediately after that be entitled to refer the Subcontract dispute to the Subcontract DAB and sub-paragraph (a) of this Sub-Clause shall no longer apply.

At any time after the expiry of the time stated or otherwise agreed in sub-paragraph (a) of this Sub-Clause:

(i) the Contractor shall be entitled to refer the Subcontract dispute to the Subcontract DAB; or

(ii) the Subcontractor shall, at his option, be entitled to refer the Subcontract dispute to the Subcontract DAB or to arbitration under Sub-Clause 20.7 [Subcontract Arbitration].

If the Contractor does not notify the Subcontractor that the Subcontract dispute involves an issue or issues which is/are also involved in a dispute between the Contractor and the Employer under the Main Contract in any Notice of Dispute given by the Contractor, or within 14 days of receiving a Notice of Dispute from the Subcontractor, either Party shall be entitled to refer the Subcontract dispute to the Subcontract DAB.

Figure 93.1 Chart comparing events leading to a Main Contract and Subcontract Dispute

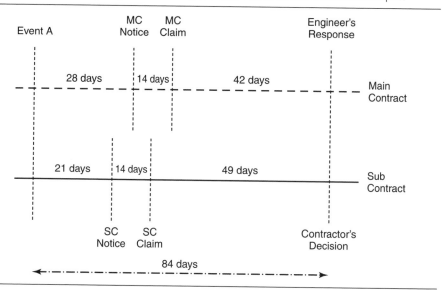

Unless the Subcontract has already been abandoned, repudiated or terminated, the Subcontractor shall proceed with the Subcontract Works in accordance with the Subcontract.

For a dispute to exist under the terms of the Subcontract it is a necessary requirement for a Party to issue a notice of such dispute to the other Party. In accordance with Sub-Clause 1.6 [*Notices, Consents, Approvals, Certificates, Confirmations, Decisions and Determinations*] such notice shall be in writing. If the Contractor issues such a notice, or within 14 days after having received such a notice from the Subcontractor, the Contractor may notify the Subcontractor that the dispute includes issues also involved in a dispute between Contractor and Employer under the Main Contract. It may be difficult to understand when a dispute under the Subcontract is declared and how the issue involving it could already be a dispute under the Main Contract. To highlight this query refer to Figure 93.1.

The chart shows the timelines applicable to an event 'A' under both the Main Contract and the Subcontract. Under the Main Contract the earliest point in time for a dispute to be validly declared would be when the Engineer responds to the submitted claim in accordance with the sixth paragraph of Sub-Clause 20.1 [*Contractor's Claims*]. The equivalent point in time under the Subcontract would be the Contractor's Decision in accordance with the second paragraph Sub-Clause 20.2(d) [*Subcontractor's Claims*]. If all involved were to use all their allotted time, the time for declaring a dispute under the Main Contract would be identical to the time under the Subcontract, that is, 84 days after the event. Hence, for issues involving a dispute under the Subcontract to be already in dispute under the Main Contract one of the following scenarios must apply.

1   The Contractor issued a claim for another event 'B' involving the same issues earlier than he received the corresponding claim 'A' from the Subcontractor. For example,

a piling subcontractor declaring a dispute on the basis of sub-soil conditions, whereas the contractor is already in dispute on something else but also involving the same sub-soil conditions.

2 The Contractor received an Engineer's response to his claim earlier than 84 days after the event 'A', and can now give a subsequent Contractor's decision for the same event 'A' to the Subcontractor.

3 The Contractor delays his decision to the Subcontractor until such time that he has received the Engineer's response under the Main Contract. This would certainly be a breach by the Contractor, but as there is no contractual sanction, it would be difficult for the Subcontractor to prove sustained damage as a result of such breach.

In the event that the issues involved in a dispute with the Subcontractor are also in dispute under the Main Contract, the Contractor has the option to notify the Subcontractor. This will initiate a series of events that are not necessarily in favour of the Subcontractor. First and foremost the dispute between the Subcontractor and the Contractor will be frozen for a period of at least 112 days (refer to Sub-Clause 20.4(a) [*Subcontract Disputes*]. The 112 days is not chosen arbitrarily, but corresponds to a combination of Subcontract Sub-Clause 20.4(b) [*Subcontract Disputes*] and Main Contract fourth paragraph of Sub-Clause 20.4 [*Obtaining Dispute Adjudication Board's Decision*]. Under the former Sub-Clause the Contractor has 28 days after a dispute is declared to refer it to the Main Contract DAB, who in turn as per the latter have 84 days to make a decision, totalling a period of 112 days. This assumes that there shall be no extension to this Main Contract DAB decision period.

In theory this puts the Contractor in a position where he will know the outcome of a Main Contract DAB decision before having to tackle the same problem with the Subcontractor. This provides a major advantage for the Contractor.

Referring back to the scenarios that must necessarily apply for a Subcontract dispute to be declared a dispute under the Main Contract (see 1–3), it can be said that scenario 1 is not common and occurs infrequently; scenario 2 will be possible theoretically, but experience has shown that almost all appointed DABs will take their allowed 84 days to reach a decision; scenario 3 is thus the most obvious option for the Contractor to force a situation where he will have the advantage of knowing the outcome to a dispute under a Main Contract DAB before discussing it with his Subcontractor. As there is no sanction for him to delay his Contractor's decision under the Subcontract (without which there cannot be a Subcontract dispute), experience has shown this to be a favourite method to be applied by most main Contractors. The result is a strong disadvantage for the Subcontractor. In reality most contractors delay almost every claim initiated by a subcontractor in its evaluation, simply to gain the advantage in later discussion after having obtained a DAB decision on the matter under the Main Contract.

Obviously all of the above assumes that the Subcontract claim issue can actually be referred to the Main Contract DAB. The truth is that any matter related to the Contract and/or the Subcontract can be referred to a Subcontract DAB, even those that have no chance of ending positively! As per the second paragraph Sub-Clause 20.4 [*Subcontract Disputes*] the Contractor shall inform the Subcontractor with reasons, if in his opinion, the Subcontract issue is involved in a Main Contract dispute. Although it is in the sole opinion of the Contractor, he can only do so if and when this issue at stake has factually become a dispute under the Main Contract, that is, after the Engineer's comments to the Contractor's Claim. There is thus again an incentive for the Contractor to act in accordance with scenario 3 above, that is, to wait until he has an Engineer's decision on the matter. If the Engineer's decision is not totally in favour of the Contractor, than he simply declares it to be a dispute under the Main Contract – who is going to challenge his opinion on this?, informs the Subcontractor accordingly, and he has gained a further possible 112 days deferral of the Subcontractor's Claim.

In this situation the Contractor has an obligation per Sub-Clause 20.4(b) [*Subcontract Disputes*] to refer the dispute to the Main Contract DAB within 28 days after a Notice of Dispute under the Subcontract. However, and again, there is no contractual sanction for a breach of this obligation. The Contractor could ostensibly delay this without much of a recourse by the Subcontractor.

The other option is to declare the Subcontract dispute related to a Main Contract dispute, but then to refrain from referring it to the Main Contract DAB at all. The Subcontract is silent on this situation. In any situation where the Contractor notifies the Subcontractor of an issue being related, the outcome is a deferral of the Subcontract claim for at least 112 days; subject of course to potential legal recourse through the courts where the Contractor has been in breach as described above. The effectiveness of such action is debatable in view of the time it takes.

Admittedly the above is a negative view of the Sub-Clause, and it is only prudent to also have a look at the other side, that is, when everything works as designed. In a situation where the Subcontractor submits a notice of dispute in accordance with the first paragraph of Sub-Clause 20.4 [*Subcontract Disputes*], but the Contractor does not declare this dispute to be related under the Main Contract (for whatever reason), the Subcontractor can, 14 days after having sent the notice of dispute, refer the matter to the Subcontract DAB. This process will be simple and will lead to a Subcontract DAB decision, that is, 84 days after referral. What happens if the Contractor realises that he made a mistake, in other words the matter is related to a dispute under the Main Contract? In this case the Contractor is still free to refer his dispute to the Main Contract DAB. We have now created the interesting situation where two independent DABs consider the same issue under the same project, potentially coming to two different decisions. A legal chaos is born.

The Sub-Clause has a multitude of conflict and dispute potential. It is in the author's view that the Sub-Clause has unbalanced risk in favour of the Contractor. Early experience in the use of the Subcontract shows that indeed main contractors realise the ample opportunities to squeeze their subcontractors into accepting less favourable terms on claims and related issues simply by playing for time and/or using the lack of sanctioned obligations for the contractor.

20.5 Appointment of the Subcontract DAB
Disputes between the Contractor and the Subcontractor shall be decided by a Subcontract DAB, which shall be jointly appointed by the Parties within 42 days after the date of a

Notice of Dispute or, where applicable, within 42 days after the date of expiry of the time stated or otherwise agreed in sub-paragraph (a) of Sub-Clause 20.4 [Subcontract Disputes]. Unless it is stated in the Appendix to the Subcontractor's Offer that it shall comprise three members, the Subcontract DAB shall comprise one suitably qualified person. If the Parties fail to agree upon the appointment of (any member of) the Subcontract DAB within 42 days after the date of a Notice of Dispute, or upon the appointment of a replacement person within 42 days after the appointed person declines or is unable to act, then the President of FIDIC or a person appointed by the President shall, upon the request of either or both Parties and after due consultation with both Parties, appoint the (member of the) Subcontract DAB. This appointment shall be final and conclusive. Each Party shall be responsible for paying one-half of the remuneration of the appointing entity or official.

The appointment of the Subcontract DAB may be terminated by mutual agreement of both Parties, but not by the Contractor or the Subcontractor acting alone. Unless otherwise agreed by both Parties, the appointment of the Subcontract DAB shall expire when the Subcontract DAB has given its decision on the dispute referred to it under Sub-Clause 20.6 [Obtaining Subcontract DAB's Decision], unless other disputes have been referred to the Subcontract DAB by that time under Sub-Clause 20.6 [Obtaining Subcontract DAB's Decision], in which event the relevant date shall be when the Subcontract DAB has also given decisions on those disputes.

In all other respects Main Contract Clause 20.2 [Appointment of the Dispute Adjudication Board] shall apply to the appointment of the Subcontract DAB, except that:

(a) the provisions relating to payment of retainer fees in the General Conditions of Dispute Adjudication Agreement, and

(b) Rules 1 to 4 of the Procedural Rules annexed to the General Conditions of Dispute Adjudication Agreement,

shall not apply.

In contrast to the Main Contract the Subcontract DAB shall be ad-hoc. It is thus appointed for a specific dispute or disputes referred to it under the terms of Sub-Clause 20.6 [*Obtaining Subcontract's DAB Decision*]. It shall be jointly appointed by the Parties within 42 days after the date of a notice of dispute in accordance with Sub-Clause 20.4 [*Subcontract Disputes*], or if applicable, within 42 days after the expiry of the deferral period in accordance with Sub-Clause 20.4(a) [*Subcontract Disputes*]. The Subcontract DAB shall by default have one member. The Parties may of course amend this to be three or more if deemed necessary. The Subcontract DAB shall be active for as long as there are specific disputes referred to it. By the time a Subcontract DAB decision is made on a dispute, and there are no other disputes referred to it by that time, the DAB appointment shall expire.

In general the provisions of Main Contract Sub-Clause 20.2 [*Appointment of the Dispute Adjudication Board*] shall apply. By implication this means that the Main Contract Appendix 'General Conditions of Dispute Adjudication Agreement' applies (refer Main Contract fifth paragraph of Sub-Clause 20.2 [*Appointment of the Dispute Adjudication Board*]). This in turn incorporates the Main Contract Annex '*Procedural Rules*' (refer to '*General Conditions of Dispute Adjudication Agreement*' clause 4(e)), with the exception of rules 1–4, which deal expressly with a standing board's obligation to visit the site regularly.

There is no express provision that prohibits the Parties from appointing the same DAB under the Main Contract and the Subcontract. However, the wisdom of such an action must be questioned. As seen previously, it is not impossible for the same dispute to be heard by both the Main Contract DAB and the Subcontract DAB. It would be almost impossible for any DAB appointed for both matters to maintain their neutral status and confidentiality towards in the one instance the claimant, and in the other instance the respondent, who by definition are the same party.

20.6 Obtaining Subcontract DAB's Decision

Either party may, subject to Sub-Clause 20.4 [Subcontract Disputes], refer a Subcontract dispute in writing to the Subcontract DAB for its decision with a copy to the other Party. The reference shall state that it is given under this Sub-Clause. In all respects, other than as stated in this Sub-Clause, Main Contract Clause 20.4 [Obtaining Dispute Adjudication Board's Decision] shall apply to the resolution of the Subcontract dispute. The reference to Main Contract Clause 20.8 [Expiry of Dispute Adjudication Board's Appointment] shall not apply.

The Subcontract DAB's decision shall be binding on both Parties unless and until it shall be revised in an amicable settlement as described below, or an arbitral award in accordance with Sub-Clause 20.7 [Subcontract Arbitration].

If either Party serves a notice of dissatisfaction with the Subcontract DAB's decision within 28 days of that decision, both Parties shall attempt to settle the Subcontract dispute amicably before the commencement of arbitration. However, unless both Parties agree otherwise, arbitration may be commenced on or after the twenty-eighth day after notice of dissatisfaction was given even if no attempt at amicable settlement has been made.

In the event that a Party fails to comply with any decision of the Subcontract DAB, then the other Party may, without prejudice to any other rights it may have, refer the failure itself to arbitration under Sub-Clause 20.7 [Subcontract Arbitration] for the purpose of obtaining an award (whether interim or other) to enforce that decision. There shall be no requirement to obtain a Subcontract DAB's decision or to attempt to reach amicable settlement in respect of this reference.

The equivalent Sub-Clause under the Main Contract (RED and PNK) is Sub-Clause 20.4 [*Obtaining Dispute Adjudication Board's Decision*], which shall apply in general under the Subcontract, safe for those provisions under this Sub-Clause that are in deviation. Obviously, the DAB is only mandated to accept referrals on actual disputes, not mere differences in opinion. However, in accordance with '*General Conditions of Dispute Adjudication Agreement*' clause 4(k) the DAB may give advice or opinions in answer to a mutual request by both Parties.

Any Subcontract DAB decision is binding on the Parties unless revised by amicable agreement or a subsequent arbitral award in accordance with Sub-Clause 20.7 [*Subcontract*

Arbitration]. Either party, may within 28 days after having received a Subcontract DAB decision, issue a notice of dissatisfaction. The Parties are then encouraged to attempt to resolve the issue amicably before proceeding to arbitration. Arbitration can commence not earlier then 28 days after a notice of dissatisfaction has been served. Interestingly the equivalent provision under the Main Contract (RED and PNK) state this period to be 56 days.

The DAB decision becomes final and binding upon the Parties if no notice of dissatisfaction has been issued within the time allowed.

20.7 Subcontract Arbitration

Unless settled amicably, any Subcontract dispute in respect of which the Subcontract DAB's decision (if any) has not become final and binding shall be finally settled under the Rules of Arbitration of the International Chamber of Commerce and Main Contract Clause 20.6 [Arbitration] shall apply to the Subcontract dispute except that the dispute may be settled by one arbitrator appointed in accordance with the Rules.

Any dispute for which a Subcontract DAB decision has been obtained, and for which a notice of dissatisfaction was issued, may be referred to arbitration. The Rules of Arbitration of the International Chamber of Commerce (ICC) and Main Contract Sub-Clause 20.6 [*Arbitration*] shall apply. With reference to Sub-Clause 20.2 [*Subcontractor's Claims*], it is highly likely that any dispute between the Contractor and the Subcontractor under the Subcontract has been considered under the Main Contract between the Contractor and the Employer. It is thus also likely that both disputes concern similar issues and reach the status of arbitration more or less at the same time. The result is two independent arbitrations running concurrent on similar issues. Apart from the hugely interesting issue concerning the outcome of these cases, there is a potential problem concerning the confidentiality on information and proceedings under arbitration conditions. It will be almost impossible for the Contractor, who is involved in both the arbitrations, to completely separate them, and to avoid using information and knowledge of the one arbitration in the other. The likelihood for subsequent arbitral awards to be overturned under law is high and should not be underestimated.

It is obvious that the wording of the various Sub-Clauses under Clause 20 [*Notices, Subcontractor's Claims and Disputes*] may be the source of numerous disputes in itself. This may be caused by decisions on similar disputes having different outcomes under the Main Contract compared with the Subcontract. This in itself is another source of potential conflict.

The Parties may prefer to have the outcome to a dispute generated under the Main Contract to also apply under the Subcontract. In other words, a DAB decision or arbitral award binds the Parties to the Subcontract as the case may be, made by the Main Contract DAB or arbitral tribunal respectively. The Subcontract Book makes provision for this option through alternative wording for Clause 20 included in the *Guidance for the Preparation of Particular Conditions of Subcontract*. This is under the heading '*Example Provisions where a Decision and/or Arbitral Award under the Main Contract Concerning a Subcontractor's Claim or an Employer's Claim shall be Binding under the Subcontract*' on pages 24 through 35.

In all cases it is advisable to seek professional and/or legal assistance when dealing with this particular Clause under the Subcontract.

Part 5

Future Updates

FIDIC Users' Guide

ISBN 978-0-7277-5856-9

ICE Publishing: All rights reserved
http://dx.doi.org/10.1680/fug.58569.527

Future Updates

Over the years FIDIC has maintained its solid name for quality services to the international construction industry through numerous services offered not only to its members, but also to the players in the industry at large. The most visible form of these services is the extensive suite of Standard Forms of Contract, also referred to as the 'Rainbow' suite of Contracts.

The history of these contracts dates back to the 1950s, when the first edition of the RED Book was published.

The following is a list of the main contract forms published by FIDIC.

- *Conditions of Contract (International) for Works of Civil Engineering Construction,* (first edition, 1957). (Due to its lengthy title and bright colour was referred to as the RED Book.)
- *Conditions of Contract (International) for Works of Civil Engineering Construction* (second edition, 1969).
- *Conditions of Contract (International) for Works of Civil Engineering Construction* (third edition, 1977).
- *Conditions of Contract for Works of Civil Engineering Construction* (fourth edition, 1987; reprinted 1992 with further amendments).
- *Conditions of Contract for Electrical and Mechanical Works* (third edition, 1987).
- *Conditions of Contract for Design Build and Turnkey* (first edition, 1995) (Orange Book).
- *Conditions of Contract for Construction* (first edition, 1999) (RED Book).
- *Conditions of Contract for Plant and Design Build* (first edition, 1999) (YLW Book).
- *Conditions of Contract for EPC/Turnkey Projects* (first edition, 1999) (Silver Book).
- *Short Form of Contract* (first edition, 1999) (Green Book).
- *Conditions of Contract for Construction MDB Harmonised* (first, second and third editions: 2005, 2006, 2010).
- *Conditions of Contract for Design, Build and Operate Projects* (first edition, 2008).
- White Book.
- Subcontract (SUB Book).
- Procurement Guide.
- Dredging.

FIDIC is conscious of the ever-changing international construction market conditions, and as such is vigilant in identifying any issues that might necessitate a re-think of the current suite of contracts published by it. It was for this reason that a special Task-Group was initiated with the mandate to research to what extent (if anything) the current suite of Contracts needed to be amended or changed, as well as identify possible needs from within the industry for new services currently not offered. As is usual for the various task-groups appointed and mandated by FIDIC it is expected that before a new or updated Standard Form of Contract is released for sale, selected industry professionals have subjected it to a rigorous review. These professionals, also referred to as 'friendly reviewers' will test these non-released versions by subjecting them to legal, commercial and other relevant standards, concepts, principles, and so on, in order to ensure the quality of product the construction industry has come to expect from FIDIC as a leading organisation. Such a friendly review usually takes anything up to a year before final release of the book is authorised.

Currently the Task-Group is working on the update (second edition) to the YLW Book. Although the Task-Group has not yet completed its mandate, the following could be seen as possible release dates of the rainbow series of Standard Forms of Contract:

Book	Possible Release Date
2nd Edition YLW	2015
2nd Edition RED	2016
2nd Edition Silver	2017
1st Edition YLW SUB	2015
1st Edition Silver SUB	2016

In addition to the Gold Book, which was created as a 'green field' project scenario, FIDIC is considering the publication of a similar book for a 'brown field' scenario. It is likely to be referred to as the 'Brown Book'. There is no information on possible release dates yet.

It should be noted however, that the above-mentioned dates are as yet unconfirmed.

Part 6

Appendices

Appendix 1
Notice Requirements for the RED and PNK Books

NOTES:
This table must be read in conjunction with the appropriate Conditions of Contract and is only intended as a guidance for when a notice may or may not be necessary. The terms have been simplified to enable the information to be tabulated.

Indicates multiple notice requirement

Issuing Notices by all Parties must:
1. be in Writing;
2. be delivered by hand, mail, courier, electronic transmission as stated in Appendix to Tender/Contract Data;
3. if delivered by hand it must have a receipt;

Ideally have the following information:
4. state that it is a Notice;
5. state the Sub-Clause(s) in which it is being made;
6. give brief description of the reason for issuing the notice.

Sub-Clause No.	Who gives Notice	To Whom?	Notification is required if	When to give notice?	Details to be included in the notice	Comments
1.8	Employer or Contractor	The other Party	(RED) errors or defects of a technical nature are discovered in documents prepared for use in executing the Works	Promptly	Details of the error or defect	
			(PNK) errors or defects are discovered in documents prepared for use in executing the Works			
1.9	Contractor	Engineer	The time for issuing drawings or instructions may delay or disrupt the work	At the time when Works are likely to be delayed	(RED) details of the drawing or instruction, why it should be issued, when it should be issued, details of the nature and amount of the delay or disruption likely	Advance Warning
					(PNK) details of the drawing or instruction, why it should be issued, when it should be issued, nature and amount of the delay or disruption likely	Advance Warning
1.9	Contractor	Engineer	Delayed drawings or instructions resulted in additional Cost or delay of the Works	As required by SC20.1 (i.e. 28 days)	As required by SC20.1 (i.e. describing the event or circumstance)	This is a further notice to the advance warning and should state that it is given in accordance with this Sub-Clause and Sub-Clause 20.1

Sub-Clause No.	Who gives Notice	To Whom?	Notification is required if	When to give notice?	Details to be included in the notice	Comments
1.13	Contractor	Appropriate authority	Notices are required by Law in relation to the execution of the Works	As required by Law	As required by Law	
2.1	Contractor	Engineer	The Contractor is delayed or incurs additional Cost as a result of a failure by the Employer to give right of access to or possession of the Site	As required by SC20.1 (i.e. 28 days)	As required by SC20.1 (i.e. describing the event or circumstance)	The notice should state that it is given in accordance with this Sub-Clause and Sub-Clause 20.1
2.4	Employer	Contractor	The Employer intends to make changes to his financial arrangements	When the situation arises	Detailed particulars	
(PNK Only) 2.4	Employer	Contractor	The Bank notifies the Borrower that the loan or credit has been suspended	When the situation arises	Detailed particulars	
2.5	Employer or Engineer	Contractor	The Employer considers himself entitled to payment under any Clause or otherwise in connection with the Contract	(RED) As soon as practicable after the Employer became aware (PNK) As soon as practicable but not longer than 28 days	Clause or basis of the claim and substantiation of the amount or extension of the Defects Notification Period	The detailed particulars may be provided at a date later than the notice, however, a summary of the matter would be usefully included
3.4	Employer	Contractor	The Employer intends to replace the Engineer	(RED) 42 days before the intended date of replacement (PNK) 21 days before the intended date of replacement	Name, address and relevant experience of the intended replacement Engineer	
3.4	Contractor	Employer	The Contractor objects to a proposed replacement Engineer	Not explicitly stated	(RED) Supporting particulars giving reasonable objection (PNK) Supporting particulars stating why Contractor considers replacement Engineer to be unsuitable	Any objection by the Contractor would need to be made immediately as time is very limited due to a relatively short notice period allowed for the Employer. There could also be local jurisdictional limitations on the period

Clause	From	To	Event	Timing	Particulars	Comments
3.5	Engineer	Contractor and Employer	The Engineer makes a determination	(RED) Not explicitly stated but must be in accordance with SC 1.3 (PNK) Within 28 days from receipt of claim unless otherwise specified	Agreement or fair determination, with supporting particulars	Shall not be unreasonably withheld as per SC 1.3
4.3	Contractor	Engineer	The Contractor's Representative intends to delegate or revoke powers, functions and authority to any person	Prior to the powers, functions and authority being delegated or revoked	Name of person, powers, functions and authority delegated or revoked	The Contractor should provide reasonable time for the Engineer to consider and respond to the request bearing in mind that alternative arrangements must be made should consent not be given
4.4	Contractor	Engineer	A Subcontractor will commence work off the Site	28 days prior to commencement of the work	Intended date of commencement and scope of work	The wording of sub-paragraph (b) requires that a notice must be issued 28 days before the intended start date for every Subcontractor. This requirement applies equally to Subcontractors employed on or off the Site.
4.4	Contractor	Engineer	A Subcontractor will commence work on the Site	28 days prior to commencement of the work	Intended date of commencement and scope of work	
4.7	Contractor	Engineer	The Contractor suffers delay and/or incurs Cost due to errors in setting out data	As required by SC20.1 (i.e. 28 days)	As required by SC20.1 (i.e. describing the event or circumstance)	The notice should state that it is given in accordance with this Sub-Clause and Sub-Clause 20.1
4.12	Contractor	Engineer	The Contractor encounters unforeseeable adverse physical conditions	As soon as practicable	Describe the physical conditions and set out the reasons why they are Unforeseeable	The notice should state that it is given in accordance with this Sub-Clause and Sub-Clause 20.1
4.16	Contractor	Engineer	There are Plant or Goods to be delivered to Site	Not less than 21 days prior to delivery	Not explicitly stated	Provide a description of the Plant or Goods and date of expected delivery

Sub-Clause No.	Who gives Notice	To Whom?	Notification is required if	When to give notice?	Details to be included in the notice	Comments
4.20	Contractor	Engineer	There are shortages, defects or defaults in any free-issue materials provided by the Employer	Promptly	Details of the shortage, defect or default in these Materials	
4.24	Contractor	Engineer	Fossils are discovered on the Site	Promptly	Not explicitly stated	The first notice should provide details of the fossils found, location, time and date when discovered. If the Contractor wishes to claim additional time and/or money he must also issue a second notice under this Sub-Clause and a further notice under Sub-Clause 20.1 followed by detailed particulars
4.24	Contractor	Engineer	The Contractor suffers delay and/or incurs Cost due to complying with the Engineer's instructions with regard to the discovery of fossils previously notified	As required by SC20.1 (i.e. 28 days)	As required by SC20.1 (i.e. describing the event or circumstance)	
5.2	Contractor	Engineer	The Contractor has reasonable objection to the employment of a nominated Subcontractor	As soon as practicable	Supporting particulars	
7.3	Contractor	Engineer	There is work to be covered up, put out of sight, or packaged for storage or transport	Whenever work is ready	Not explicitly stated	The notice should include details and location of the work which is ready for inspection
7.3	Engineer	Contractor	The Engineer receives a notice to inspect but he decides such an inspection is not required	Promptly		State that he does not require to inspect
7.4	Engineer	Contractor	The Engineer intends to attend tests?	Not less than 24 hours before the test	Not explicitly stated	Include the agreed time and place of testing in the notice
7.4	Contractor	Engineer	The Contractor suffers delay and/or incurs Cost with regard to tests	As required by SC20.1 (i.e. 28 days)	As required by SC20.1 (i.e. describing the event or circumstance)	The notice should state that it is given in accordance with this Sub-Clause and Sub-Clause 20.1
7.5	Engineer	Contractor	The Engineer finds that Plant, Materials or workmanship are defective or otherwise not in accordance with the Contract	Not explicitly stated but must be in accordance with SC1.3	Reasons for rejection	The notice should be given as soon as possible after discovering the defect

(RED) 8.1	Engineer	Contractor	When Commencement can be communicated to the Contractor	Less than 42 days after the Contractor receives the Letter of Acceptance but not less than 7 days	Commencement Date	
(PNK) 8.1	Engineer	Contractor	All precedent conditions have been fulfilled to enable Commencement Date to be notified	When the situation arises	The agreement of both Parties on fulfilment of the precedent conditions	
8.3	Engineer	Contractor	The Engineer considers that the Programme fails to meet the requirements of the Contract	Within 21 days after receiving a programme	The extent to which the Programme does not comply with the Contract	
8.3	Contractor	Engineer	The Contractor considers that there are specific probable future events or circumstances which may adversely affect the work	Promptly	Not explicitly stated	A further notice under Sub-Clause 20.1 will be required if additional Cost or delays occur
8.3	Engineer	Contractor	The Engineer considers that the Programme fails to comply with the Contract or fails to be consistant with actual progress and stated intentions	When the situation arises	The extent to which it fails to comply with the Contract	
8.6	Engineer	Contractor	The Engineer requires changes to a revised programme and supporting report submitted by the Contractor following the Engineers instruction for revised programme under SC8.3	When the situation arises	Not explicitly stated	The notice should include details of the extent to which the revised methods do not comply with the Contract
8.8	Engineer	Contractor	The Engineer wishes to suspend the Works or part of the Works	When the situation arises	The part or all the Works to be suspended	If possible the Engineer should provide reasons for the suspension with reference to SC8.9, SC8.10 and SC8.11
8.9	Contractor	Engineer	The Contractor suffers delay and/or incurrs Cost from complying with the Engineer's instructions to Suspend work	As soon as practicable	Describe the physical conditions and set out the reasons why they are Unforeseeable	The notice should state that it is given in accordance with this Sub-Clause and Sub-Clause 20.1
8.11	Contractor	Engineer	The Engineer has failed to give the Contractor permission to proceed after a suspension of part of the work for a time greater than 84 days	Within 28 days after the request	State that this is considered to be an omission of work under Clause 13	If it affects the whole of the Works, the Contractor has the option to terminate the Contract under SC16.2

Sub-Clause No.	Who gives Notice	To Whom?	Notification is required if	When to give notice?	Details to be included in the notice	Comments
9.1	Contractor	Engineer	The Contractor is ready to carry out Tests on Completion	Not less than 21 days before presumed completion date	Not explicitly stated	Include details of the Test(s) on Completion and a proposed date for the test to take place
9.2	Engineer	Contractor	The Engineer considers that the Contractor has unduly delayed the Tests on Completion	When the situation arises	Date by which the Tests must be carried out (i.e. within 21 days)	
9.2	Contractor	Engineer	If the Contractor is required to carry out Tests on Completion following a notice from the Engineer under Sub-Clause 9.2	Not explicitly stated but must be in accordance with SC1.3	Date and place of tests which must occur within 21 days of receipt of the Engineer's notice	The Contractor should give his notice as soon as possible after receiving the Engineer's Notice
9.2	Contractor	Engineer	The Contractor is delayed or incurs Cost due to unreasonable delay by the Employer when carrying out the Tests on Completion	As required by SC 10.3 & SC20.1 (i.e. 28 days)	As required by SC20.1 (i.e. describing the event or circumstance)	The notice should state that it is given in accordance with this Sub-Clause and Sub-Clause 20.1
10.1	Contractor	Engineer	The Contractor considers that the Works or Section of the Works are completed and ready for Taking Over	Not earlier than 14 days before the Works, or Sections of the Works, will be complete and ready for taking over	Not explicitly stated	The Contractor should include the date and details of the Works or Section to be taken over
10.2	Contractor	Engineer	The Contractor considers that he has incurred Costs due to the Employer taking over and/or using part of the Works	As required by SC20.1 (i.e. 28 days)	As required by SC20.1 (i.e. describing the event or circumstance)	The notice should state that it is given in accordance with this Sub-Clause and Sub-Clause 20.1
10.3	Engineer	Contractor	The Employer prevents the Contractor from carrying out Tests on Completion and the Engineer requires such Tests	When Contractor is prevented from carrying out Tests on Completion for more than 14 days	Not explicitly stated	The notice should include details of the Tests on Completion
10.3	Contractor	Engineer	The Contractor considers that he has been delayed and/or incurrs Cost due to the Employer with regard to Tests on Completion	As required by SC20.1 (i.e. 28 days)	As required by SC20.1 (i.e. describing the event or circumstance)	The notice should state that it is given in accordance with this Sub-Clause and Sub-Clause 20.1

11.4	Employer or on his behalf	Contractor	The Employer considers that the Contractor has failed to remedy a defect or damage within a reasonable time	After a reasonable time has elapsed for the Contractor to repair the damage	Date by which the defect or damage shall be remedied	
11.6	Engineer	Contractor	The Engineer requires a repetition of a test to show that the Contractor's work of remedying any defect or damage does not affect the Performance of the Works	Within 28 days after the defect or damage is remedied	Not explicitly stated	The notice should include necessary details and dates to enable the tests to be carried out
12.1	Engineer	Contractor	The Engineer requires any part of the Works to be measured	At any time	Give a reasonable time to attend	
12.1	Contractor	Engineer	The Contractor disagrees with the measurement records	No more than 14 days after being requested to examine the records	Respects in which the records are asserted to be inaccurate	Default for no timely notice is 'deemed' acceptance
12.4	Contractor	Engineer	The Contractor considers that he will incur or has incurred additional Costs due to omissions forming part of a Variation	Not stated	Supporting particulars	This is not a Contractor's claim and therefore not subject to the requirements of SC20.1. The supporting particulars, however, need to be sufficiently detailed to enable the Engineer to make his determination in accordance with SC 3.5 and C 12 (method of evaluing the additional costs)
13.1	Contractor	Engineer	(RED) the Contractor considers that he cannot readily obtain Goods required for a Variation (PNK) the Contractor considers that he cannot readily obtain Goods required for the Variation or such Variation triggers a substantial change in the sequence or progress of the Works	Promptly	Supporting particulars	
13.7	Contractor	Engineer	The Contractor considers that he has suffered (or will suffer) delay and/or incurred (or will incur) additional Cost as a result of changes in the Law	As required by SC20.1 (i.e. 28 days)	As required by SC20.1 (i.e. describing the event or circumstance)	The notice should state that it is given in accordance with this Sub-Clause and Sub-Clause 20.1

Sub-Clause No.	Who gives Notice	To Whom?	Notification is required if	When to give notice?	Details to be included in the notice	Comments
14.6	Engineer	Contractor	The Engineer has calculated an Interim Payment Certificate in an amount less than the minimum amount stated in the Appendix to Tender/Contract Data	Within 28 days of receiving the Statement	Not explicitly stated	The notice should provide details of the amount the Engineer has fairly determined and stating the minimum amount stated in the Appendix to Tender/ Contract Data
15.1	Engineer	Contractor	The Engineer considers that the Contractor has failed to carry out an obligation and must therefore remedy the failure	When the situation arises	Specified reasonable time to remedy the failure	The notice should also include details of the obligation that the Contractor has failed to carry out
15.2	Employer	Contractor	The Contractor has failed to comply with a Notice to correct; or	When the situation arises	Not explicitly stated	The notice should include full supporting particulars including assignment of any subcontract, works for the protection of life or property, safety of the Works, instructions and date for leaving the Site etc
			The Contractor has abandoned the Works; or			
			The Contractor has failed to proceed with the Works; or			
			The Contractor has failed to comply with a rejection notice; or			
			The Contractor failed to comply with Remedial Work notice; or			
			The Contractor has subcontracted the works without agreement; or			
			The Contractor is experiencing financial problems; or			
			The Contractor offered a bribe, shown favour etc			
			After termination and after completion of the Works by others, the Employer requires to release the Contractor's Equipment and Temporary Works	After completion of the Works	Details of the equipment and Temporary Works and release location (at or near the Site)	

15.5	Employer	Contractor	The Employer wishes to terminate the Contract at his convenience	At any time	Not explicitly stated	The notice should include details for the return of the Performance Security, works to protect life or property, safety of the Works, instructions and date for leaving the Site etc
(PNK) 15.6	Employer	Contractor	The Employer determines that the Contractor has been engaged in corrupt, fraudulent, collusive or coercive practises	When the situation arises	Not explicitly stated	The notice should include substantial and indisputable evidence of such activities
16.1	Contractor	Employer	The Engineer fails to certify a payment certificate or the Employer fails to provide reasonable evidence of his financial arrangements or payment has not been received	After the time of failure	Not explicitly stated	The notice should include details of failure and actions that will be taken by the Contractor (i.e. suspension or reduced rate of work)
(PNK Only) 16.1	Contractor	Employer	The Bank has suspended the loan or credit and no alternative funds are available	Not less than 7 days after the Borrower has received the suspension notification from the Bank	Not explicitly stated	The notice should include details of the suspension notification from the Bank to the Borrower and actions that will be taken by the Contractor (i.e. suspension or reduced rate of work)
16.1	Contractor	Employer	The Contractor suffers delay and/or incurs Cost as a result of suspending work (or reducing the rate of work)	As required by SC20.1 (i.e. 28 days)	As required by SC20.1 (i.e. describing the event or circumstance)	The notice should state that it is given in accordance with this Sub-Clause and Sub-Clause 20.1

Sub-Clause No.	Who gives Notice	To Whom?	Notification is required if	When to give notice?	Details to be included in the notice	Comments
16.2	Contractor	Employer	The Employer failed to provide reasonable evidence of financial arrangments within 42 days after giving notice under SC16.1	After the time of failure	Not explicitly stated	The notice should include detailed particulars and the date of termination 14 days after the notice is given
			The Engineer has failed to issue the Payment Certificate within 56 days			
			An interim payment has not been received within 98 days after the Engineer receives the Statement			
			The Employer substantially fails to perform his contractual obligations. (PNK has added requirement that the Employer's failures must adversely affect the economic balance of the Contract)			
			The Employer fails to comply with Sub-Clause 1.6 Contract Agreement			
			The Employer fails to comply with Sub-Clause 1.7 Contract Assignment?			
			There has been a prolonged suspension which affected the whole of the Works		Not explicitly stated	The notice should include detailed particulars and the date of termination which may take effect immediately
			The Employer has become bankrupt			
			(PNK only) the Contractor does not receive the letter recording an agreement of the fulfilment of condition precedents referred to in SC8.1	Not explicitly stated	Not explicitly stated	Any supporting documentation should include details of the actual letter received instructing the Commencement Date (if any)
16.2 (PNK Only)	Contractor	Employer cc Engineer	The Bank notifies the Borrower that the loan or credit has been suspended and payment has not been received within 70 days after the Engineer receives the Statement	When the situation arises	Not explicitly stated	The notice should include detailed particulars and the date of termination which must be more than 14 days after the notice is given

Clause	Giver	Recipient	Event / Situation	Timing	Particulars	Notes
17.4	Contractor	Engineer	The Contractor considers that the risks listed in Sub-Clause 17.3 have resulted in loss or damage to the Works, Goods or Contractor's Documents	Promptly	Not explicitly stated	The notice should include details of the loss or damage and any detailed particulars known at the time
17.4	Contractor	Engineer	The Contractor suffers delay and/or incurs Cost from rectifying loss or damage to the Works, Goods or Contractor's Documents	As required by SC20.1 (i.e. 28 days)	As required by SC20.1 (i.e. describing the event or circumstance)	The notice should state that it is given in accordance with this Sub-Clause and Sub-Clause 20.1
17.5	Employer or Contractor	The other Party	A claim has been made against the other Party by someone who alleges that their rights have been infringed,	Within 28 days of receiving the claim	Not explicitly stated	Failure to give notice might restrict the ability of the indemnifying Party to defend the claim
18.1	Insuring Party either Employer or Contractor	Engineer	The insuring Party has paid a premium or submitted a policy to the other Party	When submitted to the other Party	The notice should include evidence of payment and/or details of policy submitted	
18.1	Insuring Party either Employer or Contractor	The other Party	The insurer made (or attempted to make) an alteration to the policy	Promptly	Detailed particulars of the alteration (or attempted alteration)	
18.2	Contractor	Employer	The insurance cover described in Sub-Clause 18.2(d) ceased to be available at commercially reasonable terms if it is more than one year after the Base Date	When the situation arises	Supporting particulars	
19.2	Employer or Contractor	The other Party	A Party believes it is, or will be, prevented from performing any of its obligations by a Force Majeure event	Within 14 days after becoming aware or should have become aware	Specify the obligations, the performance of which is or will be prevented	**1st Notice under Force Majeure**
19.3	Employer or Contractor	The other Party	A Party believes that a Force Majeure event has ceased	When the situation arises	Specify the obligations, the performance of which are no longer prevented	**2nd Notice under Force Majeure**
19.4	Contractor	Engineer	Contractor incurs loss, damage or delay as a consequence of Force Majeure	As required by SC20.1 (i.e. 28 days)	Specify the loss or damage or delay incurred	**3rd Notice under Force Majeure**

19.6	Employer or Contractor	The other Party	A Party considers that a Force Majeure event has either prevented work continuously for a period of 84 days or has been effected for multiple periods which total more than 140 days	When the situation arises	Provide detailed particulars and the date of termination which shall take effect 7 days after notification	
19.7	Employer or Contractor	The other party	there is an event or circumstance outside the control of the Parties which makes it impossible or unlawful to complete the Contract	When the situation arises	Not explicitly stated	Provide detailed particulars and the date of termination which shall take effect immediately
20.1	Contractor	Engineer	The Contractor considers himself to be entitled to an extension of the Time and/or additional payment	As soon as practicable, and not later than 28 days	Detailed particulars to substantiate the claim	
20.4	Employer or Contractor	The other party	A Party is dissatisfied with the DAB's decision a Notice of Dissatisfaction is required	Within 28 days after receiving a decision	State that it is made pursuant to Sub-Clause 20.4, setting out the matter in dispute and giving the reason(s) for dissatisfaction	
20.4	Employer or Contractor	The other party	A DAB has failed to give its decision within 84 days a Notice of Dissatisfaction is required	Within 28 days after this period has expired	State that it is made pursuant to Sub-clause 20.4, setting out the matter in dispute and giving the reason(s) for dissatisfaction	
(RED Only) Appendix: Dispute Adjudication Agreement	Employer and Contractor	The Member of the DAB	A Dispute Adjudication Agreement has taken effect	Within six months of the Dispute Adjudication Agreement taking effect	Details of the Dispute Adjudication Agreement	

Appendix 2
Notice Requirements for the YLW Book

NOTES:

This table must be read in conjunction with the appropriate Conditions of Contract and is only intended as a guidance for when a notice may or may not be necessary. The terms have been simplified to enable the information to be tabulated.

Issuing Notices by all Parties must:

1. be in Writing;
2. be delivered by hand, mail, courier, electronic transmission as stated in Appendix to Tender;
3. if delivered by hand it must have a receipt;

Ideally have the following information:

4. state that it is a Notice;
5. state the Sub-Clause(s) in which it is being made;
6. give brief description of the reason for issuing the notice.

Indicates multiple notice requirement.

Sub-Clause No.	Who gives Notice	To Whom?	Notification is required if	When to give notice?	Details to be included in the notice	Comments
1.8	Employer or Contractor	The other Party	Errors or defects of a technical nature are discovered in documents prepared for use in executing the Works	Promptly	Details of the error or defect	
1.9	Contractor	Engineer	Delayed drawings or instructions may result in additional cost or delay of the Works	As required by SC20.1 (i.e. 28 days)	As required by SC20.1 (i.e. describing the event or circumstance)	The notice should state that it is given in accordance with this Sub-Clause and Sub-Clause 20.1
1.13	Contractor	Appropriate authority	Notices are required by Law in relation to the execution of the Works?	As required by Law	As required by Law	
2.1	Contractor	Engineer	The Contractor is delayed or incurs additional Cost as a result of a failure by the Employer to give right of access to or possession of the Site	As required by SC20.1 (i.e. 28 days)	As required by SC20.1 (i.e. describing the event or circumstance)	The notice should state that it is given in accordance with this Sub-Clause and Sub-Clause 20.1
2.4	Employer	Contractor	The Employer intends to make changes to his financial arrangements	When the situation arises	Detailed particulars	
2.5	Employer or Engineer	Contractor	The Employer considers himself entitled to payment under any Clause or otherwise in connection with the Contract	As soon as practicable after the Employer became aware	Clause or basis of the claim and substantiation of the amount or extension of the Defects Notification Period	The detailed particulars may be provided at a date later than the notice, however, a summary of the matter would be usefully included
3.4	Employer	Contractor	The Employer intends to replace the Engineer	42 days before the intended date of replacement	Name, address and relevant experience of the intended replacement Engineer	

Sub-Clause No.	Who gives Notice	To Whom?	Notification is required if	When to give notice?	Details to be included in the notice	Comments
3.4	Contractor	Employer	The Contractor objects to a proposed replacement Engineer	Not explicitly stated	Supporting particulars giving reasonable objection	Any objection by the Contractor would need to be made immediately as time is very limited due to a relatively short notice period allowed for the Employer
3.5	Engineer	Contractor and Employer	The Engineer makes a determination	Not explicitly stated but must be in accordance with SC 1.3	Agreement or fair determination, with supporting particulars	Shall not be unreasonably withheld as per SC 1.3
4.3	Contractor	Engineer	The Contractor's Representative intends to delegate or revoke powers, functions and authority to any person	Prior to the powers, functions and authority being delegated or revoked	Name of person, powers, functions and authority delegated or revoked	
4.4	Contractor	Engineer	A Subcontractor will work off the Site	28 days prior to commencement of the work	Intended date of commencement and scope of work	The wording of sub-paragraph (b) requires that a notice must be issued 28 days before the intended start date for every Subcontractor. This requirement applies equally to Subcontractors employed on or off the Site.
4.4	Contractor	Engineer	A Subcontractor will work on the Site	28 days prior to commencement of the work	Intended date of commencement and scope of work	
4.5	Contractor	Engineer	The Contractor has reasonable objection to the employment of a nominated Subcontractor	As soon as practicable	Supporting particulars	
4.7	Contractor	Engineer	The Contractor suffers delay and/or incurs Cost due to errors in setting out data	As required by SC20.1 (i.e. 28 days)	As required by SC20.1 (i.e. describing the event or circumstance)	The notice should state that it is given in accordance with this Sub-Clause and Sub-Clause 20.1
4.12	Contractor	Engineer	The Contractor encounters unforeseeable adverse physical conditions	As soon as practicable	Describe the physical conditions and set out the reasons why they are Unforeseeable	The notice should state that it is given in accordance with this Sub-Clause and Sub-Clause 20.1

Clause	From	To	Event	Time	Details	Comment
4.16	Contractor	Engineer	There are Plant or Goods to be delivered to Site	Not less than 21 days prior to delivery	Not explicitly stated	Provide a description of the Plant or Goods and date of expected delivery
4.20	Contractor	Engineer	There are shortages, defects or defaults in any free-issue materials provided by the Employer	Promptly	Details of the shortage, defect or default in these Materials	
4.24	Contractor	Engineer	Fossils are discovered on the Site	Promptly	Not explicitly stated	The first notice should provide details of the fossils found, location, time and date when discovered. If the Contractor wishes to claim additional time and/or money he must also issue a second notice under this Sub-Clause and a further notice under Sub-Clause 20.1 followed by detailed particulars
4.24	Contractor	Engineer	The Contractor suffers delay and/or incurs Cost due to complying with the Engineer's instructions with regard to the discovery of fossils previously notified	As required by SC20.1 (i.e. 28 days)	As required by SC20.1 (i.e. describing the event or circumstance)	
5.1	Contractor	Engineer	The Contractor finds any error, fault or other defect in the Employer's Requirements	Before the end of the period stated in the Appendix to Tender	Details of the error, fault or other defect	
5.1	Engineer	Contractor	After receiving a notice from the Contractor that an error, fault or other defect exists within the Employer's Requirements	Not explicitly stated but must be in accordance with SC1.3	Whether Clause 13 shall be applied	The Engineer must provide an appropriate instruction if there is a Variation or alternatively, the reasons why an experienced contractor should have been able to detect such error, fault or other defect when examining the Site
5.2	Contractor	Engineer	Where Contractor's Documents are to be submitted to the Engineer for review	According to the Employer's Requirements	A statement that the Contractor's Document is considered ready for review, ready for approval (if specified) and ready for use. The notice must also state that it complies with the Contract or give reasons if it does not comply	

Sub-Clause No.	Who gives Notice	To Whom?	Notification is required if	When to give notice?	Details to be included in the notice	Comments
5.2	Engineer	Contractor	Where a Contractor's Document fails to comply with the Contract	Within the review period (i.e. 21 days unless otherwise specified)	The extent to which the Contractor's Document fails to comply with the Contract	
5.2	Engineer	Contractor	Where a Contractor's Document is approved	Within the review period (i.e. 21 days unless otherwise specified)	With or without comments	
5.4	Contractor	Engineer	If new standards come into force after the Base Date	When the situation arises	Submit proposals for compliance if appropriate	
7.3	Contractor	Engineer	There is work to be covered up, put out of sight, or packaged for storage or transport	Whenever work is ready	Not explicitly stated	The notice should include details and location of the work which is ready for inspection
7.3	Engineer	Contractor	The Engineer receives a notice to inspect but he decides such an inspection is not required	Promptly	State that he does not require to inspect	
7.4	Engineer	Contractor	The Engineer intends to attend tests?	Not less than 24 hours before the test	Not explicitly stated	Include the agreed time and place of testing in the notice
7.4	Contractor	Engineer	The Contractor suffers delay and/or incurs Cost with regard to tests	As required by SC20.1 (i.e. 28 days)	As required by SC20.1 (i.e. describing the event or circumstance)	The notice should state that it is given in accordance with this Sub-Clause and Sub-Clause 20.1
7.5	Engineer	Contractor	The Engineer finds that Plant, Materials or workmanship are defective or otherwise not in accordance with the Contract	Not explicitly stated but must be in accordance with SC1.3	Reasons for rejection	The notice should be given as soon as possible after discovering the defect
8.1	Engineer	Contractor	When Commencement can be communicated to the Contractor	Less than 42 days after the Contractor receives the Letter of Acceptance but not less than 7 days	Commencement Date	
8.3	Engineer	Contractor	The Engineer considers that the Programme fails to meet the requirements of the Contract	Within 21 days after receiving a programme	The extent to which the Programme does not comply with the Contract	

	Contractor	Engineer	The Contractor considers that there are specific probable future events or circumstances which may adversely affect the work	Promptly	The Contractor is advised to provide as much information known at the time of issuing the notice	A further notice under Sub-Clause 20.1 will be required if additional Cost or delays occur
8.3	Engineer	Contractor	The Engineer considers that the Programme fails to comply with the Contract or fails to be consistant with actual progress and stated intentions	When the situation arises	The extent to which it fails to comply with the Contract	
8.6	Engineer	Contractor	The Engineer requires changes to a revised programme and supporting report submitted by the Contractor following the Engineers instruction for revised programme under SC8.3	When the situation arises	Not explicitly stated	The notice should include details of the extent to which the revised methods do not comply with the Contract
8.8	Engineer	Contractor	The Engineer wishes to suspend the Works or part of the Works	When the situation arises	The part or all the Works to be suspended	If possible the Engineer should provide reasons for the suspension with reference to SC8.9, SC8.10 and SC8.11
8.9	Contractor	Engineer	The Contractor suffers delay and/or incurrs Cost from complying with the Engineer's instructions to Suspend work	As soon as practicable	Describe the physical conditions and set out the reasons why they are Unforeseeable	The notice should state that it is given in accordance with this Sub-Clause and Sub-Clause 20.1
8.11	Contractor	Engineer	The Engineer has failed to give the Contractor permission to proceed after a suspension of part of the work for a time greater than 84 days	Within 28 days after the request	State that this is considered to be an omission of work under Clause 13	If it affects the whole of the Works, the Contractor has the option to terminate the Contract under SC16.2
9.1	Contractor	Engineer	The Contractor is ready to carry out Tests on Completion	Not less than 21 days before presumed completion date	Not explicitly stated	Include details of the Test(s) on Completion and a proposed date for the test to take place
9.2	Engineer	Contractor	The Engineer considers that the Contractor has unduly delayed the Tests on Completion	When the situation arises	Date by which the Tests must be carried out (i.e. within 21 days)	
9.2	Contractor	Engineer	If the Contractor is required to carry out Tests on Completion following a notice from the Engineer under Sub-Clause 9.2	Not explicitly stated but must be in accordance with SC1.3	Date and place of tests which must occur within 21 days of receipt of the Engineer's notice	The Contractor should give his notice as soon as possible after receiving the Engineer's Notice

Sub-Clause No.	Who gives Notice	To Whom?	Notification is required if	When to give notice?	Details to be included in the notice	Comments
10.1	Contractor	Engineer	The Contractor considers that the Works or Section of the Works are completed and ready for Taking Over	Not earlier than 14 days before the Works, or Sections of the Works, will be complete and ready for taking over	Not explicitly stated	The Contractor should include the date and details of the Works or Section to be taken over
10.2	Contractor	Engineer	The Contractor considers that he has incurred Costs due to the Employer taking over and/or using part of the Works	As required by SC20.1 (i.e. 28 days)	As required by SC20.1 (i.e. describing the event or circumstance)	The notice should state that it is given in accordance with this Sub-Clause and Sub-Clause 20.1
10.3	Engineer	Contractor	The Employer prevents the Contractor from carrying out Tests on Completion and the Engineer requires such Tests	When Contractor is prevented from carrying out Tests on Completion for more than 14 days	Not explicitly stated	The notice should include details of the Tests on Completion
10.3	Contractor	Engineer	The Contractor considers that he has been delayed and/or incurs Cost due to the Employer with regard to Tests on Completion	As required by SC20.1 (i.e. 28 days)	As required by SC20.1 (i.e. describing the event or circumstance)	The notice should state that it is given in accordance with this Sub-Clause and Sub-Clause 20.1
11.4	Employer or on his behalf	Contractor	The Employer considers that the Contractor has failed to remedy a defect or damage within a reasonable time	After a reasonable time has elapsed for the Contractor to repair the damage	Date by which the defect or damage shall be remedied	
11.6	Engineer	Contractor	The Engineer requires a repetition of a test to show that the Contractor's work of remedying any defect or damage may affect the Performance of the Works	Within 28 days after the defect or damage is remedied	Not explicitly stated	The notice should include necessary details and dates to enable the tests to be carried out
12.1	Employer	Contractor	When the Employer has taken over the Works and he requires to carry out Tests on Completion	As soon as reasonably practicable	The date after which the Tests on Completion shall be carried out (must be at least 21 days in advance and completed 14 days thereafter)	The notice should also include any details with the regard to the Tests on Completion
12.2	Contractor	Engineer	The Contractor is delayed or incurs Cost due to unreasonable delay by the Employer when carrying out the Tests on Completion	As required by SC20.1 (i.e. 28 days)	As required by SC20.1 (i.e. describing the event or circumstance)	The notice should state that it is given in accordance with this Sub-Clause and Sub-Clause 20.1

12.4	Employer or Engineer	Contractor	The Employer needs to specify a particular time for the Contractor to access the Site to undertake modifications following a failed Test on Completion	When the situation arises	A time that is convenient to the Employer	Failure to send such a notice may result in the Tests on Completion deemed to have passed
12.4	Contractor	Engineer	The Contractor is delayed or incurs Cost due to unreasonable delay by the Employer when permitting access to the Works to investigate a cause of failure following Tests on Completion.	As required by SC20.1 (i.e. 28 days)	As required by SC20.1 (i.e. describing the event or circumstance)	The notice should state that it is given in accordance with this Sub-Clause and Sub-Clause 20.1
13.1	Contractor	Engineer	The Contractor considers that he cannot readily obtain Goods required for a Variation, or it will reduce safety or suitability of the Works, or it will have an adverse impact on the achievement of the Schedule of Guarantees	Promptly	Supporting particulars	Upon receipt of this notice, the Engineer either cancels, confirms or varies the Instruction
13.7	Contractor	Engineer	The Contractor considers that he has suffered (or will suffer) delay and/or incurred (or will incur) additional Cost as a result of changes in the Law	As required by SC20.1 (i.e. 28 days)	As required by SC20.1 (i.e. describing the event or circumstance)	The notice should state that it is given in accordance with this Sub-Clause and Sub-Clause 20.1
14.6	Engineer	Contractor	The Engineer has calculated an Interim Payment Certificate in an amount less than the minimum amount stated in the Appendix to Tender	Within 28 days of receiving the Statement	Not explicitly stated	The notice should provide details of the amount the Engineer has fairly determined and stating the minimum amount stated in the Appendix to Tender
15.1	Engineer	Contractor	The Engineer considers that the Contractor has failed to carry out an obligation and must therefore remedy the failure	When the situation arises	Specified reasonable time to remedy the failure	The notice should also include details of the obligation that the Contractor has failed to carry out

Sub-Clause No.	Who gives Notice	To Whom?	Notification is required if	When to give notice?	Details to be included in the notice	Comments
15.2	Employer	Contractor	The Contractor has failed to comply with a Notice to correct; or The Contractor has abandoned the Works; or The Contractor has failed to proceed with the Works; or The Contractor has failed to comply with a Rejection Notice; or The Contractor failed to comply with Remedial Work notice; or The Contractor has subcontracted the works without agreement; or The Contractor is experiencing financial problems; or The Contractor offered a bribe, shown favour etc	When the situation arises	Not explicitly stated	The notice should include full supporting particulars including assignment of any subcontract, works for the protection of life or property, safety of the Works, instructions and date for leaving the Site etc
			After termination and after completion of the Works by others, the Employer requires to release the Contractor's Equipment and Temporary Works	After completion of the Works	Details of the equipment and Temporary Works and release location (at or near the Site)	
15.5	Employer	Contractor	The Employer wishes to terminate the Contract at his convenience	At any time	Not explicitly stated	The notice should include details for the return of the Performance Security, works to protect life or property, safety of the Works, instructions and date for leaving the Site etc
16.1	Contractor	Employer	The Engineer fails to certify a payment certificate or the Employer fails to provide reasonable evidence of his financial arrangements or payment has not been received	After the time of failure	Not explicitly stated	The notice should include details of failure and actions that will be taken by the Contractor (i.e. suspension or reduced rate of work)
16.1	Contractor	Employer	The Contractor suffers delay and/or incurs Cost as a result of suspending work (or reducing the rate of work)	As required by SC20.1 (i.e. 28 days)	As required by SC20.1 (i.e. describing the event or circumstance)	The notice should state that it is given in accordance with this Sub-Clause and Sub-Clause 20.1

Sub-Clause	From	To	Event	Timing		Comments
16.2	Contractor	Employer	The Employer failed to provide reasonable evidence of financial arrangements within 42 days after giving notice under SC16.1	After the time of failure	Not explicitly stated	The notice should include detailed particulars and the date of termination 14 days after the notice is given
			The Engineer has failed to issue the Payment Certificate within 56 days			
			An interim payment has not been received within 98 days after the Engineer receives the Statement			
			The Employer substantially fails to perform his contractual obligations			
			The Employer fails to comply with Sub-Clause 1.6 Contract Agreement			
			The Employer fails to comply with Sub-Clause 1.7 Contract Assignment?			
			There has been a prolonged suspension which affected the whole of the Works		Not explicitly stated	The notice should include detailed particulars and the date of termination which may take effect immediately
			The Employer has become bankrupt			
17.4	Contractor	Engineer	The Contractor considers that the risks listed in Sub-Clause 17.3 have resulted in loss or damage to the Works, Goods or Contractor's Documents	Promptly	Not explicitly stated	The notice should include details of the loss or damage and any detailed particulars known at the time
17.4	Contractor	Engineer	The Contractor suffers delay and/or incurs Cost from rectifying loss or damage to the Works, Goods or Contractor's Documents	As required by SC20.1 (i.e. 28 days)	As required by SC20.1 (i.e. describing the event or circumstance)	The notice should state that it is given in accordance with this Sub-Clause and Sub-Clause 20.1
17.5	Employer or Contractor	The other Party	A claim has been made against the other Party by someone who alleges that their rights have been infringed	Within 28 days of receiving the claim	Not explicitly stated	Failure to give notice might restrict the ability of the indemnifying Party to defend the claim
18.1	Insuring Party either Employer or Contractor	Engineer	The insuring Party has paid a premium or submitted a policy to the other Party	When submitted to the other Party	The notice should include evidence of payment and/or details of policy submitted	

Sub-Clause No.	Who gives Notice	To Whom?	Notification is required if	When to give notice?	Details to be included in the notice	Comments
18.1	Insuring Party either Employer or Contractor	The other Party	The insurer made (or attempted to make) an alteration to the policy	Promptly	Detailed particulars of the alteration (or attempted alteration)	
18.2	Contractor	Employer	The insurance cover described in Sub-Clause 18.2(d) ceased to be available at commercially reasonable terms if it is more than one year after the Base Date	When the situation arises	Supporting particulars	
19.2	Employer or Contractor	The other Party	A Party believes it is, or will be, prevented from performing any of its obligations by a Force Majeure event	Within 14 days after becoming aware or should have become aware	Specify the obligations, the performance of which is or will be prevented	**1st Notice under Force Majeure**
19.3	Employer or Contractor	The other Party	A Party believes that a Force Majeure event has ceased	When the situation arises	Specify the obligations, the performance of which are no longer prevented	**2nd Notice under Force Majeure**
19.4	Contractor	Engineer	Contractor incurs loss, damage or delay as a consequence of Force Majeure	As required by SC20.1 (i.e. 28 days)	Specify the loss or damage or delay incurred	**3rd Notice under Force Majeure**
19.6	Employer or Contractor	The other Party	A Party considers that a Force Majeure event has either prevented work continuously for a period of 84 days or has been effected for multiple periods which total more than 140 days	When the situation arises	Provide detailed particulars and the date of termination which shall take effect 7 days after notification	
19.7	Employer or Contractor	The other party	There is an event or circumstance outside the control of the Parties which makes it impossible or unlawful to complete the Contract	When the situation arises	Not explicitly stated	Provide detailed particulars and the date of termination which shall take effect immediately
20.1	Contractor	Engineer	The Contractor considers himself to be entitled to an extension of the Time and/or additional payment	As soon as practicable, and not later than 28 days	Detailed particulars to substantiate the claim	
20.4	Employer or Contractor	The other Party	A Party is dissatisfied with the DAB's decision a Notice of Dissatisfaction is required	Within 28 days after receiving a decision	State that it is made pursuant to Sub-Clause 20.4, setting out the matter in dispute and giving the reason(s) for dissatisfaction	

| 20.4 | Employer or Contractor | The other party | A DAB has failed to give its decision within 84 days a Notice of Dissatisfaction is required | Within 28 days after this period has expired | State that it is made pursuant to Sub-clause 20.4, setting out the matter in dispute and giving the reason(s) for dissatisfaction | |
| Appendix: Dispute Adjudicatuon Agreement | Employer and Contractor | The Member of the DAB | A Dispute Adjudication Agreement has taken effect | Within six months of the Dispute Adjudication Agreement taking effect | Details of the Dispute Adjudication Agreement | |

Appendix 3
Notice Requirements for the Subcontract Book

NOTES:

This table must be read in conjunction with the appropriate Conditions of Contract including those sections of the Main Contract that are applicable by statement or implication, and is only intended as a guidance for when a notice may or may not be necessary. The terms have been simplified to enable the information to be tabulated.

Issuing Notices by all Parties must:

1. be in Writing;
2. be delivered by hand, mail, courier, electronic transmission as stated in Appendix to Subcontract Offer;
3. if delivered by hand it must have a receipt;

Ideally have the following information:

4. state that it is a Notice;
5. state the Sub-Clause(s) in which it is being made;
6. give brief description of the reason for issuing the notice.

Indicates multiple notice requirement.

Sub-Clause No.	Who gives Notice	To Whom?	Notification is required if	When to give notice?	Details to be included in the notice	Comments
1.4	Subcontractor	Contractor	Another address for delivery of communications is required	When the situation arises	The new address	
1.5	Subcontractor or Contractor	The other Party	Errors or defects of a technical nature are discovered in documents prepared for executing the Subcontract Works	Promptly	Details of the error or defect	
1.7	Subcontractor or Contractor	The other Party	If a Party forms a joint venture etc	When the situation arises	Contact details of the leader having authority	
2.1	Subcontractor	Contractor	An ambiguity or discrepancy is discovered in the Subcontract, the Main Contract or during execution of the Works	When the situation arises	Details of the ambiguity or discrepancy	
3.3	Contractor	Subcontractor	The Contractor considers himself entitled to additional payment	As soon as practicable after the Contractor became aware, but not later than 28 days	Describe the event or circumstance and specify the basis of the claim	
3.3	Contractor	Subcontractor	The Contractor has made a fair decision if agreement is not reached following a Contractor's claim	The notice should not be unreasonably withheld or delayed in accordance with SC1.6	The notice should include reasons for the decision and support particulars	

Sub-Clause No.	Who gives Notice	To Whom?	Notification is required if	When to give notice?	Details to be included in the notice	Comments
3.4	Contractor	Subcontractor	The Engineer or Employer serves a notice of an Employer's claim concerning the Subcontractor on the Main Contractor	Immediately	A copy of the notice and particulars	
5.2	Subcontractor	Contractor	A Subcontractor will work off the Site	14 days prior to commencement of the work	Intended date of commencement and scope of work	The wording of sub-paragraph (b) requires that a notice must be issued 28 days before the intended start date for every Subcontractor. This requirement applies equally to Subcontractors employed on or off the Site. Hence, two separate notices for Subcontractors must be considered
5.2	Subcontractor	Contractor	A Subcontractor will work on the Site	14 days prior to commencement of the work	Intended date of commencement and scope of work	
6.1	Subcontractor	Contractor	Employer's Personnel, Contractor's Personnel etc. fails to cooperate which affects execution of the Subcontract Works	Immediately	Details of the non cooperation	The notice should state that it is given in accordance with this Sub-Clause and Sub-Clause 20.1
6.1	Subcontractor	Contractor	The Subcontractor suffers delay and/or incurs Cost due to non cooperation of Employer's Personnel, Contractor's Personnel etc	As required by SC20.1 (i.e.immediately)	As required by SC20.1 (i.e. describing the event which has occured)	
6.3	Contractor	Subcontractor	The Contractor's Subcontract Representative has not been named	Prior to the Subcontract Commencement Date	The name and particulars of the Contractor's appointed Subcontract Representative	
7.1	Subcontractor	Contractor	If there are shortages, defects or defaults in Equipment etc. provided by the Contractor	Promptly	Description of the shortage, defect or default	
7.3	Subcontractor	Contractor	If there are shortages, defects or defaults in free issue Materials provided by the Contractor	Promptly	Description of the shortage, defect or default	

8.1	Contractor	Subcontractor	When Commencement can be communicated to the Subcontractor	Not less than 14 days before Subcontract Commencement Date	The Subcontract Commencement Date	
8.6	Contractor	Subcontractor	The Contractor wishes to suspend progress of part or all of the Subcontract Works	At any time	State the reasons for the suspension	
9.1	Subcontractor	Contractor	The Subcontractor requires Subcontract Tests on Completion to be carried out	Reasonable notice	Details of the Subcontract Tests on Completion	
9.2	Subcontractor	Contractor	As per Main Contract C 9: The Subcontractor is delayed or incurs Cost due to unreasonable delay by the Employer or the Contractor when carrying out the Tests on Completion	As required by MC-SC 10.3 & SC-SC20.1 (i.e. 28 days)	As required by SC20.1 (i.e. describing the event or circumstance)	The notice should state that it is given in accordance with this Sub-Clause and Sub-Clause 20.1
10.1	Subcontractor	Contractor	The Subcontractor considers the Subcontract Works to be complete	Not earlier than 7 days before completion	Date and scope of Works to be complete	
10.1	Contractor	Subcontractor	The Contractor considers the Subcontractor's Works to be complete	21 days after receipt of Subcontractor's notice	Date on which Contractor considers the Work to be complete	
10.1	Contractor	Subcontractor	The Contractor does not consider the Subcontractor's Works to be complete	21 days after receipt of Subcontractor's notice	Provide reasons and specify work to be done to achieve completion	
12.1	Contractor	Subcontractor	Measurement of Subcontract Work is to be carried out, the Subcontractor shall be notified to attend/assist	Reasonable notice	Provide the date, time and venue and request any particulars required for the measurement	
12.1	Contractor	Subcontractor	Measurement of Subcontract Work by records is to be carried out, the Subcontractor shall be notified to attend/assist	Reasonable notice	Provide the date, time and venue for the measurement by records	
12.1	Subcontractor	Contractor	The Subcontractor disagrees with the measurement records	Within 7 days of the date of examination	The respects in which the records are inaccurate	Default for no timely notice is 'deemed' acceptance
12.1	Contractor	Subcontractor	The MC Engineer makes any determination with respect to disagreed records	Without delay	The Engineer's determination	As per SC 1.6 cannot be unreasonably withheld

Sub-Clause No.	Who gives Notice	To Whom?	Notification is required if	When to give notice?	Details to be included in the notice	Comments
12.3	Contractor	Subcontractor	The Contractor has made a fair evaluation if agreement is not reached	The notice should not be unreasonably withheld or delayed in accordance with SC1.6	The notice should include details of the evaluation and support particulars	
13.1	Subcontractor	Contractor	The Subcontractor is unable to obtain Subcontract Goods for a Subcontract Variation	Promptly	Supporting particulars	
13.4	Subcontractor	Contractor	The Subcontractor considers that he has suffered (or will suffer) delay and/or incurred (or will incur) additional Cost as a result of changes in the Law	As required by SC20.1 (i.e. immediately)	As required by SC20.1 (i.e. describing the event which has occured)	The notice should state that it is given in accordance with this Sub-Clause and Sub-Clause 20.1
15.1	Contractor	Subcontractor	The Main Contract is terminated or the Contractor and/or the Employer are released from performance under Sub-Clause 19.7 of the Main Contract, then the Contractor may terminate the Subcontract	Immediately	Not specifically stated	Provide details of the termination or release form performance
15.2	Contractor	Subcontractor	The Main Contract is terminated and the Contractor has carried out an evaluation to determine sums due to the Subcontractor	Not specifically stated	Supporting particulars	
15.5	Contractor	Subcontractor	The Subcontractor fails to carry out any obligation under the Subcontract	When the situation arises	Details of the failure and reasonable time to take remedial action	
15.6	Contractor	Subcontractor	The Subcontractor fails to comply with the requirements contained within the Main Contract Sub-Clause 15.2 (a) to (f)	When the situation arises	Give 14 days notice to terminate for (a) to (d) and immediate termination for (e) and (f)	Provide full detailed particulars of the failure within the notice
16.1	Subcontractor	Contractor	The Contractor fails to pay an amount due to the Subcontractor	When the situation arises	Give 21 days notice to suspend or reduce the rate of work describing the failure to pay	
16.1	Subcontractor	Contractor	The Subcontractor suffers delay and/or incurs Cost due to suspending or reducing the rate of work due to non payment	As required by SC20.1 (i.e.immediately)	As required by SC20.1 (i.e. describing the event which has occured)	The notice should state that it is given in accordance with this Sub-Clause and Sub-Clause 20.1

Clause	From	To	Description	When	Content	Notes
16.2	Subcontractor	Contractor	The Subcontractor may terminate the Subcontract if (a) the Subcontractor does not receive payment after giving appropriate notices under Sub-Clause 16.1, (b) the Contractor or the Employer becomes bankrupt, or (c) the Contractor substantially fails to perform his obligations under the Subcontract	When the situation arises	For (a) and (c) the termination may take effect after 14 days, for (b) the termination may take effect immediately	The notice should include the date of the intended termination, detailed particulars and supporting documents of the issues
18.2	Subcontractor	Contractor	The Subcontractor discovers inadequacies or duplication when reviewing insurances relating to the Subcontract Works	When the situation arises	Immediately	
19.1	Employer or Contractor	The other Party	As per Main Contract C 19: A Party believes it is, or will be, prevented from performing any of its obligations by a Force Majeure event	Within 14 days after becoming aware or should have become aware	Specify the obligations, the performance of which is or will be prevented	**1st Notice under Force Majeure**
19.1	Employer or Contractor	The other Party	As per Main Contract C 19: A Party believes that a Force Majeure event has ceased	When the situation arises	Specify the obligations, the performance of which are no longer prevented	**2nd Notice under Force Majeure**
19.1	Subcontractor	Contractor	As per Main Contract C 19: Subcontractor incurs loss, damage or delay as a consequence of Force Majeure	As required by SC20.1 (i.e. 28 days)	Specify the loss or damage or delay incurred	**3rd Notice under Force Majeure**
20.1	Subcontractor	Contractor	There is a requirement for the Contractor to give a notice to the Engineer or Employer under the Main Contract if it applies to the Subcontract Works	In good time to enable the Contractor to comply with the Main Contract	The notice should provide information similar to the requirements of the Main Contract	
20.1	Contractor or Subcontractor	The other Party	An event occurs or a specific probable future event or circumstance may occur which may delay execution of the Subcontract Works or Main Contract Works	Immediately	Describe the delaying event or specific probable future event and describe the Subcontract Works or the Main Contract Works which will be delayed	This is separate from the notification under SC 20.2 for intention to claim

Sub-Clause No.	Who gives Notice	To Whom?	Notification is required if	When to give notice?	Details to be included in the notice	Comments
20.1	Subcontractor	Contractor	An event occurs or a specific probable future event or circumstance may occur which may increase the Subcontract Price or Main Contract Price	Immediately	Describe the event or specific probable future event and decribe the Subcontract Works or the Main Contract Works which may incur additional costs	This is separate from the notification under SC 20.2 for intention to claim
20.2	Contractor	Subcontractor	The Subcontractor has made a claim for additional time or money for which an agreement between the Contractor and the Subcontractor has not been reached	Within 42 days after receiving the fully detailed claim from the Subcontractor	Describe the fair decision taken giving reasons and stating it is made in accordance with Sub-Clause 20.2	
20.4	Contractor or Subcontractor	The other Party	A dispute arise between the Contractor and the Subcontractor	At any time	Details of the dispute	This notice is referred to as the Notice of Dispute under this Sub-Clause
20.4	Contractor	Subcontractor	A Notice of Dispute by the Contractor or within 14 days of a Notice of Dispute by the Subcontractor involves issues also involved in a dispute between the Contractor and the Employer	When given by the Contractor, the notice should be included within the Notice of Dispute / When given by the Subcontractor, the notice should be given within 14 day of receipt of the Notice of Dispute	Reasons of his opinion that the dispute involves issue(s) that are disputed between the Contractor and the Employer	Any deferment should be detailed with actual dates, whether the dispute will be referred to the Main Contract DAB
20.4	Contractor	Subcontractor	A dispute between the Contractor and Employer arises where there is no Main Contract DAB in place	Immediately	Provide details of the referral to arbitration under Sub-Clause 20.8 of the Main Contract	
20.6	Employer or Subcontractor	The other Party	A Party is dissatisfied with the DAB's decision	Within 28 days after receiving the decision	Not stated	State that it is made pursuant to Sub-clause 20.6, setting out the matter in dispute and giving the reason(s) for dissatisfaction

The 'Guidelines for the Preparation of Particular Conditions of Subcontract' provide the following varied or alternative Sub-Clauses which contain references to notices

				Not stated	Details of the non-consent
1.9	Contractor	Subcontractor	The Employer does not consent to the Subcontract under the alternative SC 1.9	Not stated	Details of the non-consent
13.3	Contractor	Subcontractor	The Contractor fails to reach agreement with the Subcontractor for the adjustments to Price under the alternative Lump Sum Contract Sub-Clause	The notice should not be unreasonably withheld or delayed in accordance with SC1.6	Supporting particulars
15.3	Contractor	Subcontractor	The Contractor delays payment to the Subcontractor because he has not been paid under the 'pay when paid' varient Sub-Clause	As soon as is reasonably practicable but not later than the date when this payment would otherwise have become due	Provide reasons
14.12	Contractor	Subcontractor	The Contractor fails to reach agreement with the Subcontractor for the revised instalments where progress is less than expected under the alternative Lump Sum Contract Sub-Clause	The decison should not be unreasonably withheld or delayed in accordance with SC1.6	Supporting particulars
Additional Sub-Clause under Clause 17	Contractor	Subcontractor	The Parties fail to agree the liability for an event which is not the responsibility of the Subcontractor or for which he is insured against	The notice should not be unreasonably withheld or delayed in accordance with SC1.6	Reasons for the decision and supporting particulars
20.2	Contractor	Subcontractor	A Subcontractor's claim is (a) claimable under the Main Contract; (b) concerns issue(s) which are subject to the Contractor's claims under the Main Contract; or (c) concerns issues which are involved in a dispute under the Main Contract under the variant Clause 20 provisions	The notice should not be unreasonably withheld or delayed in accordance with SC1.6	Include reasons why this claim is considered to be a Related Claim	If this notification is made, the claim is thereafter considered to be a Related Claim

Sub-Clause No.	Who gives Notice	To Whom?	Notification is required if	When to give notice?	Details to be included in the notice	Comments
20.6	Contractor	Subcontractor	The Contractor considers a Notice of Dispute from the Subcontractor to involve issues within a Main Contract dispute	Within 14 days of receiving the Notice of Dispute	Include reasons why it is considered to be part of a Main Contract dispute	If there is no objection from the Subcontractor, this issue is considered to be a Related Dispute
20.8	Contractor	Subcontractor	The Main Contract DAB gives a decision on a Related Dispute	As soon as practicable but not later than 7 days	Provide the decision	

FIDIC Users' Guide
ISBN 978-0-7277-5856-9

ICE Publishing: All rights reserved
http://dx.doi.org/10.1680/fug.58569.563

Appendix 4
Contents: General Conditions for the RED Book

Contents: General Conditions

FIDIC Users' Guide
ISBN 978-0-7277-5856-9

ICE Publishing: All rights reserved
http://dx.doi.org/10.1680/fug.58569.569

Appendix 5
Contents: General Conditions for the PNK Book

Contents: General Conditions

APPENDIX: DISPUTE BOARD

GENERAL CONDITIONS OF DISPUTE BOARD AGREEMENT

ANNEX: PROCEDURAL RULES

INDEX OF SUB-CLAUSES

FIDIC Users' Guide
ISBN 978-0-7277-5856-9

ICE Publishing: All rights reserved
http://dx.doi.org/10.1680/fug.58569.575

Appendix 6
Contents: General Conditions for the YLW Book

Contents: General Conditions

FIDIC Users' Guide
ISBN 978-0-7277-5856-9

ICE Publishing: All rights reserved
http://dx.doi.org/10.1680/fug.58569.581

Appendix 7
Contents: General Conditions for the SUB Book

Contents: General Conditions

FIDIC Users' Guide
ISBN 978-0-7277-5856-9

Appendix 8
Forms and Annexes for the Red and Yellow Book

Annexes   FORMS OF SECURITIES

Acceptable form(s) of security should be included in the tender documents: for Annex A and/ or B, in the instructions to Tenderers; and for Annexes C to G, annexed to the Particular Conditions. The following example forms, which (except for Annex A) incorporate Uniform Rules published by the International Chamber of Commerce (the 'ICC', which is based at 38 Cours Albert 1er, 75008 Paris, France), may have to be amended to comply with the applicable law. Although the ICC publishes guides to these Uniform Rules, legal advice should be taken before the securities are written. Note that the guaranteed amounts should be quoted in all the currencies, as specified in the Contract, in which the guarantor pays the beneficiary.

Annex A   EXAMPLE FORM OF PARENT COMPANY GUARANTEE

[See page 3, and the comments on Sub-Clause 1.14]

Brief description of Contract ....

Name and address of Employer ....

(together with successors and assigns).

We have been informed that .... (hereinafter called the 'Contractor') is submitting an offer for such Contract in response to your invitation, and that the conditions of your invitation require his offer to be supported by a parent company guarantee.

In consideration of you, the Employer, awarding the Contract to the Contractor, we (*name of parent company*) .... irrevocably and unconditionally guarantee to you, as a primary obligation, the due performance of all the Contractor's obligations and liabilities under the Contract, including the Contractor's compliance with all its terms and conditions according to their true intent and meaning.

If the Contractor fails to so perform his obligations and liabilities and comply with the Contract, we will indemnify the Employer against and from all damages, losses and expenses (including legal fees and expenses) which arise from any such failure for which the Contractor is liable to the Employer under the Contract.

This guarantee shall come into full force and effect when the Contract comes into full force and effect. If the Contract does not come into full force and effect within a year of the date of this guarantee, or if you demonstrate that you do not intend to enter into the Contract with the Contractor, this guarantee shall be void and ineffective. This guarantee shall continue in full force and effect until all the Contractor's obligations and liabilities under the Contract have been discharged, when this guarantee shall expire and shall be returned to us, and our liability hereunder shall be discharged absolutely.

This guarantee shall apply and be supplemental to the Contract as amended or varied by the Employer and the Contractor from time to time. We hereby authorise them to agree any such amendment or variation, the due performance of which and compliance with which by the Contractor are likewise guaranteed hereunder. Our obligations and liabilities under this guarantee shall not be discharged by any allowance of time or other indulgence whatsoever by the Employer to the Contractor, or by any variation or suspension of the works to be executed under the Contract, or by any amendments to the Contract or to the constitution of the Contractor or the Employer, or by any other matters, whether with or without our knowledge or consent.

This guarantee shall be governed by the law of the same country (or other jurisdiction) as that which governs the Contract and any dispute under this guarantee shall be finally settled under the Rules of Arbitration of the International Chamber of Commerce by one or more arbitrators appointed in accordance with such Rules. We confirm that the benefit of this guarantee may be assigned subject only to the provisions for assignment of the Contract.

Date ....   Signature(s) ....

Annex B   EXAMPLE FORM OF TENDER SECURITY

[See page 3]

Brief description of Contract ....

Name and address of Beneficiary ....

.... (whom the tender documents define as the Employer).

We have been informed that .... (hereinafter called the 'Principal') is submitting an offer for such Contract in response to your invitation, and that the conditions of your invitation (the 'conditions of invitation', which are set out in a document entitled Instructions to Tenderers) require his offer to be supported by a tender security.

At the request of the Principal, we (*name of bank*) .... hereby irrevocably undertake to pay you, the Beneficiary/Employer, any sum or sums not exceeding in total the amount of .... (say: ....) upon receipt by us of your demand in writing and your written statement (in the demand) stating that:

(a)   the Principal has, without your agreement, withdrawn his offer after the latest time specified for its submission and before the expiry of its period of validity, or

(b)   the Principal has refused to accept the correction of errors in his offer in accordance with such conditions of invitation, or

(c)   you awarded the Contract to the Principal and he has failed to comply with sub-clause 1.6 of the conditions of the Contract, or

(d)   you awarded the Contract to the Principal and he has failed to comply with sub-clause 4.2 of the conditions of the Contract.

Any demand for payment must contain your signature(s) which must be authenticated by your bankers or by a notary public. The authenticated demand and statement must be received by us at this office on or before (*the date 35 days after the expiry of the validity of the Letter of Tender*) ...., when this guarantee shall expire and shall be returned to us.

This guarantee is subject to the Uniform Rules for Demand Guarantees, published as number 458 by the International Chamber of Commerce, except as stated above.

Date .... Signature(s) ....

Annex C   EXAMPLE FORM OF PERFORMANCE SECURITY - DEMAND GUARANTEE

[See comments on Sub-Clause 4.2]

Brief description of Contract ....

Name and address of Beneficiary ....

.... (whom the Contract defines as the Employer).

We have been informed that .... (hereinafter called the 'Principal') is your contractor under such Contract, which requires him to obtain a performance security.

At the request of the Principal, we (*name of bank*).. hereby irrevocably undertake to pay you, the Beneficiary/Employer, any sum or sums not exceeding in total the amount of .... (the 'guaranteed amount', say: ....) upon receipt by us of your demand in writing and your written statement stating:

    (a)   that the Principal is in breach of his obligation(s) under the Contract, and

    (a)   the respect in which the Principal is in breach.

[Following the receipt by us of an authenticated copy of the taking-over certificate for the whole of the works under clause 10 of the conditions of the Contract, such guaranteed amount shall be reduced by .... % and we shall promptly notify you that we have received such certificate and have reduced the guaranteed amount accordingly.] [1]

Any demand for payment must contain your [minister's/directors'] [1] signature(s) which must be authenticated by your bankers or by a notary public. The authenticated demand and statement must be received by us at this office on or before (*the date 70 days after the expected expiry of the Defects Notification Period for the Works*) .... (the 'expiry date'), when this guarantee shall expire and shall be returned to us.

We have been informed that the Beneficiary may require the Principal to extend this guarantee if the performance certificate under the Contract has not been issued by the date 28 days prior to such expiry date. We undertake to pay you such guaranteed amount upon receipt by us, within such period of 28 days, of your demand in writing and your written statement that the performance certificate has not been issued, for reasons attributable to the Principal, and that this guarantee has not been extended.

This guarantee shall be governed by the laws of .... and shall be subject to the Uniform Rules for Demand Guarantees, published as number 458 by the International Chamber of Commerce, except as stated above.

Date .... Signature(s) ....

---

[1] *When writing the tender documents, the writer should ascertain whether to include the optional text, shown in parentheses [ ]*

Annex D   EXAMPLE FORM OF PERFORMANCE SECURITY - SURETY BOND

[See comments on Sub-Clause 4.2]

Brief description of Contract ....

Name and address of Beneficiary ....

.... (together with successors and assigns, all as defined in the Contract as the Employer).

By this Bond, (*name and address of contractor*) ....
(who is the contractor under such Contract) as Principal and (*name and address of guarantor*)
.... as Guarantor are irrevocably held and firmly bound
to the Beneficiary in the total amount of .... (the 'Bond Amount', say:
.....................................) for the due performance of all such Principal's obligations and liabilities
under the Contract. [Such Bond Amount shall be reduced by .... % upon the issue of the taking-
over certificate for the whole of the works under clause 10 of the conditions of the Contract.][1]

This Bond shall become effective on the Commencement Date defined in the Contract.

Upon Default by the Principal to perform any Contractual Obligation, or upon the occurrence of any of
the events and circumstances listed in sub-clause 15.2 of the conditions of the Contract, the Guarantor
shall satisfy and discharge the damages sustained by the Beneficiary due to such Default, event or
circumstances,[2] However, the total liability of the Guarantor shall not exceed the Bond Amount.

The obligations and liabilities of the Guarantor shall not be discharged by any allowance of time or
other indulgence whatsoever by the Beneficiary to the Principal, or by any variation or suspension
of the works to be executed under the Contract, or by any amendments to the Contract or to the
constitution of the Principal or the Beneficiary, or by any other matters, whether with or without the
knowledge or consent of the Guarantor.

Any claim under this Bond must be received by the Guarantor on or before (*the date six months after
the expected expiry of the Defects Notification Period for the Works*) .... (the 'Expiry
Date'), when this Bond shall expire and shall be returned to the Guarantor.

The benefit of this Bond may be assigned subject to the provisions for assignment of the Contract,
and subject to the receipt by the Guarantor of evidence of full compliance with such provisions.

This Bond shall be governed by the law of the same country (or other jurisdiction) as that which
governs the Contract. This Bond incorporates and shall be subject to the Uniform Rules for
Contract Bonds, published as number 524 by the International Chamber of Commerce, and words
used in this Bond shall bear the meanings set out in such Rules.

Wherefore this Bond has been issued by the Principal and the Guarantor on (*date*) ....

Signature(s) for and on behalf of the Principal ....

Signature(s) for and on behalf of the Guarantor ....

[1] *When writing the tender documents, the writer should ascertain whether to include the optional text,
shown in parentheses [ ]*

[2] *Insert:* [and shall not be entitled to perform the Principal's obligations under the Contract.]
    Or: [or at the option of the Guarantor (to be exercised in writing within 42 days of receiving the claim
         specifying such Default) perform the Principal's obligations under the Contract.]

Annex E   EXAMPLE FORM OF ADVANCE PAYMENT GUARANTEE

[*See comments on Sub-Clause 14.2*]

Brief description of Contract   ....

Name and address of Beneficiary ....

.... (whom the Contract defines as the Employer).

We have been informed that .... (hereinafter called the 'Principal') is your contractor under such Contract and wishes to receive an advance payment, for which the Contract requires him to obtain a guarantee.

At the request of the Principal, we (*name of bank*) .... hereby irrevocably undertake to pay you, the Beneficiary/Employer, any sum or sums not exceeding in total the amount of .... (the 'guaranteed amount', say:..) upon receipt by us of your demand in writing and your written statement stating:

(a) that the Principal has failed to repay the advance payment in accordance with the conditions of the Contract, and

(b) the amount which the Principal has failed to repay.

This guarantee shall become effective upon receipt [of the first instalment] of the advance payment by the Principal. Such guaranteed amount shall be reduced by the amounts of the advance payment repaid to you, as evidenced by your notices issued under sub-clause 14.6 of the conditions of the Contract. Following receipt (from the Principal) of a copy of each purported notice, we shall promptly notify you of the revised guaranteed amount accordingly.

Any demand for payment must contain your signature(s) which must be authenticated by your bankers or by a notary public. The authenticated demand and statement must be received by us at this office on or before (*the date 70 days after the expected expiry of the Time for Completion*) .... (the 'expiry date'), when this guarantee shall expire and shall be returned to us.

We have been informed that the Beneficiary may require the Principal to extend this guarantee if the advance payment has not been repaid by the date 28 days prior to such expiry date. We undertake to pay you such guaranteed amount upon receipt by us, within such period of 28 days, of your demand in writing and your written statement that the advance payment has not been repaid and that this guarantee has not been extended.

This guarantee shall be governed by the laws of .... and shall be subject to the Uniform Rules for Demand Guarantees, published as number 458 by the International Chamber of Commerce, except as stated above.

Date .... Signature(s) ....

Annex F EXAMPLE FORM OF RETENTION MONEY GUARANTEE

[See comments on Sub-Clause 14.9]

Brief description of Contract ....

Name and address of Beneficiary ....

.... (whom the Contract defines as the Employer).

We have been informed that .... (hereinafter called the 'Principal') is your contractor under such Contract and wishes to receive early payment of [part of] the retention money, for which the Contract requires him to obtain a guarantee.

At the request of the Principal, we *(name of bank)* .... hereby irrevocably undertake to pay you, the Beneficiary/Employer, any sum or sums not exceeding in total the amount of .... (the 'guaranteed amount', say: ....) upon receipt by us of your demand in writing and your written statement stating:

(a) that the Principal has failed to carry out his obligation(s) to rectify certain defect(s) for which he is responsible under the Contract, and

(b) the nature of such defect(s).

At any time, our liability under this guarantee shall not exceed the total amount of retention money released to the Principal by you, as evidenced by your notices issued under sub-clause 14.6 of the conditions of the Contract with a copy being passed to us.

Any demand for payment must contain your signature(s) which must be authenticated by your bankers or by a notary public. The authenticated demand and statement must be received by us at this office on or before *(the date 70 days after the expected expiry of the Defects Notification Period for the Works)* .... (the 'expiry date'), when this guarantee shall expire and shall be returned to us.

We have been informed that the Beneficiary may require the Principal to extend this guarantee if the performance certificate under the Contract has not been issued by the date 28 days prior to such expiry date. We undertake to pay you such guaranteed amount upon receipt by us, within such period of 28 days, of your demand in writing and your written statement that the performance certificate has not been issued, for reasons attributable to the Principal, and that this guarantee has not been extended.

This guarantee shall be governed by the laws of .... and shall be subject to the Uniform Rules for Demand Guarantees, published as number 458 by the International Chamber of Commerce, except as stated above.

Date............................... Signature(s) ....

Annex G   EXAMPLE FORM OF PAYMENT GUARANTEE BY EMPLOYER

[See page 17: Contractor Finance]

Brief description of Contract   ....

Name and address of Beneficiary   ....

.... (whom the Contract defines as the Contractor).

We have been informed that .... (whom the Contract defines as the Employer and who is hereinafter called the 'Principal') is required to obtain a bank guarantee.

At the request of the Principal, we *(name of bank)* .... hereby irrevocably undertake to pay you, the Beneficiary/Contractor, any sum or sums not exceeding in total the amount of .... (say: ....) upon receipt by us of your demand in writing and your written statement stating:

   (a)   that, in respect of a payment due under the Contract, the Principal has failed to make payment in full by the date fourteen days after the expiry of the period specified in the Contract as that within which such payment should have been made, and

   (b)   the amount(s) which the Principal has failed to pay.

Any demand for payment must be accompanied by a copy of *[list of documents evidencing entitlement to payment]* .... in respect of which the Principal has failed to make payment in full.

Any demand for payment must contain your signature(s) which must be authenticated by your bankers or by a notary public. The authenticated demand and statement must be received by us at this office on or before *(the date six months after the expected expiry of the Defects Notification Period for the Works)* .... when this guarantee shall expire and shall be returned to us.

This guarantee shall be governed by the laws of .... and shall be subject to the Uniform Rules for Demand Guarantees, published as number 458 by the International Chamber of Commerce, except as stated above.

Date................................. Signature(s) ....

LETTER OF TENDER

NAME OF CONTRACT:

TO:

We have examined the Conditions of Contract, Specification, Drawings, Bill of Quantities, the other Schedules, the attached Appendix and Addenda Nos .... for the execution of the above-named Works. We offer to execute and complete the Works and remedy any defects therein in conformity with this Tender which includes all these documents, for the sum of (in currencies of payment) ....
....
or such other sum as may be determined in accordance with the Conditions of Contract.

We accept your suggestions for the appointment of the DAB, as set out in Schedule.......................

> [We have completed the Schedule by adding our suggestions for the other Member of the DAB, but these suggestions are not conditions of this offer].*

We agree to abide by this Tender until .... and it shall remain binding upon us and may be accepted at any time before that date. We acknowledge that the Appendix forms part of this Letter of Tender.

If this offer is accepted, we will provide the specified Performance Security, commence the Works as soon as is reasonably practicable after the Commencement Date, and complete the Works in accordance with the above-named documents within the Time for Completion.

Unless and until a formal Agreement is prepared and executed this Letter of Tender, together with your written acceptance thereof, shall constitute a binding contract between us.

We understand that you are not bound to accept the lowest or any tender you may receive.

Signature .... in the capacity of ....

duly authorised to sign tenders for and on behalf of ....
....

Address: ....

....

Date: ....

* If the Tenderer does not accept, this paragraph may be deleted and replaced by:

> We do not accept your suggestions for the appointment of the DAB. We have included our suggestions in the Schedule, but these suggestions are not conditions of this offer. If these suggestions are not acceptable to you, we propose that the DAB be jointly appointed in accordance with Sub-Clause 20.2 of the Conditions of Contract.

APPENDIX TO TENDER

[Note: with the exception of the items for which the Employer's requirements have been inserted, the following information must be completed before the Tender is submitted]

Item	Sub-Clause	Data
Employer's name and address	1.1.2.2 & 1.3	
Contractor's name and address	1.1.2.3 & 1.3	
Engineer's name and address	1.1.2.4 & 1.3	
Time for Completion of the Works	1.1.3.3	_____ days
Defects Notification Period	1.1.3.7	365 days
Electronic transmission systems	1.3	
Governing Law	1.4	
Ruling language	1.4	
Language for communications	1.4	
Time for access to the Site	2.1	_____ days after Commencement Date
Amount of Performance Security	4.2	_____ % of the Accepted Contract Amount, in the currencies and proportions in which the Contract Price is payable
Normal working hours	6.5	
Delay damages for the Works	8.7 & 14.15(b)	_____ % of the final Contract Price per day, in the currencies and proportions in which the Contract Price is payable
Maximum amount of delay damages	8.7	_____ % of the final Contract Price
If there are Provisional Sums: Percentage for adjustment of Provisional Sums	13.5(b)	_____ %

Initials of signatory of Tender _____

If Sub-Clause 13.8 applies:
Adjustments for Changes in Cost;
Table(s) of adjustment data . . . .   13.8 . . . .   for payments each
month/[*YEAR*] in _____ (*currency*)

Coefficient; scope of index	Country of origin; currency of index	Source of index; Title/definition	Value on stated date(s)* Value	Date
a= 0.10 Fixed				
b=_____ Labour				
c=_____				
d=_____				
e=_____				

* These values and dates confirm the definition of each index, but do not define Base Date indices

Total advance payment . . . .   14.2 . . . .   ____% of the Accepted Contract Amount

Number and timing of instalments . . . .   14.2 . . . .

Currencies and proportions . . . .   14.2 . . . .   ____ % in _____
                                                                        ____ % in _____

Start repayment of advance payment .   14.2(a)   when payments are _____ %
                                                          of the Accepted Contract Amount
                                                          less Provisional Sums

Repayment amortisation of advance
payment . . . .   14.2(b) . . . . .   ____ %

Percentage of retention . . . .   14.3 . . . .   ____ %

Limit of Retention Money . . . .   14.3 . . . .   ___ % of the Accepted Contract Amount

If Sub-Clause 14.5 applies:
Plant and Materials for payment
when shipped en route to the Site . .   14.5(b) . . . . .   _____ [list]
                                                                _____ [list]

Plant and Materials for payment
when delivered to the Site . . . .   14.5(c) . . . . .   _____ [list]
                                                                    _____ [list]

Minimum amount of Interim Payment
Certificates . . . .   14.6 . . . .   ___ % of the Accepted Contract Amount

If payments are only to be made in a currency/currencies named on the first page of the Letter of Tender:

Currency/currencies of payment . . . .   14.15 . . . . . .   as named in the Letters of Tender

Initials of signatory of Tender _____

If some payments are to be made in a currency/currencies not named on the first page of the Letter of Tender:

Currencies of payment 14.15

Currency Unit	Percentage payable in the Currency	Rate of exchange: number of Local per unit of Foreign
Local: _____ [name]		1.000
Foreign: _____ [name]		
_____ [name]		

Periods for submission of insurance:
(a) evidence of insurance . . . . 18.1 . . . .   ___ days
(b) relevant policies . . . . 18.1 . . . .   ___ days

Maximum amount of deductibles for
insurance of the Employer's risks . . . . 18.2(d) . . . . . _____

Minimum amount of third party
insurance . . . . 18.3 . . . . _____

Date by which the DAB shall be appointed . 20.2 . . . . 28 days after the Commencement Date

The DAB shall be . . . . 20.2 . . . . *Either:*
____ One sole Member/adjudicator
Or:
____ A DAB of three Members

Appointment (if not agreed) to be
made by . . . . 20.3 . . . . The President of FIDIC or a person appointed by the President

If there are Sections:
Definition of Sections:

Description (Sub-Clause 1.1.5.6)	Time for Completion (Sub-Clause 1.1.3.3)	Delay Damages (Sub-Clause 8.7)

[*In the above Appendix, the text shown in italics is intended to assist the drafter of a particular contract by providing guidance on which provisions are relevant to the particular contract. This italicised text should not be included in the tender documents, as it will generally appear inappropriate to tenderers.*]

Initials of signatory of Tender _____

In the Yellow Book the Appendix to Tender is changed to suit the Sub-Clauses:

■ the period for notifying unforeseeable errors, faults and defects in the Employer's Requirements is added for Sub-Clause 5.1;

■ the date by which the DAB shall be appointed is deleted for Sub-Clause 20.2.

CONTRACT AGREEMENT

This Agreement made the _____ day of _____ 20 _____

Between _____ of _____ (hereinafter called 'the Employer') of the one part,
and _____ of _____ (hereinafter called 'the Contractor') of the other
part

Whereas the Employer desires that the Works known as _____ should be executed by
the Contractor, and has accepted a Tender by the Contractor for the execution and completion of
these Works and the remedying of any defects therein.

The Employer and the Contractor agree as follows:

1. In this Agreement words and expressions shall have the same meanings as are respectively
   assigned to them in the Conditions of Contract hereinafter referred to.

2. The following documents shall be deemed to form and be read and construed as part of this
   Agreement:

   (a) The Letter of Acceptance dated _____

   (b) The Letter of Tender dated _____

   (c) The Addenda nos. _____

   (d) The Conditions of Contract

   (e) The Specification

   (f) The Drawings, and

   (g) The completed Schedules.

3. In consideration of the payments to be made by the Employer to the Contractor as
   hereinafter mentioned, the Contractor hereby covenants with the Employer to execute and
   complete the Works and remedy any defects therein, in conformity with the provisions of the
   Contract.

4. The Employer hereby covenants to pay the Contractor, in consideration of the execution and
   completion of the Works and the remedying of defects therein, the Contract Price at the
   times and in the manner prescribed by the Contract.

In Witness whereof the parties hereto have caused this Agreement to be executed the day and
year first before written in accordance with their respective laws. _____

SIGNED by: _____   SIGNED by: _____

for and on behalf of the Employer in the presence   for and on behalf of the Contractor in the presence
of   of

Witness: _____   Witness: _____
Name: _____   Name: _____
Address: _____   Address: _____
Date: _____   Date: _____

In the Yellow Book the Contract Agreement is changed to suit the requirements of the Contract: At paragraph 2(e) 'The Specification' is replaced by 'Employer's Requirements'; at 2(f) 'Drawings' are replaced by 'completed Schedules'; at 2(g) 'completed Schedules' are replaced by 'Contractor's proposal'. This confirms that the Contractor's proposal, even if approved, does not supersede the Employer's Requirements. If particular items in the Contractor's proposal are intended to have priority then they must be confirmed as Variations.

FIDIC Users' Guide
ISBN 978-0-7277-5856-9

ICE Publishing: All rights reserved
http://dx.doi.org/10.1680/fug.58569.601

Index

Index by page numbers. Flow charts and Tables are indicated by *italic page numbers*. Definitions by initial capitals (except 'day' and 'year')+**emboldened page numbers**. 'RB' means '1999 Red Book' [Construction], 'YLW' is '1999 Yellow Book' [Plant and Design-Build], and 'PNK' is 'Pink Book' [2010 Multilateral Development Banks Harmonised Conditions of Contract for Construction], 'SUB' is [2011 Conditions of Subcontract for building and engineering works designed by the Employer]